J. F. Bonneville J. L. Dietemann

Radiology of
The Sella Turcica

With the Collaboration of
J. C. Demandre G. Didierlaurent C. Edus P. Gresyk M. Pion
N. Quantin T. Taillard

Illustrations by M. Gaudron
Translation Reviewed by I. Moseley

With a Foreword by J. L. Vezina, a Preface by A. Wackenheim
and a Historical Review by J. Metzger

With 370 Figures in 693 Separate Illustrations

Springer-Verlag Berlin Heidelberg New York 1981

Jean François Bonneville, M.D.

Professor of Radiology, Head of Department of Neuroradiology,
University Hospital of Besançon, 2, Place Saint Jacques, F-25030 Besançon

Jean Louis Dietemann, M.D.

Department of Neuroradiology, University Hospital of Strasbourg
1, Place de l'Hôpital, F-67091 Strasbourg

The cover picture shows a lateral view of the normal sella turcica

ISBN-13: 978-3-642-67788-5 e-ISBN-13: 978-3-642-67786-1
DOI: 10.1007/978-3-642-67786-1

Softcover reprint of the hardcover 1st edition 1981

Reproduction of the figures: Gustav Dreher GmbH, Stuttgart

2127/3130-543210

To my family
To my staff

J. F. B.

To my family

J. L. D.

Foreword

Master of all endocrine activity and executive organ of one's quality of life, the pituitary gland is tightly lodged in the "turkish saddle." As a bony container, the sella turcica is to the hypophysis what the skull is to the brain; it can therefore be looked upon as a little vault within the cranial vault. Just as the cranium is moulded by the growth of the brain, so is the sella fashioned by its content. It becomes locally enlarged in response to expanding intrasellar lesions, and it tends to return to its original size and shape upon their removal or destruction.

Pituitary adenomas have in the past been diagnosed upon enlargement of the sella turcica. In the past decade, as a direct result of interdisciplinary cooperation, we have learned that tiny adenomas, the immediate cause of some cases of acromegaly, amenorrhea–galactorrhea syndrome, or Cushing's disease, can exist with minimal or no observable effect on the size of the sella. The breakthrough started when radioimmunoassay, as a new method of accurately measuring specific hormonal output, indicated selective pituitary oversecretions in patients with normal-sized sellae. Neurosurgeons highly skilled in the transsphenoidal approach with the surgical microscope were obliged to operate on some of these patients and confirmed the presence of tiny oversecreting adenomas in their pituitary glands. Their challenge to the neuroradiologist, to come up with a means of diagnosing these tumors, resulted in scrupulous studies of the sella turcica with emphasis on subtle changes of its floor. Hypocycloidal tomography confirmed the neurosurgical findings, and this simple method has become the tool of choice for preoperative diagnosis of microadenomas having reached the threshold size of 4 mm. Diagnosis of smaller lesions awaits the refinements of high resolution coronal computerized tomography.

From the point of view of the radiologist, the sella turcica therefore serves two diagnostic purposes. As part of the sphenoid bone, provides direct evidence of systemic or specific bony diseases. As a bony container of the hypophysis, it displays indirect signs of a variety of intrasellar of intracranial lesions. The medical literature has widely reported on bony landmarks of the sella and their variations in health and disease since Oppenheim's pioneering work in 1901. Bonneville and Dietemann include in their book the feature points of these reports along with their own observations. In addition, they set forth the concept of the sella as a mirror of the pathology of the pituitary gland. The material is presented in a cartesian topographical order and it is convincingly illustrated. They are to be complimented for bringing to the literature this new radiologic textbook on the sella turcica. Their work is a splendid contribution to medical learning.

Montreal, Canada, September 1980 *Jean Lorrain Vezina*

Preface

This book by Prof. J.F. Bonneville and Dr. J.L. Dietemann fills a gap in the radiologic literature, for the sellar area has never previously been so accurately studied. Over the past few years we have had the opportunity of listening to lectures and of reading papers on the subject by J.F. Bonneville. His contributions to the field include several very original views and new approaches, and the reputation that he has already earned in France will doubtlessly extend beyond our borders as a result of this remarkably conceived and well-documented work.

It might be appropriate to point out that J.F. Bonneville belongs to the transitional generation of radiologists who succeeded in integrating and dealing with computerized tomography, and J.L. Dietemann, to the new generation for whom computerized tomography is a part of routine training.

The Strasbourg neuroradiologic school is honored by J.F. Bonneville's friendship, and I hope that the ties between the neuroradiologic teams of Strasbourg and Besançon remain productive in the field of scientific research and become even closer in the future.

A. Wackenheim

Professor and Chairman of Diagnostic Radiology
University of Strasbourg

Preface

Historical Review

My good friend J.F. Bonneville has kindly asked me to contribute to his book on the sella turcica and to review the main stages in the development of radiologic knowledge of this region.

Naturally, the starting point for this review is Roentgen, to whom the technique owes its name. The way to radiologic assessment of the sella was opened by clinicians and pathologists: in 1886, Pierre Marie described acromegaly, and one year later, Minkowski related this to the existence of a pituitary tumor. It was in 1902, as Fischgold and Bull recalled in their excellent history of neuroradiology*, that Antoine Béclère described the corresponding radiologic manifestations: "a notable increase in the anteroposterior diameter of the pituitary fossa, whose margins appear thickened, and which gives the general impression of a hemispherical cup," together with irregular thickening of the skull bones and exaggerated development of the frontal and maxillary sinuses.

Another pioneer was Arthur Schuller, of Vienna, in whose work on the skull base, published in 1905, is to be found an analysis of the effects of cerebral tumors on the appearance of the sella turcica. It was also Schuller who advised the surgeon Hirsch to try the nasal approach (through the sphenoid sinus) to pituitary tumors.

Subsequently the work of Haas (1925), David and Dilenge (1957), and particularly that of du Boulay (1957) and Mahmoud (1958) has increased our knowledge of the radiology of the sella turcica. More recently, Vezina (1974) emphasized the possibility of detecting intrasellar microadenomas at an early stage; we also were working along these lines at La Pitié (1975).

Contrast examination of the sellar region started when Dandy injected air or oxygen by the lumbar route, for the study of the intracranial cerebrospinal fluid spaces. Gas encephalography has given valuable information about the sella and its neighboring structures. The use of iodinated contrast instead of gas was a much later development.

Egaz Moniz carried out the first cerebral angiogram with sodium iodine solution in 1927, after numerous experimental trials. This method permits analysis not of the sella itself but of the neighboring structures and is of great value for meningiomas of the middle cranial fossa and certain aneurysms. The techniques of subtraction developed in 1935 by Ziedses des Plantes and unrecognized for a number of years, and of tomography, by which the skull may be studied in a series of transverse or sagittal planes, must be mentioned. The latter technique combined with pneumoencephalography, has been of inestimable value for the study of pituitary tumors. It would also be quite wrong not to give credit to the technical advances made by Lysholm, firstly with his head-holding device

---

* Fischgold, H., Bull, J.: A Short History of Neuroradiology. VIIIth Symposium Neuroradiologicum, September 1967.

(1931) and subsequently, with Schoenander, in enabling skull radiographs to be obtained in a number of different projections.

Finally, computerized tomography, has been developed by the physicists Hounsfield and McCormack, and put into clinical use by Ambrose. It is useful for the study of the skull and its contents and for the sellar region particularly when used in combination with contrast media. This technique also allows us to demonstrate the curious anomaly which Busch in 1951 called the "empty sella" as a result of his detailed anatomic studies of the diaphragma sellae; the term empty sella is, however, inexact, since the pituitary, more or less compressed, is still detectable.

This brief review will, I hope, serve to indicate the great value of, on one hand, clinicoradiologic correlation and, on the other, advances in physics (and more recently computer science) for the study of the skull, particularly the sella turcica.

La Pitié, Paris, September 1980 *J. Metzger*

Introduction

It may be asked whether a monograph on plain radiography of the sella turcica is needed in 1980, that is, in the CT era. Progress in endocrinology and biochemistry and the early demonstration of pituitary dysfunction make it essential for the radiologist to be able to assess the sella accurately and to detect minor variations in its size. Detailed knowledge of the range of morphological variations which simulate pathological changes is indispensable, and for this reason normal variants and pictures which may give rise to confusion are extensively illustrated here alongside the classic pathological appearances. At the end of this work, the reader will also find a chapter presented as a series of exercises allowing him to test his knowledge and to familiarize himself with possible pitfalls. Angiographic and cisternographic findings are frequently non-specific in cases of parasellar pathology; furthermore, they are well documented elsewhere and are therefore not considered here. The authors are of the opinion that because of the relatively recent introduction of high resolution computerised tomography and its continuing refinement, the time is not yet ripe for an ambitious monograph on the sella turcica, which would include all the features of this technique. Nevertheless, in an up-to-date chapter they have collected the main results of CT, illustrating the marvellous possibilities of this exciting new technique. The bibliography is exhaustive: the reader will find essential bibliographical references with the authors' names mentioned directly in the text and can also consult a complete numerical bibliography at the end of each chapter. The most recent references concerning CT scanning of the sella have been added at the end of the bibliography.

Besançon, September 1980 *J.F. Bonneville · J.L. Dietemann*

Contents

Chapter 1

Embryology of the Sellar Region

Chapter 2

Anatomy of the Sellar Region

Chapter 6

Intrasellar Pathology

Chapter 7

Suprasellar Pathology

Chapter 8

Presellar Pathology

Chapter 9

Parasellar Pathology

Chapter 10

Retrosellar Pathology

Chapter 11

Infrasellar Pathology

Chapter 12

Sella Turcica in Raised Intracranial Pressure and Hydrocephalus

Chapter 13

Generalized Diseases and Changes in the Sella Turcica

Chapter 1 Embryology of the Sellar Region

A. Development of the Sphenoid Bone[1]

The cranial skeleton develops in three stages: the membranous stage (desmocranium), the cartilaginous stage (chondrocranium), and the stage of ossification (osteocranium). This development begins at the paraxial mesoderm, which envelopes the cranial end of the notochord (Arey 1965). The skull derives from two different origins: first, from ossification of the cartilaginous skull (bones of the skull base), and second, from direct bone formation in the membranous skull (bones of the skull vault). In fact, the pterygoid processes and the greater wings of the sphenoid have a composite origin, both cartilaginous and membranous.

When considering the development of the sphenoid bone it is necessary to distinguish four stages (Augier 1931; Kier 1968):

I. Membranous Stage

The membranous skull is formed by the development of mesodermal somites which grow and differentiate from the front backward and from the inside outward, beginning at the notochord.

During the 1st month of fetal life, the membranous skull is formed by only a triangular plate (the basal plate) which has an anterior apex that reaches the pituitary region. Its base is posteriorly located and corresponds to the foramen magnum. This plate then extends forward to reach the ethmoid region, and surrounds the pharyngeal diverticulum of Rathke. The mesenchyme of the basal plate surrounds the cranial nerves, forming the primitive basal foramina.

II. Cartilaginous Stage

Beginning in the 2nd month of fetal life, the chondrocranium develops caudocranially around the notochord. It derives from numerous chondrification centers.

The sphenoid region is divided into two parts:

The *posterior part* (hypophyseal or postsphenoid portion) embodies the pituitary gland and gives rise to the osseous postsphenoid. In a fetus of 40 mm, postsphenoid cartilage extends backward without demarcation with the occipital cartilage. The pituitary fossa is limited posteriorly by the dorsum sellae and anteriorly by the tuberculum sellae. Laterally, postsphenoid cartilage is continuous with the alisphenoid (future greater wing of the sphenoid).

The *anterior part* (prehypophyseal or presphenoidal portion) extends from the tuberculum sellae to the cribiform plate of the ethmoid bone. The osseous presphenoid derives from this anterior portion. In a fetus of 40 mm, the presphenoidal region consists of a wide surface, the hypochiasmatic plate, in front of which the future planum sphenoidale is located. Laterally, the presphenoid is continuous with the orbitosphenoid. The latter is composed of two crura which mark the boundary of the optic canal (the crus anterius and the crus posteriorus of the orbitosphenoid) (Kodama 1971).

1 Refs. [100, 288, 495, 497, 509, 510, 511, 512]

III. Stage of Ossification

Ossification of the sphenoid is complex, being derived from 18 or 19 ossification centers (Augier 1931). In the following section the different ossification centers and their respective ossifying periods are discussed first, then their pre- and postnatal development.

1) Ossification Centers, Ossifying Periods, and Fusion

Schematically, the sphenoid is formed by the fusion of six different regions: the presphenoid, the postphenoid (basisphenoid), the orbitosphenoid, the alisphenoid, the pterygoid centers, and the ossicles of Bertin.

The development of the sphenoid body is of particular interest; it takes place essentially by the fusion of the pre- and postsphenoid; however, the orbitosphenoid also participates in the formation of the sphenoid body, notably in its presphenoid portion (Fig. 1.1).

The alisphenoid, which produces the greater wings of the sphenoid and like the pterygoid centers does not contribute to the formation of the sphenoid body, will not be discussed further. The ossicles of Bertin (future sphenoid sinus) will be described briefly.

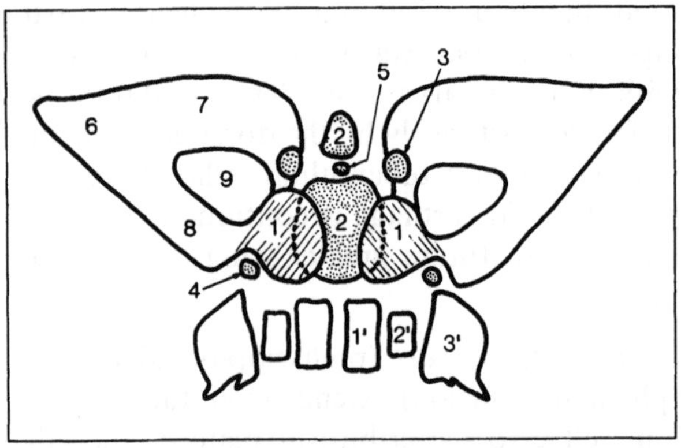

Fig. 1.1. The ossification centers of the presphenoid, postsphenoid, and orbitosphenoid. Modified from Kodama (1965) and Sasaki and Kodama (1973).

Presphenoid centers:
1 Main centers
2 Medial corporeal center
3 Anterior accessory center
4 Posterior accessory center
5 Middle accessory center
6 Orbitosphenoid, with its anterior crus, *7;* posterior crus, *8;* and optic canal, *9*

Postsphenoid centers:
1' Medial postsphenoid center
2' Lateral postsphenoid center
3' Alisphenoid center

a) Postsphenoid Centers

These centers, of which there are four, have been studied by Augier (1931), Kodama (1971) and Sasaki and Kodama (1973). They appear below the pituitary gland, within the postsphenoidal cartilage. There is one pair of medial centers and another of lateral centers. These centers begin to appear in the fetus at the end of the 3rd month (9 cm stage). They are always present in embryos larger than 13 cm (Fig. 1.1).

The union of these four centers takes place in one of two ways (Sasaki and Kodama 1973): either by joining together of the two medial centers followed by fusion with the lateral centers (80% of cases); or by joining together of a lateral center with a medial center on one side, followed by fusion of the two medial centers (20% of cases). This fusion occurs between the end of the 5th month (23.5 cm) and the beginning of the 6th month (26.5 cm). When the postsphenoid is formed it fuses with the surrounding structures.

With the orbitosphenoid, this fusion starts at the beginning of the 6th month (26.5 cm) and is always complete in the fetus of 41 cm.

With the presphenoid, this fusion begins laterally toward the end of the 6th month (28 cm) and is complete in the fetus of 46 cm.

With the alisphenoid fusion starts at the beginning of the 9th month and is always complete by the 10th month after birth.

The medial centers of the postsphenoid form the sellar floor and the dorsum sellae (i.e., all the medial osseous structures extending from the tuberculum sellae to the sphenooccipital synchondrosis). The lateral centers of the postsphenoid form the lingula sphenoidalis and the carotid sulcus.

b) Presphenoid Centers

These nine centers (consisting of two main and seven accessory centers) have been well described by Kodama (1965) and Kier (1968). As will be seen, these centers appear much later in fetal life than the postsphenoid centers (Fig. 1.1).

The two main centers are symmetric and situated near the posterior crus of the orbitosphenoid; they appear in the fetus of between 18 and 29.5 cm (4th to 6th month).

The two anterior centers are symmetric and situated anterior to the main centers, medial to the upper root of the future lesser wing of sphenoid; they appear in the fetus of between 24.5 cm (end of the 5th month) and the 9th month.

The two posterior centers, symmetric and located immediately behind the main centers, may appear in the fetus of 31 cm (approx. 7th month), but are often not ossified during fetal life.

A middle accessory center, situated in the midline between the two parts of the medial corporeal center, may appear in a fetus of 34 cm (8th month). It is often absent, however, since the center is ossified at birth in only 30% of cases.

The medial corporeal center consists of two parts and is deeply situated. It appears in the fetus of between 8th and the 9th month.

Fusion of these presphenoid centers and those of the opposite side with the orbitosphenoid and postsphenoid centers takes place progressively:

Orbitosphenoid. The main presphenoid centers unite at an early stage with the posterior crus of the orbitosphenoid (between 20 and 25 cm). The main centers then fuse with the anterior crus of the orbitosphenoid (between 26.5 cm and the 9th month).

Basisphenoid. This fusion occurs in the fetus between 31.5 cm and the 9th month.

Presphenoid of the opposite side. The medial corporeal centers unite with the two main centers which will unite in the midline. These fusions begin in the fetus of 31.5 cm and end in the 9th month.

The presphenoid centers form the presphenoid part of the sphenoid body. The two main centers, being together in the midline, constitute the tuberculum sellae. The chiasmatic sulcus is formed by the union of the anterior accessory centers.

c) Orbitosphenoid

This merits particular attention. It forms the lesser wing of the sphenoid and the optic canal; however, it also participates in the formation of both the presphenoid portion of the body of the sphenoid and the planum sphenoidale. As Kodama (1965) has shown, ossification of the orbitosphenoid begins in the posterolateral part of the optic canal at an early stage, i.e., in the embryo of 12.5 cm (Fig. 1.1). This ossification extends around the future optic canal. In the embryo of 30 cm, the orbitosphenoid (medial part of the future upper root of the lesser wing of the sphenoid) fuses with the main presphenoid center to form the optic canal. The orbitosphenoid unites with the medial corporeal center toward the 8th month of fetal life, and thus contributes to the formation of the body of the sphenoid. The two orbitosphenoids develop toward the midline, overlapping the presphenoid, and fuse to form the planum sphenoidale. Fusion in the midline of the orbitosphenoids ends during the 1st year after birth.

d) Alisphenoid Centers

These are paired and symmetric, appearing in the lateral part of the ala temporalis, lateral to the foramen rotundum, in the embryo of 4 cm. These centers unite with the postsphenoid centers, this process beginning in the fetus of 9 months and ending during the 10th month after birth.

e) Pterygoid Centers

These centers are paired and symmetric. They appear precociously in the embryo of 30 mm. They develop upward to form the medial pterygoid lamina and the vaginal processes. Finally, these centers unite with the alisphenoid and the postsphenoid.

f) Ossicles of Bertin

The ossicles of Bertin are paired and symmetric. Each is formed by the union of one, two, or three centers appearing between the 4th and 6th months. At birth, these two centers join and form a triangular plate near the sphenoidal rostrum. On the nasal side, this plate is hollowed by the future sphenoid sinus. The ossicles of Bertin do not fuse completely with the sphenoid body until puberty.

2) Prenatal Development of the Pre- and Postsphenoid Ossification Centers

In the preceding section, the different ossification centers and their development were discussed. The present section concerns three particular aspects:

Fusion between the pre- and postsphenoid
Ossification of the dorsum sellae
The craniopharyngeal canal

a) Fusion Between the Pre- and Postsphenoid

This fusion gives rise to the sphenoid body. It begins in the fetus of 31.5 cm and is finished during the 10th month of intrauterine life (Kodama 1965). Fusion initially takes place laterally (Kier 1968). A presphenoid cartilaginous zone thus continues to exist, bounded posteriorly by the anterior margin of the postsphenoid, laterally by the union of the pre- and postsphenoid, and anteriorly by both the main presphenoid ossification centers and the medial corporeal center. This presphenoid region will form the subjacent region of the tuberculum sellae. In fact, this cartilaginous region constitutes a column of cartilage extending from the upper to the lower surface of the sphenoid body. This column ossifies progressively during the last 3 months of fetal life by posterior and medial growth of the presphenoid centers. At birth, only the upper part of this column is ossified and it remains fragile. The remainder of the cartilaginous column ossifies more or less quickly during the 1st year of postnatal life. This cartilaginous zone is termed the intersphenoid synchondrosis.

b) Ossification of the Dorsum Sellae

Although the postsphenoid ossifies during early fetal life, the dorsum sellae remains cartilaginous until birth and only ossifies after birth by extension of the sphenoid body. It is completely formed at about 4 years.

c) Craniopharyngeal Canal[2]

This canal, which at the cartilaginous stage transmits the pharyngeal pedicle of the pituitary gland, normally disappears very early in the chondrocranium period (at the 8th week of fetal life), being occluded by growth of the surrounding cartilage. In the fetus of 9 months, two medial canals develop, extending from the upper to the lower surface of the sphenoid (Kodama 1965). The anterior orifice is located in or slightly in front of the tuberculum sellae, whereas the posterior is behind it. The anterior orifice represents the intersphenoid synchondrosis and closes during the 1st year of life. The posterior orifice represents the upper tip of a medial canal crossing the entire sphenoid body. It is this canal which has wrongly been named the "craniopharyngeal canal." Indeed, this canal is formed during the 9th month of fetal life (Kodama 1965). Studies carried out by Lowman et al. (1966) show that this canal never contains remnants of pituitary tissue but always an artery (which comes from arteries of the pituitary capsule) and a vein. These vessels supply the sphenoid body. It is sometimes possible to find other canals, situated laterally, which also contain vessels (Harwood-Nash and Fitz 1976). The vascular canal usually closes during the 1st year of life. In exceptional cases it may persist in the adult but without any pathologic significance. Lowman et al. (1966) have suggested that it would be preferable to call this the "intrasphenoidal canal."

IV. Postnatal Development of the Basisphenoid

1) Postsphenoid

The dorsum sellae ossifies progressively by extension of the sphenoid body until the age of 4 years.

2) Presphenoid

In the postnatal development of the presphenoid, the planum sphenoidale is of paramount

---

2 Refs. [18, 20, 151, 240, 257, 513, 582, 732, 889, 909]

interest. At birth, the planum sphenoidale does not exist; it is formed during the 1st year of postnatal life by fusion in the midline of the medial portions of the two orbitosphenoids. The orbitosphenoid overlaps the upper surface of the presphenoid. The posterior boundary of the orbitosphenoid, forming the planum, is the limbus sphenoidale (Kodama 1971).

In addition, to its orbitosphenoidal origin, the planum sphenoidale present a second important characteristic: its posterior growth. During the 1st year of postnatal life, the planum develops backward and progressively overlaps the presphenoid; yet it remains separated from the subjacent presphenoid. Fusion of the two structures occurs during the first 6 years of postnatal life. The posterior boundary of the planum sphenoidale (future limbus sphenoidale) is separate from the subjacent sphenoid in 18% of children aged less than 6 years (Kier 1968). That part of the presphenoid which is not covered by the planum sphenoidale constitutes the chiasmatic sulcus. As Kier (1968) has demonstrated, the respective lengths of the planum sphenoidale and the chiasmatic sulcus depend on posterior development of the planum; thus, a long planum sphenoidale is found with a short chiasmatic sulcus and vice versa.

B. Development of the Sphenoid Sinus[3]

Examination of the embryologic development of the sphenoid sinus facilitates understanding of the septa of the sphenoid sinus, as found in the adult.

I. Prenatal Development

The primitive sphenoid sinus is formed by an outgrowth of the nasal mucous membrane in the nasal capsule, and it appears between the 9th and 10th week of fetal life. The size of this posterior recess does not vary during intrauterine life. Toward the 5th month of fetal life, the primitive sphenoid sinus is surrounded by several ossicles (of Bertin) which will fuse about the 9th month. At birth, there is thus a small mucous cavity (several mm in diameter) connected to the nasopharynx and located within the ossicles of Bertin. This primitive sphenoid sinus is completely separated from the sphenoid body at this stage.

II. Postnatal Development

About the 4th year of postnatal life, the ossicles of Bertin, containing the primitive sphenoid sinus, unite with the presphenoid. Pneumatization of the presphenoid begins, which then extends toward the postsphenoid and laterally toward the greater wings of the sphenoid. At puberty, the sphenoid sinus reaches its final size. However, after disappearance of the spheno-occipital synchondrosis (at about 18 years), the sphenoid sinus may extend toward the occipital bone. Formation of bone sphenoid septa can be explained by the absence of resorption in the fusion zones of the different pre- and postsphenoid ossification centers (Cope 1917; Congdon 1920; Hammer and Rådberg 1961). These fusion zones are, in fact, more resistant. Resistance in the area of fusion between the pre- and postsphenoid explains the absence, in certain cases, of extension of the sinus toward the postsphenoid, corresponding to the presellar type. When the sinus penetrates the postsphenoid, the area of fusion between pre- and postsphenoid forms the transverse septum. Correlation of Kodama's fetal studies (1965) concerning presphenoid ossification centers with anatomic findings confirms Cope's theory (1917): the presphenoid derives from a medial group and a lateral group of ossification centers; in practice the sphenoid sinus is most commonly found to have three cells and two septa.

In the postsphenoid formed by two postsphenoid centers there are usually two cells. However, there may be lateral extensions, separated from the postsphenoid cavities by the fusion zone between the medial and lateral postsphenoid centers. Cope's diagrams (1917) illustrate the different varieties of septa. It is

3 Refs. [100, 171, 174, 417, 914, 949, 988]

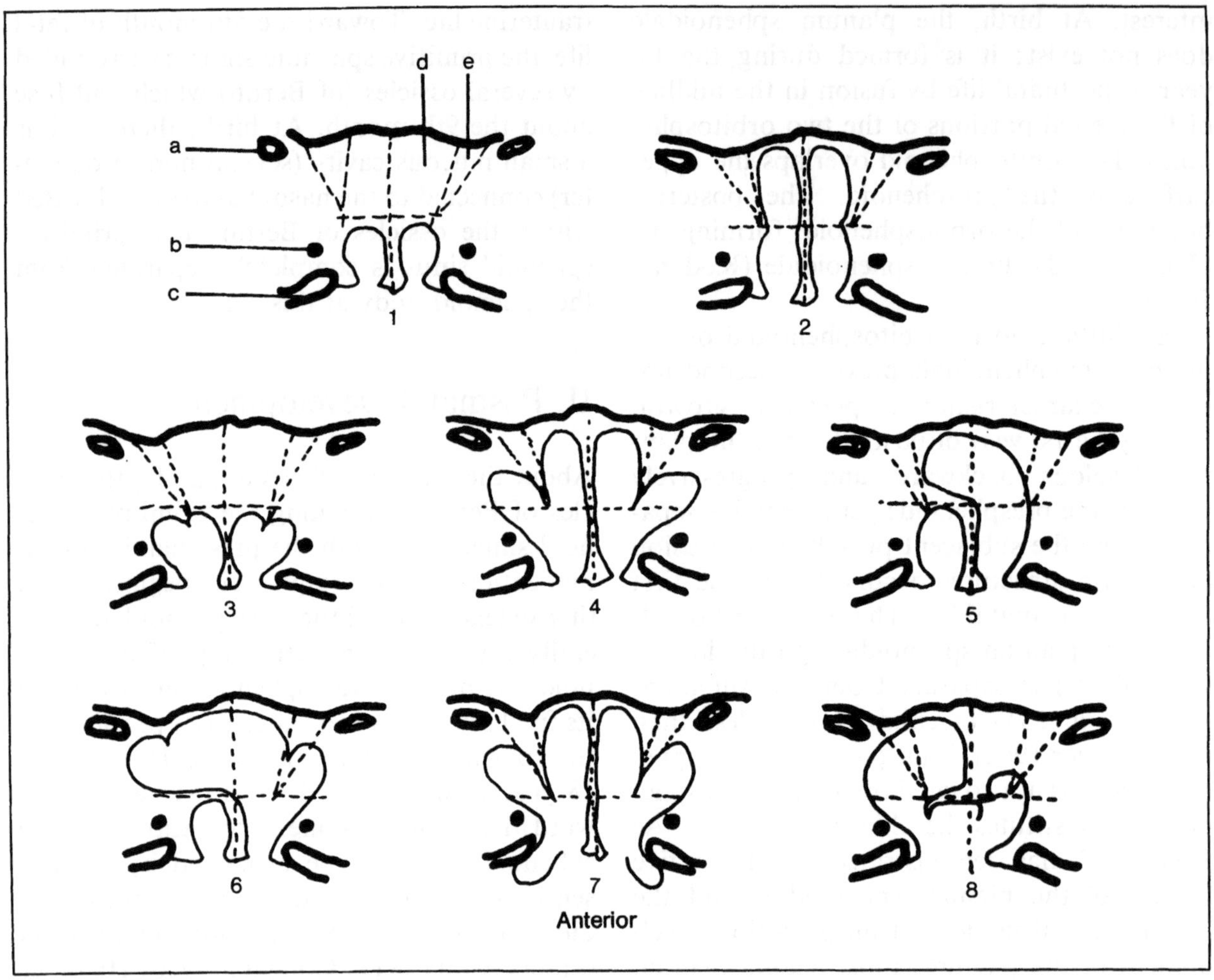

Fig. 1.2. The development of the sphenoid sinus and of the septal variations (Cope 1917).

a Foramen ovale
b Foramen rotundum
c Superior orbital fissure
d Basipostsphenoid center
e Lingula center

1 Presphenoid sinus
2 Pre- and postsphenoid sinus
3 Presphenoid sinus with small lateral recesses
4 Postsphenoid sinus with marked lateral recesses
5 and *6* Unilateral extension of the sphenoid sinus toward the postsphenoid
7 Pre- and postsphenoid sinus with lateral and anterior recesses
8 Extension of the lateral recess toward the postsphenoid

necessary to note, however, that this diagram is based on an erroneous conception concerning the number of presphenoid ossification centers and is only valid, therefore, for the postsphenoid (Fig. 1.2).

C. Development of the Pituitary Gland

The pituitary gland has a double origin: the anterior lobe derives from the ectoderm of the stomodeum (Rathke's pouch), and the posterior neural lobe of the hypophysis derives from the neuroectoderm of the floor of the forebrain. Development of the pituitary takes place during the first 4 months of fetal life (Arey 1965).

I. Neurohypophysis

Behind Rathke's pouch there is an outgrowth of the floor of the forebrain, which forms the neurohypophysis and the pituitary stalk.

II. Adenohypophysis

In embryos of 8 mm, a recess in the roof of the stomodeum forms Rathke's pouch (Fig. 1.3). This outgrowth becomes constricted by the surrounding mesenchyme and forms a vesicle in the embryo of about 15 mm. The pharyngohypophyseal pedicle continues to exist until the 8th week, at which time it disappears completely. However, there are frequently some remnants of the lower part of the pharyngohypophyseal pedicle below the sphenoid or even in the sphenoid body. These remnants are, respectively, the pharyngeal hypophysis and the parahypophysis. During the 3rd and 4th months of fetal life, the pharyngeal pouch acquires its characteristic structure (Arey 1965) (Fig. 1.4). Thus the cavity of the pharyngeal pouch is transformed into a "residual lumen." The anterior wall of the pouch thickens and differentiates into adenohypophyseal glandular tissue. The posterior wall, located between the residual lumen and the neurohypophysis, does not develop, and constitutes the pars intermedia in the adult. The pars tuberalis is formed by the extension of adenohypophyseal glandular tissue along the anterior aspect of the pituitary stalk. In the pars tuberalis, residual pharyngeal epithelial cells corresponding to the upper part of the hypophyseal duct may be found.

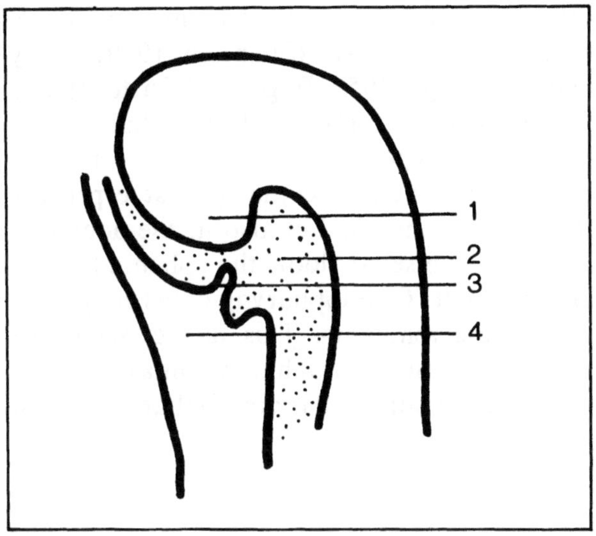

Fig. 1.3. The development of Rathke's pouch (embryo of 8 mm).
1 Forebrain
2 Mesoderm
3 Outgrowth in the roof of the stomodeum forming Rathke's pouch
4 Stomodeum

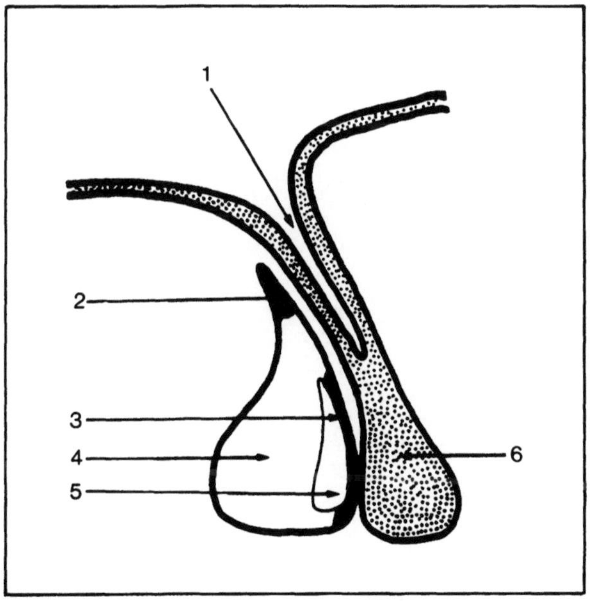

Fig. 1.4. Pituitary gland of human embryo (3 months).
1 Infundibular recess of the third ventricle
2 Pars tuberalis
3 Pars intermedia (derived from the posterior wall of Rathke's pouch)
4 Adenohypophysis (derived from the anterior wall of Rathke's pouch)
5 Residual lumen
6 Neurohypophysis

III. Capsule of the Pituitary Gland

Both the anterior and posterior lobes of the hypophysis are surrounded by a capsule which extends toward the pituitary stalk, where it joins the intracranial pia mater. As Ciric (1977) has recently shown, this capsule derives from mesoderm between the neuroectodermal recess and Rathke's pouch. This same mesoderm also gives rise to the pia mater.

D. Main Anomalies in the Fetal Development of the Sellar Region[4]

I. In the Postsphenoid

Lack of fusion of the postsphenoid centers in the midline. This causes a sagittal cleft in

4 Refs. [271, 369, 876, 886].

the sphenoid body (Harwood-Nash and Fitz 1976). [For the persistent intrasphenoidal canal (see Chap. 1.A.III.2c).]

II. In the Pre- and Orbitosphenoid

Lack of fusion of the planum sphenoidale with the chiasmatic sulcus. This causes an unfused planum sphenoidale, which must be distinguished radiologically from a fracture of the planum (Smith and Kier 1971).

Absence of development of the optic strut of the lesser wing of the sphenoid: the optic foramen is then in direct communication with the superior orbital fissure.

III. In the Pituitary Gland and the Pituitary Stalk

Pars intermedia: the residual lumen is lined by cuboidal or columnar ciliated epithelium, which sometimes contains mucus-secreting cells. Epithelial cysts result from slow secretion into these cysts (Shanklin 1949; Berry and Schlesinger 1959; Fager and Carter 1966).

Pars tuberalis: islands of epithelial cells of malpighian type corresponding to remnants of the hypophyseal duct may persist, especially in the pars tuberalis but also in the adenohypophysis, the sellar floor, the sphenoid body, and below it. Craniopharyngiomas may arise from these remnants. This fact explains the topography of craniopharyngiomas and the frequency of suprasellar craniopharyngiomas.

Chapter 2 Anatomy of the Sellar Region

A. Descriptive Anatomy of the Sellar Region

I. Sella Turcica

The sellar region includes the sella turcica itself and the presellar region, which extends from the tuberculum sellae to the cribriform plate of the ethmoid bone.

The sella turcica is a central depression in the upper aspect of the body of the sphenoid bone. Posteriorly, the sella turcica is bounded by the dorsum sellae, laterally by the carotid sulcus, and anteriorly by the tuberculum sellae.

The presellar region includes the chiasmatic sulcus and the planum sphenoidale (Figs. 2.1–2.3).

1) Dorsum Sellae

This is the posterior wall of the sella turcica. The dorsum sellae is rectangular in shape and tilted anteriorly. Its height, width, and thickness are variable, the average height being 7 mm, the average width 20 mm, and the thickness varying from 2 to 7 mm (Mahmoud 1958). Its median portion is much thinner than the lateral portions as a result of the depression on the anterior aspect of the dorsum sellae which contains the neurohypophysis. It is, however, rare for a foramen to be present (Le Double 1903).

The posterior aspect of the dorsum sellae may present a horizontal transverse depression which corresponds to the groove for the basilar sinus. (The basilar sinus connects the posterior portions of the two cavernous sinuses.)

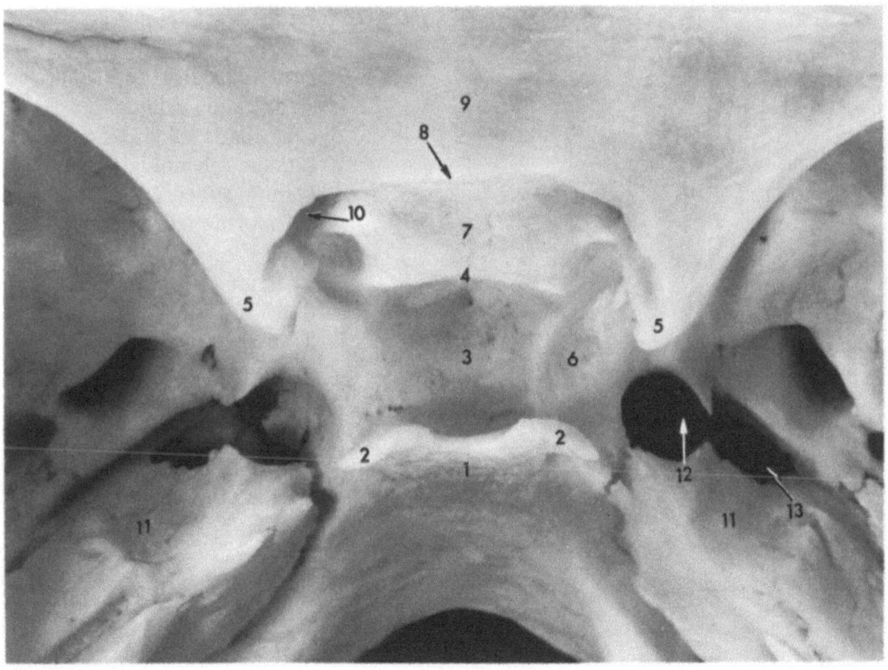

Fig. 2.1. Endocranial surface of the skull base (seen from above).
1 Dorsum sellae
2 Posterior clinoid processes
3 Floor of the sella turcica
4 Tuberculum sellae
5 Anterior clinoid processes
6 Carotid sulcus
7 Chiasmatic sulcus
8 Limbus sphenoidale
9 Planum sphenoidale
10 Cranial opening of the optic canal
11 Trigeminal impression
12 Cranial opening of the carotid canal
13 Foramen lacerum

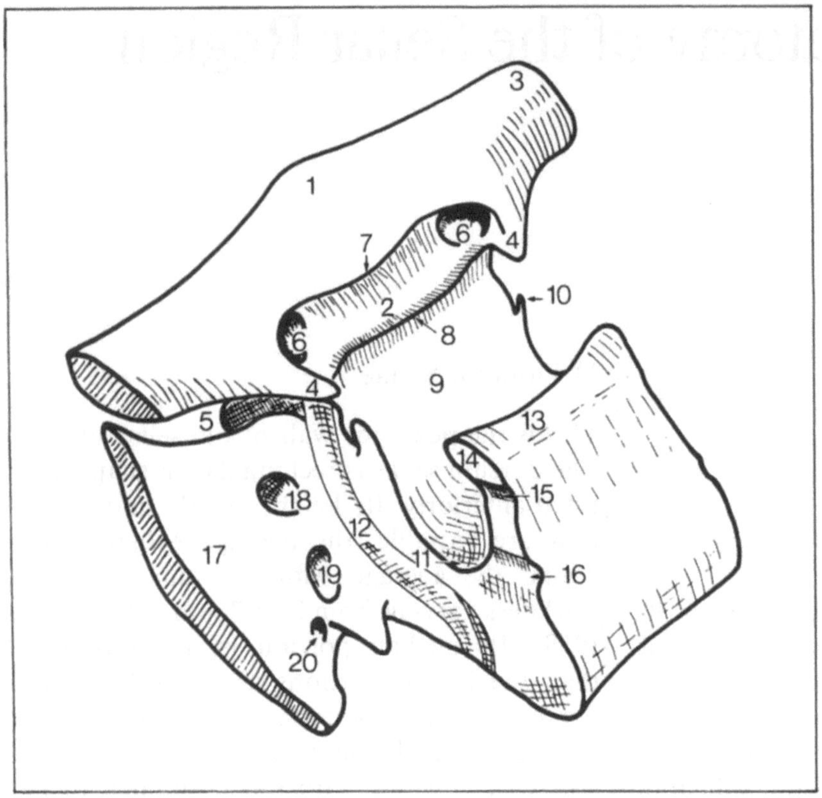

Fig. 2.2. Diagram of sella turcica (modified from Perlemuter and Waligora 1974).
1 Planum sphenoidale
2 Chiasmatic sulcus
3 Lesser sphenoid wing
4 Anterior clinoid processes
5 Superior orbital fissure
6 Cranial opening of the optic canal
7 Limbus sphenoidale
8 Tuberculum sellae
9 Anterior wall of the sella turcica
10 Middle clinoid process
11 Floor of the sella turcica
12 Carotid sulcus
13 Dorsum sellae
14 Posterior clinoid process
15 Notch for the oculomotor nerve
16 Notch for the inferior petrosal sinus
17 Greater sphenoid wing
18 Foramen rotundum
19 Foramen ovale
20 Foramen spinosum

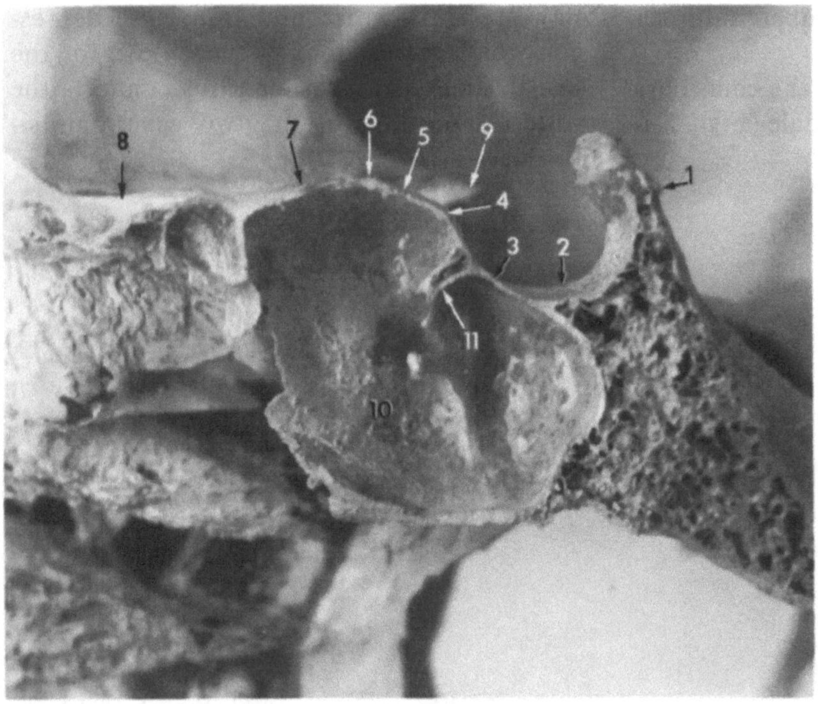

Fig. 2.3. Midline sagittal section of skull base.
1 Dorsum sellae
2 Floor of the sella turcica
3 Anterior wall of the sella turcica
4 Tuberculum sellae
5 Chiasmatic sulcus
6 Limbus sphenoidale
7 Planum sphenoidale
8 Cribiform plate of ethmoid bone
9 Anterior clinoid process
10 Sphenoid sinus
11 Transsphenoidal crest (remaining portion of the barrier between the pre- and postsphenoid)

The posterior clinoid processes form the lateral extensions of the superior edge of the dorsum sellae and give attachment to the fixed margin of the tentorium cerebelli.

At the lateral edge of the dorsum sellae a notch corresponding to the imprint of the sixth cranial nerve may be found in some cases (Pribram and du Boulay 1971).

2) Floor of the Sella Turcica

The sellar floor extends from the dorsum sellae to the tuberculum sellae; laterally, it is bounded by the carotid sulcus. On frontal sections the floor appears flat, although it may present a slight upward or downward concavity. The floor of the sella is formed by dense, resistant cortical bone, yet it is thin, notably anteriorly, where its thickness is 1 mm or less in 82% and less than 0.5 mm in 40% of specimens, according to Renn and Rhoton (1975).

3) Anterior Wall of the Sella Turcica

This is the anterior and upward extension of the sellar floor and is anteriorly bounded by the tuberculum sellae. The middle clinoid processes, when present, form the lateral limits of the anterior wall.

4) Tuberculum Sellae

This separates the sella turcica from the chiasmatic sulcus. The tuberculum sellae is more or less prominent and affects the ease with which a transfrontal approach to the pituitary fossa may be carried out (Renn and Rhoton 1975).

5) Middle Clinoid Processes

These are two condensations of bone which laterally limit the anterior wall of the sella turcica. They are situated below the tuberculum sellae and medial to the internal carotid artery.

6) Anterior Clinoid Processes

These are the posteromedial extensions of the lesser wings of the sphenoid and are situated lateral and posterior to the optic canal, and lateral to the internal carotid artery. They are variable in size and shape, and may be asymmetric.

There is frequently a bony bridge between the middle and anterior clinoid processes forming the caroticoclinoid canal (Le Double 1903; Bergland et al. 1968; Lang 1977). In 6% of cases there is also a bony bridge between the anterior and posterior clinoid processes (Bergland et al. 1968). The anterior clinoid processes give attachment to the anterior end of the free border of the tentorium cerebelli.

7) Carotid Sulcus

This is the groove on the lateral aspect of the sphenoid bone for the internal carotid artery. The superior margin of this groove forms the lateral limit of the sellar floor. The groove can bulge into the posterolateral portion of the sphenoid sinus. In 4% of normal cases there exists a bony defect in this groove and the internal carotid artery is separated from the sphenoid sinus only by the mucous membrane. In 66% of all cases the bony wall of the carotid sulcus is less than 1 mm thick and is consequently less resistant (Renn and Rhoton 1975).

II. Presellar Region

From behind forward the presellar region is formed by the chiasmatic sulcus and the planum sphenoidale, separated by the limbus sphenoidale.

1) Chiasmatic Sulcus

The chiasmatic sulcus is bounded posteriorly by the tuberculum sellae and anteriorly by the limbus sphenoidale. Laterally it extends toward the optic canals. It is generally a flat surface. When the sphenoid sinus is poorly pneumatized the chiasmatic sulcus may present a slight upward concavity, while in a well-pneumatized sinus the chiasmatic sulcus may present an upward convexity. It generally has a slight backward and downward tilt yet may be horizontal, or even vertical.

2) Planum Sphenoidale

This is a flat bony plate situated between the two lesser sphenoid wings, anteriorly bounded by the cribiform plate and posteriorly by the limbus sphenoidale. A midline crest extends along the upper aspect of the planum sphenoidale and is more pronounced at its anterior extremity.

3) Limbus Sphenoidale

This is a transverse bony crest which extends between the anteromedial portions of the two anterior clinoid processes. It separates the planum sphenoidale from the chiasmatic sulcus. The limbus sphenoidale is of variable size. Its prominence depends on the angle between the planum and the chiasmatic sulcus, and on the shape of the chiasmatic sulcus. It is well defined when the planum – chiasmatic sulcus angle is perpendicular and when the chiasmatic sulcus is flat and/or upwardly concave. Conversely, it is less well-defined when the planum and the chiasmatic sulcus form a straight line.

III. Ligaments and Unusual Ossifications of the Sellar Region

1) Interclinoid Ligaments

These join the anterior and posterior clinoid processes on each side. There may also be ligaments joining the middle clinoid processes to the anterior and/or posterior clinoid processes. All these ligaments may ossify.

2) Petroclinoid Ligaments

These extend on either side from the upper edge of the petrous apex (medial to the Gasserian ganglion) to the lateral aspect of the dorsum sellae (below the posterior clinoid processes). The sixth cranial nerve crosses under this ligament, which often ossifies.

3) Unusual Ossifications of the Sellar Region

a) Sellar Spine (Lang 1977)

This is a bony intrasellar spine arising in the midline from the inferior portion of the anterior aspect of the dorsum sellae. It is directed anterosuperiorly (Fig. 5.83).

b) Trigeminal Bridge

This is formed by ossification of the dura mater overlying the Gasserian ganglion. It ends near the dorsum sellae without actually reaching it.

c) Os Supra Petrosum of Meckel

This is a pea-sized cartilage located laterally and above the internal carotid opening (Currarino and Weinberg 1974; Lang 1977).

B. Relationships Between the Sella Turcica and Surrounding Structures

I. Structures Above the Sella Turcica

1) Dura Mater of the Sella Turcica

The floor of the sella turcica is completely covered by dura mater. A horizontal dural fold forms the roof of the sella turcica, extending from the dorsum sellae to the tuberculum sellae; it is called the diaphragma sellae (or sellar diaphragm). Laterally, the sella turcica is closed by a vertical dural fold which separates the pituitary fossa from the cavernous sinus. The pituitary gland is therefore enclosed, except at the diaphragma sellae where a central circular opening transmits the pituitary stalk (Fig. 2.4 and Fig. 2.5).

It should be noted that numerous variations of the diaphragma sellae are possible and are frequently encountered. These variations play an important role in the pathogenesis of numerous sellar diseases. The variations involve the following:

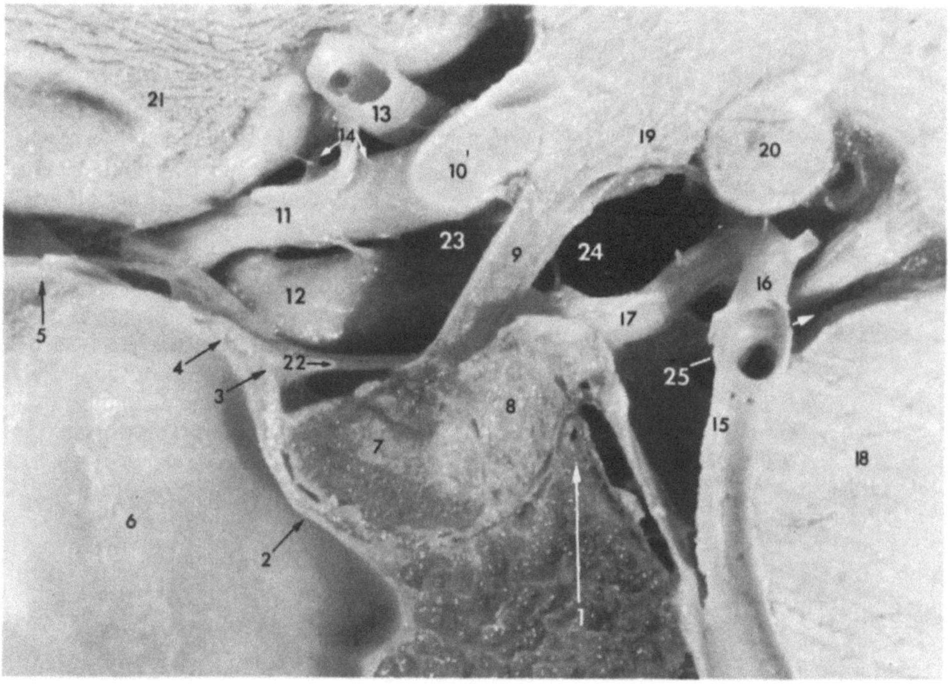

Fig. 2.4. Midsagittal section of the sellar region (Courtesy, H. Duvernoy).

1 Dorsum sellae
2 Anterior wall of the sella
3 Tuberculum sellae
4 Chiasmatic sulcus
5 Planum sphenoidale
6 Sphenoid sinus
7 Adenohypophysis
8 Neurohypophysis
9 Pituitary stalk
10 Optic Chiasm
11 Optic nerve
12 Internal carotid artery
13 Anterior cerebral artery
14 Chiasmal branch of the internal carotid artery
15 Basilar artery
16 Posterior cerebral artery
17 Posterior communicating artery
18 Pons
19 Hypothalamus
20 Mamillary body
21 Frontal lobe
22 Diaphragma sellae
23 Prechiasmatic cistern
24 Postchiasmatic cistern
25 Interpeduncular cistern

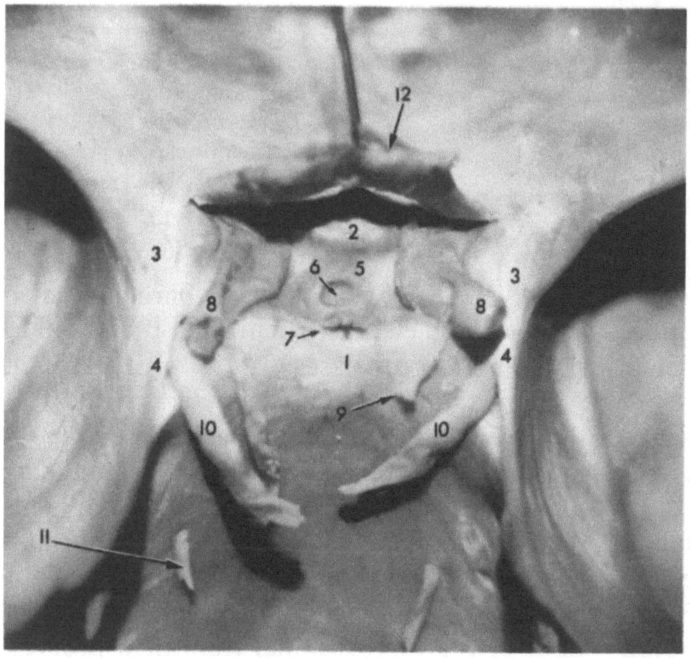

Fig. 2.5. Dura mater of the sellar region (seen from above) (Courtesy, C. Maillot).

1 Dorsum sellae
2 Tuberculum sellae
3 Anterior clinoid processes
4 Free edge of the tentorium
5 Diaphragma sellae
6 Small opening in the center of the diaphragma sellae
7 Pituitary stalk
8 Supraclinoid segment of internal carotid siphon
9 Posterior communicating artery
10 Oculomotor nerve
11 Abducent nerve
12 Optic chiasm

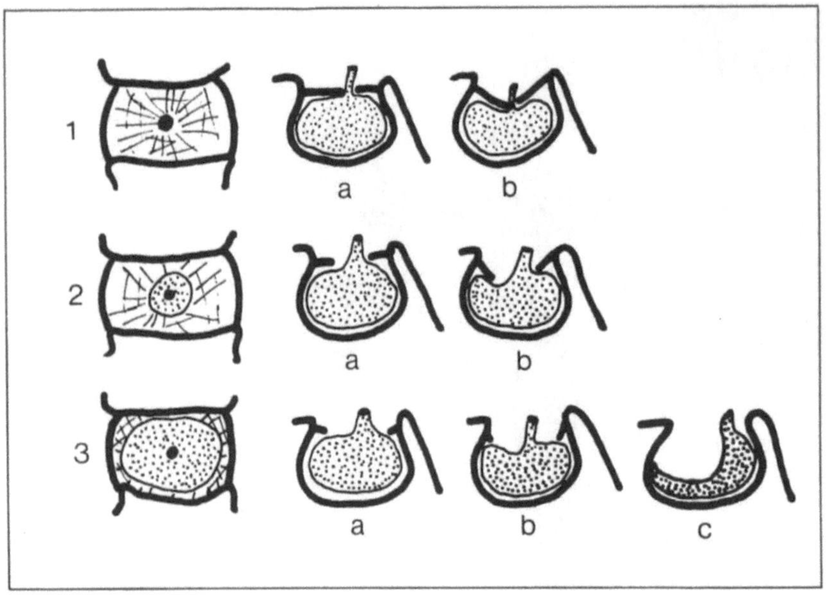

Fig. 2.6. Diagrams showing the types of diaphragma sellae deficiency (Busch 1951). Three types of opening are described: type 1 (41.9%) with only a small central opening that transmits the pituitary stalk; type 2 (37.6%) with a greater opening (3 mm); type 3 (20.5%) with a large central opening and a persistent dural rim of 2 mm in width. The diagram demonstrates the consequence of the diaphragma sellae deficiency on the shape of the pituitary gland.
Type 1: *a* (38.4%), *b* (3.5%);
Type 2: *a* (23.3%), *b* (14.3%);
Type 3: *a* (8.8%), *b* (6.2%), *c* (5.5%)

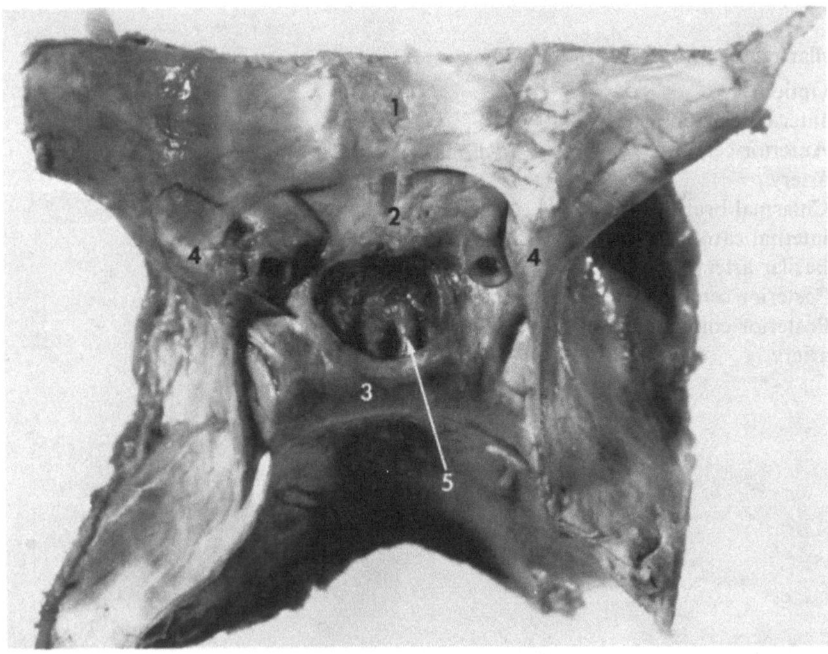

Fig. 2.7. Large opening in the diaphragma sellae (seen from above) (Courtesy, N. Heldt). The pituitary fossa seems to be empty, however, the pituitary gland is, in fact, flattened against the anterior aspect of the dorsum sellae.
1 Planum sphenoidale
2 Chiasmatic sulcus
3 Dorsum sellae
4 Anterior clinoid process
5 Pituitary stalk

1. Attachments; usually the diaphragma is attached to the upper edge of the dorsum sellae and on the tuberculum sellae. This attachment may be a few millimeters lower.
2. Thickness; usually the diaphragma sellae is thinner at the center and thicker at its periphery; it is often very thin throughout.
3. Opening (Fig. 2.6); numerous anatomic studies have been devoted to the shape of the diaphragma sellae (Schaeffer 1924; Busch 1951; Bergland et al. 1968; Renn and Rhoton 1975). These show that the diaphragma sellae is only completely closed around the pituitary stalk in 42% of cases. In 56% of cases, it is more than 5 mm in diameter. In 20% of the specimens studied by Busch, the diaphragma sellae was formed solely by a dural ring 2 mm in width. The consequences of this diaphragmatic deficiency are very important, because it allows intrasellar herniation of the suprasellar subarachnoid spaces, the pulsations of the cerebrospinal fluid then progres-

sively flatten the pituitary gland. This is confirmed by Busch's discovery that such a pituitary gland was flattened against the bottom of the sella turcica in 5.5% of his specimens. This anatomic fact explains the origin of the so-called primary empty sella turcica (Fig. 2.7, and Fig. 6.1).

The dura mater of the sella turcica contains intercavernous venous connections. These connections are essentially three in number (Renn and Rhoton 1975) (Fig. 4.21).

1. The most important is the basilar sinus, which is situated in the dura mater overlying the posterior surface of the dorsum sellae. It is present in 82% of specimens. The basilar sinus receives on each side the superior and inferior petrosal sinuses. It sometimes creates a groove on the posterior aspect of the dorsum sellae.

2. The anterior intercavernous sinus is situated anterior to the pituitary gland in 76% of specimens. Usually it is contained by the diaphragma sellae, but in 10% of cases it lies in the dura mater covering the anterior wall of the sella turcica.

3. The posterior intercavernous sinus has been found in 32% of specimens. It is situated behind the pituitary gland and is contained either by the diaphragma sellae or by the dura mater covering the anterior aspect of the dorsum sellae.

2) Pituitary Gland

The pituitary is an endocrine gland made up of two parts (adenohypophysis and neurohypophysis), and lying within the sella turcica. It is connected to the anterior part of the third ventricle by the pituitary stalk (Fig. 2.4).

a) Description

The adenohypophysis develops from Rathke's pouch (ectodermal diverticulum). The adenohypophysis synthesizes and releases seven different hormones: adrenocorticotropin (ACTH), thyrotropin (TSH), gonadotropins (FSH and LH), prolactin hormone (PRL), melanocyte stimulating hormone (MSH), and somatotropin (STH).

The neurohypophysis is of neurectodermal origin and is connected directly to the floor of the third ventricle by the pituitary stalk. It secretes vasopressin (ADH) and ocytocin, although these are produced within the hypothalamus, and are only released by the neurohypophysis.

The pituitary stalk connects the hypothalamus to the neurohypophysis. It is directed obliquely upward and backward, passing behind the optic chiasm. Its average length is around 5 mm. The pituitary stalk contains the portal circulation from the hypothalamic region to the pituitary gland (Popa and Fielding 1930).

b) Weight

The weight of the pituitary gland varies greatly. Rasmussen (1924) found that it varied for males between 350 and 800 mg, and for females between 450 and 900 mg; this may double during pregnancy.

c) Shape

The shape also varies greatly. Usually it is shaped like a chick-pea and has the following dimensions: height 5 mm, length 10 mm, and width 15 mm. The pituitary gland fills only 80% of the sellar volume, the remaining 20% being occupied by dura mater and intercavernous sinuses (Di Chiro and Nelson 1962).

The greater the age, the greater the alterations in shape. Two factors are important in modification of the shape: (1) deficiency of the diaphragma sellae, which leads to a flattening of the pituitary gland, and (2) intrasellar extensions of the intracavernous portion of the internal carotid artery, whose loops, found particularly in elderly patients, lead to lateral compression of the pituitary gland.

3) Suprasellar Vascular and Nervous Structures (Fig. 2.4)

a) Optic Chiasm

The optic chiasm is a flat quadrangular bundle of nerve fibers: Its anterolateral angles are continuous with the optic nerves, and the posterolateral angles with the optic tracts.

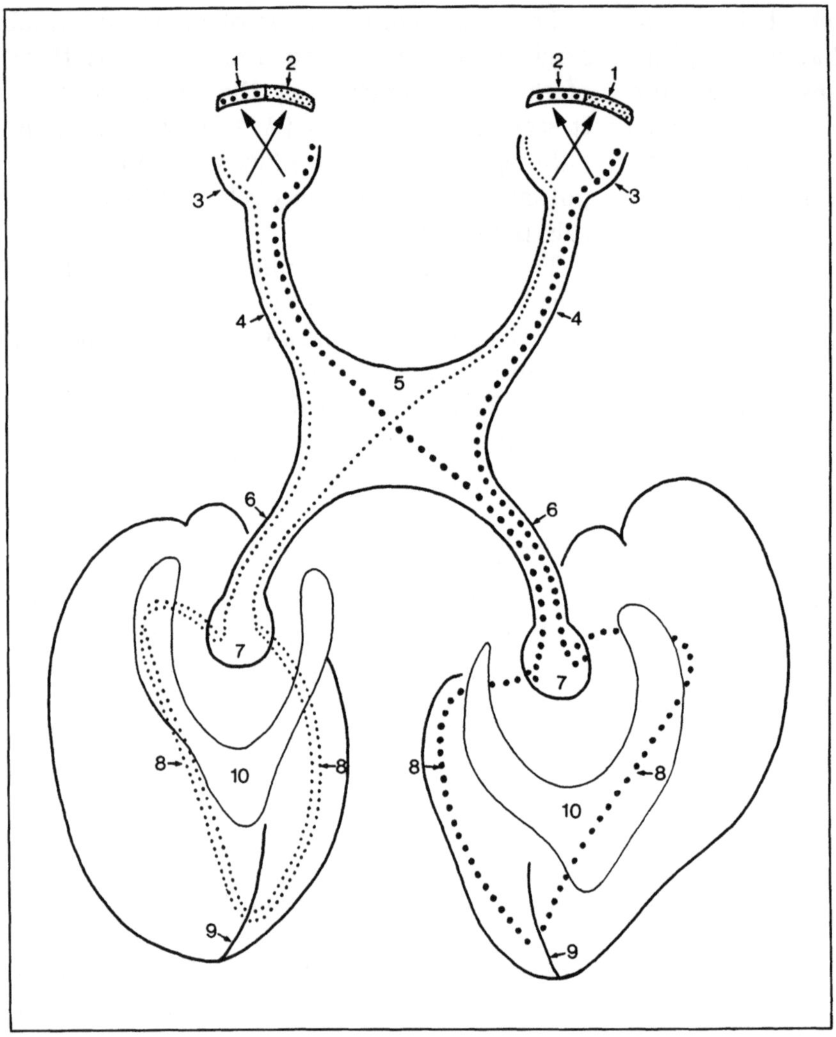

Fig. 2.8. Diagram of the visual pathways.
1 Temporal visual field
2 Nasal visual field
3 Temporal half of the retina
4 Optic nerve
5 Optic chiasm
6 Optic tract
7 Lateral geniculate nucleus
8 Optic radiation
9 Calcarine sulcus
10 Lateral ventricle

The average dimensions of the optic chiasm are: width 10 mm, length 6 mm, and thickness 2 mm. According to several authors (Schaeffer 1924; Bergland et al. 1968; Renn and Rhoton 1975), the optic chiasm overlies the diaphragma sellae in 80% of cases. In 9% it overlies the tuberculum sellae (prefixed chiasm) and in 11% overlies the dorsum sellae (postfixed chiasm). A postfixed chiasm is associated with long optic nerves, while with a prefixed chiasm the optic nerves are short. The angle between the two optic nerves varies from 50° to 80°. The space delimited posteriorly by the anterior edge of the optic chiasm and laterally by the two optic nerves is called the prechiasmatic space. It has to be taken into consideration during the transfrontal surgical approach to the pituitary gland. A small prechiasmatic space renders such an approach difficult. The topographical variations of the chiasm explain atypical visual field defects occurring with pituitary tumors. The spatial arrangement of nerve fibers within the optic chiasm is as follows (Fig. 2.8). The nerve fibers coming from the temporal half of each retina run in the lateral part of the optic chiasm and pass toward the ipsilateral optic tract. The nerve fibers coming from the nasal half cross the midline. Those originating from the superior quadrant of the nasal retina pass through the posterior part of the optic chiasm, while those originating from the inferior quadrant pass through the anterior part; both pass toward the controlateral optic tract.

b) Third Ventricle

The anterior part of the third ventricle is situated above and behind the optic chiasm and presents its two anterior recesses, the optic recess lying in front of the infundibular recess.

c) Upper Relations

The olfactory tracts, the anterior perforated substance, and the inferior aspect of the frontal lobes constitute the upper relations of the presellar region.

The branches of the circle of Willis which form the upper relations of the sella turcica are essentially the anterior cerebral artery (in its horizontal portion), the anterior communicating artery, and the posterior communicating artery (Fig. 2.4). The anterior cerebral artery arises from the internal carotid artery, slightly above and lateral to the optic chiasm and courses anteriorly and medially, crossing above the optic chiasm or optic nerve. The anterior communicating artery runs transversaly above the optic chiasm. The level of the crossing depends on the position of the optic chiasm and the degree of tortuosity of the horizontal portion of the anterior cerebral artery. The internal carotid artery ends 2 cm above the posterior clinoid process, lateral to the optic chiasm.

d) Subarachnoid Spaces

The suprasellar vascular and nervous elements are located in the suprasellar subarachnoid space. The arachnoid mater does not penetrate the sella turcica when the diaphragma sellae is completely closed. The arachnoid only covers the upper half of the pituitary stalk. However, as Key and Retzius (1875) and Ferner (1960) have demonstrated, the arachnoid and consequently the suprasellar subarachnoid space, may penetrate the pituitary fossa when there is a defect in the diaphragma sellae.

II. Lateral Structures: the Cavernous Sinus

The cavernous sinuses are paired, symmetric structures lying on either side of the sphenoid bone. They are venous lakes traversed by the carotid siphon, the oculomotor nerves (III, IV, VI), and the ophthalmic branch of the fifth cranial nerve. The walls of the cavernous sinus are dural and periosteal. The confluences are divided into three groups (anterior, inferior, and posterior).

1) Walls of the Cavernous Sinus

The studies of Taptas (1949), Bonnet (1955), and Patouillard and Vanneuville (1972) have demonstrated that the cavernous sinus is formed by a dural fold and that consequently only its superior, medial, and lateral walls are dural while the inferior wall is periosteal only (Figs. 2.9, 2.10).

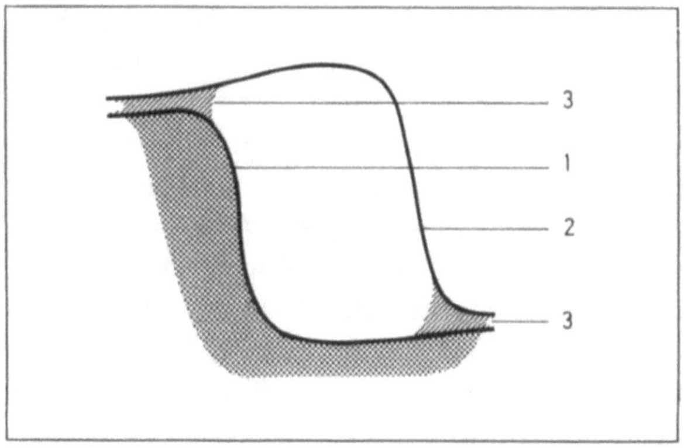

Fig. 2.9. Diagram showing the dural fold of the cavernous sinus.
1 Periosteal bone of the sphenoid body and greater wing
2 Superior and lateral dural wall of the cavernous sinus
3 Junction of the dura mater of the periosteal bone

2) Contents of the Cavernous Sinus

The cavernous sinus contains venous blood brought or drained by the three anterior, inferior, and posterior confluences. As shown by Bergland et al. (1968), the size of the sinus varies greatly; it may be large or completely obliterated by its vascular and nervous contents (Fig. 2.10).

The cavernous sinus contains the following: *The internal carotid artery,* which courses medially in the carotid sulcus below the level of the sellar floor. In elderly patients the carotid arteries are often tortuous and come to

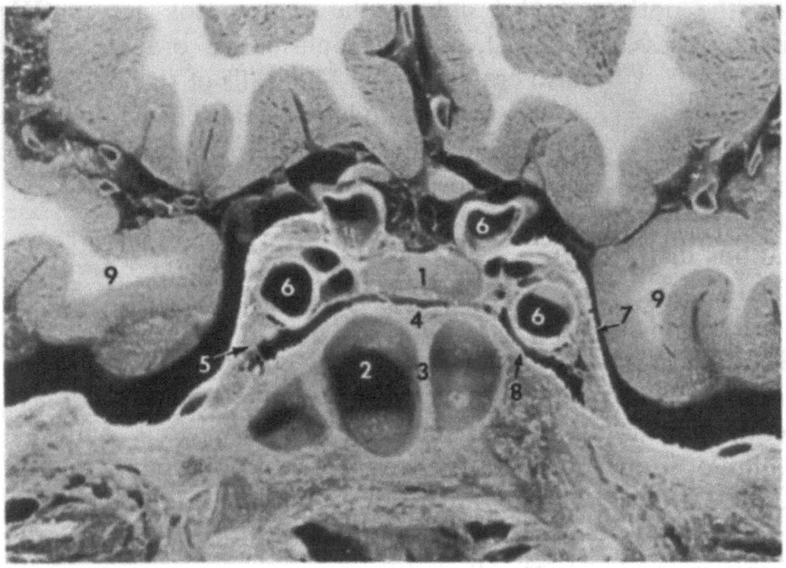

Fig. 2.10. Frontal section of the pituitary fossa (Atlas anatomique Sandoz)
1 Pituitary gland
2 Sphenoid sinus
3 Intersinusal septum
4 Sellar floor
5 Cavernous sinus
6 Internal carotid artery
7 Lateral dural wall of the cavernous sinus
8 Carotid sulcus
9 Temporal lobe

lie lateral to the pituitary gland, from which they are separated by only a thin dural layer. *The sixth cranial nerve* (abducent nerve), which runs lateral to the internal carotid artery. *The third and fourth cranial nerves and the ophthalmic division of the fifth,* which, surrounded by their own meningeal layer, lie laterally against the lateral dural wall of the cavernous sinus. *Between these different neurovascular elements* numerous fibrous trabeculae are to be found.

3) Afferent and Efferent Veins of the Cavernous Sinus

The afferent veins are anterior; they comprise the superior and inferior ophthalmic veins, the sphenoparietal sinus, and the uncinate vein.

The efferent veins are posterior (superior and inferior petrosal sinuses) and inferior (toward the pterygoid plexus). The superior petrosal sinus runs in the attached margin of the tentorium cerebelli and enters the transverse sinus. The inferior petrosal sinus courses in a groove between the temporal and occipital bones to end in the jugular bulb. The inferior efferent veins drain to the pterygoid plexus through the foramen ovale, the foramen lacerum, and the foramen of Vesalius.

III. Structures Below and Anterior: the Sphenoid Sinus

The relations between the sella turcica and the sphenoid sinus are most important; in fact the shape of the planum sphenoidale, chiasmatic sulcus, and sellar floor is directly related to the morphology of the sphenoid sinus.

1) General Shape and Size of the Sphenoid Sinus

The size of the sphenoid sinus is variable. Hammer and Rådberg (1961) described three anatomic types of sphenoid sinus: conchal (3%), presellar (11%), and sellar (86%).

2) The Septa in the Sphenoid Sinus

It is possible to distinguish an intersinusal septum, a transverse septum, a sagittal lateral septum, and often certain unclassifiable septa. As has been seen in Chap. 1 (Fig. 1.2), these septa correspond to the fusion areas of the pre- and postsphenoid ossification centers (Cope 1917).

a) Intersinusal Septum

Sometimes median in its anterior portion, this septum becomes lateral at its posterior extremity, where it is frequently attached to the carotid prominence (Fig. 2.10).

b) Transverse Septum

This represents the pre- and postsphenoid fusion area, and is attached to the anterior wall of the sella turcica. This septum may be complete but is often reduced to a transverse crest.

c) Sagittal Lateral Septum

This represents the fusion area of the postsphenoid center with the lingual center on either side. The posterior attachment of this septum is often on the anterior part of the carotid groove.

d) Unclassifiable Septa

Numerous transverse or sagittal septa are often found, especially in the presphenoid. These septa have an embryologic explanation.

Using Cope's theory concerning the formation of septa and Kodama's description of ossification centers, it is possible to understand the presence of numerous septa corresponding to the fusion areas of the main and accessory presphenoid centers. The variable growth of these centers could explain the numerous variations in the topography of these septa.

3) Important Variations in the Walls of the Sphenoid Sinus

Three important vascular or nervous structures (the intracavernous internal carotid artery, the optic nerve, and the upper two divisions of the trigeminal nerve) are usually separated from the sphenoid sinus by its bony wall, but may be in contact with the mucous membrane of the sinus, as the result of bony defects in the floor of the optic canal, carotid sulcus, or lateral wall of the sphenoid sinus. It has been noted (Schaeffer 1924; Renn and Rhoton 1975) that mucous diverticula extend from the sphenoid sinus to the cavernous sinus through these bony defects.

4) Recessus Sphenoethmoidalis

A case of an exceptional relationship between the sella turcica and the recessus sphenoethmoidalis has been described by Pecina and Novoselac (1970), whereby the anterior wall of the sella turcica and the chiasmatic sulcus were in direct relationship with the nasal cavities. In this case, the recessus sphenoethmoidalis, which is normally located between the sphenoid sinus posteriorly and the ethmoid sinus anteriorly, was hypertrophic.

IV. Posterior and Anterior Structures

Posteriorly, the sella turcica is related to the basilar artery, the proximal portions of the posterior cerebral arteries, the anterior aspect of the cerebral peduncles, the posterior perforated substance, and the pons; all these structures are enclosed by the interpeduncular fossa and pontine cistern. Anteriorly, the sella turcica is related to the cribiform plate of the ethmoid bone.

C. Vascular Supply of the Sellar Region

I. Dura Mater and Osseous Structures

1) Sellar Floor

The supply comes essentially from the capsular arteries, which arise from the medial wall of the internal carotid artery. These may be divided into the anterior branches, which arise from the supracavernous segment of the internal carotid artery and supply the anterior part of the diaphragma sellae, and the inferior branches, which arise from the intracavernous segment, and supply the dura mater of the sellar floor (McConnell 1953; Faure et al. 1972). The inferior hypophyseal artery, arising from the meningohypophyseal trunk, supplies a part only of the dura mater of the sellar floor and diaphragma sellae. The clival branch of the meningohypophyseal trunk supplies the dura mater of the dorsum sellae, which is also supplied by branches arising from the ascending pharyngeal artery and from the vertebral artery (Djindjan and Merland 1978).

The vascular supply of the diaphragma sellae (Lee et al. 1978) comes from the inferior hypophyseal artery, for the posterior quadrant, from branches arising from the intracavernous internal carotid artery, for the right and left lateral quadrants, and from anterior capsular arteries for the anterior quadrant.

2) Cavernous Sinus

The dura mater is supplied mainly by the inferior cavernous sinus artery, which also supplies the dura mater of the middle cranial fossa and has connections with branches arising from the middle meningeal artery (Parkinson 1965), the accessory meningeal artery (Lasjaunias 1975), and the artery of the foramen rotundum (Djindjan and Merland 1978). The posterior part of the dura mater of the cavernous sinus is supplied by the clival branch of the ascending pharyngeal artery.

3) Presellar Region

The dura mater is supplied by right and left posterior ethmoid arteries arising from the ophthalmic artery, and by the supracavernous branches of the internal carotid artery.

The bone is supplied by branches arising from the dural arteries.

II. Pituitary Gland

The origin of the vascular supply is twofold: the superior hypophyseal group to the adenohypophysis and the inferior hypophyseal arteries to the neurohypophysis.

1) Superior Hypophyseal Group

Numerous arterial branches arise on each side of the supraclinoid internal carotid artery, before its bifurcation. The extent of these arteries has been assiduously studied by Duvernoy (1958). They give rise to the hypophyseal portal vessels. There probably exist in man, as in numerous mammals, connections between arteries of the adenohypophysis and of the neurohypophysis.

2) Inferior Hypophyseal Arteries

One such vessel is found on either side; it is the medial terminal branch of the meningohypophyseal trunk, arising from the first part of the carotid siphon. The meningohypophyseal trunk divides into three branches (Parkinson 1965): the tentorial artery, the dorsal meningeal artery, and the inferior hypophyseal artery. This latter runs medially and bifurcates into superior and inferior branches which supply mainly the neurohypophysis. These branches anastomose with the controlateral branches.

III. Optic Chiasm

The chiasm is well supplied by numerous branches arising from neighboring arteries (the anterior cerebral and anterior communicating arteries, the superior hypophyseal arteries, and the internal carotid artery). This has been described fully by Bergland and Ray (1969).

D. Innervation of the Sellar Region

The nerve supply of the bone and dura mater of the sellar region comes from the Gasserian ganglion and the three divisions of the trigeminal nerve. The nerve supply of the planum sphenoidale comes from the ophthalmic nerve.

Chapter 3 Radiographic Techniques

Radiologic examination of the sella turcica requires frontal and lateral projections of the highest technical quality, which are sometimes complemented by tomography.

A. Plain Radiography

Routine radiographs are essential and must be of perfect quality. They allow not only a study of the morphology of the sella turcica but also an assessment of the state of the lamina dura.

I. Equipment and Film Quality Factors

It is possible to obtain a good radiograph of the sella turcica using ordinary X-ray tables; the ideal apparatus is obviously the skull radiographic unit, whether it be isocentric or not. Tilting fluoroscopic tables allow control of the projection by television fluoroscopy but, of course, this should not be abused.

In general, a 0.6 mm focal spot and a focus-film distance of 1 m are used. The grid must have a high ratio of about 12 to 1, and the intensifying screens must be of a high definition type. A rather high exposure time (0.5–1 s) is preferable for a kilovoltage between 60 and 80 kV. In addition, one generally prefers a film of high contrast in order to obtain in the normal radiographs, in both frontal and lateral projections, a white line corresponding to the floor of sella.

The cortical bone of the pituitary fossa appears much more clearly on a high contrast film than on a mild contrast film (Fig. 3.3). Even minimal demineralization of the sellar floor will be detected more easily on a high contrast film.

II. Projections

1) The Two Basic Projections for the Sella Turcica

a) Lateral Skull Projections

Radiologic examination of the sella turcica always includes a lateral film of the skull. Coned views of the sella turcica are never of superior quality but on the contrary sometimes mask an imperfect projection. In addition, a single lateral projection is always sufficient in studying the sella turcica. Lateral radiography of the sella turcica is performed in a sitting, supine, or prone position. Criteria for a perfect lateral projection include superimposition of the orbital roofs, of the floors of the middle cranial fossae and of the external auditory meatus. In practice, we insist on perfect superimposition of the two orbital roofs; this criterion is a sine qua non. If it is not fulfilled, one must consider the radiograph of the sella turcica to be quite uninterpretable (Fig. 3.1).

b) Inclined Posteroanterior Projection (Caldwell's View)

Obtaining this film is of major importance. The posteroanterior projection is preferable and is taken in either a sitting or prone position, with the forehead and nose touching the film, and caudal angulation of the central ray of 22° (Fig. 3.2). The frontal projection must meet two essential criteria:

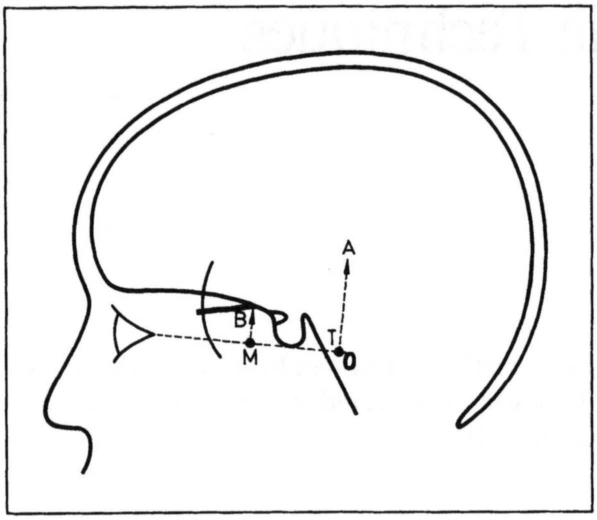

Fig. 3.1. Diagram showing the centering points. Routine lateral projection of the skull. Centering point (*A*) is located 4 cm above the tragus (*T*) on a line perpendicular to the orbitomeatal line. For magnified coned view of the sella turcica, centering point (*B*) is located 2 cm above the midpoint (*M*) of the orbitomeatal line

Fig. 3.2a, b. Inclined posteroanterior projection (Caldwell's view).
a Positioning: forehead and nose touch the film, the central ray making an angle of 22° craniocaudally.
b Result: in the correct projection the sellar floor (*1*) appears 1.5 cm beneath the planum sphenoidale (*2*)

▽

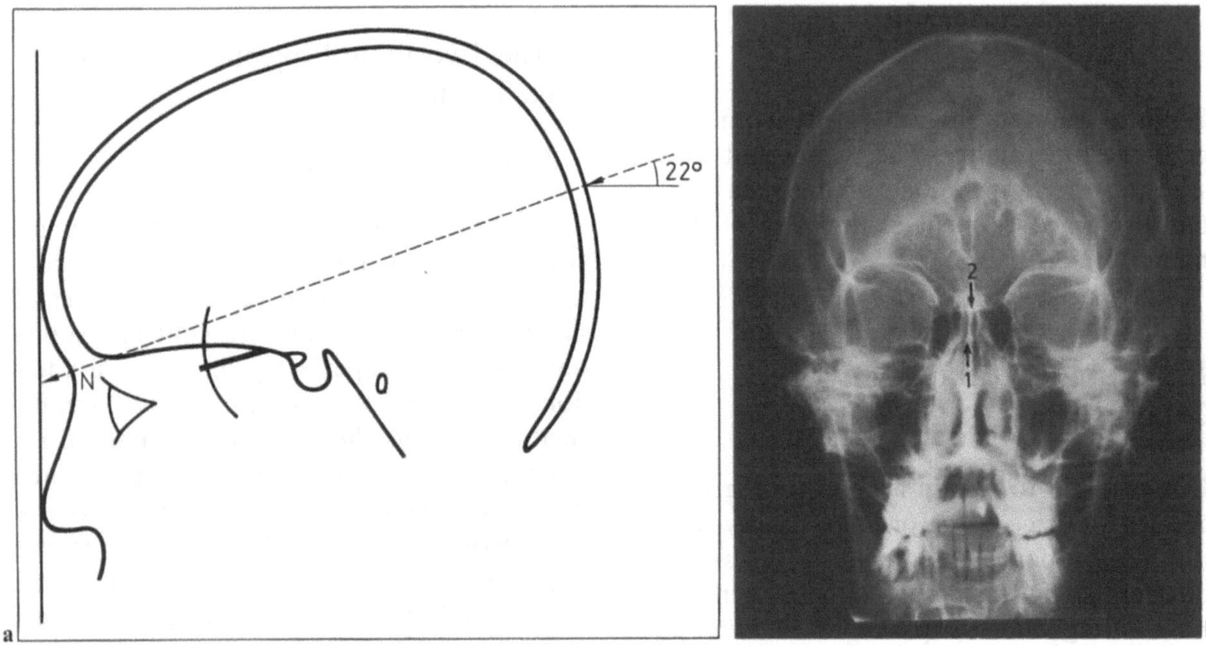

1) The head must be straight: the innominate line and inner table of the squama of the temporal bone must be equidistant on the two sides.

2) Angulation of the central ray must be correct: the floor of the sella turcica must project 1.5 cm below the planum sphenoidale and the upper margins of the petrous pyramids should be in the lower quadrant of the orbit. These projection criteria may be verified with television fluoroscopy.

Under certain conditions (trauma, uncooperative patient), anteroposterior projections can be used, in which case the anterior structures (orbits, superior orbital fissures, paranasal sinuses) are enlarged and slightly blurred.

2) Direct Magnified Views

Only X-ray tubes with a 0.3 mm or 0.1 mm focal spot are used for frontal and lateral magnified views. The beam is centered 2 cm above the middle of the orbitomeatal base line (Fig. 3.2). These focal spots may be used to give a three- to fourfold enlargement of the sella turcica; on films taken with direct magnification, precise study of the bony structures and in particular the lamina dura is possible.

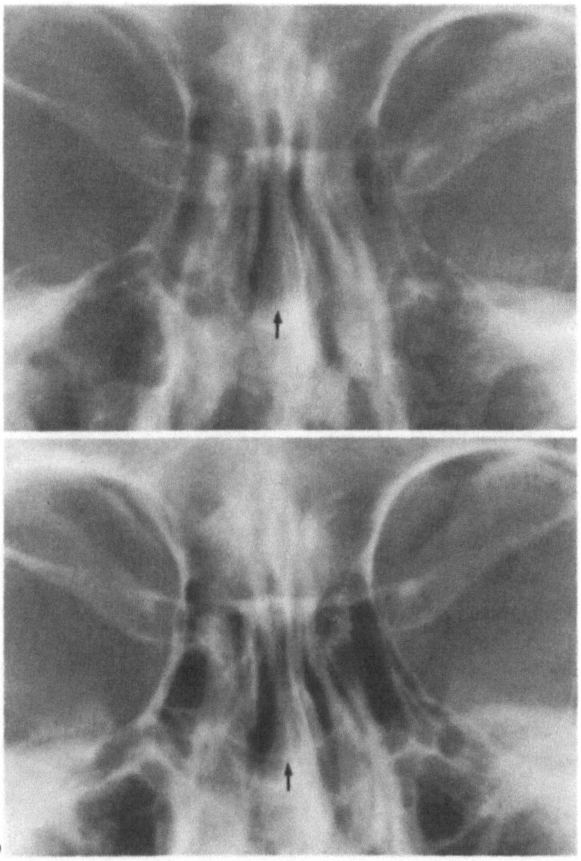

a

b

Fig. 3.3a, b. Sharpness and film characteristics. Planum sphenoidale and floor of the sella appear more clearly in **b** than in **a** (*arrows*).
a Medium contrast film
b High contrast film

However, the markedly divergent nature of the X-ray beam gives rise to nonpathologic double floors.

Recently, a new electron gun has made possible good magnified views of the sella turcica and permits a precise study of the lamina dura of the sella turcica (Fig. 4.8b).

3) Other Projections

a) Variations of Caldwell's View

In certain cases, the floor of the sella is not visualized on the frontal view. Visualization of the floor on frontal projections is possible only if the central beam is tangential to the floor. In cases in which the floor is not shown, the direction of the floor must be evaluated on the lateral projection, and the inclination of the X-ray beam modified accordingly.

b) Anteroposterior Half-Axial Projection (Towne's View)

This projection is taken in the sitting or supine positions, with the orbitomeatal line perpendicular to the film and the central ray is directed craniocaudally, forming an angle of 25°–30° with the orbitomeatal line. The dorsum sellae and the posterior clinoid processes then project into the posterior half of the foramen magnum; the anterior clinoid processes are seen in some cases. To project the dorsum sellae into the foramen magnum exactly, Etter (1961) suggested that the basal angle be taken into account.

c) Posteroanterior Half-Axial Projection

This projection is the opposite of the preceding one, the ray being posteroanterior; it also shows projection of the dorsum sellae over the foramen magnum.

d) Axial Projection (Base or Submentovertical View)

The axial projection is taken in a sitting or supine position, the head being fully extended with support at the top of the skull. The central ray, directed caudocranially, is centered on the midpoint of the biauricular line, forming an angle of 100° with the orbitomeatal line; the latter is parallel to the film.

In the subaxial projection, the central ray is perpendicular to the orbitomeatal line. In the hyperaxial projection, the central ray is directed caudocranially, forming an angle of 120° with the orbitomeatal line.

This projection need not be routine, but it is sometimes useful and it complements the two basic projections. Indeed, it gives a good image of the sphenoid wings and the sphenoid sinus. The axial view may be avoided when there is cerebrovascular insufficiency, and it is obviously contraindicated in the case of cervical spine trauma.

e) Special Projections

Numerous circumstances may lead to particular projections being obtained to study structures around the sella turcica (superior orbital fissures, optic canals, apex of the petrous bone). The unilateral projection of the superior orbital fissure demonstrates clearly its medial margin; the optic canal view (Hartman and Gilles 1959) projects the optic canal perpendicular to the film; and finally, Stenver's view demonstrates clearly the petrous bone apex.

B. Tomography

Tomography eliminates superimposed structures and permits a serious study of the quality of the cortical bone of the sella turcica.

I. Tomographic Devices

Pluridirectional tomography is the technique chosen for precise analysis of the sella cortical bone; however, many radiographic units perform only linear tomography.

1) Linear Blurring Movement

Linear tomography has two main disadvantages: on the one hand, the relatively great thickness of the sections (a minimum of 2–3 mm) and on the other hand, imperfect effacement of the structures adjacent to the focal plane, thus creating superimposed blurs (Fig. 3.4). These two adverse factors restrict considerably the efficiency of linear blurring movement in the study of the sella turcica. Precise study of the cortical bone of the sella turcica requires thin tomographic sections of about 1 mm, with perfect effacement, that pluridirectional tomography alone can produce. If linear tomography only is possible, use of a focal spot smaller than 1 mm is necessary. The plane of section must be perpendicular to the floor of the sella. A high section angle of 30°–50° has to be used to reduce the thickness of the section (Fig. 3.5).

2) Pluridirectional Blurring Movement

There are several types of pluridirectional blurring movement: circular, elliptical, spiral, and hypocycloidal; however, the hypocycloidal movement is best suited to the study of the sellar region. This movement gives very thin sections, about 1 mm thick, but requires a long exposure time of 6 s, necessitating absolute immobility of the patient. Production of false images and incomplete blur related to pluridirectional blurring movement are rare. However, one may find:

Poor visibility of the sellar floor due to an extensive sphenoid sinus (Fig. 5.86).

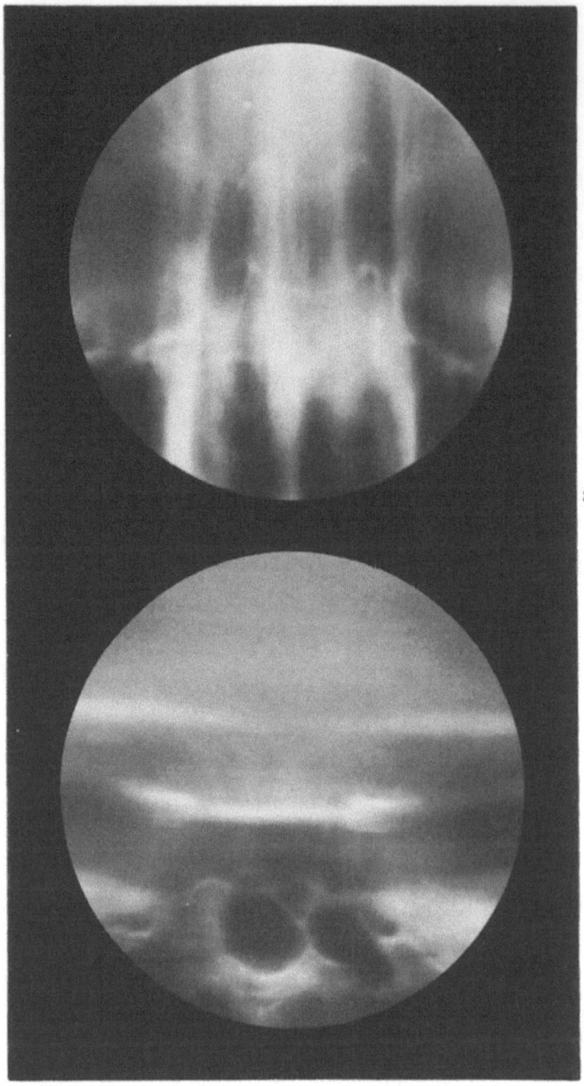

a

b

Fig. 3.4a, b. Tomographic artifacts in linear tomography.
a Linear vertical tomogram through sellar floor. Incomplete blurring of the walls of the nasal cavities produces vertical artifacts.
b Linear transverse tomogram: transverse artifacts related to incomplete blurring of sphenoid ridges overlie the sellar area

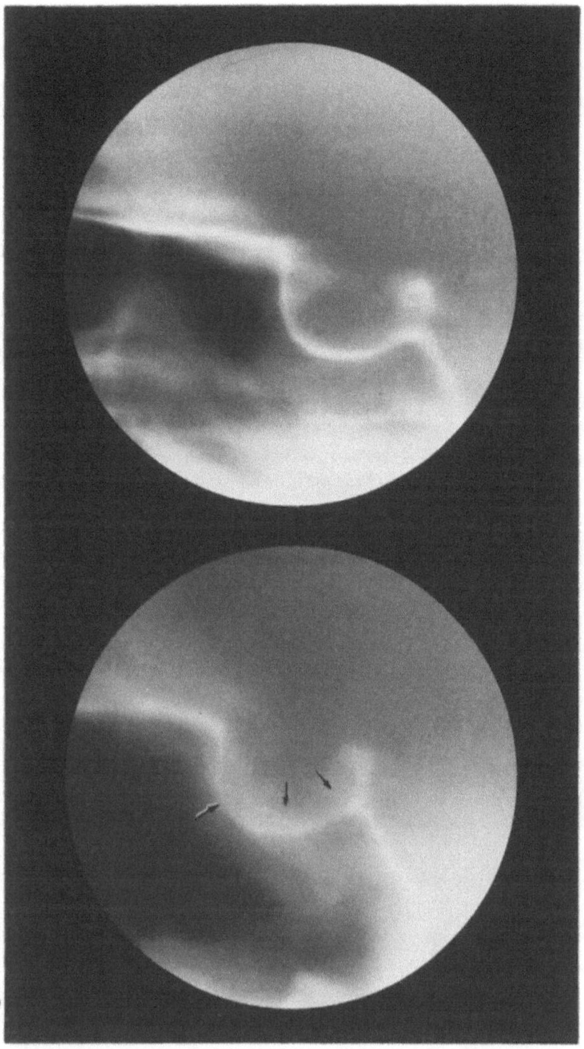

a

b

Fig. 3.5a, b. Lateral linear tomogram a and polytomogram b of a microadenoma are compared. On the linear tomogram, it is impossible to confirm thinning of the lamina dura of the floor of the sella. The polytomogram clearly demonstrates a localized erosion of the cortical bone of the anterior wall of the sella *(black and white arrow)*. False shadows doubling bony structures are noted *(black arrows)*

False clouding of the sphenoid sinus when a lateral tomographic section is tangential to a sagittal septum of the sinus.
False shadows surrounding very dense structures (Fig. 3.5).

There are two types of pluridirectional tomograms:

Constant magnification apparatus (Massiot-Philips Polytom).
Variable magnification apparatus.

Tomographic study of the sella turcica often requires comparison with previous radiographs. We, therefore, prefer the Polytom.

II. Tomographic Projections

Three projections are used in tomography of the sella: frontal, lateral, and (rarely) axial. Frontal tomograms are of major importance.

1) Lateral Tomograms

Such tomograms are preferably taken in the prone position, with the head turned. The beam is centered 2 cm above the midpoint of the orbitomeatal line. With the first section passing through the right carotid sulcus, 12 sections at intervals of 1 mm are obtained, thus allowing a millimetric study of the floor of the sella between the two carotid sulci. In certain cases, more lateral sections are necessary for visualization of the anterior clinoid processes (Fig. 3.6). If there is any doubt concerning a fluid level within the sphenoid sinus, sections in the erect position are required.

2) Frontal Tomograms

The patient is in the prone position, with forehead and nose touching the film. The first section passes through the planum sphenoidale. We obtain six localized sections at intervals of 2 mm (Fig. 4.12).

3) Axial Tomograms

Axial tomograms are used to study the sphenoid sinus, the anterior clinoid processes, the optic canals, or the lesser sphenoid wings. To obtain both the anterior clinoid processes and the optic canals on the same section, it is necessary to use the hyperaxial projection. This requires a considerable extension of the head, the central ray forming an angle of 120°–125° with the orbitomeatal line. Millimetric sections are then obtained upward from the orbitomeatal line.

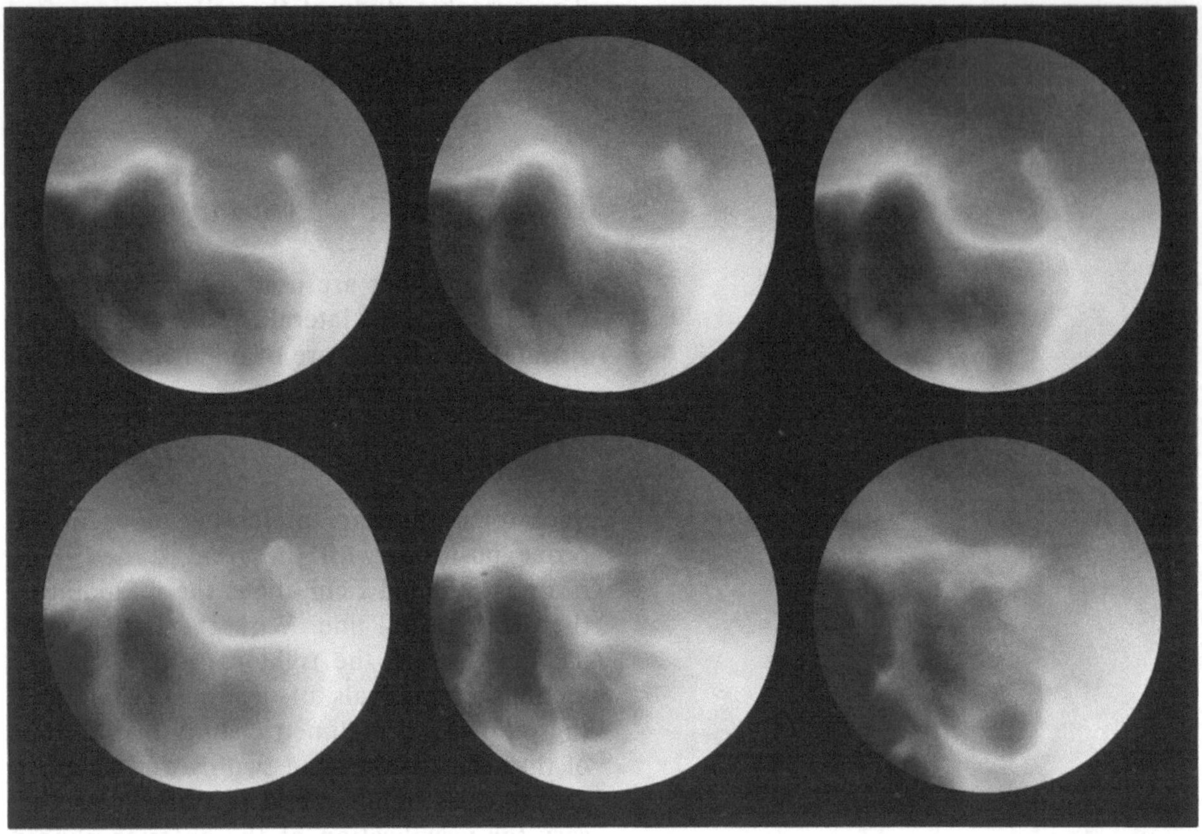

Fig. 3.6. Lateral coned tomograms of sella turcica. Six sections allow a millimetric study of the lamina dura from midline to carotid sulcus (study of the entire floor needs twelve sections at intervals of 1 mm)

III. When is Tomography Required?

Clinical, laboratory, or plain radiographic findings can lead to tomography.

1) Clinical or Laboratory Findings

If the clinical or laboratory findings strongly suggest a possible pituitary microadenoma, polytomograms must be obtained, even if standard films are normal (amenorrhea-galactorrhea syndrome with hyperprolactinemia, acromegaly, Cushing's syndrome).

If the clinical or laboratory findings are less suggestive than in the above supposition (e.g., amenorrhea or galactorrhea without hyperprolactinemia), then a magnified roentgenogram of the sella turcica with a very fine focal spot (electron gun if possible) could be sufficient after plain films.

2) Radiologic Signs

Poor visibility of the sellar floor on the frontal projection requires a tomographic study of the sella turcica. The same is true for all appearances of a double floor where the lower line appears less well mineralized than the upper one; and also for any abnormality of the cortical bone found on plain films. Of course, any sella turcica which appears pathologic on standard films must be studied by tomography, in order to confirm the diagnosis and determine the lateralization or extension of the lesion. It is obvious that if the lesion is very large and the sella turcica enlarged, tomographic sections can be performed every 3 or even every 5 mm.

Chapter 4 Radiologic Anatomy

A. Radiologic Anatomy of the Sella Turcica and of the Presellar Region

I. Children [7]

A radioanatomic study of the normal sellar region in the child must take into account both the pre- and postnatal development of the sphenoid body (Chap. 1).

We have divided our radioanatomic study into three stages:

The sellar region during the 1st year of life;
The sellar region from 1 to 5 years;
The sellar region from 6 years to adulthood.

In children younger than 6 years of age, the sella turcica is studied essentially on lateral projections (the sellar floor is, in fact, difficult to discern on frontal projections due to the absence of the sphenoid sinus). Tomography is of no value in children.

1) Sellar Region During the First Year of Life

The sella turcica is a regular cup of varying dimensions, the dimensions have been evaluated by several authors (Kovacs 1934; Silverman 1957). The dorsum sellae is totally cartilaginous at birth and begins its ossification at the 1st year of life. The anterior clinoid processes are present and often very large (Fig. 4.2).

At birth, fusion of the presphenoid with the postsphenoid is not complete. A column of cartilage separating the pre- and postsphenoid appears on the lateral projection as a lucent gap, the intersphenoid synchondrosis. This starts at the tuberculum sellae and extends downward and forward as far as the undersurface of the sphenoid (Fig. 4.1). This synchrondrosis has been found in 64% of the cases during the 1st month (Shopfner et al. 1968).

Moreover, the presellar region is quite incomplete: the planum sphenoidale begins to develop during the 1st year of life (Fig. 4.2); only the chiasmatic sulcus is present at birth. During the 1st month, the chiasmatic sulcus often presents a slight upward convexity (Fig. 4.1). The planum sphenoidale extends progressively backward. The extent of posterior development of the planum determines the respective lengths of the chiasmatic sulcus and the planum sphenoidale in adults. During its posterior development, the planum progressively overlaps the chiasmatic sulcus. The posterior margin of the planum (which corresponds to the future limbus sphenoidale) does not always join the chiasmatic sulcus, which accounts for the formation of a notch, which may persist several years (Fig. 4.3) or even into adulthood (Fig. 5.47), between these two structures.

At birth, the roof of the optic canal is the most superior sphenoid structure of the sellar region. Before formation of the planum sphenoidale, the roof of the optic canal must not be mistaken for the planum (Kier 1968). Moreover, the chiasmatic sulcus usually appears long when posterior growth of the planum is not completed (Fig. 4.4).

All these particular features of postnatal presphenoid development give a characteristic

7 Refs. [2, 18, 20, 57, 121, 130, 154, 178, 201, 257, 278, 304, 316, 326, 400, 449, 495, 496, 497, 508, 519, 582, 625, 752, 872, 928, 929]

radiologic appearance to the sellar region, with a false aspect of "excavation" of the presellar region (Fig. 4.1).

This excavated appearance of the presellar region is often called a "J" or "omega" shape normal sella. In fact, this appearance is clearly different from that seen with chiasmatic gliomas or certain craniopharyngiomas. Actually, this excavated appearance of the presellar region is due to incomplete development of this region during the 1st year of life: the absence of the planum, the high position of the roof of the optic canal, and the presence of a chiasmatic sulcus with slight upward convexity all contribute to this excavated appearance of the presellar region. Precise description of the way in which the anatomic structures are deformed in both normal and pathologic cases is greatly preferable to the terms J and omega sella turcica, which are often misinterpreted in the literature.

It should be noted that, during the 1st year, the sphenoid body often present a normal dense area of ossification (Harwood-Nash and Fitz 1976) (Fig. 4.5).

A central intrasphenoidal channel, referred to by many authors as the craniopharyngeal canal, extends between the lowest point of the floor of the sella and the sphenoid rostrum. It may sometimes be identified on a lateral radiograph (Fig. 4.1). This canal is median, vertical, or oriented downward and forward. Many authors believe that this canal corresponds to the traces of the stomodeal end of Rathke's pouch. Lowman et al. (1966) do not share this opinion and believe that the canal appears toward the end of fetal life and contains vessels supplying the basipostsphenoid. This canal has been found on only 1% of skull radiographs of newborns and generally disappears during the 1st year of life. However, in exceptional cases, it may persist in adult life (Fig. 4.6).

Sometimes, the inferolateral wall of the optic canal, the optic strut, appears in newborns as a thin, dense, curve line below the chiasmatic sulcus (Fig. 4.2).

2) Sellar Region from One to Four Years

The sellar cup grows very gradually. The dorsum sellae finishes its ossification about the age of 4 (Berger et al. 1976). The interphenoid synchondrosis has always disappeared by the age of 3. The presellar region has all its elements. Later, its shape will be modified by the development of the sphenoid sinus.

3) Sellar Region from Four Years to Adulthood

The region of the sella develops gradually. Basisphenoid pneumatization begins about the age of 4. Fusion of the ossicles of Bertin (sphenoid concha) containing the primary sphenoid sinus with the sphenoid body will allow extension of the sphenoid sinus toward the basisphenoid. This fusion takes place at about the age of 4 (Fujioka and Young 1978).

Growth of the sphenoid sinus is linear from 4–15 years of age. In one-third of cases, the sinus is well developed at 15 and does not continue to develop, but in 50% there is still moderate growth after the age of 15 (Hinck and Hopkins 1965) (Fig. 4.7). Sphenoid sinus development is indeed quite variable (see Chap. 5). When it is extensive, the shape of both the planum sphenoidale and the chiasmatic sulcus may be modified.

During this period, the sella turcica grows as a whole. Only the anterior clinoid processes show as further development, because they have already reached their final size at the age of 5. The spheno-occipital synchondrosis closes between the ages of 12 and 18 (Irwin 1960).

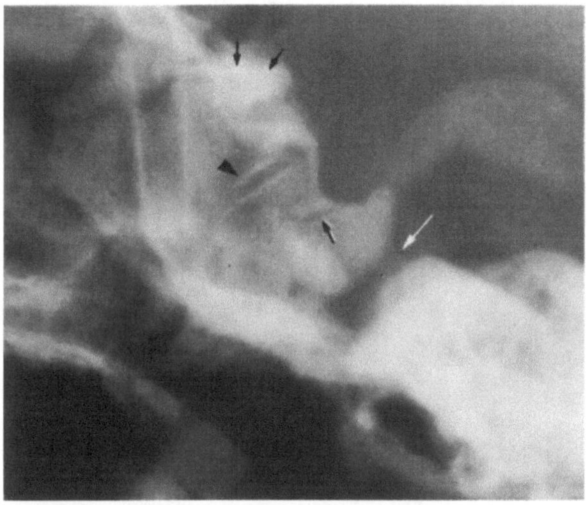

Fig. 4.1. Normal 1-week-old infant; lateral view. The optic canal roof *(black arrows)* is the most superior structure of the sphenoid region and gives a false appearance of an excavated presellar region. Note the normal presence of the intersphenoid synchondrosis *(arrowhead)* and of the spheno-occipital synchondrosis *(white arrow)*. Note the persistent craniopharyngeal canal *(black and white arrow)*

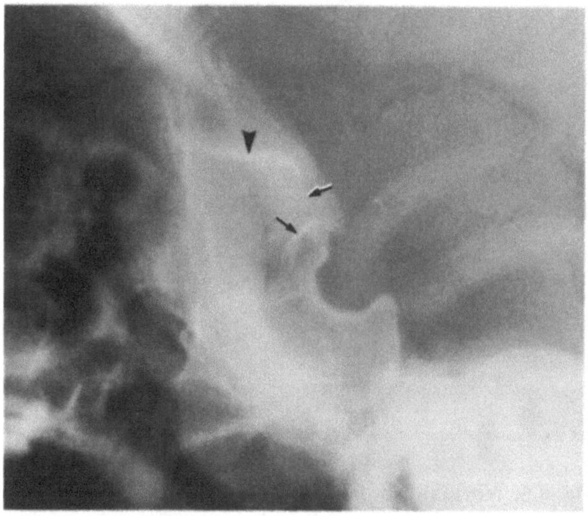

Fig. 4.2. Normal 5-month-old infant; lateral view. The planum sphenoidale *(arrowhead)* is formed; the chiasmatic sulcus presents a normal upward concavity *(black and white arrow)*. Note the projection of the optic strut just below the chiasmatic sulcus *(black arrow)*

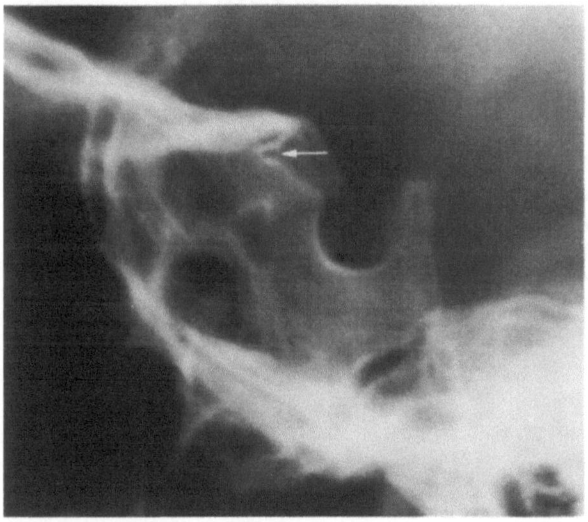

Fig. 4.3. Normal 6-year-old child. The lateral roentgenogram shows a notch corresponding to the absence of fusion of the limbus sphenoidale with the chiasmatic sulcus *(white arrow)*

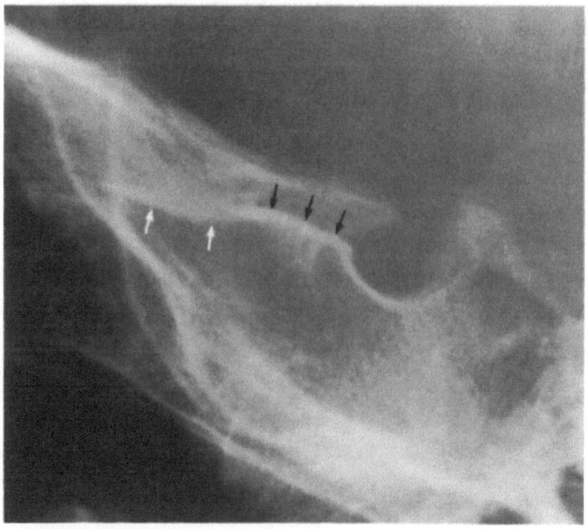

Fig. 4.4. Normal 2-year-old child. Note the presence of a short planum sphenoidale *(white arrows)* and of a long chiasmatic sulcus *(black arrows)*

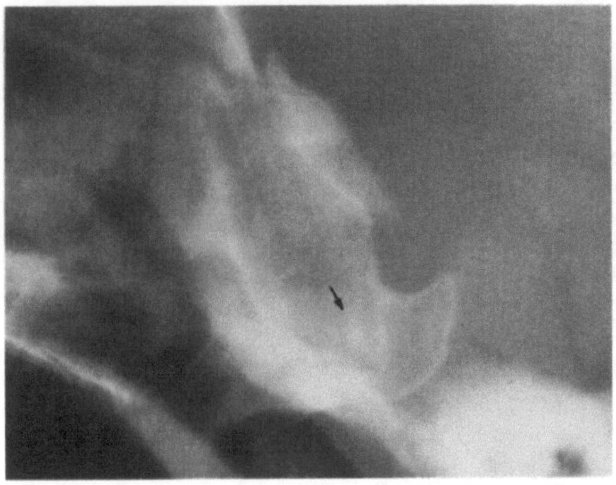

Fig. 4.5. Normal 1-month-old infant; lateral view. Note the dense nidus of bone below the sellar floor *(arrow)*

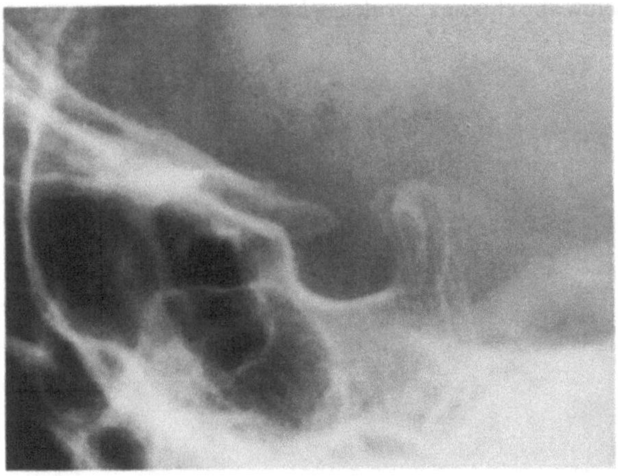

Fig. 4.7. Normal 14-year-old boy. Note the normal pneumatization of the sphenoid bone

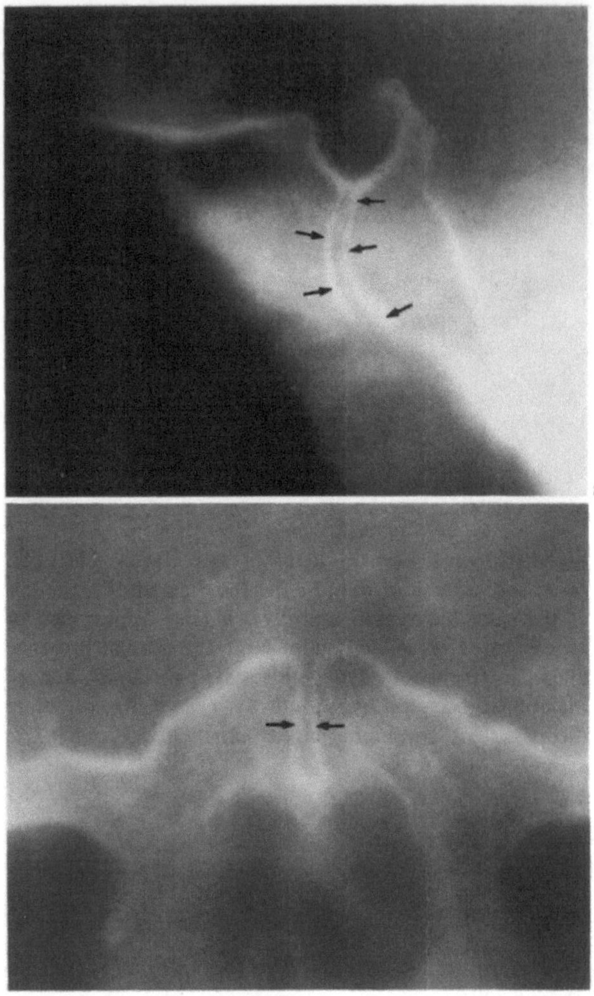

a

b

Fig. 4.6 a, b. 19-year-old male.
a Lateral tomogram
b Frontal tomogram.
Note the persistent craniopharyngeal canal extending medially from the sellar floor to the inferior aspect of the sphenoid bone *(arrows)*

II. Adults[8]

1) Lateral Projection

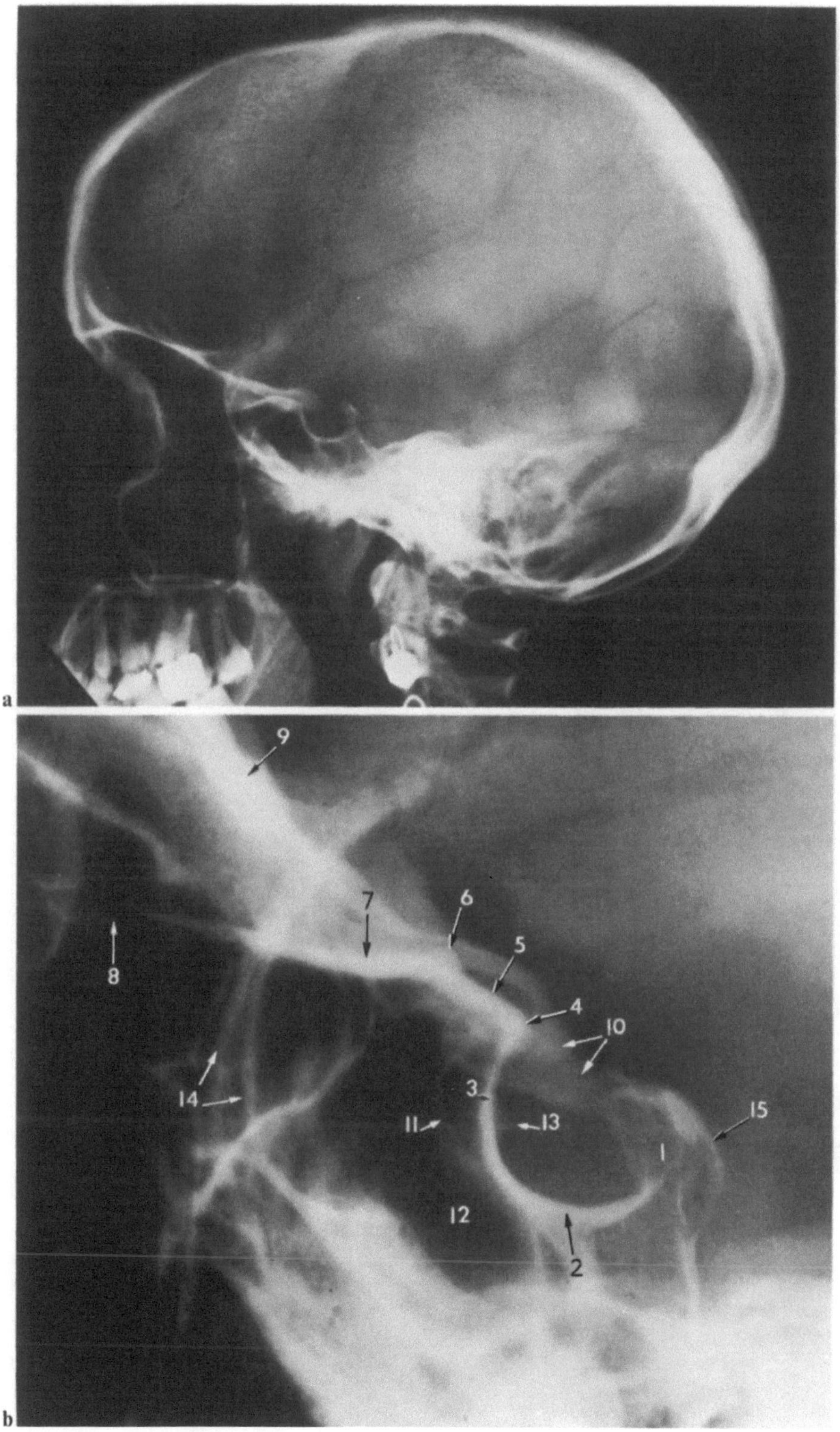

Fig. 4.8
a Lateral projection
b Magnified lateral projection
1 Dorsum sellae
2 Floor of the sella turcica
3 Anterior wall of the sella turcica
4 Tuberculum sellae
5 Chiasmatic sulcus
6 limbus sphenoidale
7 Planum sphenoidale
8 Cribriform plate of the ethmoid bone
9 Orbital roofs
10 Anterior clinoid processes
11 Carotid sulcus
12 Sphenoid sinus
13 Middle clinoid process
14 Anterior walls of the middle cranial fossae
15 Calcified petroclinoid ligament

8 Refs. [74, 75, 85, 109, 115, 135, 154, 178, 203, 224, 230, 242, 248, 255, 266, 276, 284, 301, 312, 317, 325, 361, 362, 451, 500, 558, 747, 748, 752, 762, 763, 789, 790, 892]

1) Lateral Projection (continued)

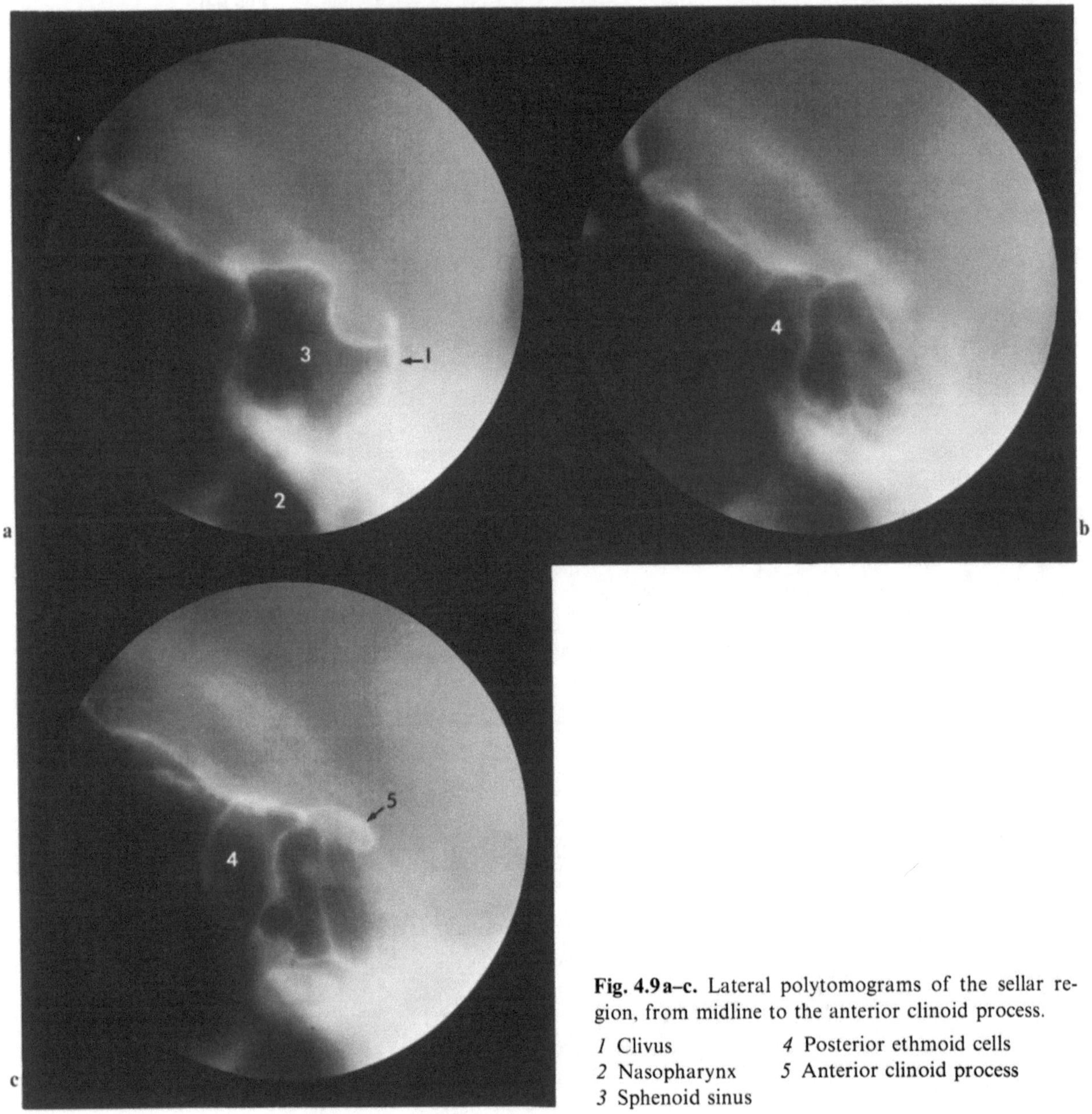

Fig. 4.9 a–c. Lateral polytomograms of the sellar region, from midline to the anterior clinoid process.

1 Clivus
2 Nasopharynx
3 Sphenoid sinus
4 Posterior ethmoid cells
5 Anterior clinoid process

2) Frontal Projection

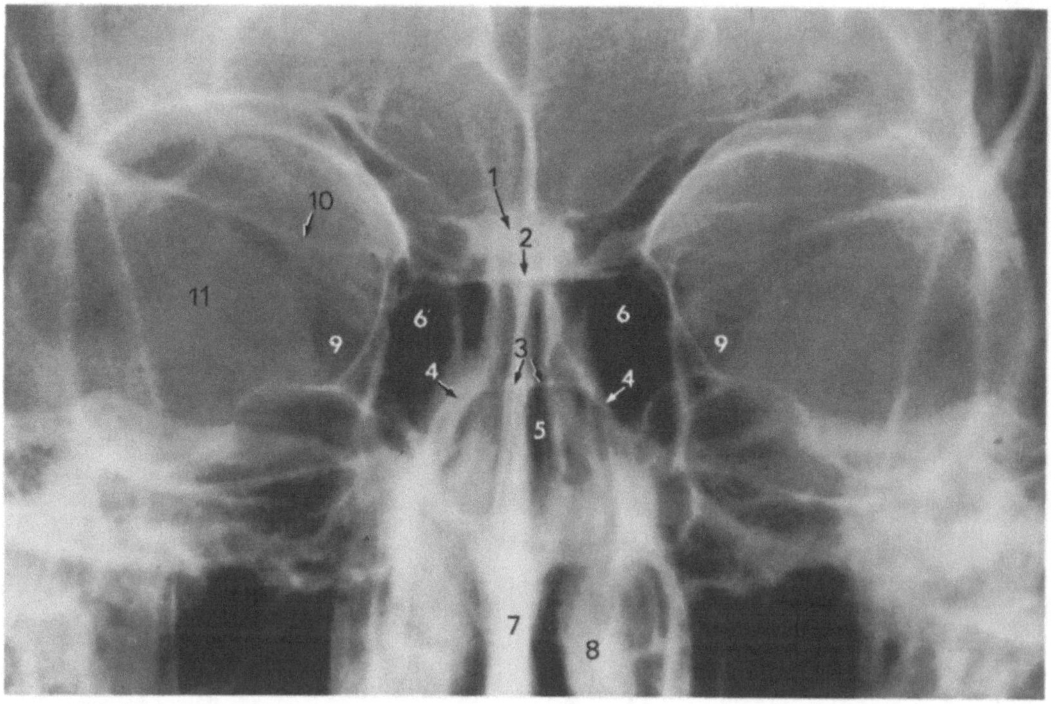

Fig. 4.10. Posteroanterior projection of the sella turcica.

1 Crista galli
2 Planum sphenoidale
3 Floor of the sella turcica
4 Carotid sulcus
5 Sphenoid sinus
6 Ethmoid cells
7 Nasal septum
8 Inferior turbinate
9 Superior orbital fissure
10 Lesser wing of the sphenoid
11 Greater wing of the sphenoid

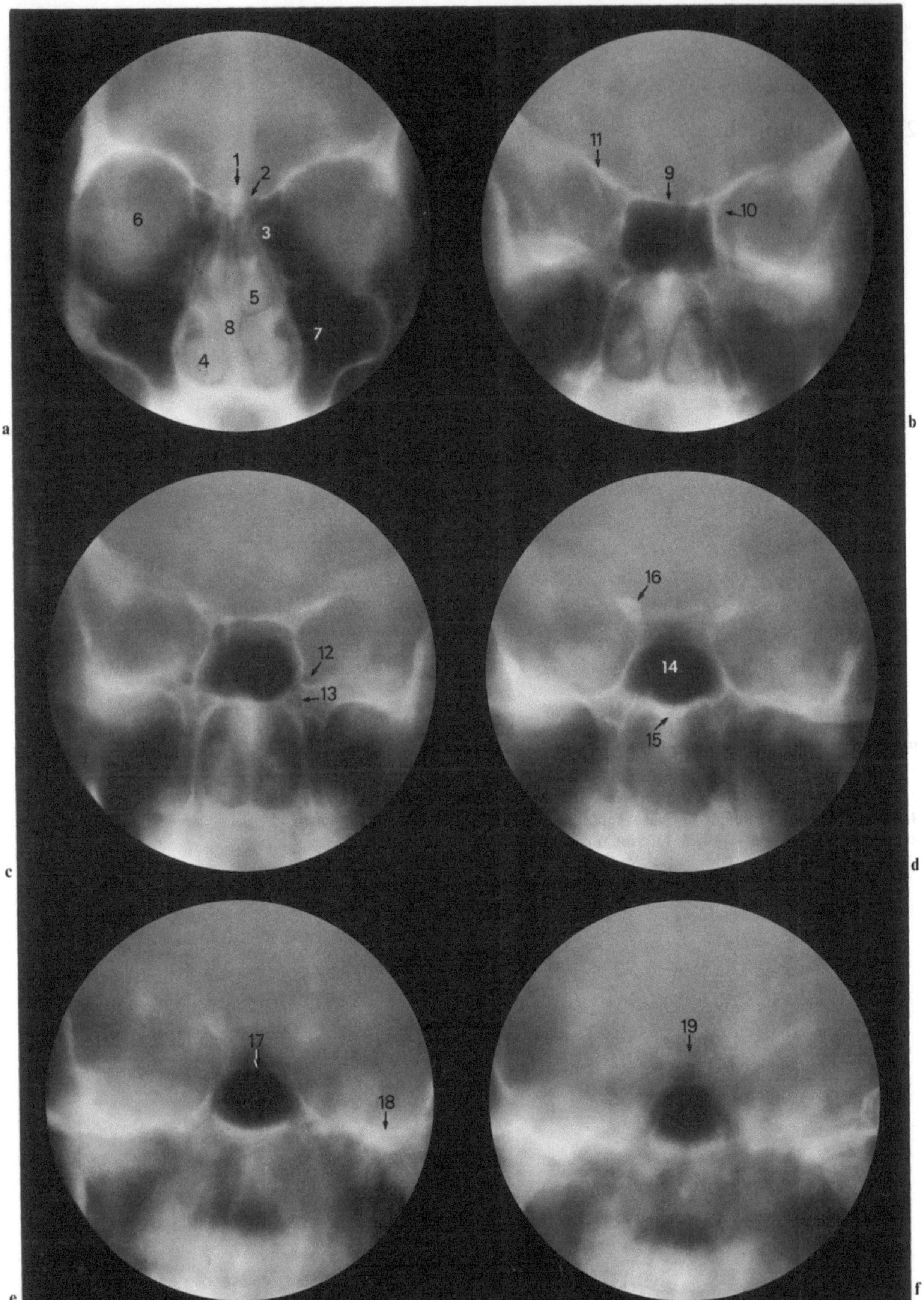
1
2
3
4
5
6
7
8
9
10
11
12
13
14
15
16
17
18
19
a
b
c
d
e
f

Fig. 4.12. Coned millimetric frontal polytomograms.

1 Anterior clinoid process
2 Floor of the sella
3 Carotid sulcus
4 Sphenoid sinus
5 Intersinusal septum

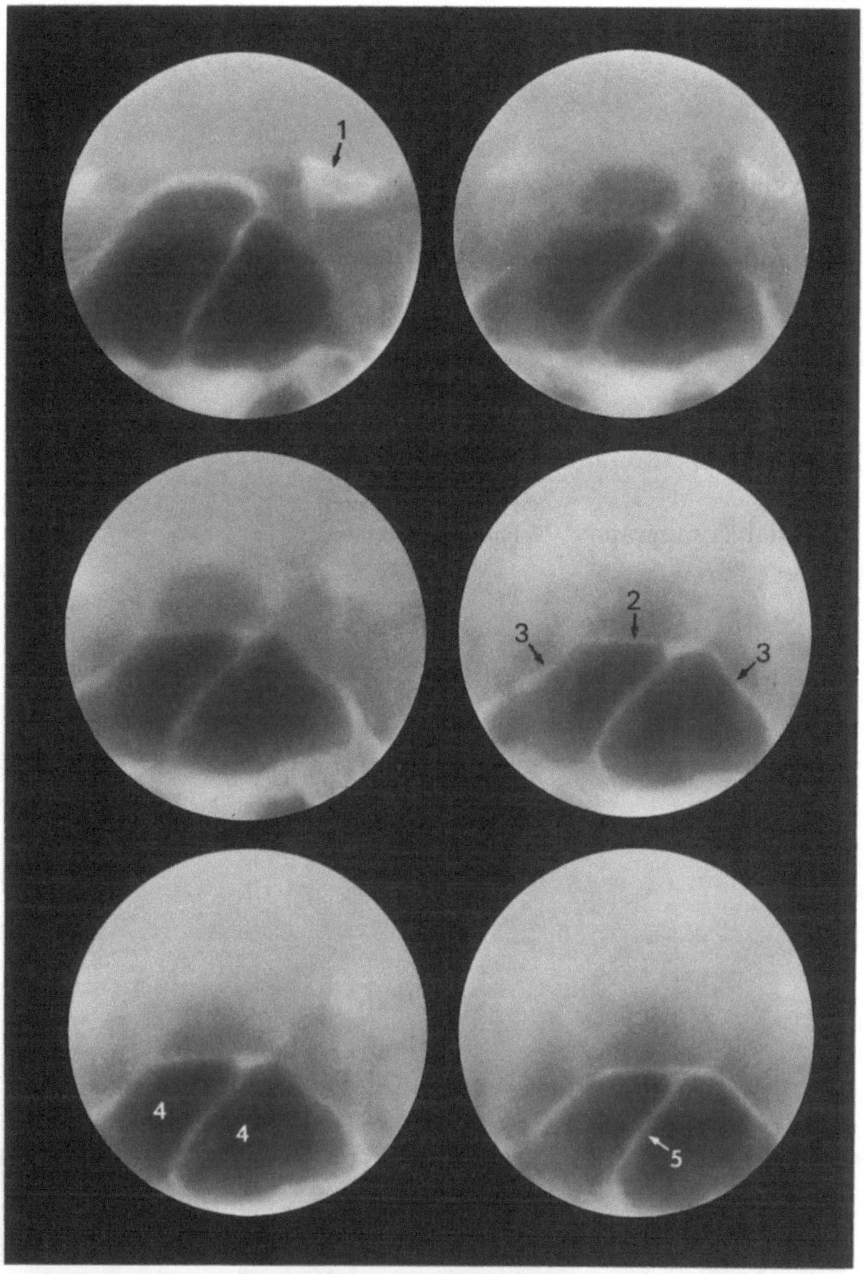

◁ **Fig. 4.11a–f.** Frontal polytomograms, from crista galli to dorsum sellae.

1 Crista galli
2 Cribriform plate
3 Ethmoid cells
4 Inferior turbinate
5 Middle turbinate
6 Orbit
7 Maxillary sinus
8 Nasal septum
9 Planum sphenoidale
10 Superior orbital fissure
11 Lesser sphenoid wing
12 Foramen rotundum
13 Pterygoid canal
14 Sphenoid sinus
15 Sphenoid rostrum
16 Anterior clinoid process
17 Floor of the sella turcica
18 Floor of the middle cranial fossa
19 Dorsum sellae

3) Half Axial View

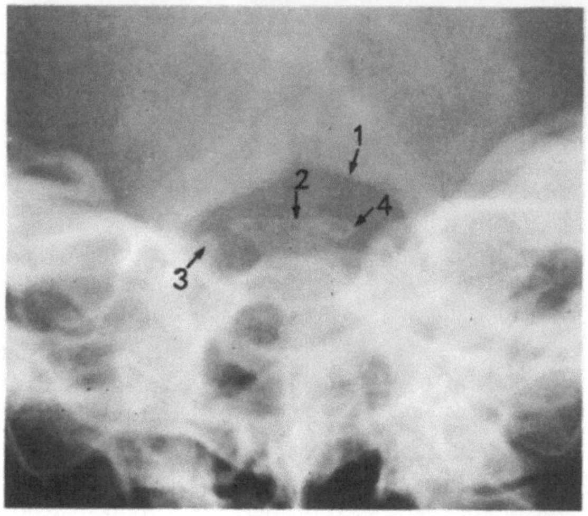

Fig. 4.13. Half axial view (Towne's view).

1 Posterior margin of the foramen magnum
2 Dorsum sellae
3 Anterior clinoid process
4 Posterior clinoid process

4) Axial View

a) Standard View

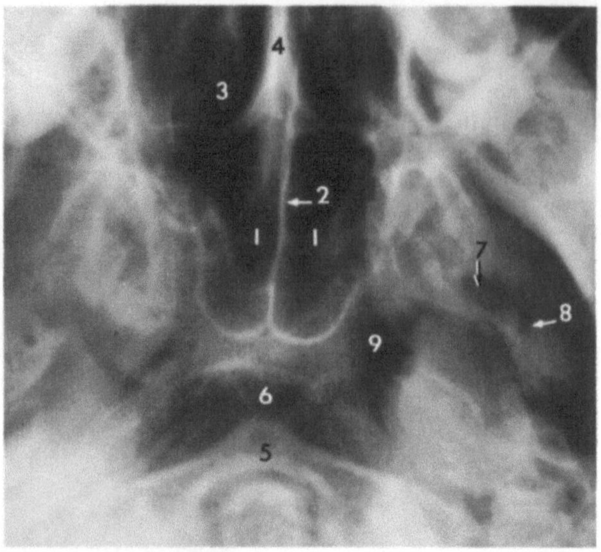

Fig. 4.14. Axial projection.

1 Sphenoid sinus
2 Intersinusal septum
3 Ethmoid sinus
4 Nasal septum
5 Anterior arch of atlas
6 Clivus
7 Foramen ovale
8 Foramen spinosum
9 Foramen lacerum

b) Axial Tomography

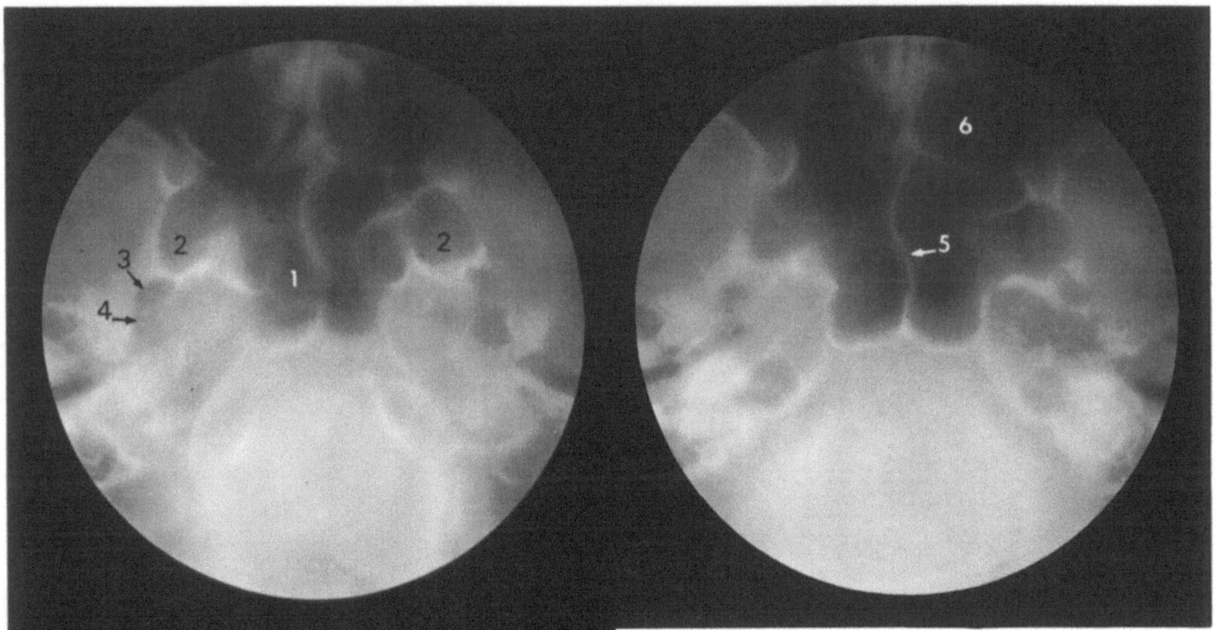

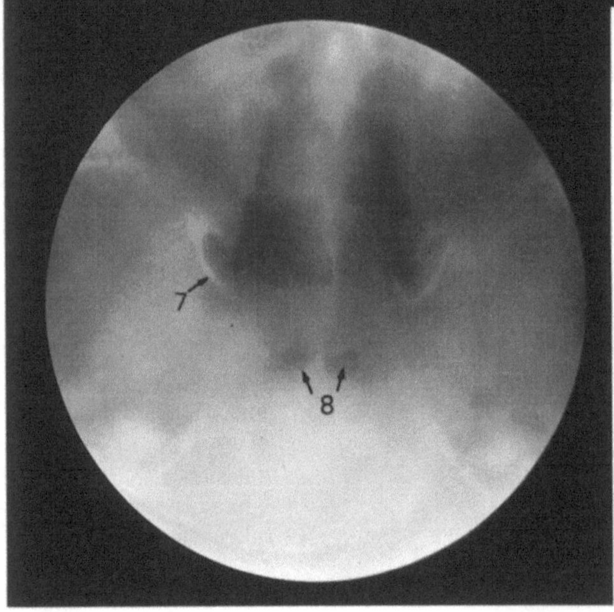

Fig. 4.15. Axial tomograms, from the floor of the middle cranial fossa to the anterior clinoid processes.

1 Sphenoid sinus
2 Lateral extension of the sphenoid sinus toward the greater wings of the sphenoid
3 Foramen ovale
4 Foramen spinosum
5 Intersinusal septum
6 Ethmoid cells
7 Pneumatized anterior clinoid process
8 Pneumatized dorsum sellae

5) Other Projections

a) Optic Canal

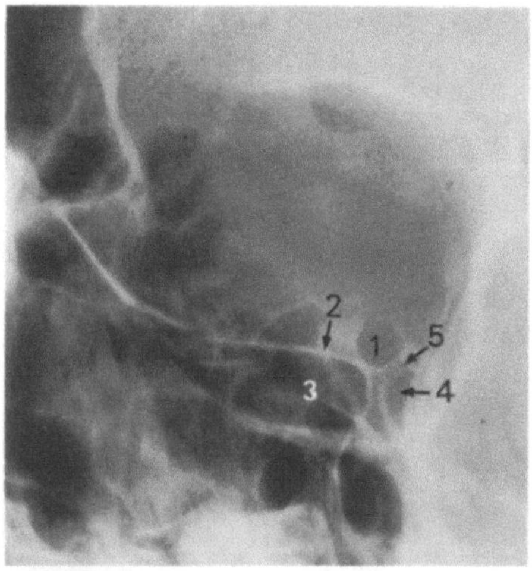

Fig. 4.16. Optic canal view.
1 Optic canal
2 Planum sphenoidale
3 Sphenoid sinus
4 Superior orbital fissure
5 Optic strut

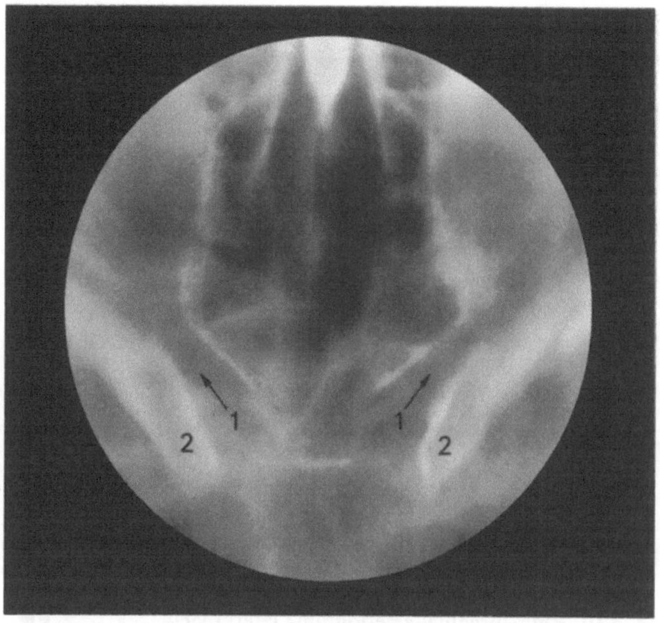

Fig. 4.17. Axial tomogram of the optic canal.
1 Optic canal *2* Anterior clinoid process

b) Superior Orbital Fissure

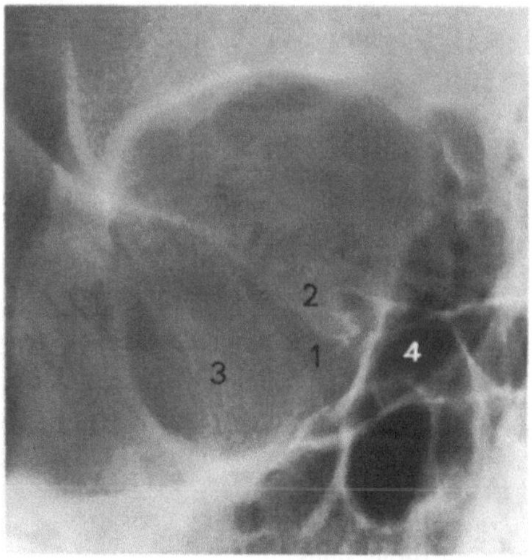

◁ **Fig. 4.18.** Unilateral view of the superior orbital fissure (Brunetti's view).
1 Superior orbital fissure
2 Lesser wing of the sphenoid
3 Greater wing of the sphenoid
4 Sphenoid sinus

B. Regional Radiologic Anatomy

I. Vascular Anatomy

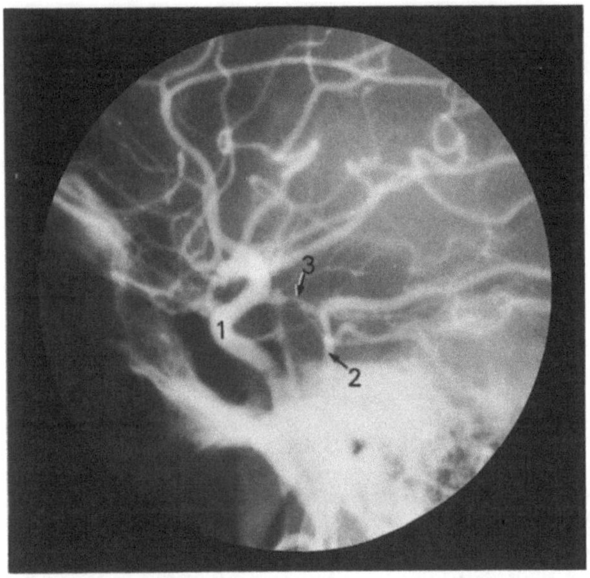

Fig. 4.19. Right brachial arteriogram; lateral view.

1 Internal carotid siphon
2 Basilar artery
3 Posterior communicating artery

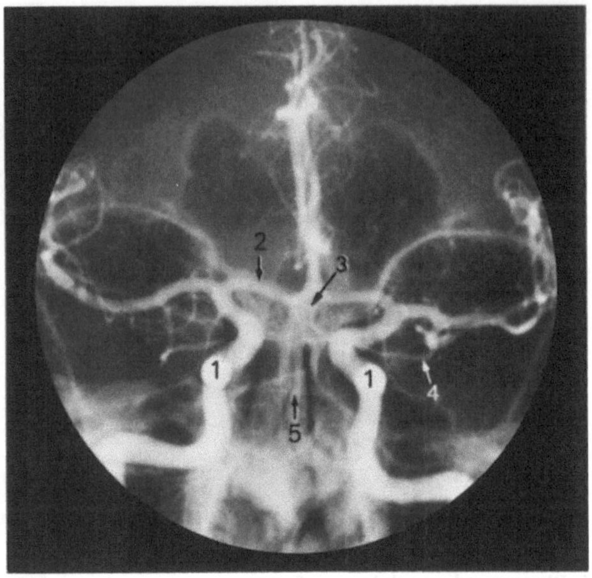

Fig. 4.20. Bilateral internal carotid arteriogram; frontal view.

1 Internal carotid siphon
2 Anterior cerebral artery
3 Anterior communicating artery
4 Ophthalmic artery
5 Floor of the sella

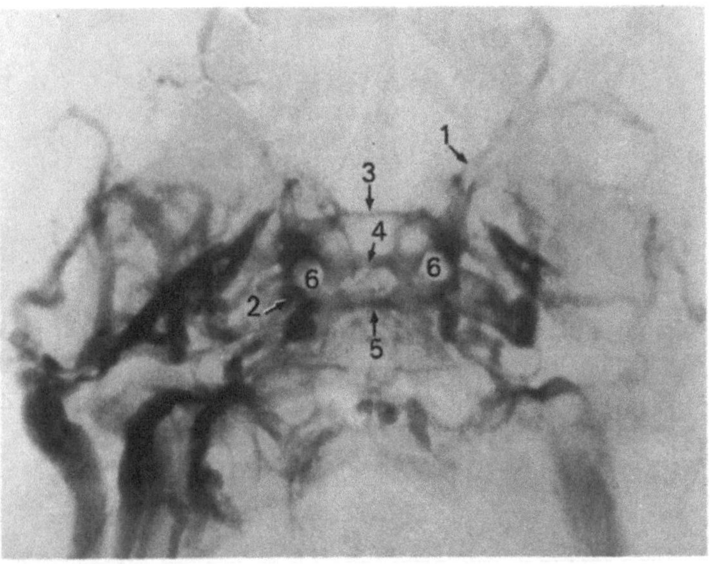

Fig. 4.21. Normal cavernous sinus phlebogram; axial view (Courtesy, D. Doyon).

1 Superior ophthalmic vein
2 Cavernous sinus
3 Anterior intercavernous sinus
4 Posterior intercavernous sinus
5 Basilar sinus
6 Internal carotid artery

II. Ventricular and Cisternal Anatomy

a

b

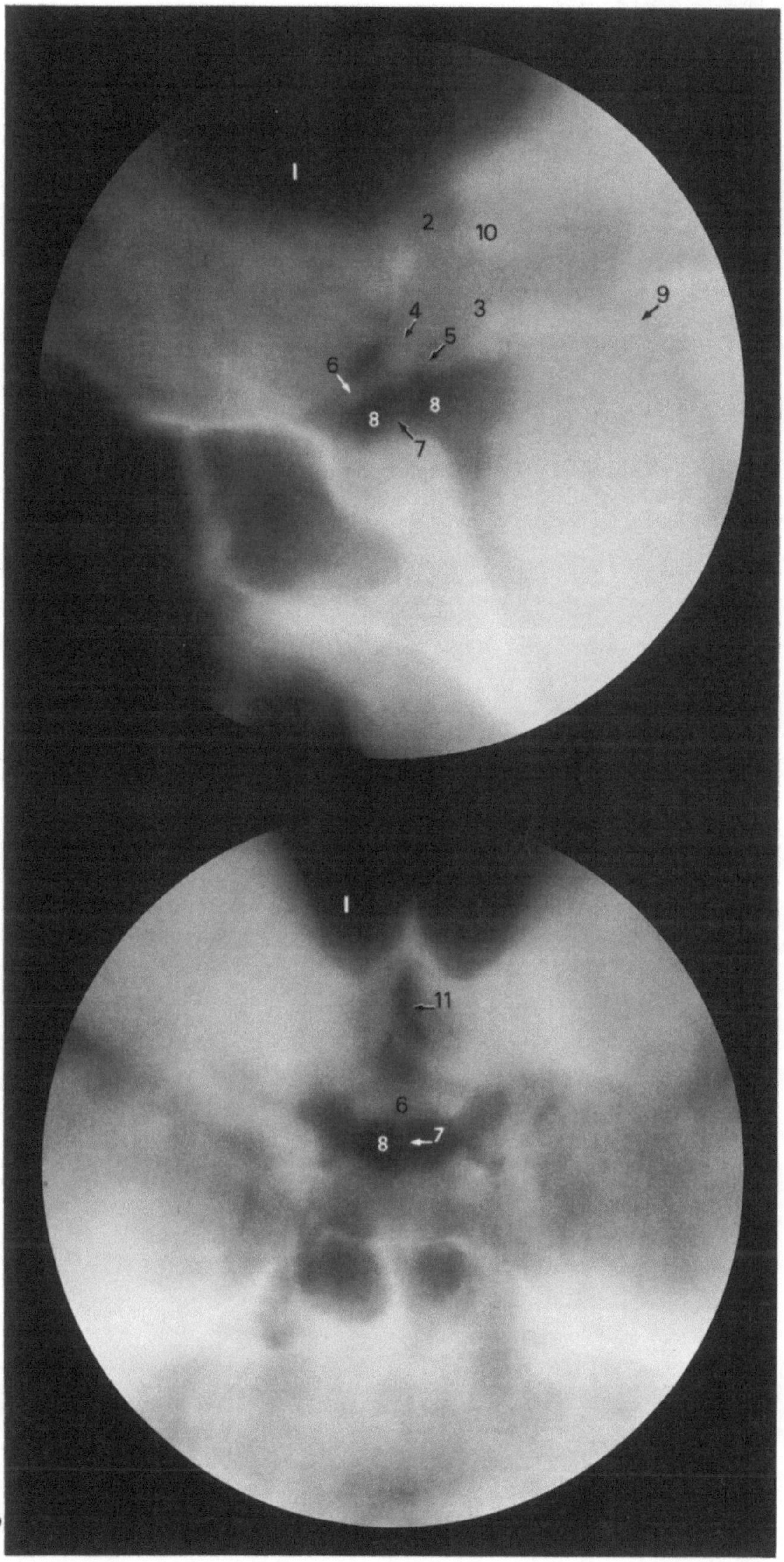

Fig. 4.22 a, b. Pneumoencephalogram.
a Midsagittal polytomogram
b Frontal polytomogram
1 Frontal horn
2 Foramen of Monro
3 Third ventricle
4 Optic recess
5 Infundibular recess
6 Optic chiasm
7 Pituitary stalk
8 Suprasellar cistern
9 Aqueduct of sylvius
10 Massa intermedia
11 Cistern of the lamina terminalis

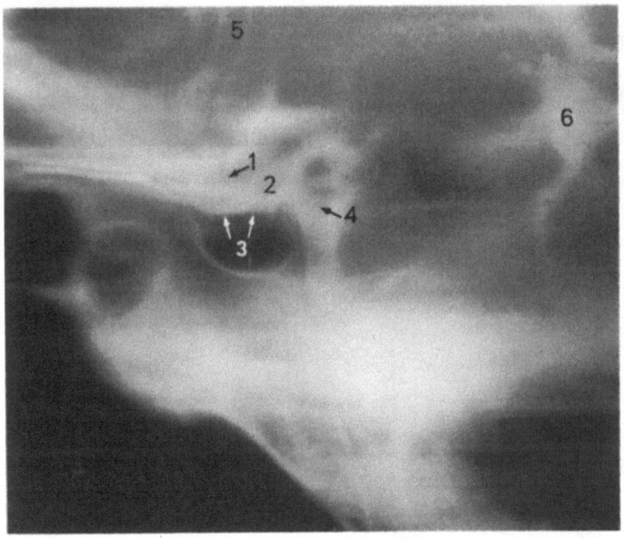

Fig. 4.23. Metrizamide cisternography; lateral tomogram.

1 Optic chiasm
2 Pituitary stalk
3 Diaphragma sellae
4 Basilar artery
5 Anterior cerebral artery
6 Quadrigeminal cistern

Chapter 5 Variations and Normal Limits[9]

A. Variations in the General Appearance of the Sella Turcica

The general configuration of the sella turcica varies considerably from one individual to another. It is necessary to distinguish its numerous variations from pathologic abnormalities.

I. Variations in Shape

Variations in the shape of the sella turcica on the frontal projection will be discussed together with those of the sellar floor.

On the lateral projection the sella turcica may appear as oval (58%), round (25%), or flat (17%) (Camp 1924) (Fig. 5.1). However, all intermediate shapes may also be found. The configuration of the sella turcica as seen in the lateral projection depends on skull shape (Fig. 5.2). In dolichocephalic and African subjects in particular, the sella is quite elongated from front to back, whereas in brachycephalic subjects, the sella appears round, having a smaller anteroposterior diameter. The sella is generally referred to as "open" or "closed" depending on the size of the aditus (the distance between the tuberculum sellae and the tip of the dorsum) (Fig. 5.3).

In the closed sella, the dorsum usually shows a definite forward tilt and must therefore be distinguished from the bridged sella, whose appearance is given by ossification of the interclinoid ligaments.

9 Refs. [41, 59, 71, 97, 115, 116, 181, 203, 246, 266, 272, 317, 352, 363, 364, 365, 366, 367, 429, 431, 521, 536, 537, 539, 617, 726, 737, 893, 927, 932].

II. Variations in Size[10]

Much has been written about the dimensions of the sella turcica, with numerous descriptions of methods used to determine the length, height, width, surface area, or volume of the pituitary fossa (Martinez-Farinas 1967; Lorenz 1949; Haas 1954; Oon 1963). We have not found them to be useful, and it would seem that the large number of suggested methods only serves to indicate their individual inadequacy. Those methods which attempt to determine linear measurements or areas fail to take into consideration the width of the pituitary fossa, while those which estimate the volume refer to the width of the sella turcica, whose evaluation is distinctly difficult due to the fact that the sella does not possess bony lateral limits. In addition, the extreme range of normal values is such that they are of no assistance in evaluating a minimal change in the volume of the pituitary fossa. It is also noteworthy that the volume measured in such a case merely gives an approximation of the volume of the pituitary gland, which itself occupies about 80% of the hypophyseal fossa. Today, determining its precise dimensions can only be of interest in monitoring the evolution of a pathologic sella turcica, and we prefer in such cases to use tomography with constant magnification since it enables assessment of possible changes directly and reliably by comparison of the tomograms. Measurements of the sella turcica may, however, be quite useful to the young radiologist in training, since measurement can to a certain degree be used as a substitute for lack of experience.

10 Refs. [58, 223, 228, 368, 604, 697, 698, 778, 875, 967].

The length (L) of the sella turcica, i.e., the largest anteroposterior diameter, measures 5–16 mm, an average of 10.5 mm, under normal conditions of projection. Any length exceeding 17 mm will always be considered abnormal, according to Taveras and Wood (1964).

The depth (D) is the longest perpendicular line drawn from the presumed level of the diaphragma sellae to the floor. The diaphragma sellae extends from the tuberculum sellae to the top of the dorsum sellae. The depth generally varies from 4–12 mm, with an average of 8 mm. A sella whose depth exceeds 14 mm must be considered as abnormally deep.

The surface area of the sella on the lateral view can be calculated by multiplying the length by the depth, or even more precisely by using graph paper; however, if the surface area of a sella turcica exceeds 130 mm^2, it is considered abnormal (Mahmoud 1958).

The above data must be used with care, remembering that, owing to the variability in width of the pituitary fossa, a sella turcica whose area on a lateral view appears small may indeed have a large volume; conversely, a sella turcica which seems large in lateral view may have a normal volume if its floor is narrow. It is for this reason that Di Chiro (1960) introduced the notion of this third dimension.

The width (W) represents the width of the floor of the sella turcica as seen on the posteroanterior projection. It varies from 10–15 mm, but can only be regarded as pathologic beyond 19 mm.

The volume (V) can be calculated by applying the following formula: $V=\frac{1}{2}L \times D \times W$. The extreme values are 233 and 1092 mm^3, with an average of 594 mm^3 (Di Chiro and Nelson 1962).

In our opinion, when one is seeking for an early pituitary tumor (microadenoma), the measured values have little practical value, due to the large range of normal values. Conversely, when pathologic changes are obvious, measurement of the sella turcica loses all diagnostic value. It is for this reason that we usually estimate the dimensions of the hypophyseal fossa by eye alone. We agree with du Boulay and El Gammal (1966) that assessment of the dimensions of the sella turcica is of no value in showing a volume increase in the hypophysis, but that on the contrary, contour changes and especially a localized cortical thinning are much better indications of the presence of a possible intrasellar tumor.

In rejecting numerical measurement of the sella turcica, we are obviously introducing the notion of subjectivity in our assessment of sellar dimensions; therefore, it is possible to find "large" nonpathologic sellae and "small" nonpathologic sellae (Fig. 5.4). The latter, which are present without any pathologic context and notably no hypopituitarism, could be considered by some as being related to hyperpneumatization of the sphenoid sinus.

III. Unusual Configurations

1) Lack of Visibility of the Floor of the Sella on the Routine Posteroanterior Projection

The fact that the floor of the sella may not be seen on the usual posteroanterior projections ("frontal projection of the sella turcica") merits attention and raises the suspicion of blurring due to increased intracranial pressure, even in the absence of other bony changes. In certain cases, however, where the anterior wall of the sella slopes markedly, the floor is short and does not therefore appear on the frontal view in the usual way, i.e., as a dense linear opacity (Fig. 5.5).

2) Vertical Chiasmatic Sulcus

The slope of the chiasmatic sulcus with respect to the planum sphenoidale varies considerably and can reach an angle of 90° (Pribram and du Boulay 1971). In this case, the limbus sphenoidale appears very prominent, the chiasmatic sulcus and the anterior wall of the sella forming one continuous line; the tuberculum sellae is less well defined that usual. This general appearance must not be confused with that of an enlarged sella turcica (Figs. 5.6, 5.7, 5.48, 5.49).

3) Bridged Sella Turcica

Uni- or bilateral ossification of interclinoid and laterosellar ligaments changes the general appearance of the sella turcica on the lateral view (Figs. 5.62, 5.63).

4) Sella Turcica with Thin Cortical Bone

In certain cases, the entire bony margin of the sella turcica is thin and could suggest pathologic thinning, especially if the dimensions of the sella turcica are at the upper limit of normal (Fig. 5.8). This appearance may be considered as physiologic when the whole of the skull base and particularly the presellar area and the orbital roofs present the same appearance, i.e., very thin cortical bone whose density is conserved without blurring.

5) Small Sella Turcica on One Side

In such cases, one-half of the pituitary fossa is too small due to an upward bulging of the sphenoid sinus, usually limited to one side. On the lateral projection, a double floor can be seen, the lower line of which indicates the normal hemisella, while the upper line shows the hemisella whose volume is reduced (Fig. 5.11). Diagnosis is easy on the frontal view. The floor of the sella lies too high on one side and is too close to the planum sphenoidale (Fig. 5.10). This must not be mistaken for depression of the other side by an intrasellar tumor. Frontal and lateral tomograms may be useful in confirming the diagnosis of this morphological abnormality (Fig. 5.9).

IV. Normal Limits

1) Empty Sella in the Early Stage

The diagnosis of any empty sella turcica is nearly always possible on plain films. In general, the sella is enlarged vertically, its floor presenting a regular and symmetric concavity (sellar floor *en cuvette* on frontal projection). Changes in the cortical bone of the sellar floor vary according to the pneumatization of the sphenoid sinus (see Chap. 6). However, intermediate patterns between typical empty sellae turcicae and normals, which often make radiologic diagnosis quite difficult, have never been recorded. The sellae which we have designated as early empty sellae are those which present an intrasellar expansion of the subarachnoid spaces (confirmed by pneumography, metrizamide cisternography, or CT), but in which the cerebrospinal fluid pulsations have not yet enlarged the sella turcica. These early empty sellae obviously correspond anatomically to an absent or deficient diaphragma sellae.

In general, they are associated with a poorly aerated sphenoid sinus (presellar type sinus) and the sellar floor is in contact with the sphenoid cancellous bone. In this particular case, it appears that the floor is more easily modified by pulsations of the cerebrospinal fluid than when it is in contact with the sphenoid sinus itself. The radiologic appearance of these early empty sellae turcicae in their early stage is quite revealing (Figs. 5.12, 5.13).

In the frontal projection, the floor presents a regular and slight upward concavity. In the lateral projection, the appearance is that of a very characteristic double floor: opposite a poorly pneumatized sphenoid, the sellar floor itself seems doubled, giving the appearance of two parallel lines, the upper one being partly obscured and the lower one very distinct, as if traced with a pencil. These two lines join to form one line at the anterior wall. However, they may in some cases continue to be doubled as far as the dorsum sellae.

Precise changes in the cortical bone of the sellar walls, better defined on tomograms, will be thoroughly discussed in the chapter dealing with the empty sella turcica (Chap. 6).

In the absence of any endocrine or visual signs, the foregoing morphological abnormalities may be regarded as changes at the physiologic boundary, related to an anatomic variation, the deficiency of the diaphragma sellae. Nevertheless, they require radiologic monitoring.

2) Sella Turcica and Multiparity

Radiologists have different opinions regarding possible changes of the sella turcica in multiparous patients. For some, multiparity does not induce enlargement of the sella turcica, while others (Taveras and Wood 1976) believe that it often causes an increase in the size of the pituitary fossa. Still others maintain that multiparity is indeed an essential factor in the formation of an empty sella turcica. It can be concluded that all of these authors' opinions are correct, in that the influences of multiparity on the sella vary considerably.

Moreover, we would like to draw attention to two interesting points. Multiparity usually does not change the radiologic pattern of the sella turcica. In certain cases, the appearance of the sella turcica is changed in a very characteristic way. The sella is moderately enlarged; its anterior wall and floor appear quite irregular and mamillated, presenting several very limited depressions as if "each pregnancy had left its mark on the sella turcica walls." In the lateral projection, these multiple depressions often show a highly irregular double floor, i.e., the two lines forming the double floor are not parallel but rather tend to converge or even to intersect (Figs. 5.14–5.16).

3) Sella Turcica and the Internal Carotid Artery

The general appearance of the sella turcica may be modified by the course, size, and pulsations of the carotid siphons (Engels 1958). Within its presellar course, the internal carotid artery sometimes runs in a deep vertical groove, notching the lateral surface of the sphenoid bone, which is shown clearly in a lateral projection as a translucent vertical band. This is seen particularly in older patients or in those with arterial hypertension (Figs. 5.17, 5.21 d, e). The internal carotid artery can simply thin the lateral wall of the sphenoid sinus (Fig. 5.18).

At the junction of the pre- and juxtasellar segments, the carotid siphon may leave its trace on the lateral portion of the base of the dorsum sellae and on the part adjacent to the sellar floor, thus creating a localized change or erosion which can simulate increased intracranial pressure. It is especially in the presence of a double carotid siphon that these changes occur (Fig. 5.19).

A wide carotid groove in its juxtasellar segment may modify the general appearance of the sella turcica in both frontal and lateral projections: the sellar floor itself appears narrow and extends from one or both sides by a more or less wide platform, either slightly sloping, or horizontal and displaced downward with regard to the sellar floor (Fig. 5.21 a, b, c). Such an appearance indicates a more medial position of the internal carotid artery than ordinarily seen (Fig. 5.22). In such cases a concave semilunar notch with thinning of the cortical bone may still be seen in the anterosuperior wall of the sella turcica; this must not be considered as pathologic (Fig. 5.20).

Exceptionally, the carotid siphon can lie within the sella turcica. This abnormality can cause the sella turcica to become enlarged or may merely demineralize its floor (Anderson 1976). The diagnosis can be made using plain films and frontal tomograms which show one or several localized depressions in the sellar floor, corresponding to the imprint of the intrasellar carotid siphon (Figs. 5.23, 5.24).

It is obvious that calcification of the carotid siphons also changes the general appearance of the sellar region.

4) Sella Turcica in Old Age

With age, osteoporosis first blurs electively the dorsum sellae and sometimes the posterior portion of the floor. Thinning of the lamina dura occasionally gives the impression of an enlarged sella. In the frontal view, the sellar floor almost always remains clearly visible (Fig. 5.25), which allows a differentiation of changes due to osteoporosis from those related to increased intracranial pressure in which the sellar floor is affected at an early stage. Calcified carotid siphons are frequently projected on the sellar area in older patients.

5) Sella Turcica and Craniostenosis

In oxycephalic patients, premature closure of the coronal suture leads to a shortness of the anterior cerebral fossa, while the middle cerebral fossa is abnormally deep. As a result, the sella turcica appears lower than usual in the lateral projection. In addition, the sella is often deformed and enlarged both anteroposteriorly and vertically (Fig. 5.26). In these cases, cisternography or CT generally demonstrates an empty sella.

Changes in the visual field can also occur in oxycephalic patients. In order to avoid a false diagnosis of pituitary adenoma, for example, it is extremely important to have a good knowledge of the typical appearance of the sella turcica in this form of craniostenosis.

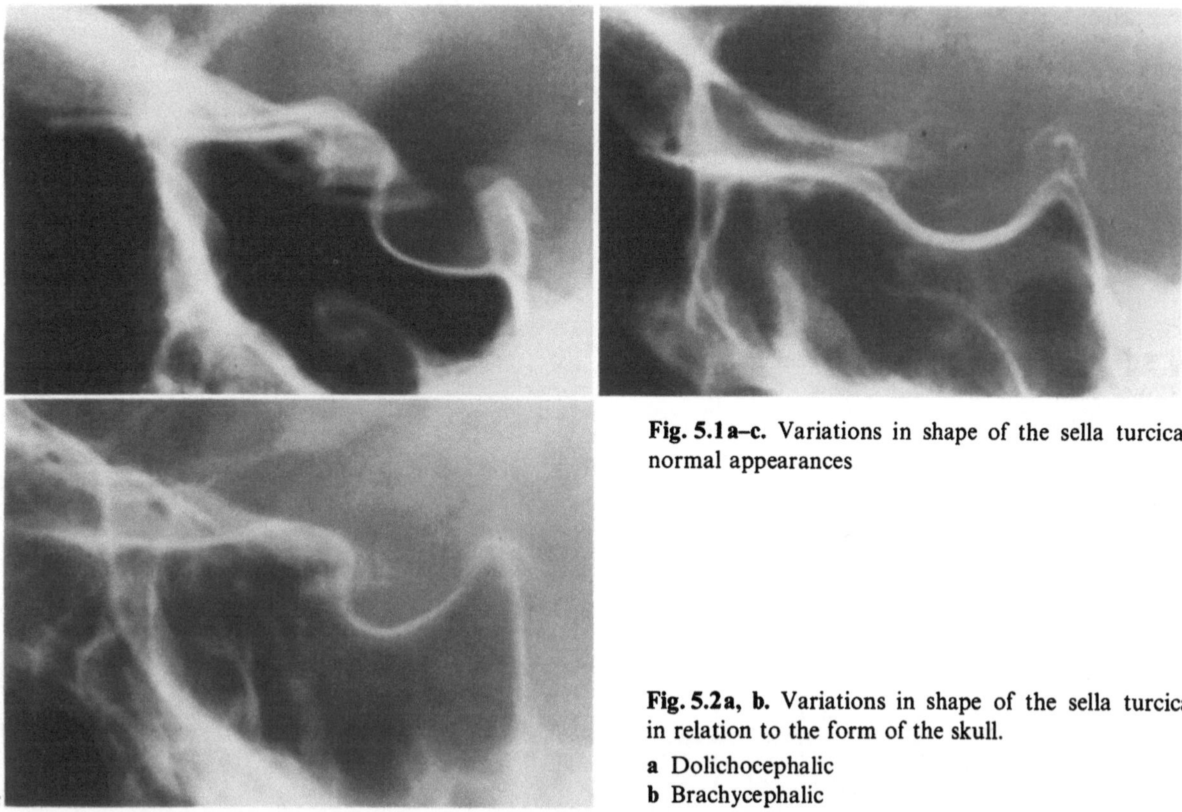

Fig. 5.1a–c. Variations in shape of the sella turcica: normal appearances

Fig. 5.2a, b. Variations in shape of the sella turcica in relation to the form of the skull.
a Dolichocephalic
b Brachycephalic

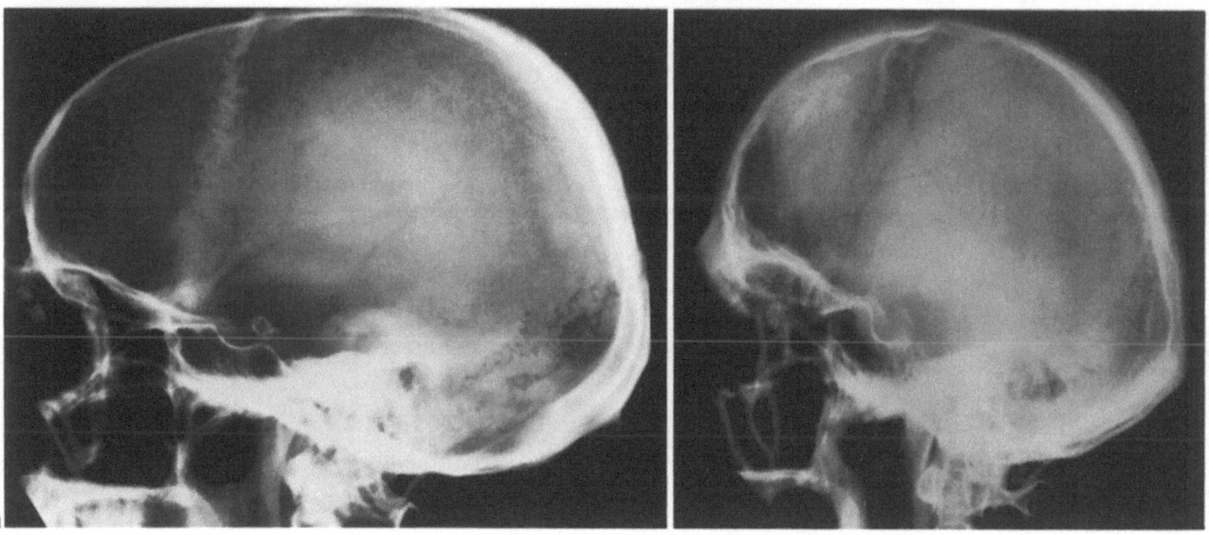

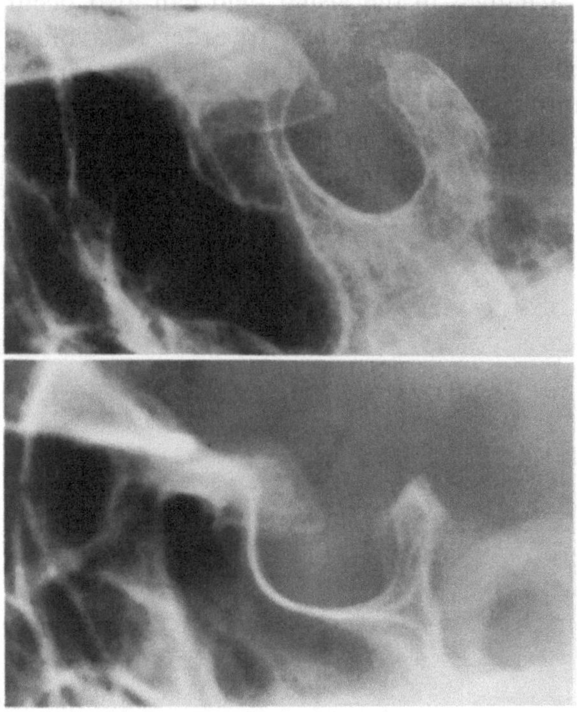

Fig. 5.3a, b. Variations of the sellar opening.
a Closed sella
b Open sella

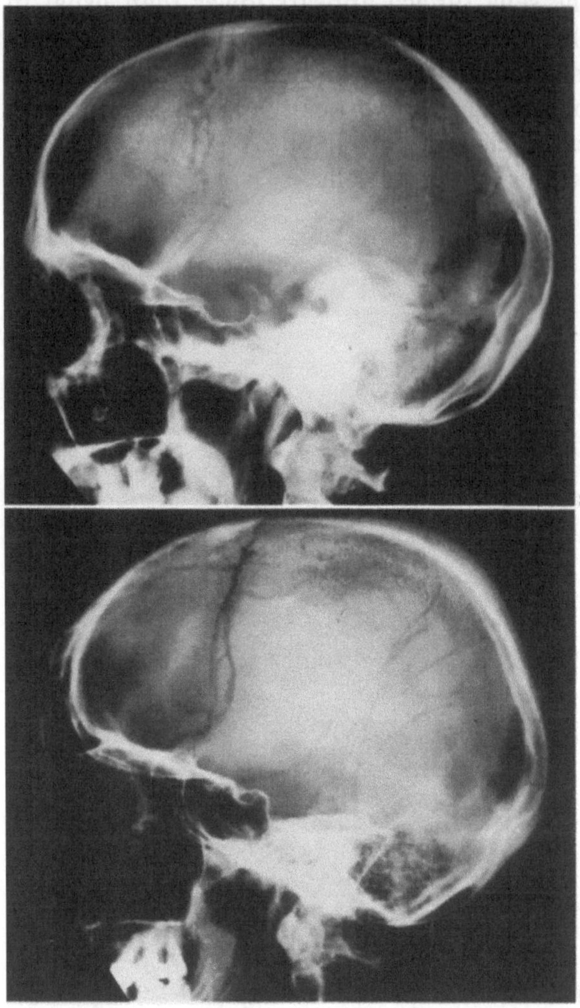

Fig. 5.4a, b. Variations in size of the sella turcica.
a Large normal sella
b Small normal sella

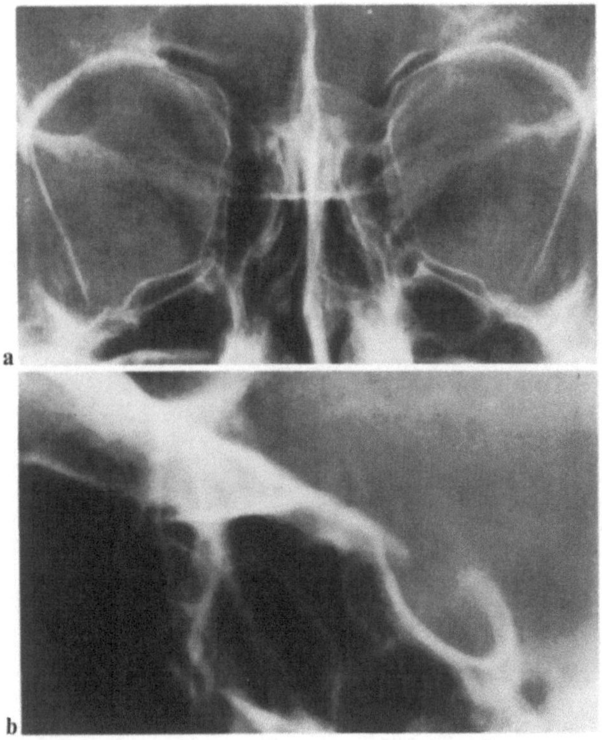

Fig. 5.5
a Frontal projection of the sella turcica: the floor is not visible, this appearance is not, however, pathologic.
b Lateral projection: the anterior wall slopes markedly and the floor of the sella is short

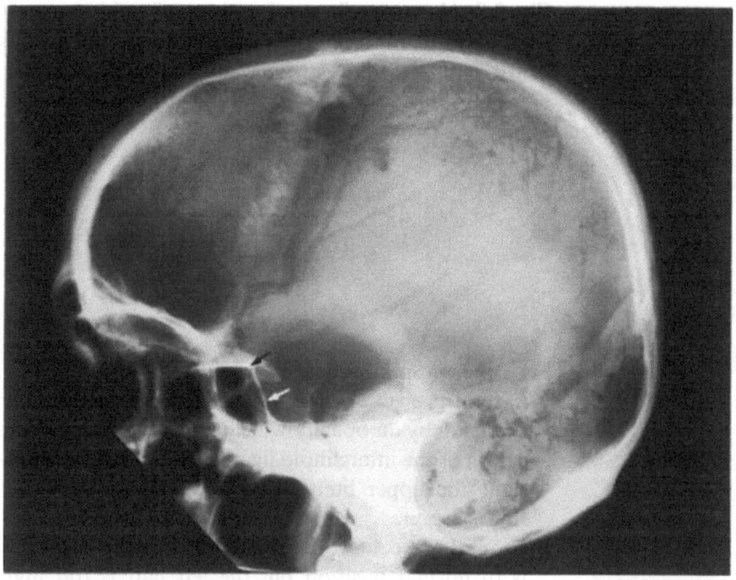

Fig. 5.6. Verticalization of the chiasmatic sulcus, giving a false appearance of an enlarged sella turcica. Limbus sphenoidale *(black arrow)*; tuberculum sellae *(white arrow)*

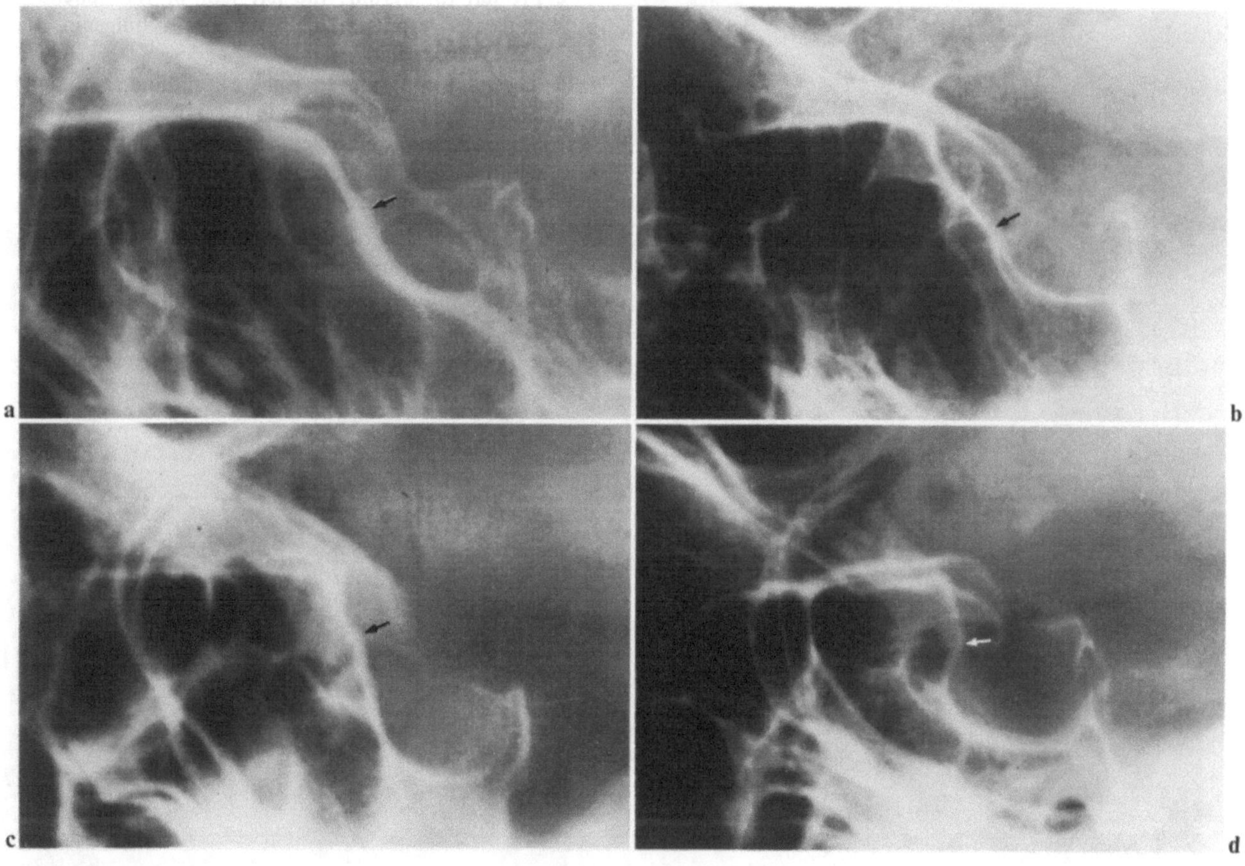

Fig. 5.7a–d. Marked angulation between the chiasmatic sulcus and the planum sphenoidale; at the limit the chiasmatic sulcus is vertical (tuberculum sellae, *arrow*)

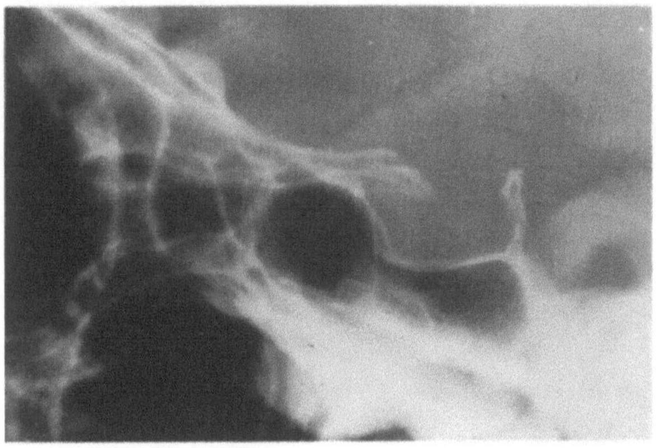

◁ **Fig. 5.8.** Normal sella turcica: generally thin appearance of the orbital roofs, the presellar region, the walls of the pituitary fossa, and the dorsum sellae

Fig. 5.9a–e. Small sella on one side.

a Lateral projection: the sella turcica seems partially occupied by an opacity which could simulate calcification of the interclinoid ligaments, but in fact represents the upper limit of the left half of the sellar floor *(arrows)*.

b Frontal projection: the right half of the sella turcica is in normal position but the left half is too high and too close to the planum sphenoidale *(arrows)*.

c Frontal tomogram: the left half of the sphenoid sinus is overpneumatized, causing a bulge in the floor of the sella.

d Sagittal tomogram to right of midline: the right side of the pituitary fossa is normal.

e Sagittal tomogram to left of midline: flat hemisella, without excavation; the left side of the floor *(arrow)* is almost at the level of the planum sphenoidale

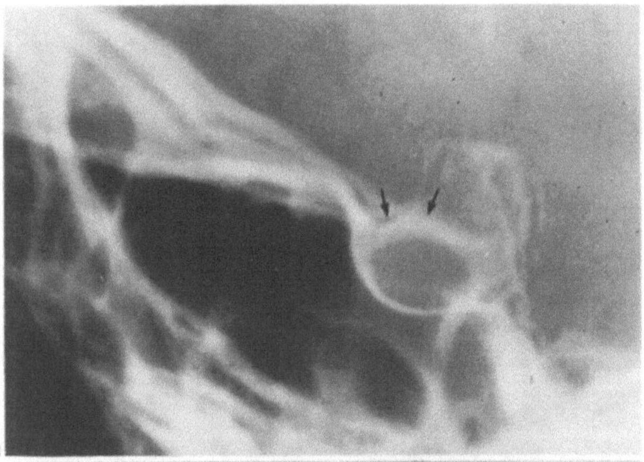

a

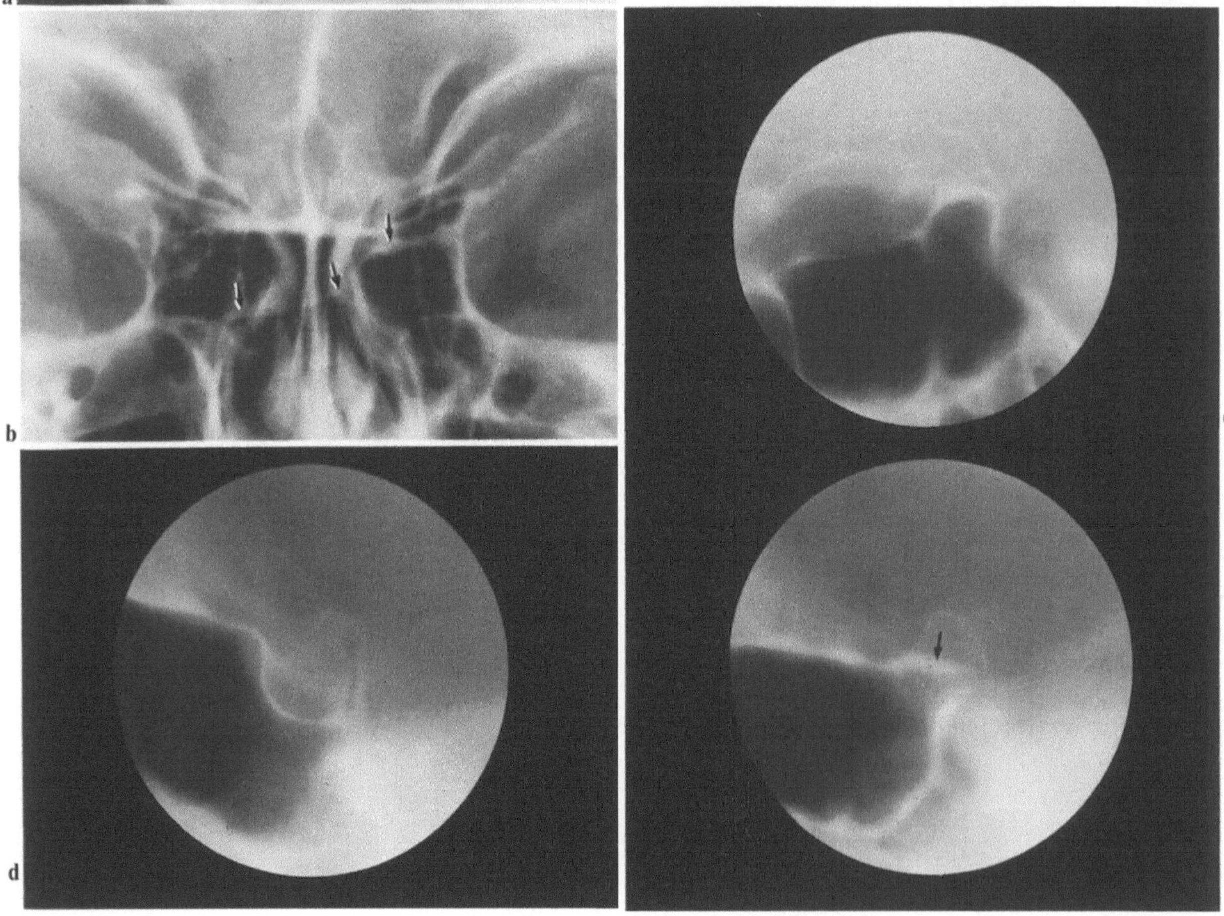

b c d e

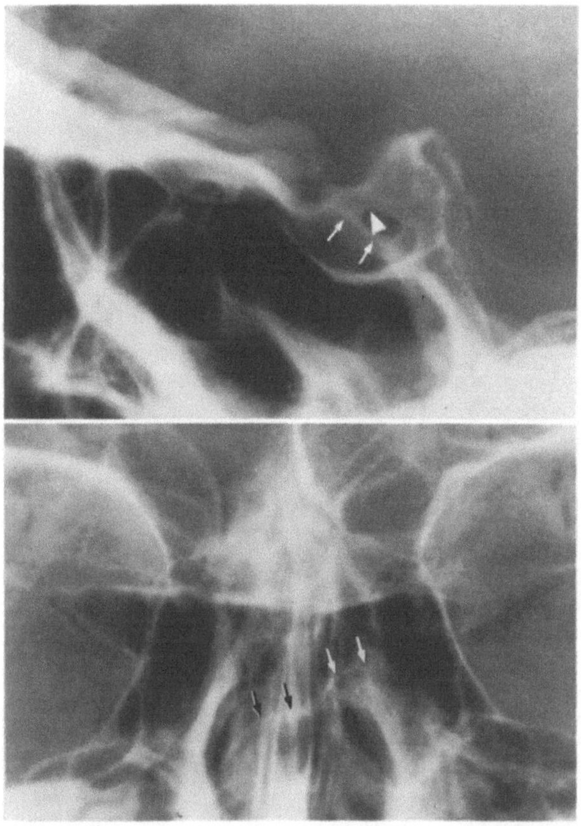

a b

Fig. 5.10a, b. Small sella on one side: the left half of the floor shows upward bulging due to extensive pneumatization of one half of the sphenoid sinus *(white arrows)*. The right side is normal *(black and white arrows)*. Note a calcified ligament between the left middle and posterior clinoid processes, giving the appearance of a perforated sella *(arrowhead)*

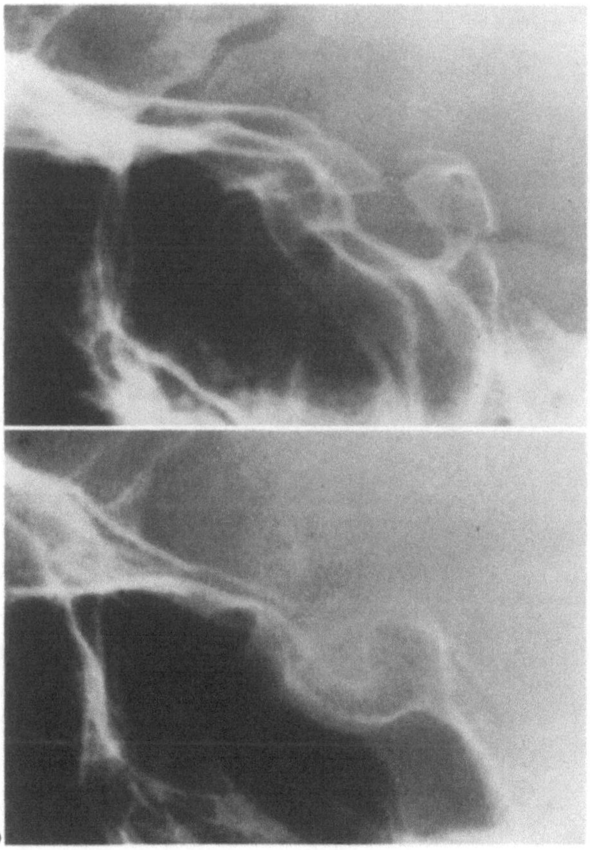

a b

Fig. 5.11a, b. Small sella on one side: in each case, overpneumatization of one-half of the sphenoid sinus causes considerable asymmetry of the floor of the sella. One side of the sella turcica is of normal size, while the other is too small

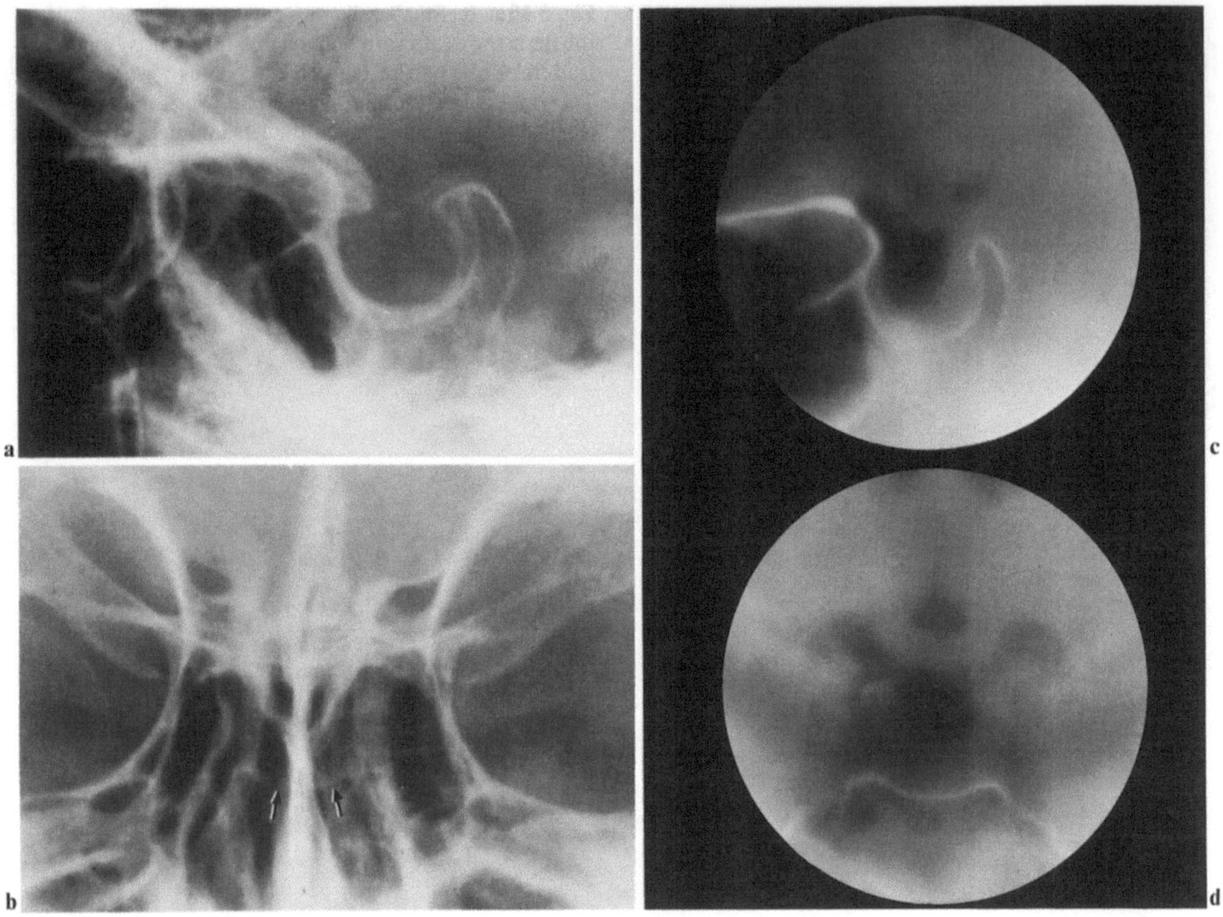

Fig. 5.12 a–d (caption see opposite page)

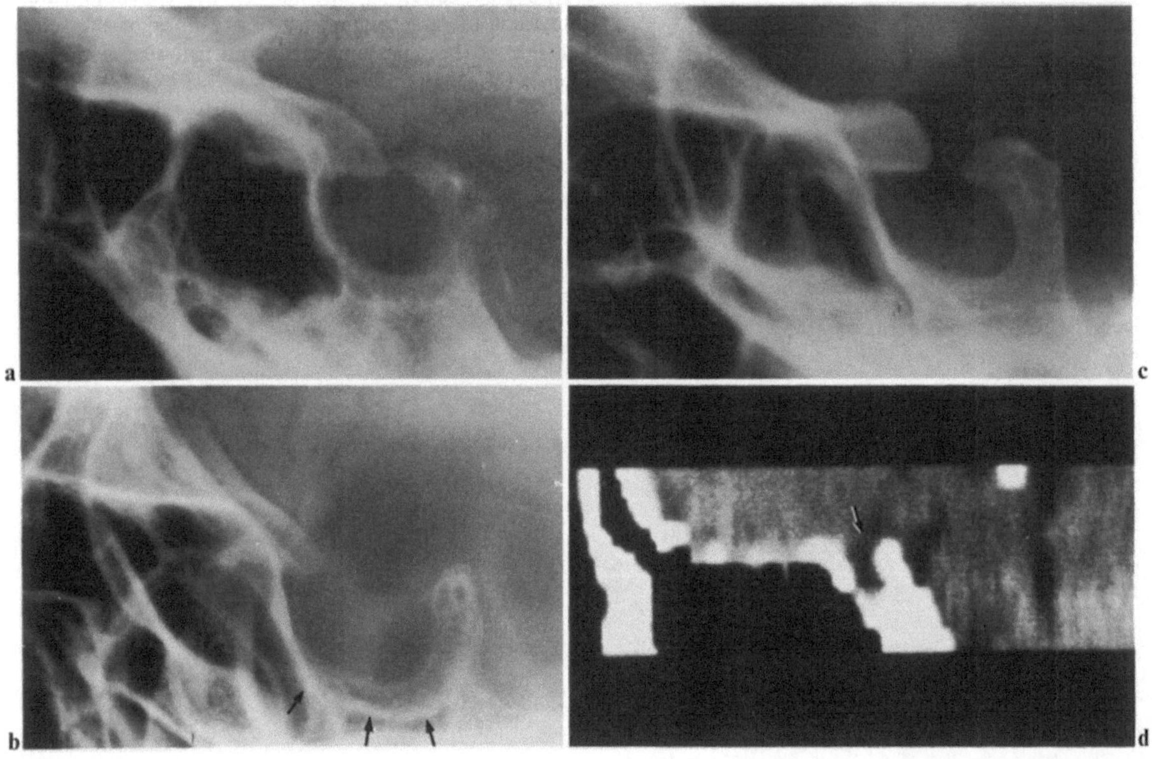

Fig. 5.13 a–d (caption see opposite page)

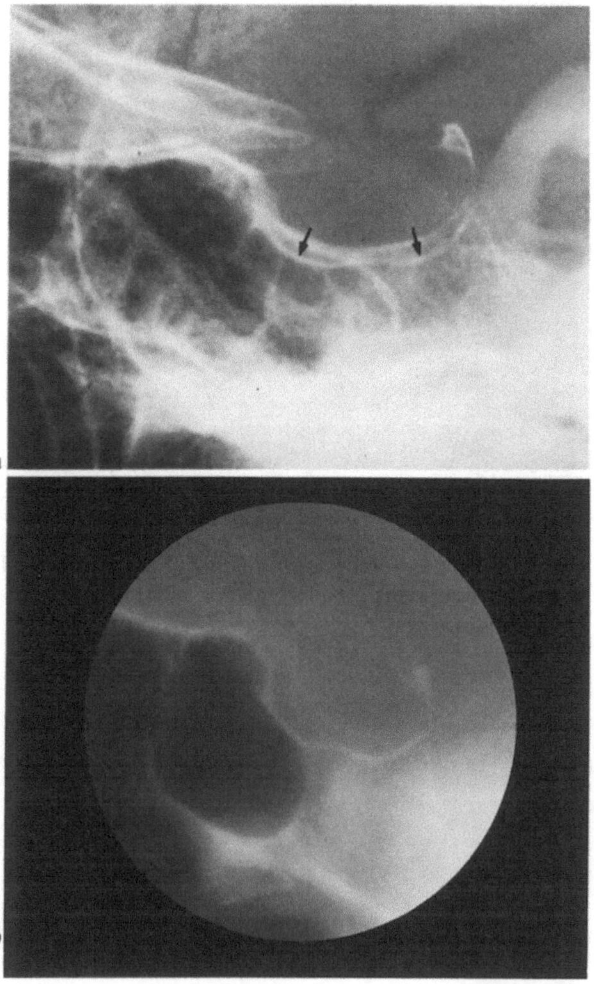

a

b

Fig. 5.14a, b. Sella turcica in multiparity (eight pregnancies). **a** Lateral projection: the floor shows a characteristic double outline, formed by two nonparallel lines which approach one another and then diverge *(arrows)*. This effect is produced by an irregular and mamillated floor, clearly shown by tomography **b**

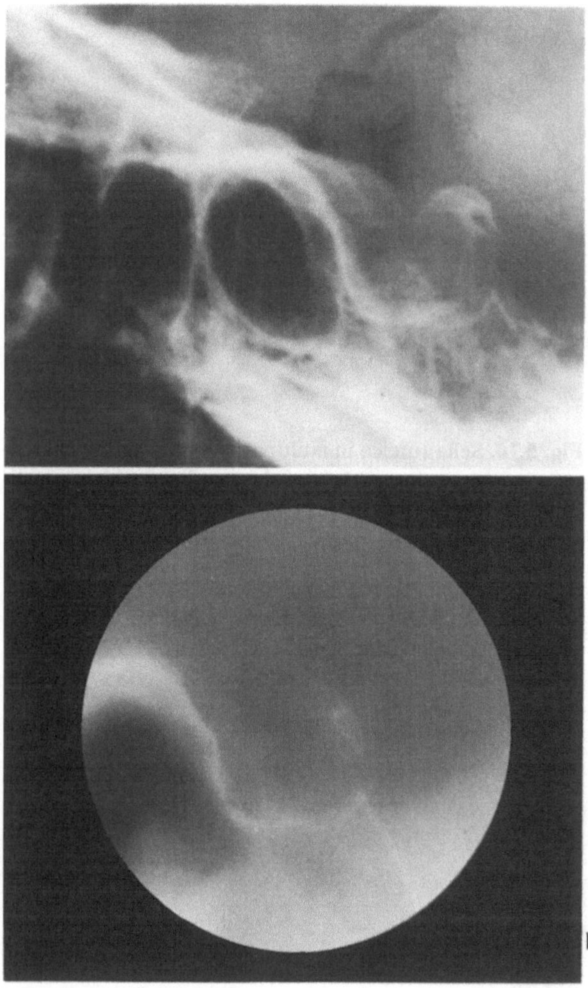

a

b

Fig. 5.15a, b. Sella turcica in multiparity (eleven pregnancies); irregular, mamillated appearance of the floor, especially apparent with tomography. Sphenoid pneumatization is of presellar type

---

Fig. 5.12a–d. Early empty sella.

a Lateral projection: no change in the size of the sella turcica; slight double floor with parallel margins and a clear lower line. Note that pneumatization of the sphenoid body is presellar.

b Frontal projection: slight, regular, upward, concavity of the floor of the sella *(arrows)*. Note the slight but very regular upward concavity of the floor of the sella, the cortical bone of which is perfectly distinct.

c, d Midsagittal and frontal pneumotomograms: intrasellar air confirms the presence of enlarged intrasellar subarachnoid space

◁ **Fig. 5.13**

a–c Three further examples of early empty sellae; on the lateral projection, the size of the sella remains virtually normal. Note in each case the parallel character of the double floor whose upper line is obscured and whose lower line is very well defined *(arrows)*.

d Early empty sella: CT sagittal reconstruction demonstrates presence of CSF within the sella *(arrow)*

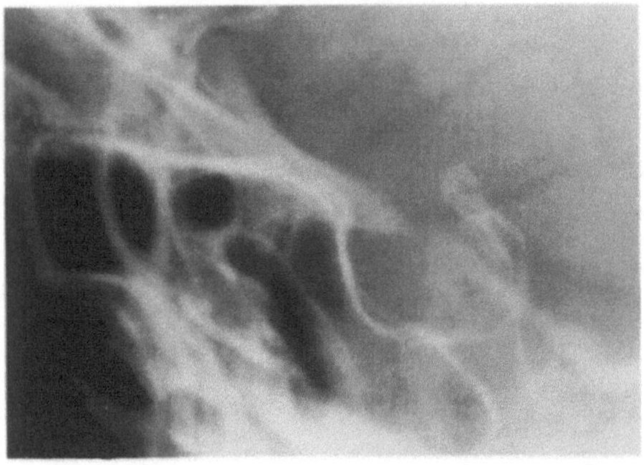

Fig. 5.16. Sella turcica in multiparity, showing the characteristic intersecting pattern of the two lines which constitute the double floor

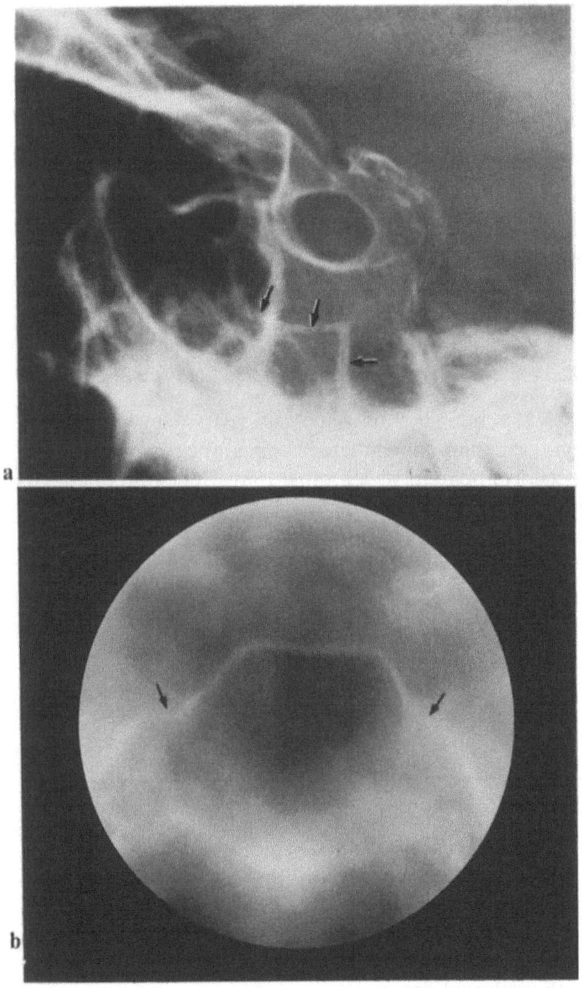

a

b

Fig. 5.17a, b. Carotid sulcus.

a Lateral projection: very pronounced imprint of the carotid sulcus *(arrows)*.

b Frontal tomogram through the sella: the carotid sulci are wide, thus creating a "stairlike" appearance in the lateral walls of the sphenoid *(arrows)*

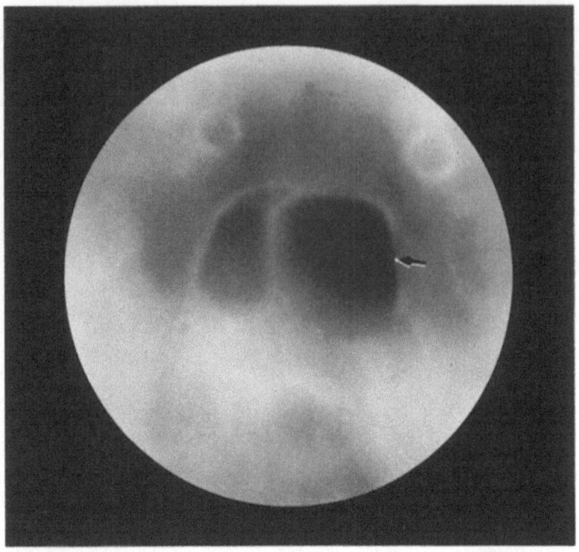

Fig. 5.18. Frontal tomogram: thinning of the lateral wall of the sphenoid sinus *(arrow)* by a partially calcified internal carotid artery

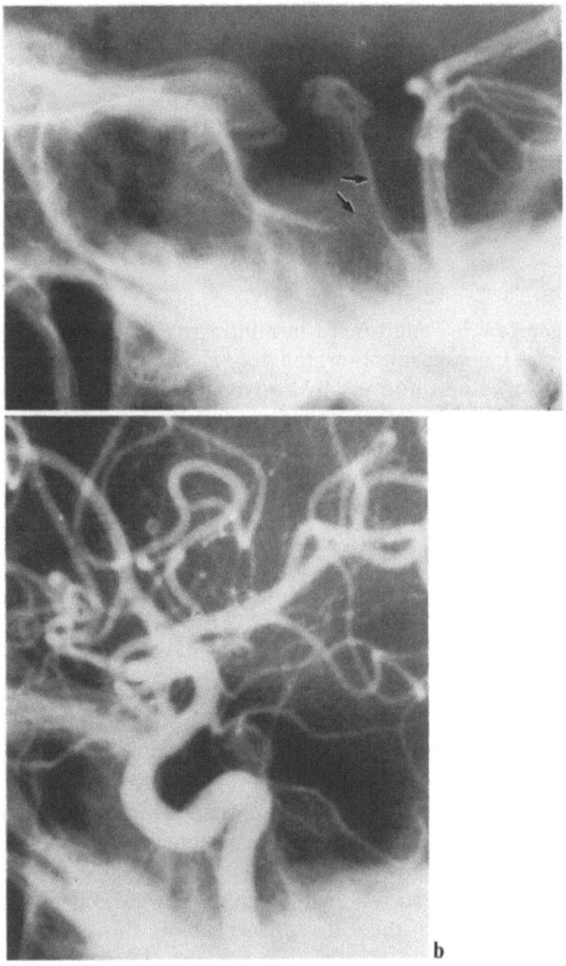

a

b

Fig. 5.19a, b. Depression of the posterior part of the floor of the sella, extending to the lower part of the dorsum sellae **a** (*arrows*), related to an ectatic carotid siphon (shown in **b**)

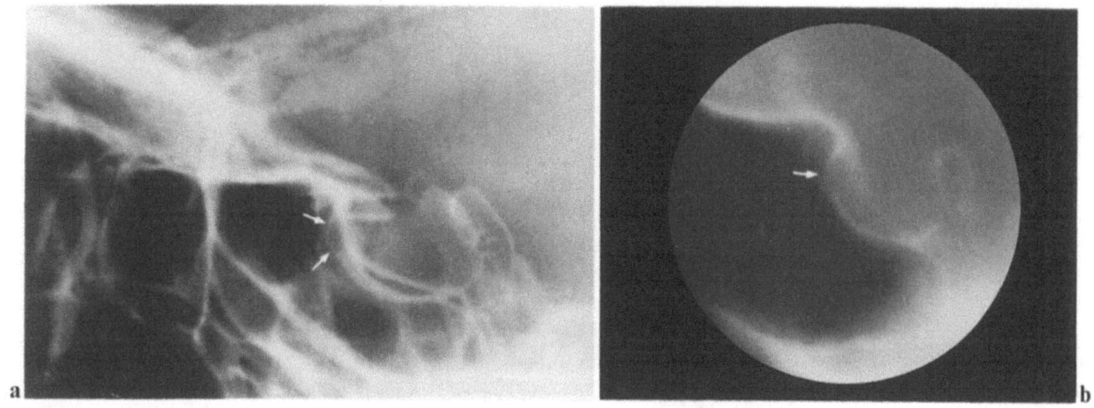

Fig. 5.20. a Lateral view and **b** tomogram: concave semilunar notch in the upper part of the anterior wall of the sella, with thinning of the cortical bone; imprint of a medially lying carotid siphon *(arrows)*

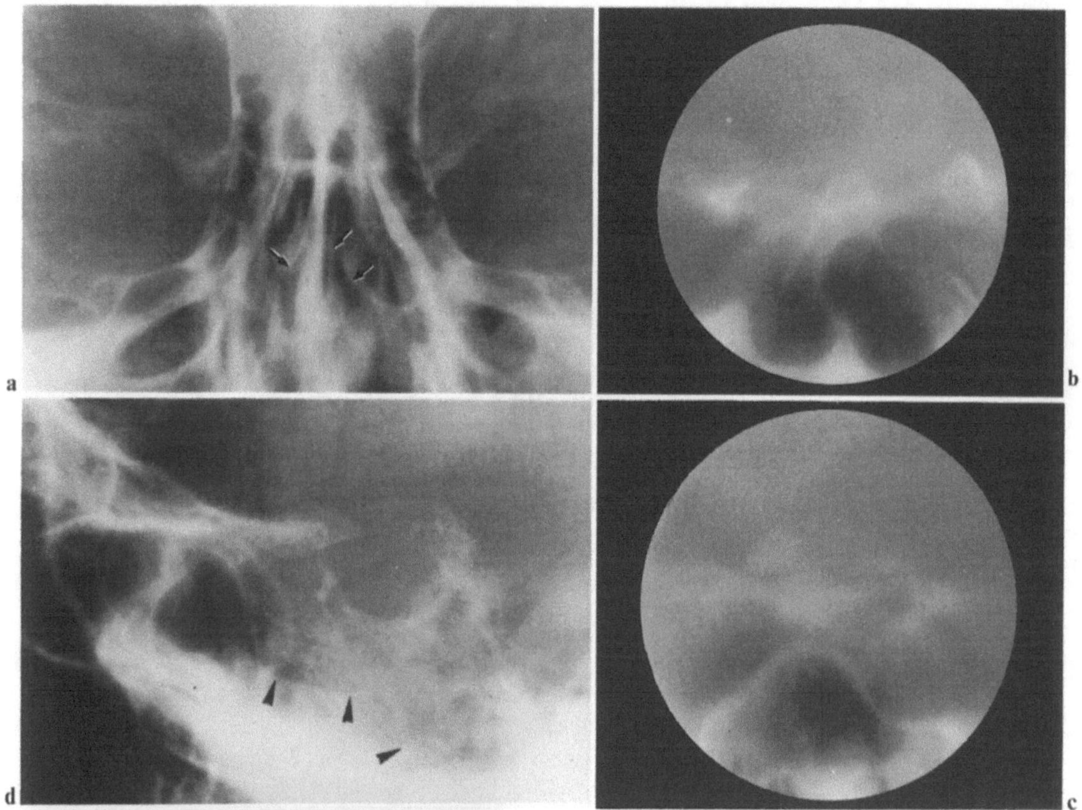

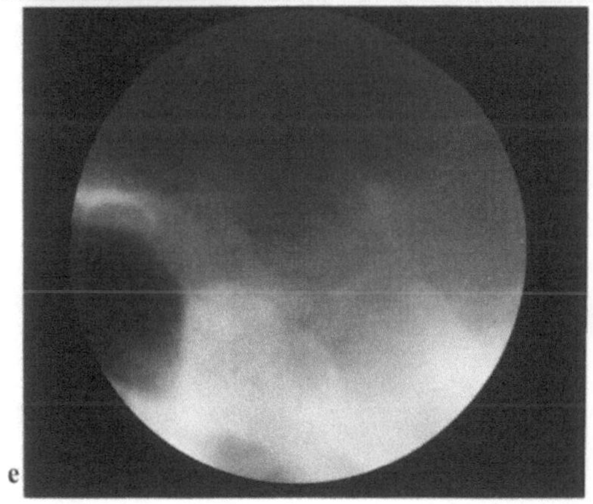

Fig. 5.21a–e. Carotid sulcus.

a Frontal projection: narrow floor of the sella, extended on each side by the wide carotid sulci *(arrows)*, especially on the left.

b, c Frontal tomograms passing through the anterior clinoid processes and the floor of the sella.

d Lateral projection: very large imprint of the carotid sulcus *(arrowheads)*.

e Corresponding lateral tomogram

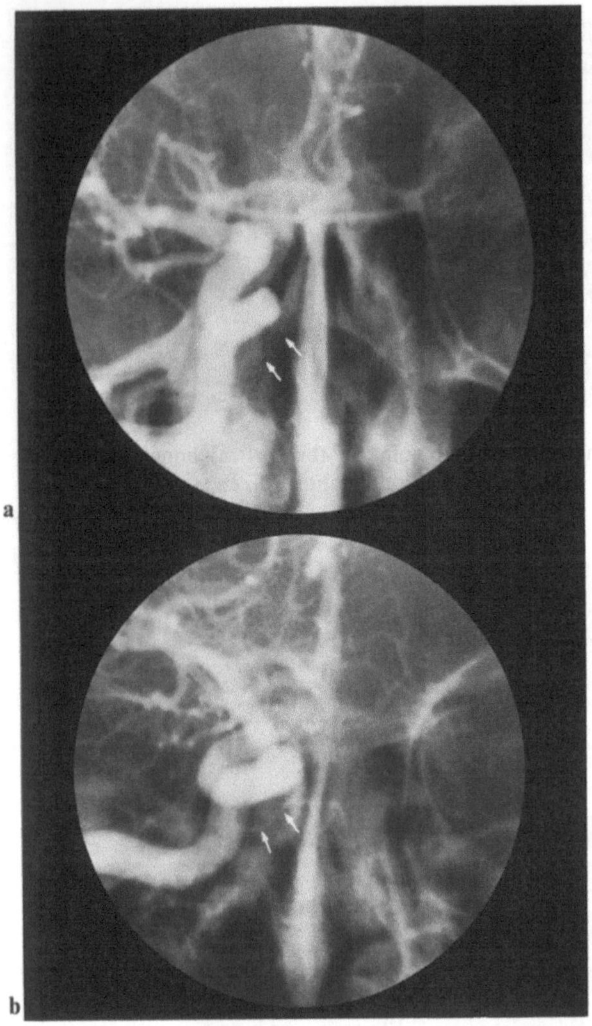

Fig. 5.22. **a** Two intrasellar carotid siphons *(arrows)*; note in **b** the narrowness of the floor of the sella and the shape of the carotid sulcus

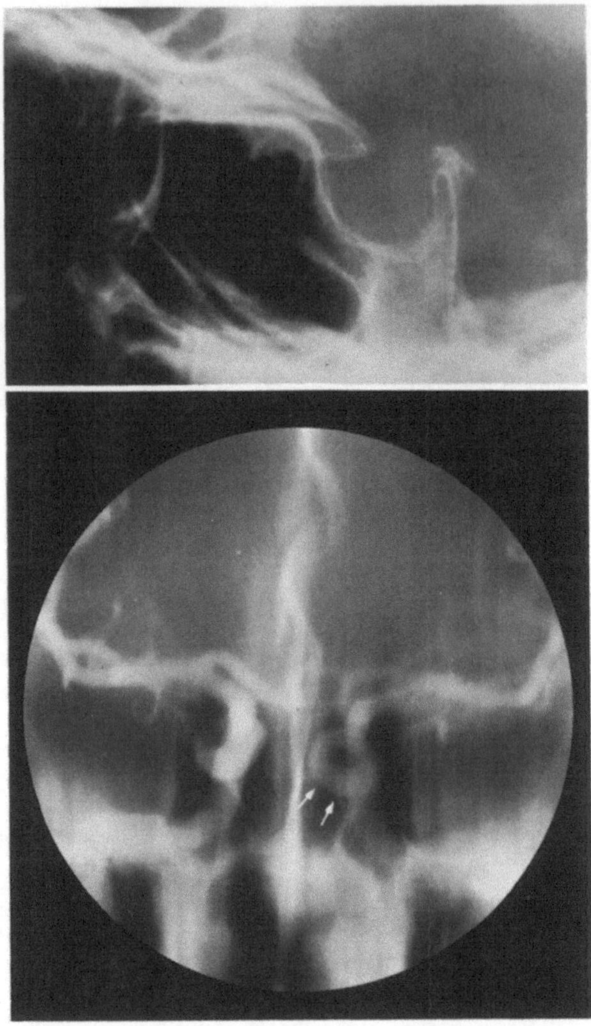

Fig. 5.23a, b. Intrasellar siphon.

a Very irregular appearance of the floor of the sella, which is doubled.

b Bilateral frontal angiotomogram: right and left ectatic intrasellar siphons; the left siphon almost reaches the midline *(arrows)*

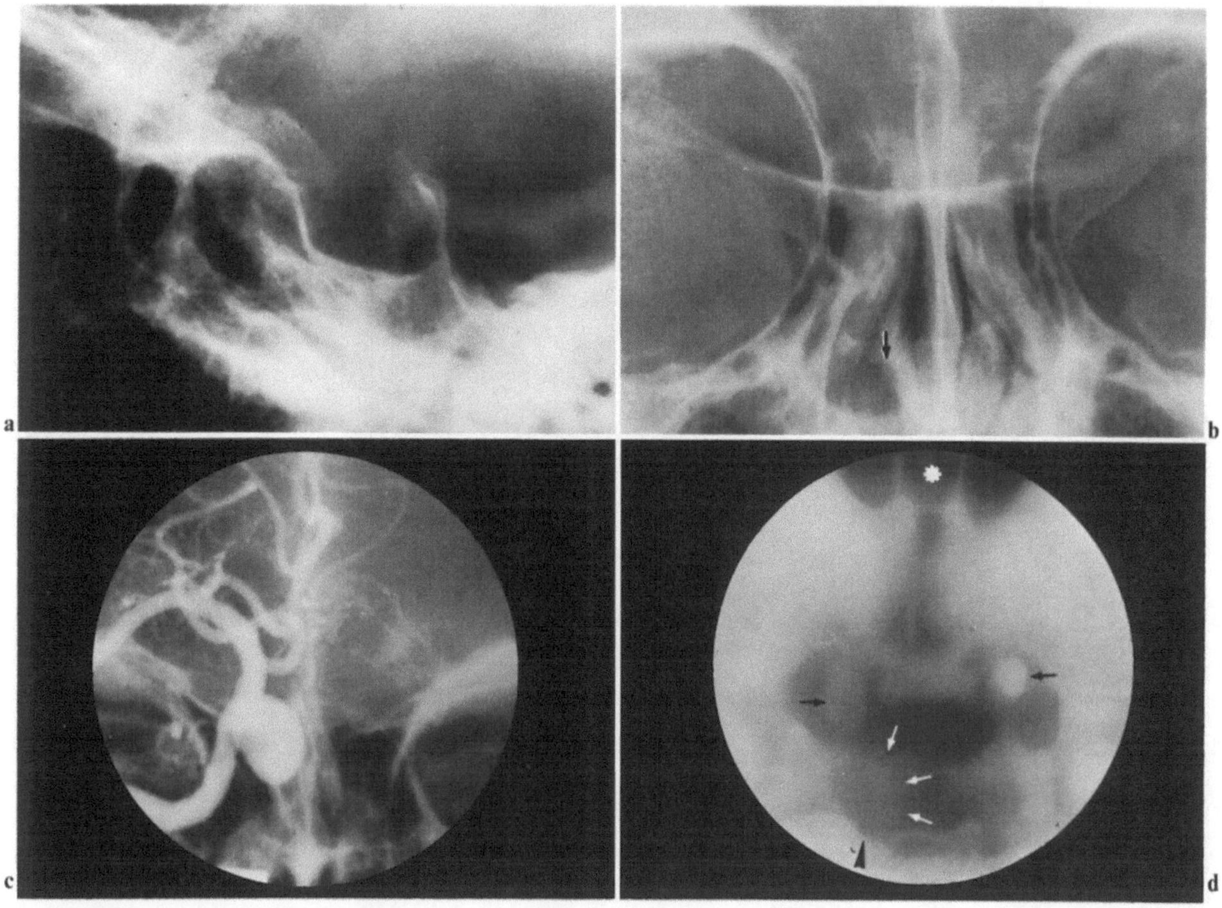

△

Fig. 5.24 a–d. Intrasellar siphon.

a Lateral projection of the sella turcica.

b Frontal projection of the floor of the sella: localized depression of the right side of the floor *(arrow)*.

c Frontal angiogram: intrasellar carotid siphon corresponding to the depression of the floor.

d Frontal pneumotomogram through the sella: imprint of the carotid siphon on the floor. The air outlines the supraclinoid carotid siphons *(black arrows)* and the right intrasellar siphon *(white arrows)*, which lies close to the depression of the floor *(arrowhead)*. Cavum septi pellucidi *(white star)*

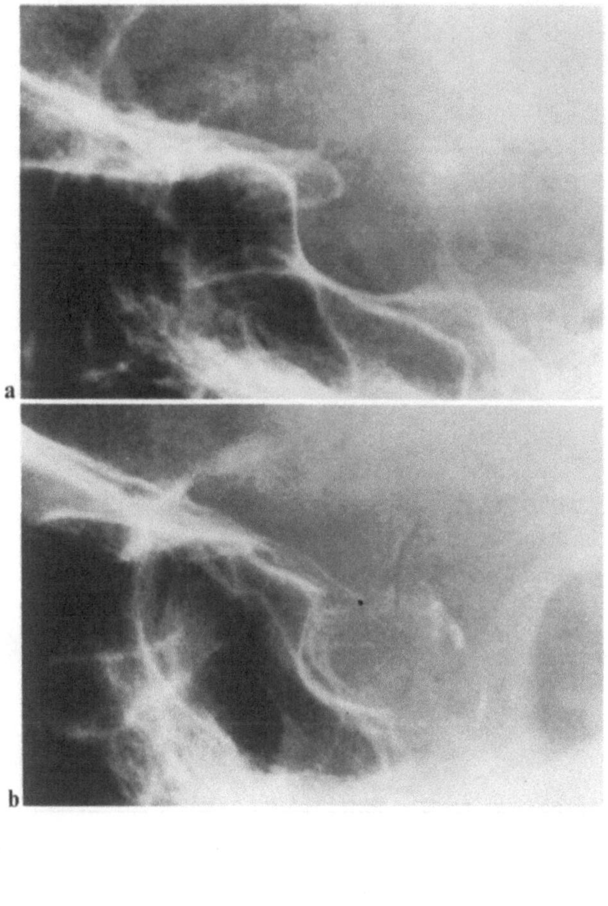

Fig. 5.25 a, b. Two examples of osteoporosis in elderly subjects, with considerable demineralization of the dorsum sellae and the posterior part of the floor of the sella

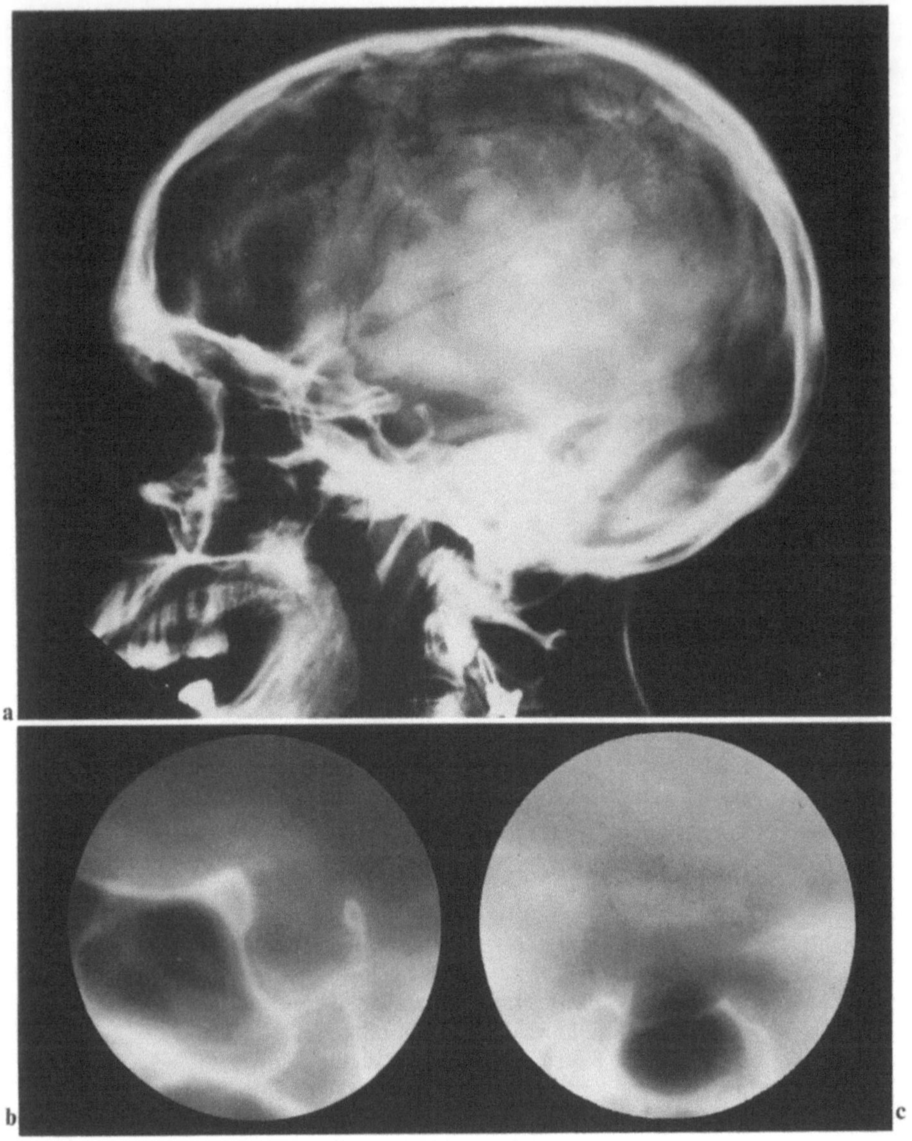

Fig. 5.26a–c. Oxycephaly.

a The sella turcica appears roughly triangular and seems to be embedded in the skull base.

b, c Sagittal and frontal tomograms: hollowed appearance of the floor of the sella

B. Variations in Different Anatomic Structures

I. Variations in the Sphenoid Sinus [11]

The variations of the sphenoid sinus are of great importance since they may affect the morphology of the presellar area, the general configuration of the sella turcica itself, and especially its floor.

Pneumatization in the sphenoid bone develops from front to back, until puberty (Fujioka and Young 1978). However, pneumatization in the dorsum sellae and posterior clinoid processes can continue until the age of 20 (Vidic and Stom 1968).

Three types of sinus are generally described in terms of their anteroposterior extension (Fig. 5.27) (Hammer and Rådberg 1961):

The conchoidal type ("conchal") in which the sphenoid is only slightly pneumatized (3%);

The presellar type, in which pneumatization does not extend beyond the tuberculum sellae (11%);

The sellar type, where the sinus extends at least as far posteriorly as the floor of the sella (86%).

Pneumatization of the sphenoid may be quite extensive, the sinus occupying the pterygoid processes, the lesser and greater wings of the sphenoid (Fig. 5.28), the floor of the middle fossa (Fig. 5.29), and the clivus. We have even observed pneumatization of an occipitalized anterior arch of the atlas (Fig. 5.30).

The right and left sphenoid sinuses are usually separated by an intersphenoidal septum, the anterior portion of which is most often in the midline. More posteriorly, it slopes laterally, sometimes joining the carotid groove (Fig. 5.31) (Kinnman 1977). Pneumatization may in fact be very asymmetric, in which case the translucency of the sinus appears heterogeneous in a lateral projection. This could be mistaken for pathologic clouding (Fig. 5.32). The cavity of the sinus is often divided by other septa in anteroposterior, transverse, horizontal, or oblique directions (Fig. 5.33); the septa may be incomplete, forming crests (Fig. 5.34).

All of the foregoing septa are well shown in frontal and submentovertical tomograms. When transphenoidal surgery is to be performed, such a study is of fundamental importance.

Surgeons must have a good understanding of several important facts. Following a septum in a more or less longitudinal direction does not necessarily lead to the floor of the sella, but may lead laterally to the carotid groove. The possibility of a vertical chiasmatic sulcus must be recognized so the tuberculum sellae is not mistaken for the limbus sphenoidale. Surgical perforation of the skull base above the tuberculum, i.e., above the diaphragma sellae, leads to a cerebrospinal fluid fistula (Fig. 5.49) (Rothman S.L.G. et al. 1976).

The optic eminence is a bulging of the optic canal toward the sphenoid sinus at the upper portion of its lateral wall (Fig. 5.35). This becomes particularly clear when pneumatization extends into the lesser wings. The wall of the sinus around the optic eminence is thus very fragile and must of course be strictly respected.

II. Variations in the Presellar Region

The presellar region, situated in the middle of the anterior cranial fossa, extends from the posterior portion of the cribriform plate to the tuberculum sellae. From front to back it is composed of the planum sphenoidale, separated from the chiasmatic sulcus by the limbus sphenoidale. Too often neglected in everyday radiodiagnosis, the presellar region is of particular interest, and therefore a detailed analysis of these anatomic variations is of some importance (Lee 1976; Bonneville et al. 1979). The planum sphenoidale must be recognized under any circumstances since, its morphology being constant, it constitutes a reference of

11 Refs. [230, 231, 246, 276, 376, 377, 581, 595, 655, 678, 715, 722, 754, 760, 977, 987, 993, 1043]

normality for the skull base. Conversely the morphology of the chiasmatic sulcus is highly variable.

In frontal projections of the floor of the sella, the planum sphenoidale has always been shown as a linear, horizontal plate of cortical bone joining the upper edge of the lesser sphenoid wings across the midline. In fact, the frontal image of the presellar region depends on the position of the chiasmatic sulcus with regard to the planum and on the obliquity of the planum.

One can conclude, therefore, that the presellar region takes many different forms (Figs. 5.36, 5.37):

A horizontal line (Figs. 5.36a and 5.37a);

A line with a slight upward concavity (Figs. 5.36b, 5.37b);

A line pointing upward in the midline which corresponds to the optic crest of the anterior portion of the planum, presenting a median ridge (Figs. 5.36c, 5.37c);

A double line corresponding to the superimposition of a+c (Figs. 5.36a+c; 5.37d).

A double line, one part of which is horizontal and the other curved, corresponding to the images of the planum and the chiasmatic sulcus (Figs. 5.36d, e, 5.37e, f);

A butterfly-shaped image (Figs. 5.36f, 5.37g);

A linear and very narrow image (Figs. 5.36g, 5.37h).

Other appearances of the presellar region in the frontal projection are more exceptional; they also change the appearance of the region in the lateral projection:

Upward convexity of the planum in a frontal projection in cases where sphenoid pneumatization is very pronounced (Figs. 5.36h, 5.38a); however, this pattern must only be regarded as physiologic after the possibility of regional pathology, notably mengioma of the planum, has been excluded.

A slanted or staircase-shaped planum in the asymmetric skull base (Figs. 5.36i, j, 5.38b, 5.46).

Finally, it is possible that the planum would not appear on the frontal projection if the X-ray beam was not parallel to the planum. This is the case when the planum is very oblique, and its direction nearly parallel with that of the orbital roofs (Figs. 5.38c, 5.36k).

When studying the presellar region, the lateral projection is the most important. It is necessary to identify the chiasmatic sulcus, which varies in length, shape, and direction; and the planum sphenoidale, a dense plate whose thickness, less than 2 mm, is constant from front to back. It extends slightly further forward than the greater wings of the sphenoid. The constant, regular characteristics of the planum are such that it is an extremely valuable normal reference point (Fig. 5.39). Nevertheless, it must be remembered that the planum sphenoidale can appear short when the chiasmatic sulcus is long (Fig. 5.40). Most often, the chiasmatic sulcus forms an angle of about 160° with the planum. In some cases, the planum and chiasmatic sulcus form a straight line. It is nearly always possible to identify them, however, because their structure is different (Fig. 5.41). In other cases, the planes of the chiasmatic sulcus and the planum sphenoidale are parallel, but the former is slightly displaced downward relative to the former, in which case, the limbus always appears prominent (Fig. 5.42).

In contrast to the planum sphenoidale, the morphology of the chiasmatic sulcus is occasionally affected by pneumatization of the sphenoid sinus (Pribram and du Boulay 1971); a convex chiasmatic sulcus may be seen with a very well pneumatized sphenoid sinus, for example, and a concave chiasmatic sulcus with a poorly pneumatized sinus (Fig. 5.43).

Upward convexity of the planum sphenoidale related merely to hyperpneumatization of the underlying sphenoid is rare (Fig. 5.44a) (Lombardi et al. 1968). In general, the morphology of the planum is not affected by pneumatization of the sphenoid bone (Fig. 5.44b); similarly a planum whose thickness exceeds 2 mm, even if it is regular, must be examined carefully for a possible meningioma. Additional investigations would be necessary in order to rule out this possibility (Fig. 5.45). A double line planum in the lateral projection is not pathologic and usually corresponds to a staircaselike appearance on the frontal view (Fig. 5.46). A break, slanting downward and forward between the planum and the chiasmatic sulcus, can be seen in rare cases where these structures are in line (Smith and Kier 1971). This abnormality, called an unfused planum, must not be mistaken for a fracture (Fig. 5.47). Verticalization of the chiasmatic sulcus is the most interesting abnormality: in such cases, the chiasmatic sulcus forms an angle of about 90° with the planum sphenoidale and is in line with the anterior wall of the sella (Fig. 5.48). The limbus sphenoidale is very prominent, while the tuberculum sellae seems absent. It is important to differentiate this pattern from pathologic enlargement of the pituitary fossa. A vertical chiasmatic sulcus often appears unusually long, and makes the planum appear shorter; it may not reach the level of the greater wings of the sphenoid.

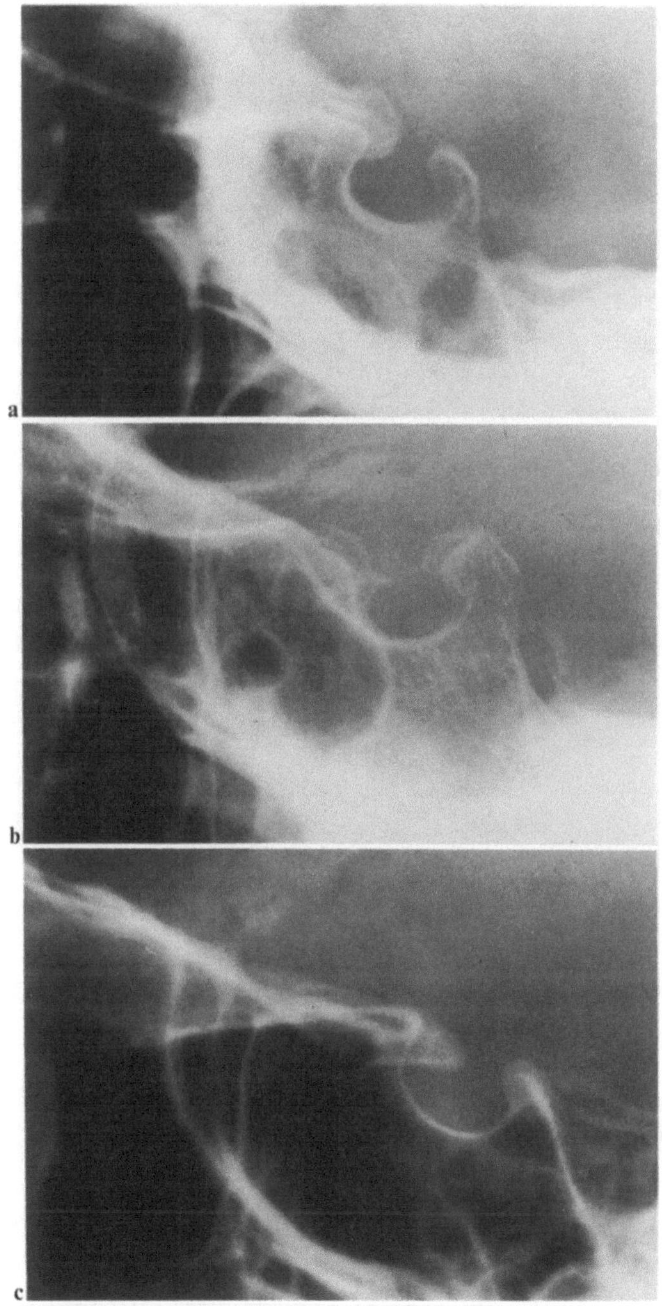

Fig. 5.27a–c. The three main types of sphenoid sinus pneumatization.
a Conchal; **b** presellar; **c** sellar

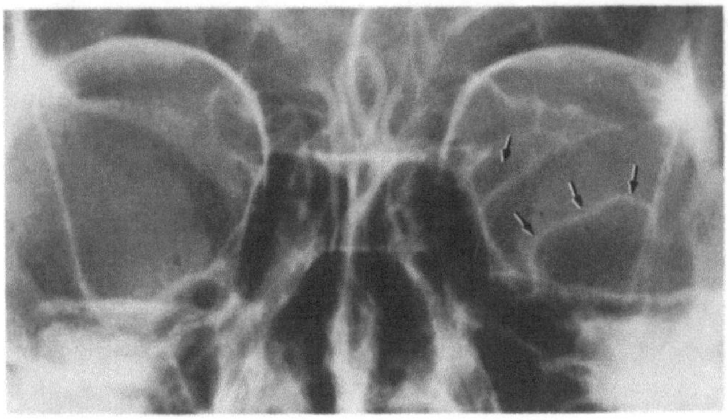

Fig. 5.28. Extensive pneumatization of the sphenoid sinus; pneumatization of the lesser and greater wings of the sphenoid *(arrows)*

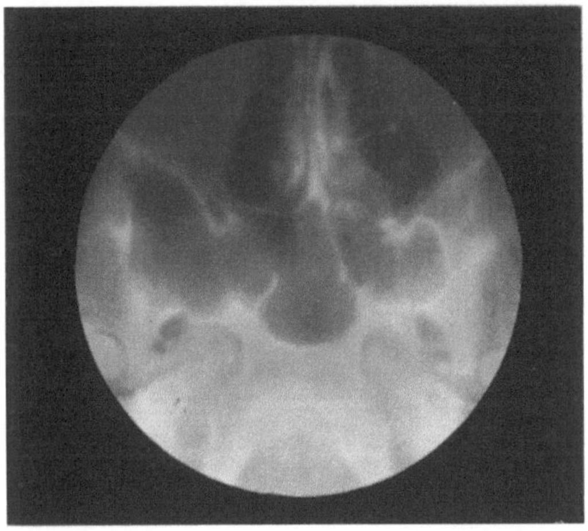

Fig. 5.29. Axial tomography: extensive pneumatization of the sphenoid and considerable pneumatization of the greater wing of the sphenoid on the right and of the floors of both middle fossae

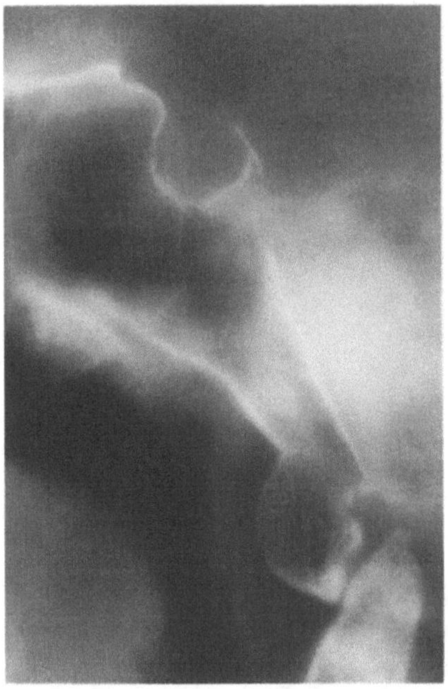

Fig. 5.30. Midline tomogram: unusual pneumatization of the clivus and of an occipitalized anterior arch of atlas

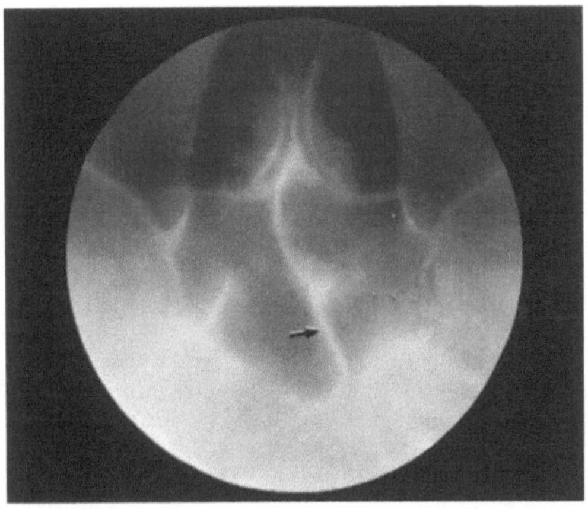

Fig. 5.31. Axial tomogram: good visibility of the intersinusal septum, which slopes laterally backward *(arrow)*

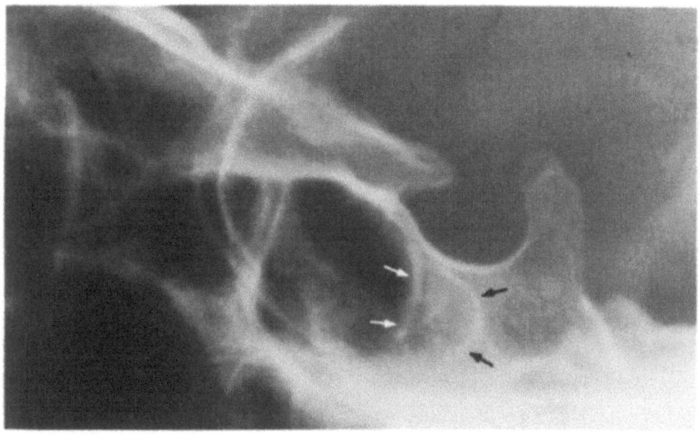

Fig. 5.32. Asymmetric pneumatization of the sphenoid below the anterior part of the floor of the sella, creating a false appearance of clouding in the posterior part of the sinus *(arrows)*

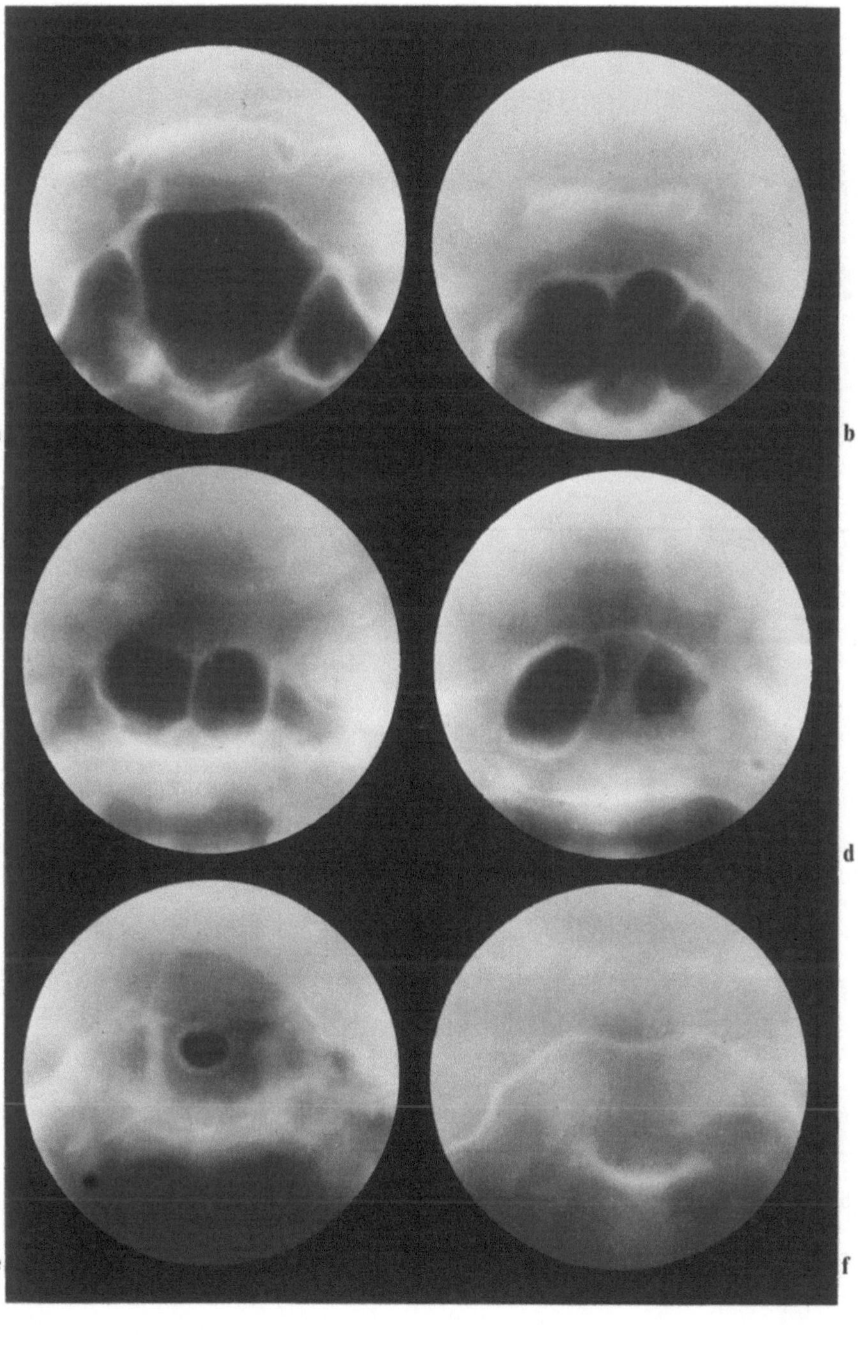

Fig. 5.33 a–f. Frontal tomograms: different patterns of sphenoid pneumatization

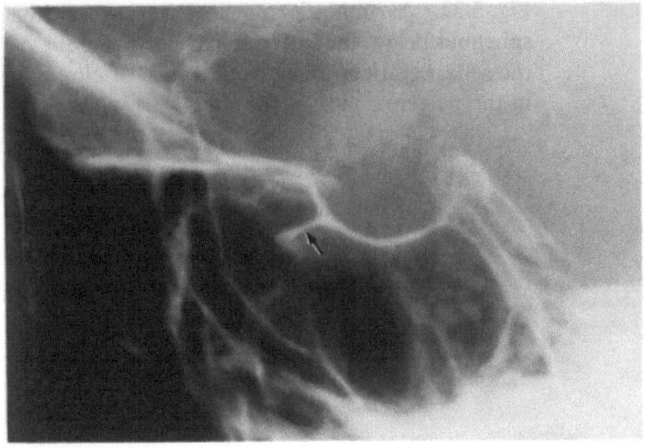

Fig. 5.34. Incomplete septum or crest of the sphenoid sinus *(arrow)*

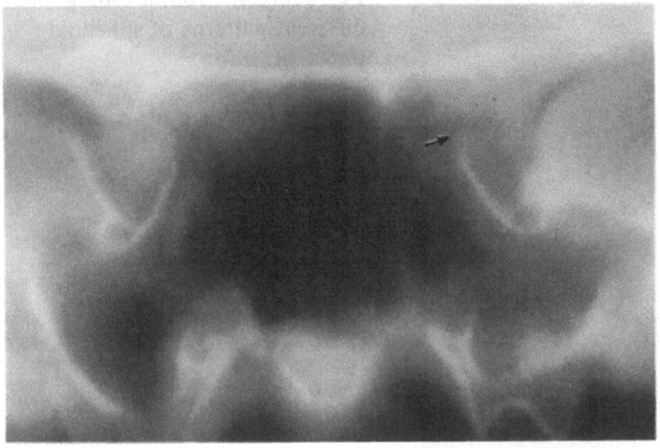

Fig. 5.35. Frontal tomogram through the optic canals. Note the very thin lateral wall of the sphenoid sinus opposite the optic canal *(arrow)*

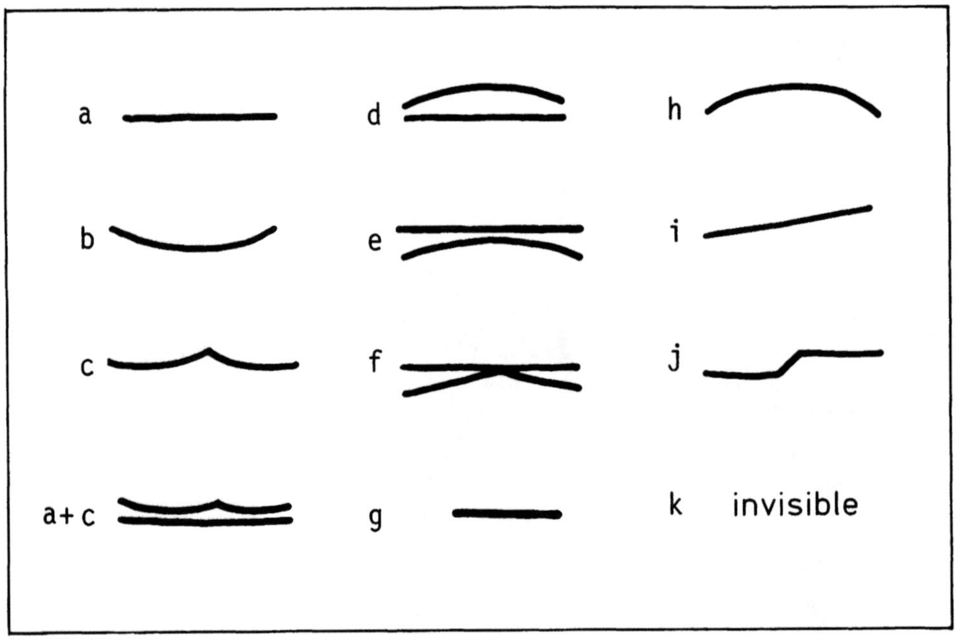

Fig. 5.36. Different appearances of the planum sphenoidale on frontal projection

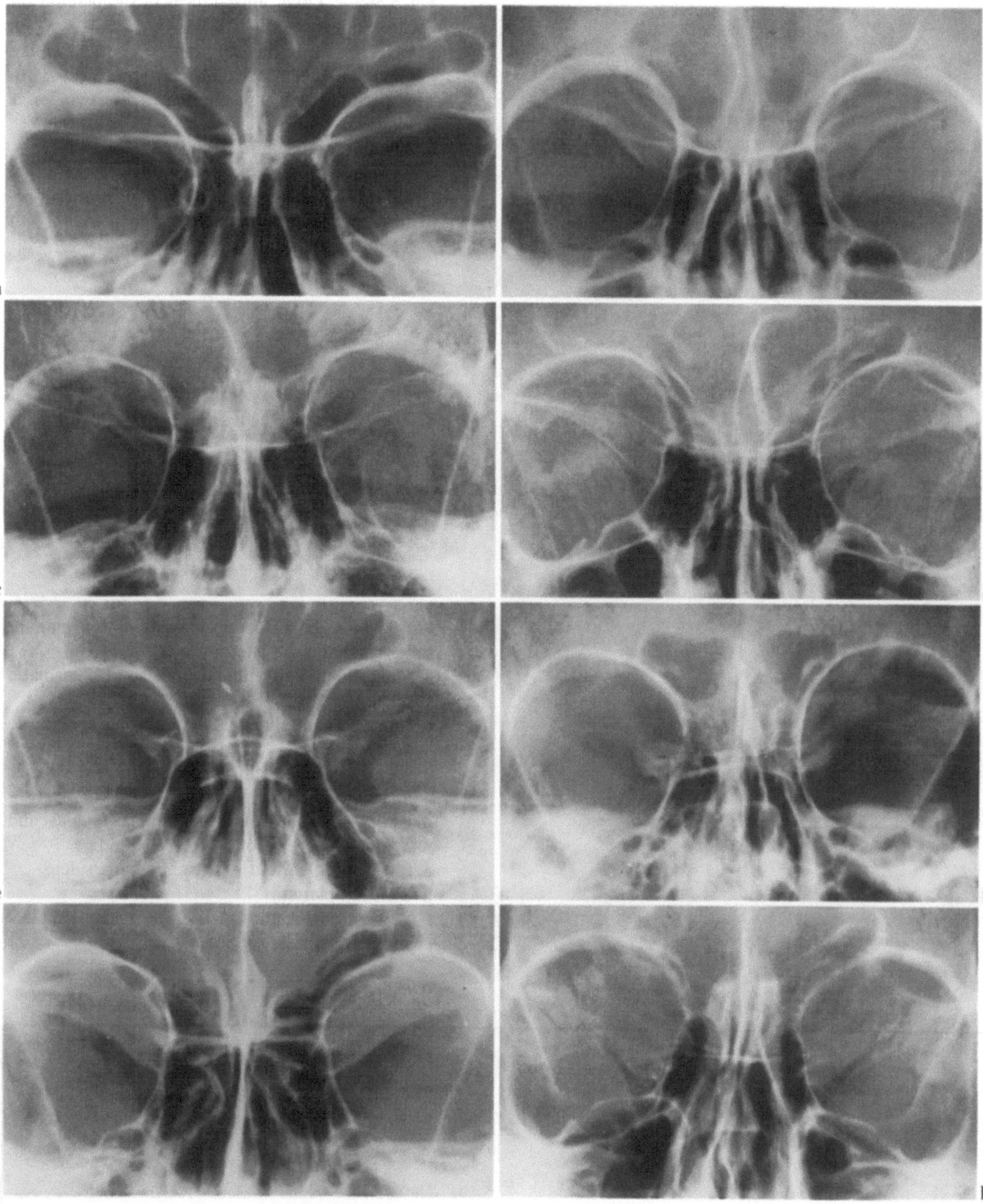

a b c d e f g h

Fig. 5.37 a–h. Different forms of the planum sphenoidale on frontal projection. Note that the floor of the sella is visible in every case

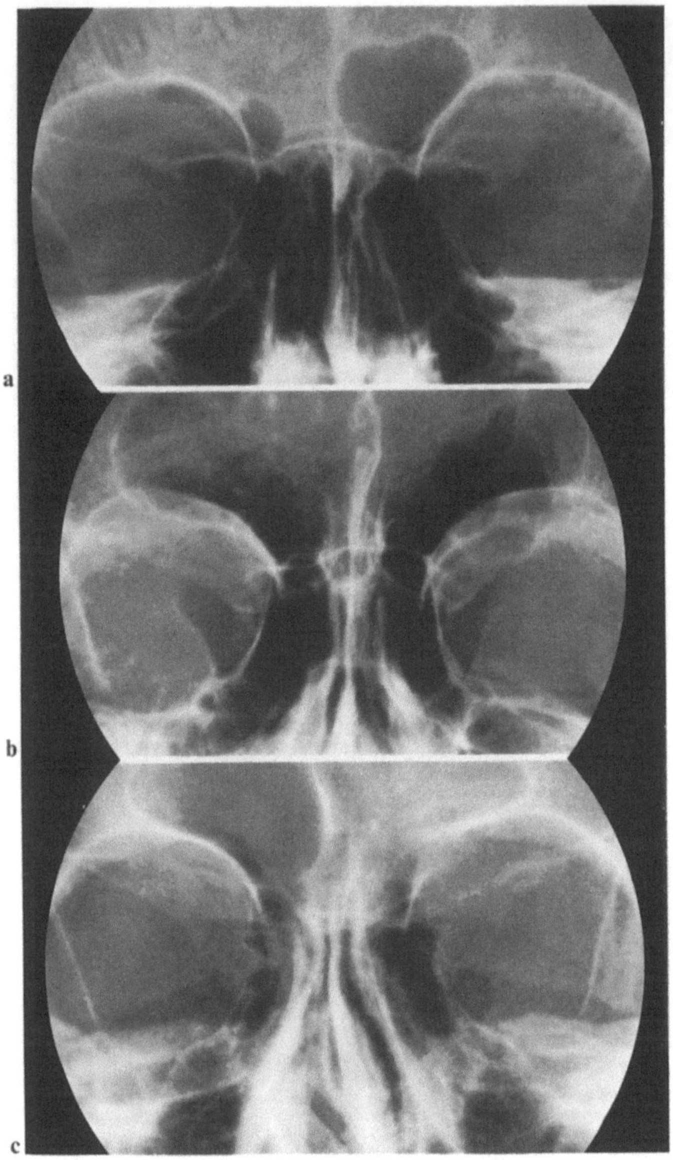

a

b

c

Fig. 5.38a–c. Other appearances of the planum on the frontal projection.

a Large upward convexity of the planum, where sphenoid pneumatization is pronounced, with, in particular, pneumatization of the two lesser wings.

b Asymmetric planum, "stairlike" appearance.

c Lack of visibility of the planum on the usual frontal view. This appearance is most often produced when the direction of the planum with regard to the skull base is unusual

Fig. 5.39a–i. Normal appearances of the planum sphenoidale on the lateral projection.

a Long planum sphenoidale and short chiasmatic sulcus.

b Short planum sphenoidale and long chiasmatic sulcus.

c Concavity of the chiasmatic sulcus and absence of sphenoid pneumatization.

d Upper convexity of the chiasmatic sulcus and hyperpneumatization of the sphenoid.

e, f, g Sloping variant of the chiasmatic sulcus

h Unfused planum

i Vertical chiasmatic sulcus, which may simulate an enlarged sella turcica

▽

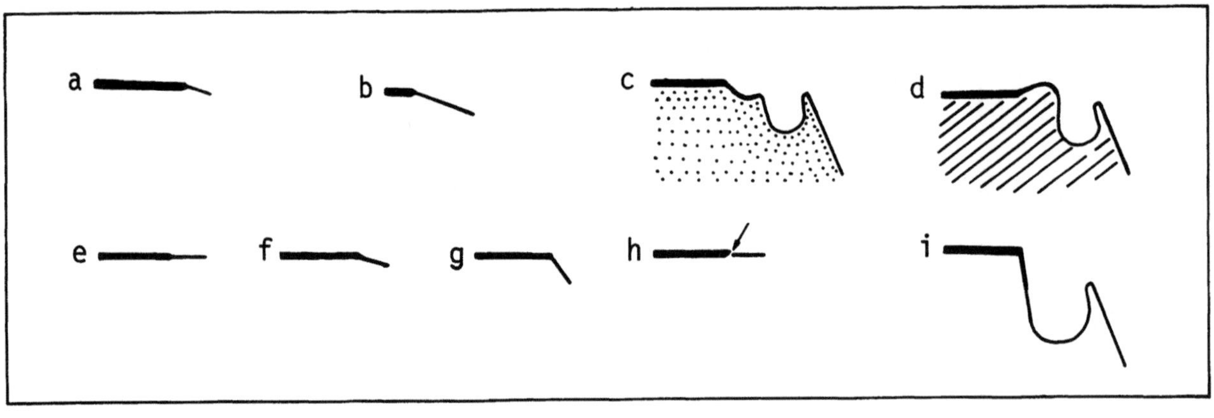

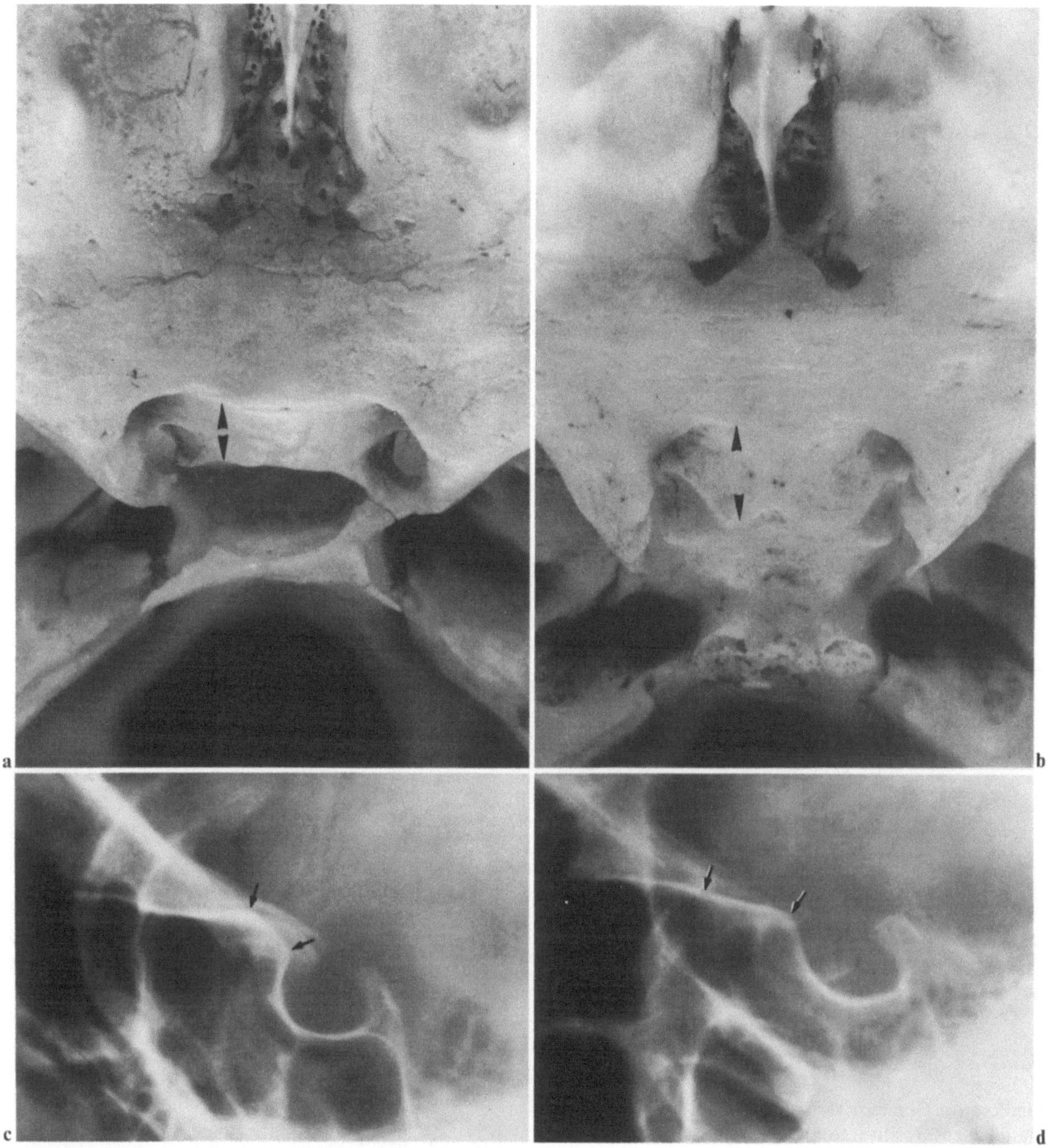

Fig. 5.40 a–d. Variations in length of the chiasmatic sulcus *(arrows* and *arrowheads)*.
a, c Short chiasmatic sulcus
b, d Long chiasmatic sulcus

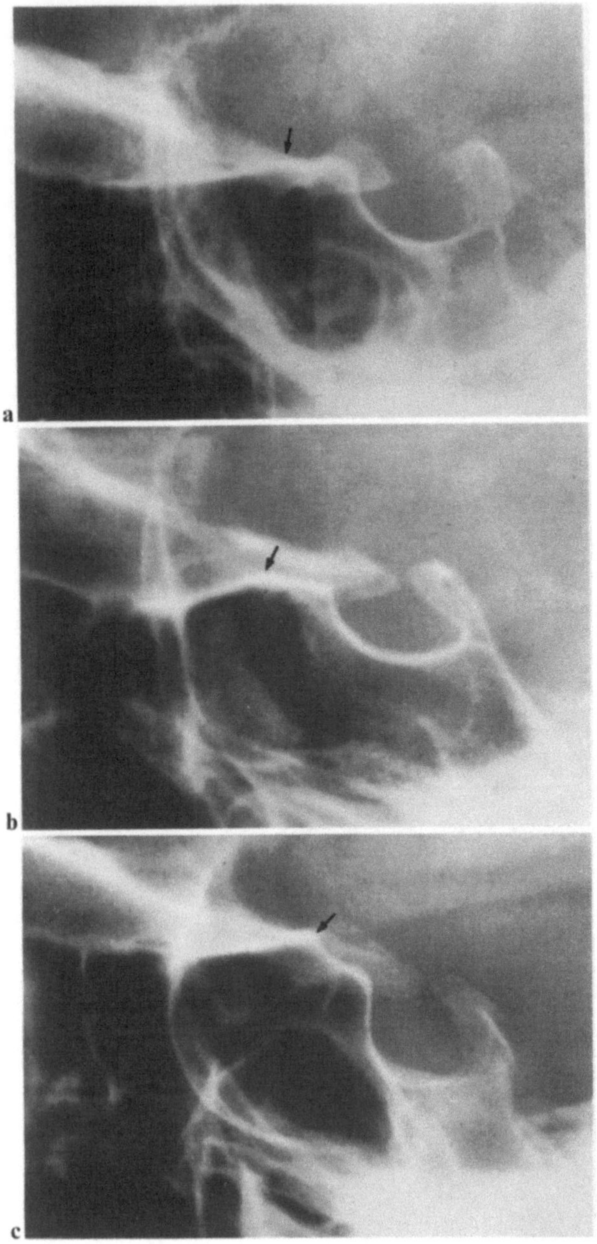

Fig. 5.41 a–c. Common variations in the slope of the chiasmatic sulcus relative to the planum sphenoidale. The chiasmatic sulcus is always identifiable; limbus sphenoidale *(arrow)*

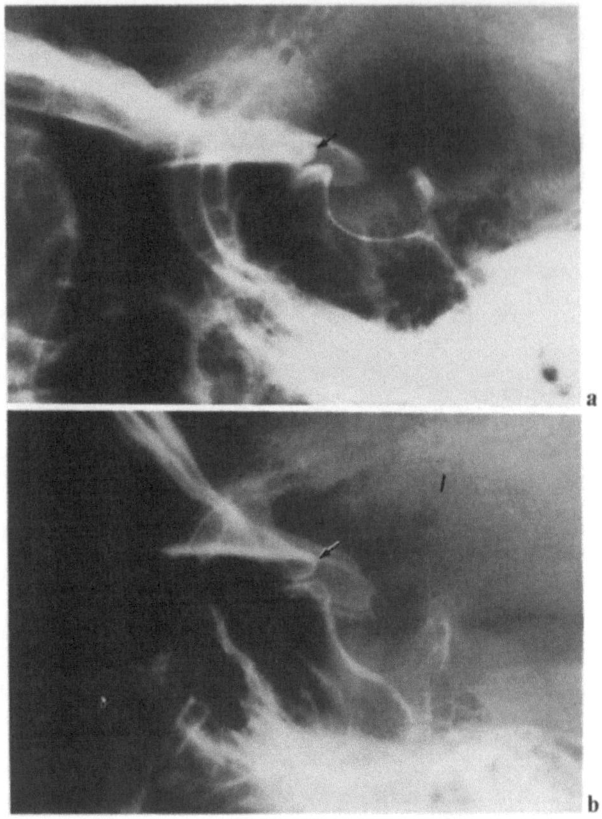

Fig. 5.42 a, b. Two examples of a well-defined limbus sphenoidale *(arrow)*

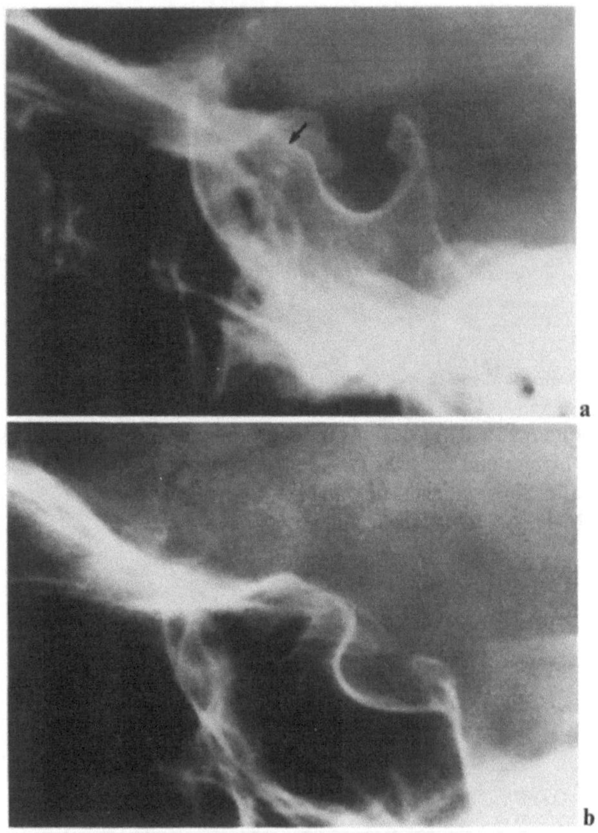

Fig. 5.43 a, b. Variations in shape of the chiasmatic sulcus with sphenoid pneumatization.

a Concave chiasmatic sulcus above a poorly pneumatized sphenoid *(arrow)*.

b Convex chiasmatic sulcus above a well-pneumatized sphenoid sinus

Fig. 5.44
a Slight upper convexity of the planum sphenoidale above a well-developed sphenoid sinus *(arrows)*. This appearance is not pathologic.
b As a rule, the planum sphenoidale remains flat, even when pneumatization of the sinus is extensive

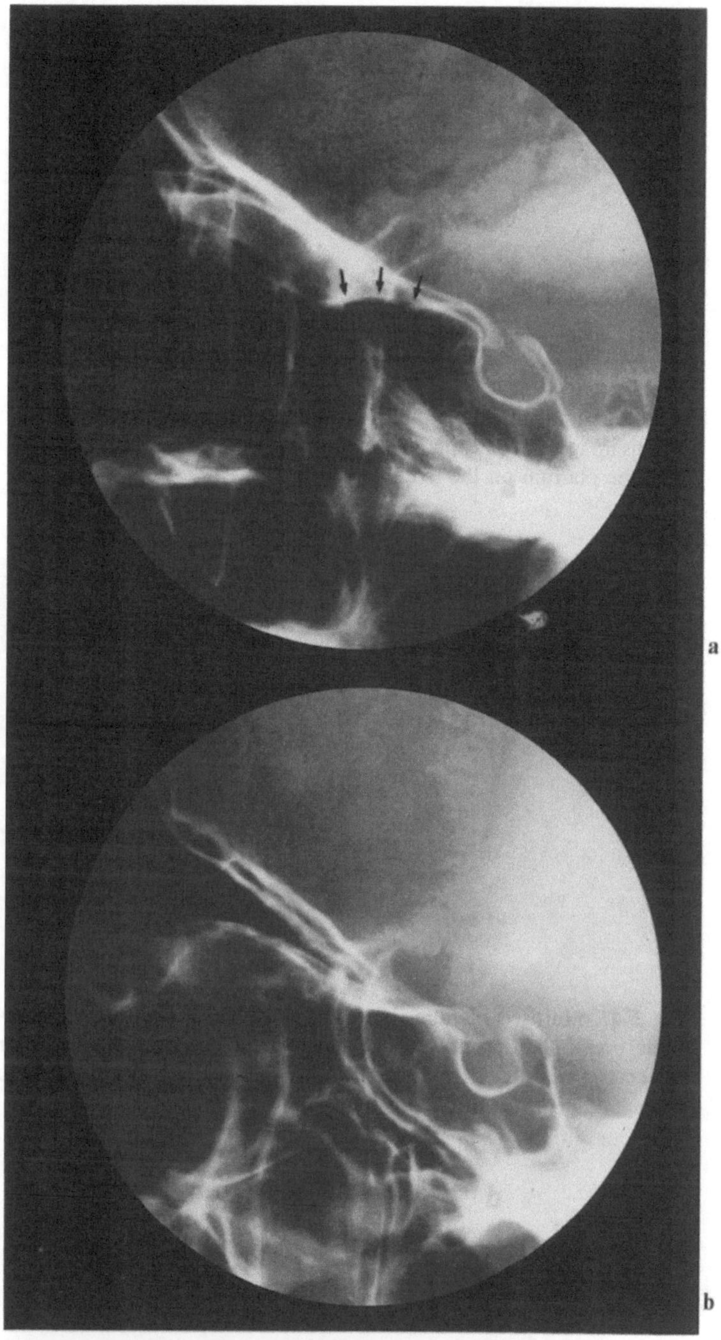

a

b

Fig. 5.45. Thick planum sphenoidale throughout its length. This pattern may be normal, but requires additional investigation

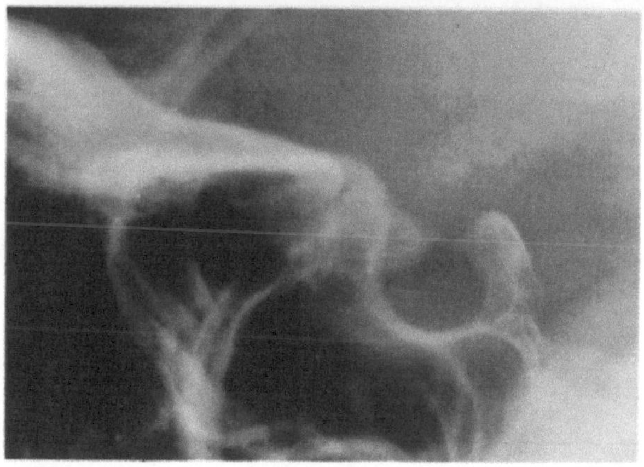

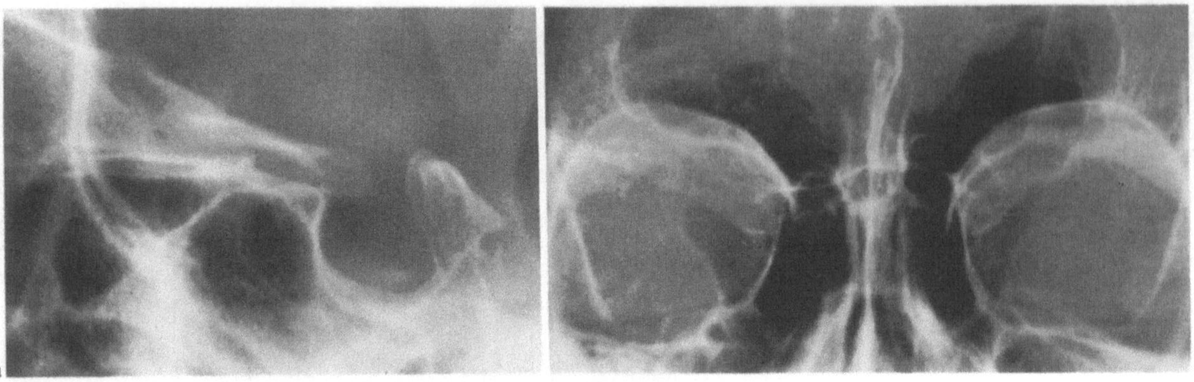

a b

Fig. 5.46
a Double image of the planum on the lateral projection.
b This image can be explained by the stairlike appearance of the planum on the frontal view

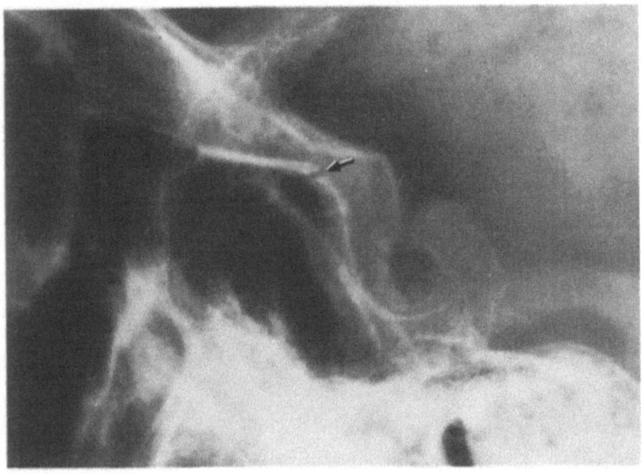

Fig. 5.47. Unfused planum. Note the presence of an oblique break slanting downward and forward, separating the chiasmatic sulcus from the planum; *(arrow)* this anatomic variant must not be mistaken for a fracture

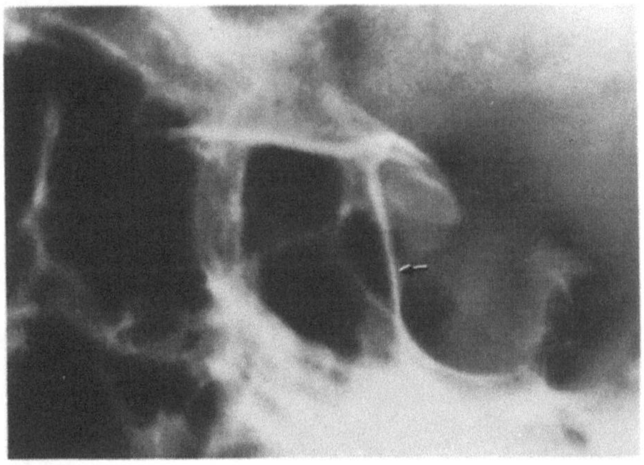

Fig. 5.48. Vertical chiasmatic sulcus giving the illusion of a large sella turcica; in fact, the sella turcica is normal. Tuberculum sellae *(arrow)*

III. Variations in the Sella Turcica Itself

1) Tuberculum Sellae, Anterior and Middle Clinoid Processes

The tuberculum sellae marks the anterior boundary of the pituitary fossa. It is situated at the right-angle which normally separates the chiasmatic sulcus from the anterior wall of the sella turcica.

In lateral projection, the anterior clinoid processes generally appear superimposed on the tuberculum sellae, which nevertheless always remains easily identifiable. Care must be taken, however, not to confuse it with the limbus sphenoidale, especially when the chiasmatic sulcus slopes markedly, even to the extent of being vertical with regard to the planum sphenoidale, or with possible middle clinoid processes. Inability to recognize this could cause errors in calculating the size of the pituitary fossa. The tuberculum sellae is used when drawing Welcker's angle or Twining's line. It constitutes the anterior attachment of the diaphragma sellae and can, therefore, be clearly demonstrated by pneumoencephalography (Fig. 5.49) or metrizamide cisternography.

On each side of the sella turcica, the anterior clinoid processes form the posterior extension of the lesser wings of the sphenoid; they seem to extend the orbital roofs posteriorly. In the lateral projection, they frequently occupy a large portion of the area of the sella turcica (Fig. 5.50). In evaluating by eye alone the dimensions of the hypophyseal fossa, it is, of course, necessary to disregard the superimposed anterior clinoid processes since they constitute very lateral structures in relation to the pituitary fossa. In addition, their morphology being very variable (Fig. 5.51), they can appear dumpy, hooked, thin, pneumatized, rudimentary, or even absent. Very often they appear asymmetric in both length and height (Fig. 5.52) (Bruneton et al. 1979). Short, finely pointed anterior clinoid processes do not necessarily indicate increased intracranial pressure. If this morphological abnormality, often bilateral, is an isolated finding, without demineralization of the walls of the sella turcica or the planum sphenoidale, it must be regarded as a simple variation from the normal (Fig. 5.53). Sagittal, paramedial, and especially frontal tomograms allow a careful examination of the anterior clinoid processes. The carotid siphons come into contact with their medial borders.

The middle clinoid processes appear as two small triangular elevations located at each side of the sella turcica below and medial to the anterior clinoid processes (Fig. 5.54). On either side they extend the synostotic crest described by anatomists as the lower margin of the anterior intercavernous sinus groove, which is seen only exceptionally on skull films (Fig. 5.55). Middle clinoid processes are often clearly visible on direct magnified views. The internal carotid arteries pass immediately lateral to them. Dural calcification is often present, adjacent to the middle clinoid processes, leading to the formation of the caroticoclinoid canal.

2) Dorsum Sellae and Posterior Clinoid Processes

The dorsum sellae is the posterior limit of the sella turcica. It is composed of a plate of bone presenting an anterior concavity which, in a lateral projection, generally gives an image of three lines from front to back: the anterior line represents the right and left lateral parts of the dorsum sellae; the intermediate line is the posterior boundary of the sella turcica in the midline; and the posterior line forms the posterior limit of the dorsum itself. The dorsum sellae varies in its morphology (Fig. 5.56): it can appear thick, massive, pneumatized, or even short. The cortical bone of the anterior surface of the dorsum sellae, in the midline, is smooth; it is generally slightly thinner than the cortical bone of the floor of the sella. The dorsum sellae itself sometimes appears so thin that its pattern on a midline sagittal tomogram may simulate that of increased intracranial pressure (Fig. 5.57); there may even be a foramen at its center (Pribram and du Boulay 1971) (Fig. 5.58). The posterior surface of the dorsum is uneven due to the insertion of calcified dural folds, and because of circular or triangular notches whose edges are sometimes condensed (Fig. 5.59), indicating the passage of the third nerve and the petrosal sinuses on the lateral aspect of the dorsum sellae. A persistent trigeminal artery may rarely run in a groove on the lateral aspect of the dorsum sellae (Fig. 13.10).

The general line of the dorsum usually prolonges that of the clivus. It therefore slants upward and forward. However, it may slant upward and backward, thus creating the image of the open sella, which should be regarded as normal if there is no other sign of bone erosion in the region (Fig. 5.60). A congenital anomaly of ossification giving an apparently eroded dorsum sellae, has been recently reported (Penkrot and Bures 1979).

The posterior clinoid processes are the superolateral angles of the dorsum sellae: they generally extend forward beyond the level of the dorsum sellae. Highly variable in their morphology, they frequently appear asymmetric. Unilateral hypoplasia of one of the clinoid processes or of the adjacent part of the dorsum may give rise to a truncated hemidorsum (Fig. 5.61).

The dorsum sellae and the posterior clinoid processes are clearly seen on frontal tomograms or in the Towne's view, which shows very distinctly the frequent asymmetry of these structures, their often irregular pneumatization, and the lateral notches created by the oculomotor nerves or petrosal sinuses.

3) Calcification of the Ligaments and the Dura Mater of the Sellar Region

The dura mater stretched on each side of the sella between the anterior and posterior clinoid processes may calcify to form the interclinoid ligaments. The appearances thus created is that of a "bridged sella." Uni- or bilateral calcification of the interclinoid ligaments, which constitute the lateral margins of the pituitary fossa, does not influence its volume in any way (Figs. 5.62, 5.63).

The dura folds, arising from the middle clinoid processes and connecting them to the anterior and posterior clinoid processes, may also calcify (Fig. 5.64). One rarely sees calcification of the diaphragma sellae itself or of the dura mater of the floor of the sella turcica.

Calcification of the petroclinoid ligaments occurs very frequently after the age of 20. In the lateral projection, the appearance is that of a dense line extending from the posterior clinoid processes to the petrous apex. The petroclinoid ligaments can be partially or totally calcified (Fig. 5.66). They sometimes appear as a dense opacity with a slightly bulbous outline and must not be confused with pathologic calcification. In elderly patients with an osteoporotic dorsum, dense calcification of the petroclinoid ligaments must not lead to a false diagnosis of anteroposterior enlargement of the sella (Fig. 14.50). In the frontal view, parasellar calcification of the petroclinoid ligaments, which can be unilateral, is projected on the wide part of the superior orbital fissure (Fig. 5.67). Limited calcification of the free edge of the tentorium can also be seen, having the same appearance as the petroclinoid ligaments but occupying a higher position (Fig. 5.68).

Plaquelike calcification of the anterior part of the free edge of the tentorium may rarely simulate atheromatous calcification of the internal carotid artery (Fig. 5.65).

The os suprapetrosum of Meckel is a small dural calcification which on the frontal projection lies immediately above the petrous apex and on the lateral view, is projected on the spheno-occipital synchondrosis (Fig. 5.69) (Currarino and Weinberg 1974).

The dura mater lining the lateral endocranial margin of the superior orbital fissure can also calcify (Bonneville et al. c, to be published). This calcification, which is nearly always bilateral and larger than that of the petroclinoid ligaments, is easily recognized on axial tomograms. The anteromedial edge is linear, whereas the posterolateral border is irregular (Figs. 5.70, 5.71).

4) Floor of the Sella

We define the floor of the sella as both the anterior wall and the floor of the pituitary fossa itself, as the bed of the adenohypophysis. Variations of the floor are numerous and may in certain cases simulate pathologic changes such as those of a pituitary tumor (Swanson and du Boulay 1975; Banna et al. 1978; Metzger et al. 1975).

a) Variations of the Morphology of the Floor of the Sella in the Frontal View

In the frontal view of the sella, the floor appears as a dense horizontal line 10–15 mm in length, projected about 10 mm below the planum sphenoidale. It is of great importance that this line be visible in the projection as it is a fundamental criterion of normality of the skull radiograph. If the floor of the sella is invisible in the usual projection, in the absence of any pathologic condition, this is usually due to an anteroposterior shortness of the floor itself with pronounced obliquity of the anterior wall (Fig. 5.5). The floor of the sella can be visualized by changing the direction of the X-ray beam (Fig. 5.72); it is seen more clearly when the sphenoid sinus extends below it (Fig. 5.73a). It is nevertheless visible in the presence of a presellar sinus when it overlies the cancellous bone (Fig. 5.73b). In such cases the floor of the sella often presents a slight upward concavity in the frontal projection. Conversely, in the routine frontal projection of the sella, the floor rarely appears convex (Fig. 5.73c). This appearance may be simulated by a narrow floor extended on either side by a carotid sulcus in medial position (Fig. 5.73d). In the frontal view, the sellar floor can also be slightly sloping (Fig. 5.73c). If the difference in level between the two lateral boundaries of the floor does not exceed 2 mm, then it is considered a normal variation. In fact, it is clear that a small lateralized intrasellar tumor can give the same pattern. When the floor of the sella is sloping, the cortical bone must be carefully examined, if necessary by tomography, seeking possible erosive changes. Of course, a sloping floor may be the source of a double floor in the lateral projection. The floor of the sella may also be seen as a broken line at the attachment of one or several septa of the sphenoid sinus (Fig. 5.73f). A difference of level in the floor may sometimes coexist with some asymmetry of pneumatization of the sphenoid sinus. Finally, we have already noted the particular pattern of the floor in both the small sella on one side (Fig. 5.9) and the early empty sella (Fig. 5.12).

b) Morphological Variations of the Floor of the Sella in the Lateral Projection

Aside from the double floor patterns related to the presence of a small sella on one side or of an early empty sella, it may be stated that any floor which is not strictly horizontal in the frontal view gives the appearance of a double floor in the lateral projection. The presence of a double floor must always be very carefully explained, drawing a correlation between its frontal and lateral appearances (Figs. 5.74, 5.76). In general, it is possible that a double floor should be normal, particularly if its lower margin is as thick, or a fortiori, thicker than the upper margin (Fig. 5.75).

A double floor can be created by a poor lateral projection (Fig. 5.77); conversely, an apparently single floor on an unsatisfactory projection can still be abnormal. A false double floor may result from the juxtaposition of the sellar floor and the cortical bone of the sphenoid sinus, when a narrow band of cancellous bone is present between the sella and the sinus (Figs. 5.78, 5.79). Moreover, if the floor is not horizontal, triple or even quadruple floors may be shown. Visibility of the carotid sulcus may also give the appearance of a double floor (Fig. 14.64). In such cases, the lower margin, representing the carotid sulcus, extends backward and vertically downward instead of running up toward the dorsum as in double floors related to anatomic variations or an abnormality of the floor of the sella.

c) Particular Aspects of the Floor of the Sella

The usual configuration of the floor is sometimes modified locally over 1 or 2 mm by small depressions or by reliefs which have not all been explained anatomically and must not, therefore, be regarded as pathologic. Ordinarily these images are not accompanied by localized modifications of the cortical bone (Fig. 5.80). When they are located in the central area of the floor, they may represent either an extension of the intrasellar end of the craniopharyngeal canal or a large vascular channel.

Another frequent observation is that of a depression limited to the posterior part of the sellar platform, which sometimes extends to the lower part of the dorsum sellae, forming the fossula hypophyseos or neurohypophyseal bed. This normal depression often appears on the most posterior frontal cuts and must not be considered pathologic (Figs. 5.80, 14.22a, b).

Limited reliefs may give a varied, uneven appearance to the sellar floor, such as localized blister of the sphenoid sinus (Fig. 5.82) or sellar spine (Fig. 5.83) (Lang 1977; Dietemann and Bonneville to be published).

d) Variations of the Cortical Bone of the Floor of the Sella

It is of fundamental importance to examine the cortical bone forming the walls of the hypophyseal fossa. It would appear to be the case that an intrasellar lesion generally thins the adjacent cortical bone before bulging occurs, therefore, well before the sella turcica appears enlarged. The value of studying the cortical bone has recently been contested, however, because of the great variability of the lamina dura (Swanson and du Boulay 1975). In fact, careful examination or the cortical bone using direct magnification and possibly polytomography is extremely important in interpreting any radiograph of the sella turcica.

The thickness of the cortical bone of the floor of the sella varies according to the degree of pneumatization of the sphenoid sinus. Composed of compact bone bordering a pneumatized zone, the cortical bone has traditionally been considered more resistant than the cancellous bone opposite a nonpneumatized portion of the sphenoid (Mahmoud 1958). This increased resistance is due to the constitution of the compact bone and to the fusion of two layers of cortical bone since both the floor of the sella and the sphenoid sinus have their own cortical bone. Actually, greater resistance of the floor of the sella opposite the sinus, verified during the course of various pathologic processes, is not necessarily represented radiologically by greater thickness. On the contrary, in most cases, the cortical bone of the floor seems thinner opposite the sinus; but the reverse is also possible (Fig. 5.84).

Whatever the pneumatization of the sphenoid, the floor of the sella may appear evenly thick or evenly thin in a continuous manner; homogeneous thinness of the sellar floor, if it is not accompanied by a bulging of the fossa walls, does not suggest an intrasellar tumor. Similarly, thinning of the sellar floor, beginning or ending abruptly without transition, is not pathologic. This also applies in the case of limited thickening of the floor (Fig. 5.84d).

Localized thinning of the floor does not indicate an intrasellar tumor either, if tomography demonstrates that the cortical bone is thinned on both sides or that only the lower surface is thinned (Fig. 5.85). False thinning may be suggested by tomographic artifacts (Fig. 5.86).

Insertion of septa of the sphenoid sinus usually creates localized thickening of the cortical bone and, in exceptional cases, depression of the floor of the sella (Fig. 5.87). Persistence of a small area of cancellous bone below the floor of the sella modifies the appearance of the lamina dura (Fig. 5.88).

In rare cases, changes in the cortical bone of the floor are very difficult to interpret, particularly if the clinical findings raise the possibility of a pituitary microadenoma. The diagnosis between normal and pathologic thinning cannot, therefore, be reached without further investigation (Fig. 5.89).

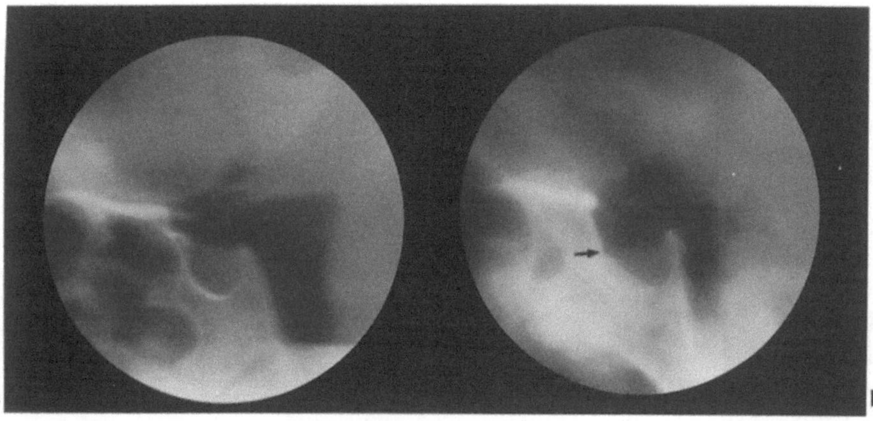

a b

Fig. 5.49 a, b. Midsagittal pneumotomograms.
a Usual appearance, the anterior part of the diaphragma is attached to the tuberculum sellae.
b Vertical chiasmatic sulcus; the position of the tuberculum sellae *(arrow)* is perfectly defined, as is the anterior attachment of the diaphragma sellae

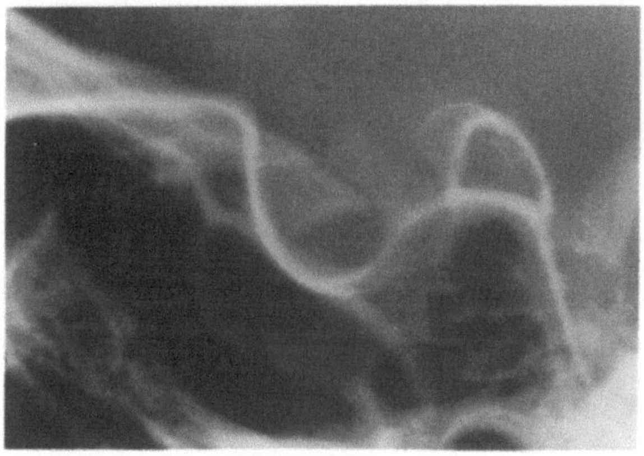

Fig. 5.50. Considerable asymmetry of both the anterior and posterior clinoid processes, which overlie most of the area of the sella

Fig. 5.51 a–d. Different types of normal anterior clinoid processes
▽

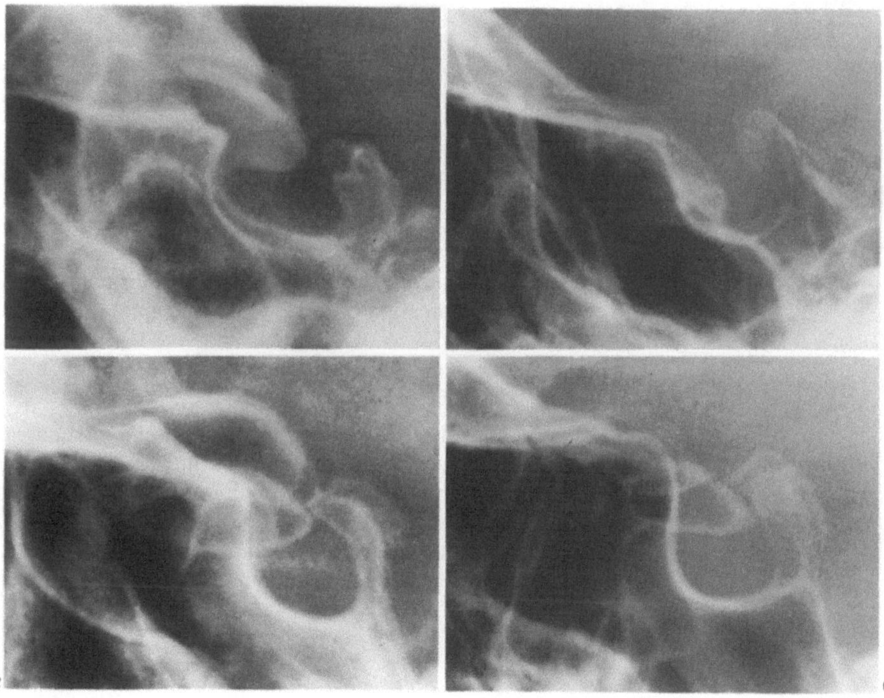

a b c d

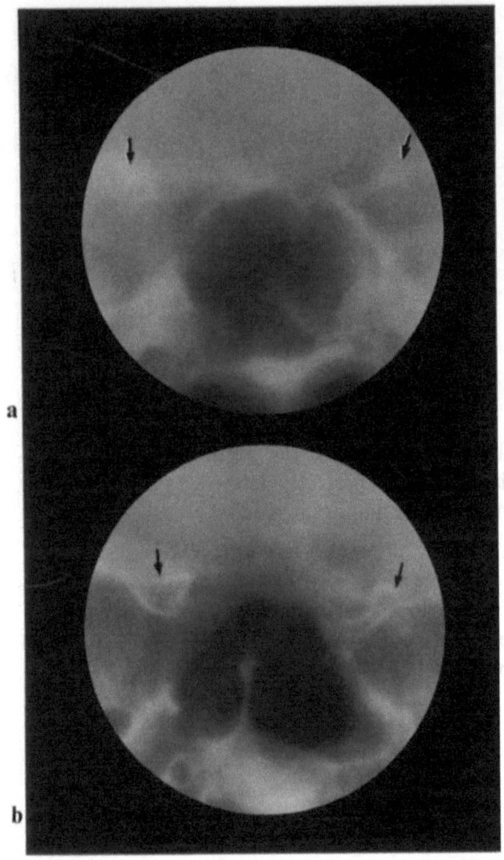

Fig. 5.52a, b. Frontal tomogram of the sphenoid: asymmetry of the anterior clinoid processes *(arrows)*

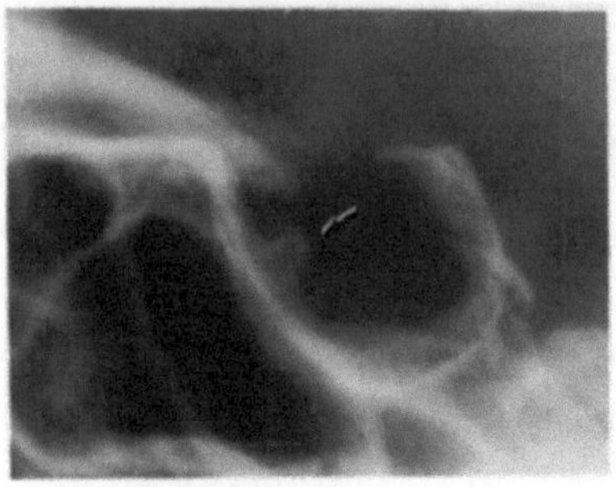

Fig. 5.54. Middle clinoid process *(arrow)*

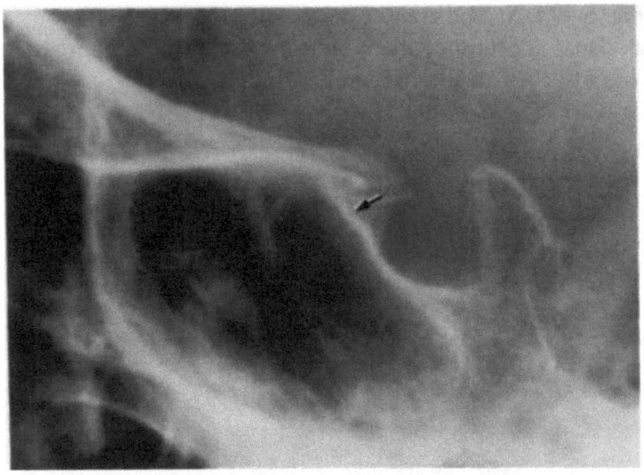

Fig. 5.55. Limited depression subjacent to the tuberculum sellae: groove of the intercavernous sinus *(arrow)*?

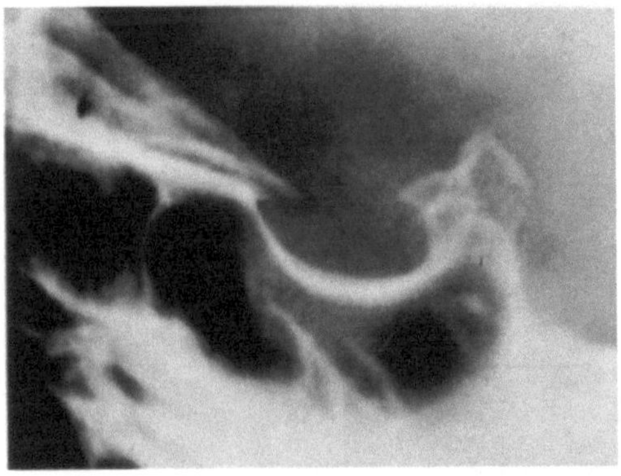

◁ **Fig. 5.53.** Very short, finely pointed anterior clinoid processes: this appearance must be considered normal if there is no other indication of increased intracranial pressure

Fig. 5.57. a The line representing the middle portion of the anterior surface of the dorsum sellae *(arrow)* nearly reaches the posterior surface of the clivus *(arrowhead)*. As a result, **b,** a midsagittal tomogram, shows a very thin middle portion of the dorsum sellae, which should not be interpreted as an indication of increased intracranial pressure ▷

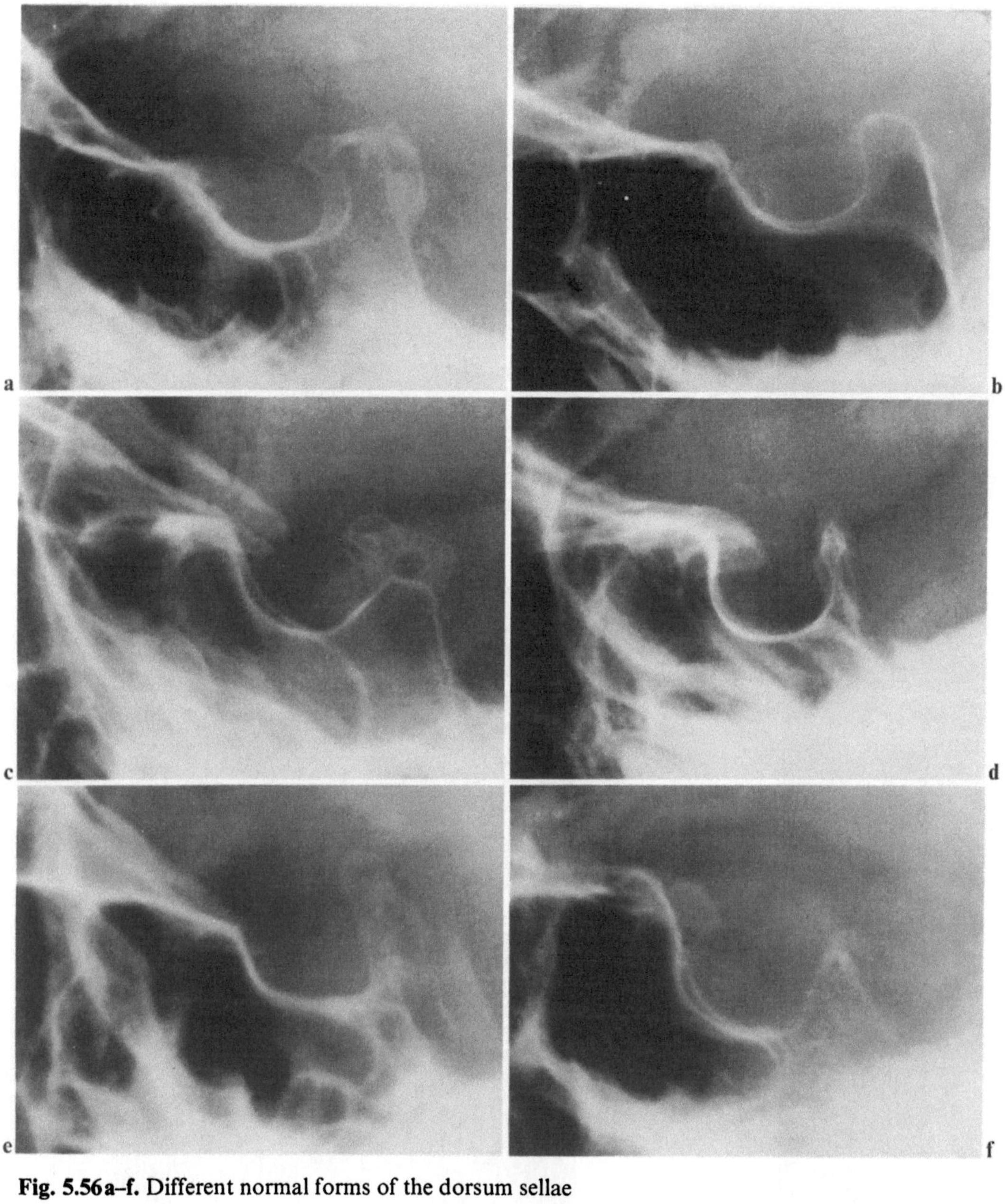

Fig. 5.56 a–f. Different normal forms of the dorsum sellae

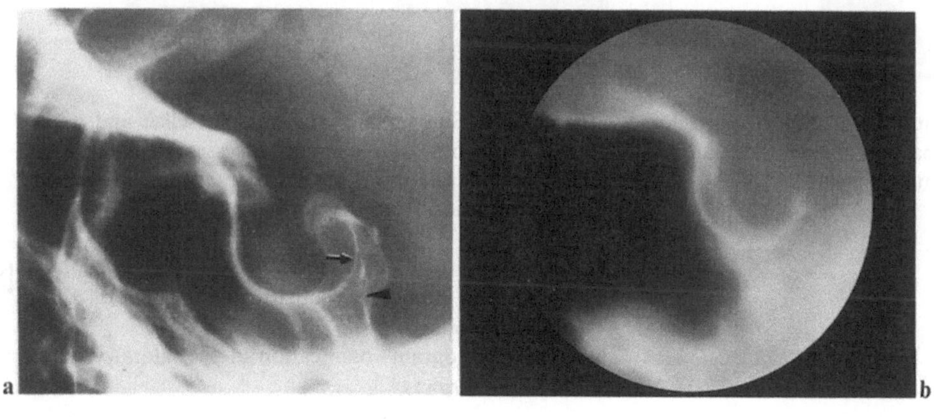

Fig. 5.57 a, b
(caption see opposite page)

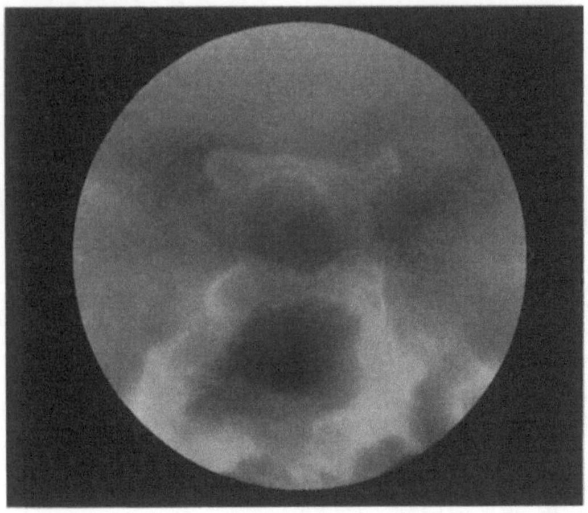

Fig. 5.58. Frontal tomography of the sphenoid through the dorsum sellae: appearance of a round hole in the dorsum sellae

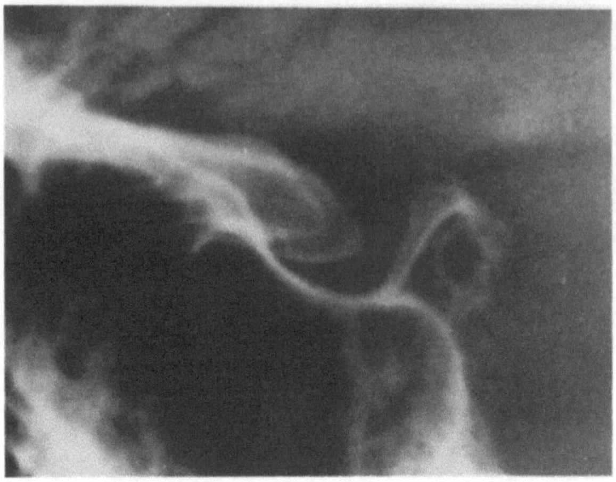

Fig. 5.60. Open sella turcica: the dorsum sellae slants upward and backward. This appearance must be considered as normal if there is no sign of bone erosion

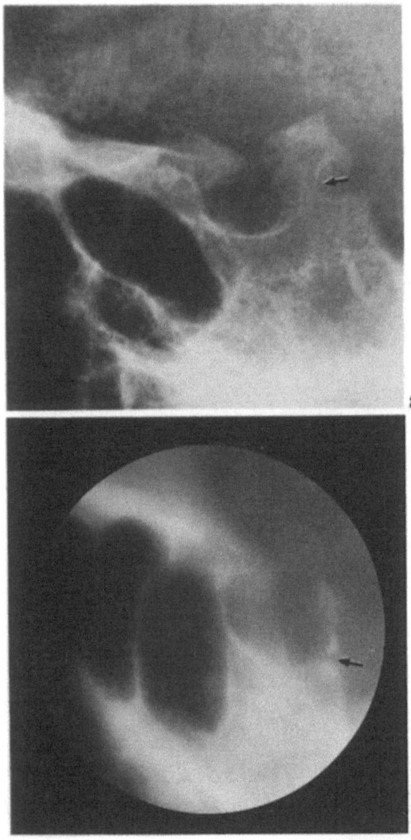

Fig. 5.59 a, b. The posterior surface of the dorsum sellae shows notches indicating the passage of both the oculomotor nerve and the petrosal sinus *(arrows)*

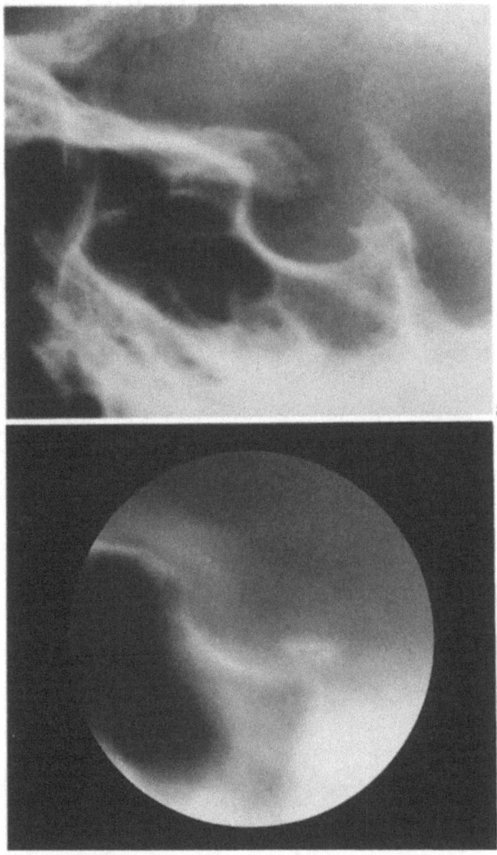

Fig. 5.61 a, b. Shortness of the dorsum sellae and asymmetry of the posterior clinoid processes: at one of the lateral boundaries the dorsum sellae appears truncated; this is a variation from the normal

Fig. 5.63
- **a** Closed sella turcica: calcification of the lateral dural fold of the sella
- **b** Frontal tomogram: ossification of the lateral walls of the sella *(arrows)*

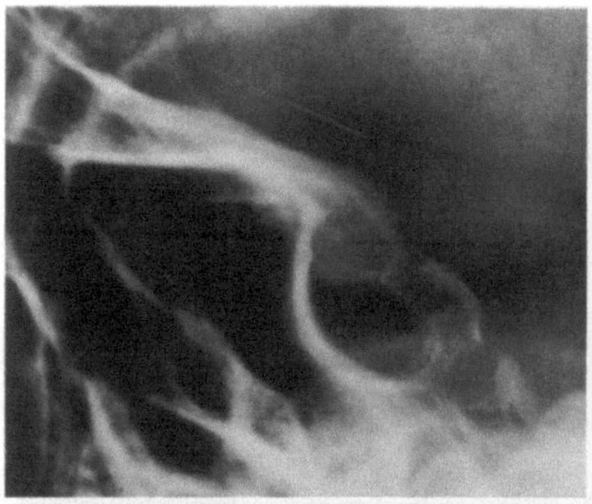

a

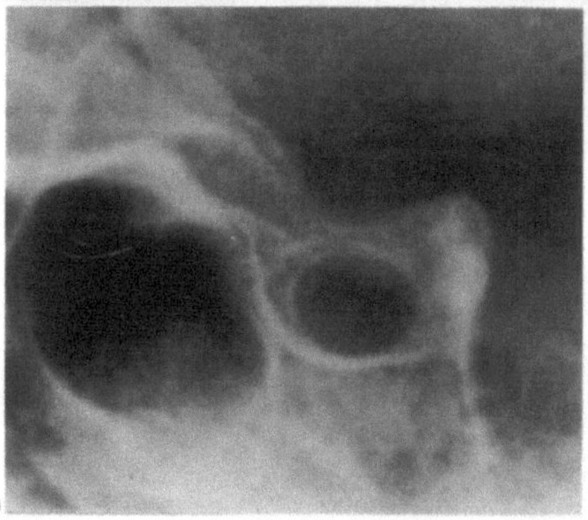

b

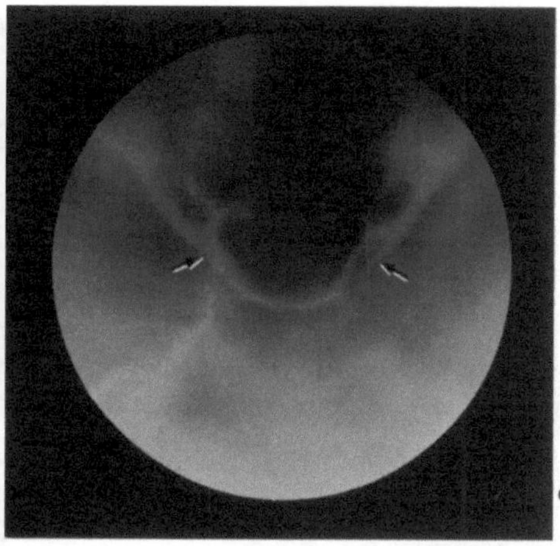

c

Fig. 5.62
- **a** Closed sella: the anterior and posterior clinoid processes are in contact.
- **b** Closed sella turcica: calcification of the interclinoid ligaments.
- **c** Axial tomogram: the sella turcica itself remains unchanged by the existence of interclinoid ligament calcifications *(arrows)*

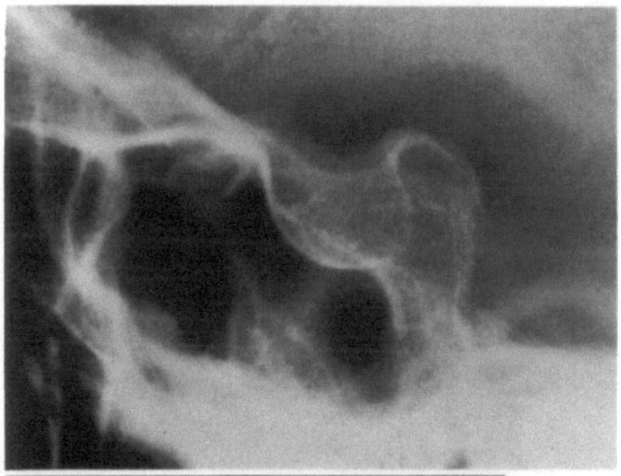

a

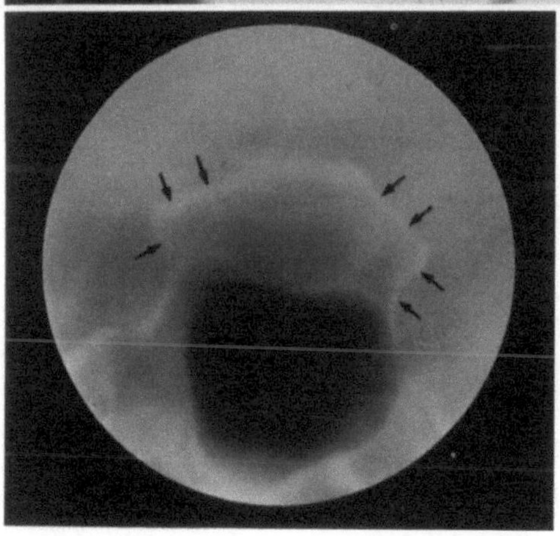

b

Fig. 5.63 a, b

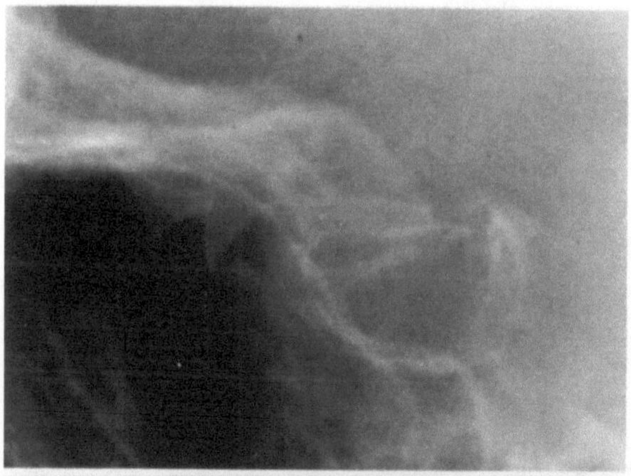

a

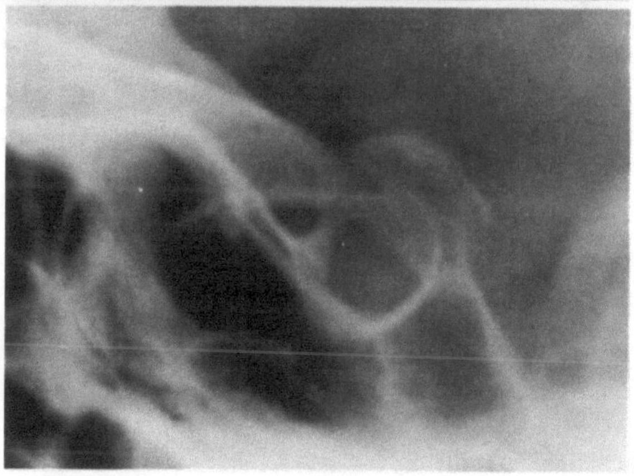

b

Fig. 5.64 a, b. Calcification of the dural folds connecting the anterior, middle, and posterior clinoid processes

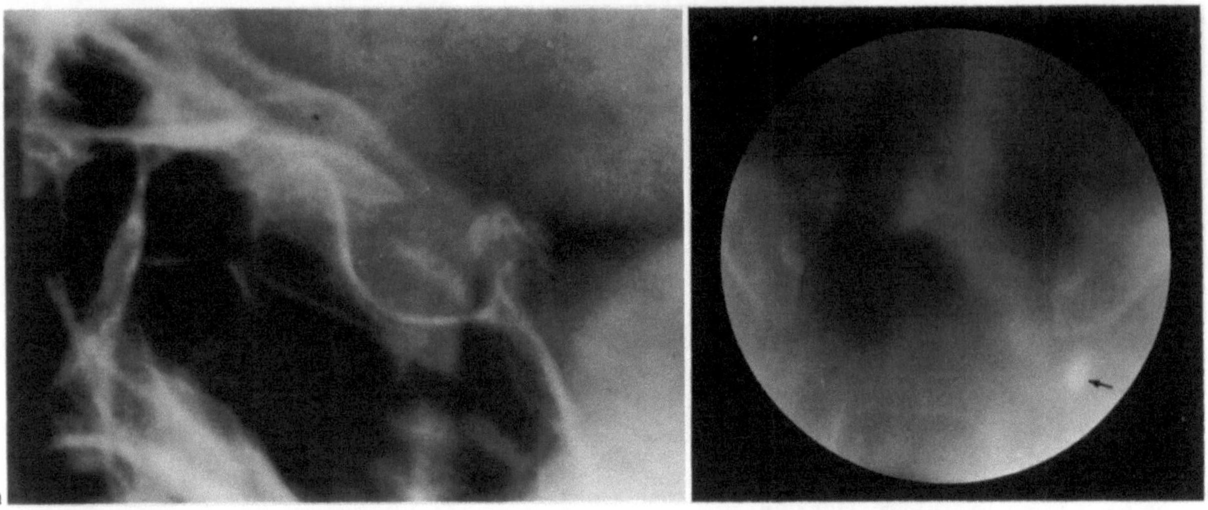

Fig. 5.65a, b (caption see opposite page)

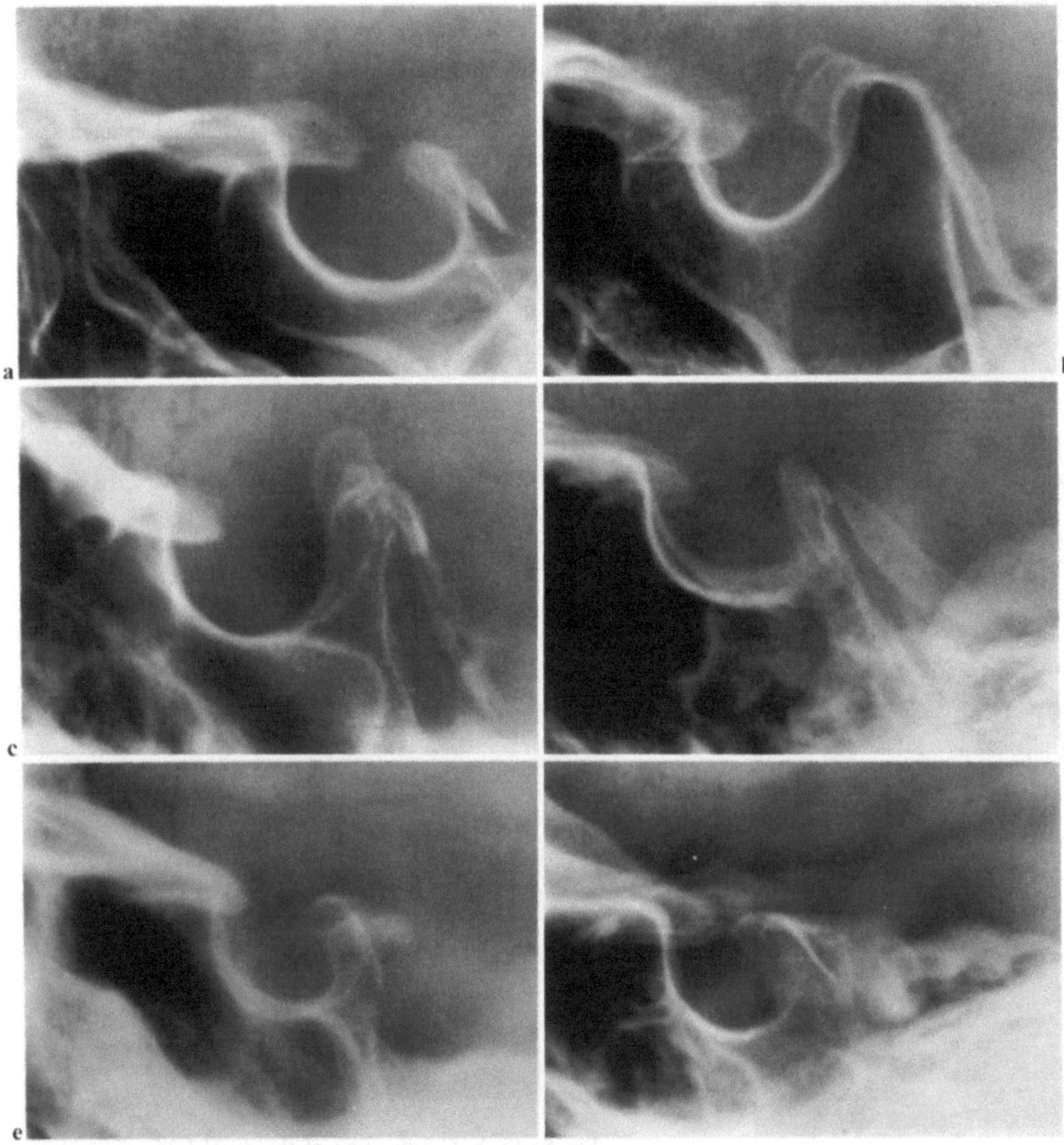

Fig. 5.66a–f. Different patterns of calcification of the petroclinoid ligaments

◁ **Fig. 5.65a, b.** Calcification of the anterior portion of the free edge of the tentorium.
a Lateral projection: projection on the sella turcica of an elongated opacity slanting downward and backward, simulating carotid siphon calcification.
b Axial tomography: calcification of the anterior part of the free edge of the tentorium *(arrow)*

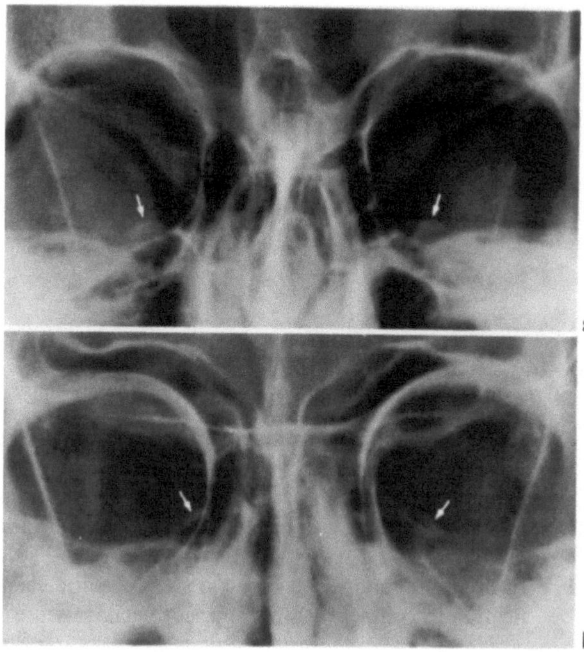

a

b

▷ **Fig. 5.67a, b.** Appearances of calcification of the petro-clinoid ligaments on the frontal view *(arrows)*

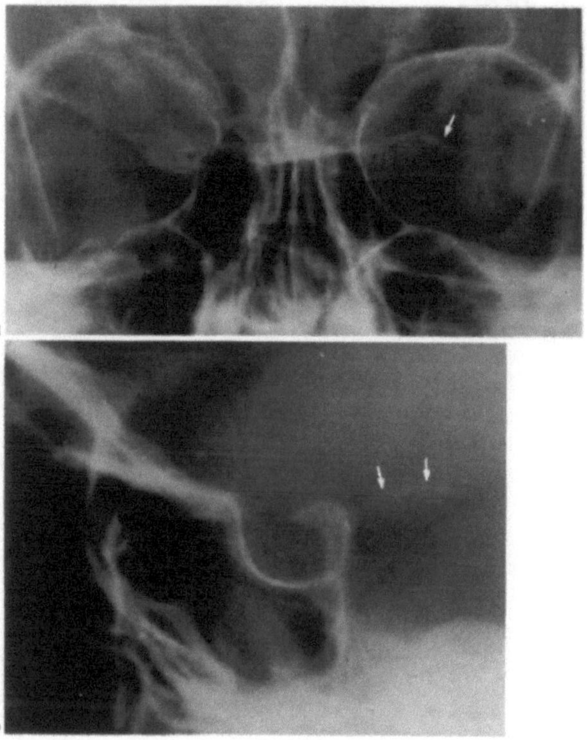

a

b

Fig. 5.68a, b. Limited calcification of the free edge of the tentorium

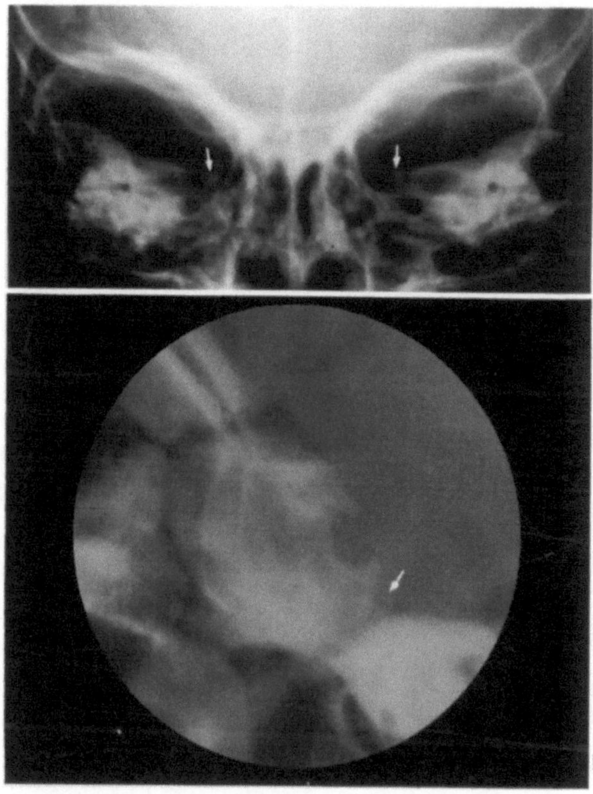

a

b

Fig. 5.69a, b. Os suprapetrosum of Meckel *(arrows)*

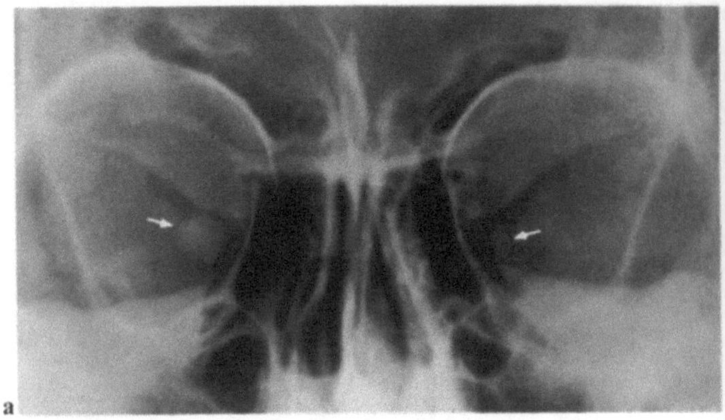

a

Fig. 5.70a–c. Calcification of the dura mater lining the superior orbital fissure.

a Frontal view: asymmetric, rounded calcifications partially occupying most of the superior orbital fissure *(arrows)*.

b, c Frontal tomograms showing two areas of dural calcification in contact with the superior orbital fissure

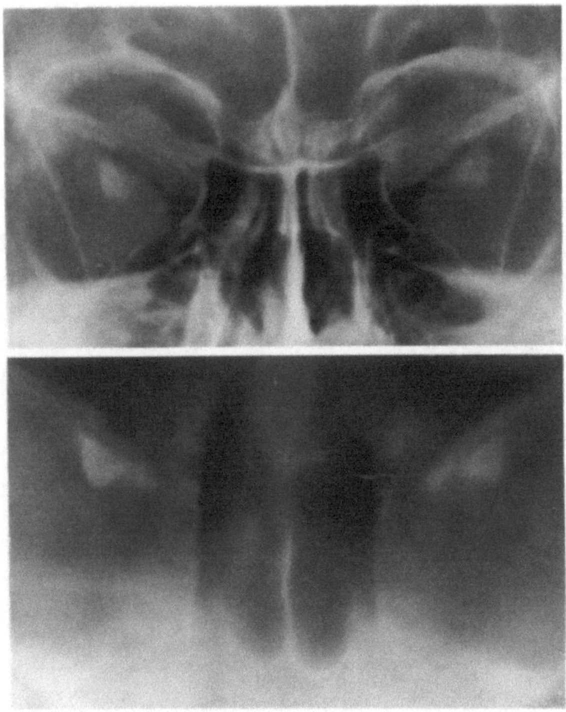

a

b

Fig. 5.71 a, b. Dural calcification lining the superior orbital fissures.

a Frontal view: dense, triangular calcification lining the lateral part of the two superior orbital fissures.

b Axial tomogram: the anterior surface of the dural calcification is strictly parallel to the greater wing of the sphenoid. The posterior surface is irregular

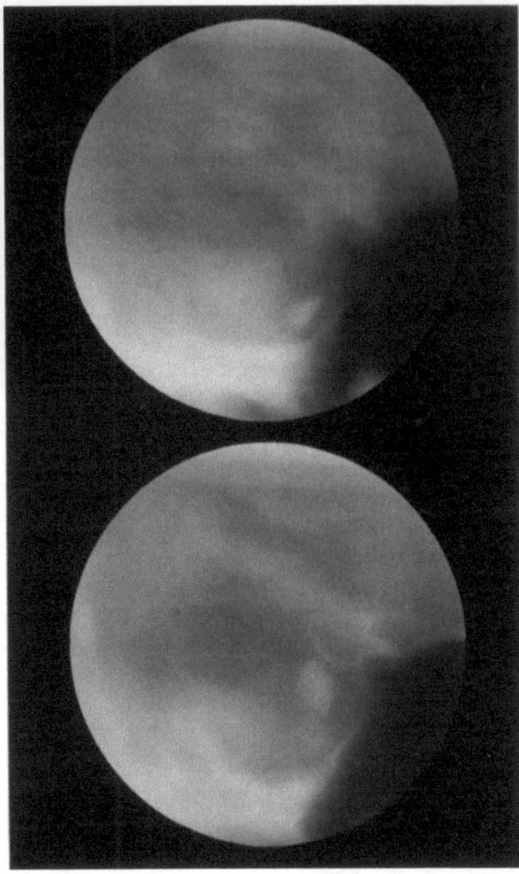

b

c

Fig. 5.73a–f. Different patterns of the floor of the sella, on the frontal view. ▷

a Well-developed sphenoid sinus

b Absence of sphenoid pneumatization

c Slight upward convexity of the floor

d Narrow floor and internal carotid sulci

e Slight lateral sloping of the floor

f Broken floor at the point of insertion of the sagittal septa of the sphenoid sinus

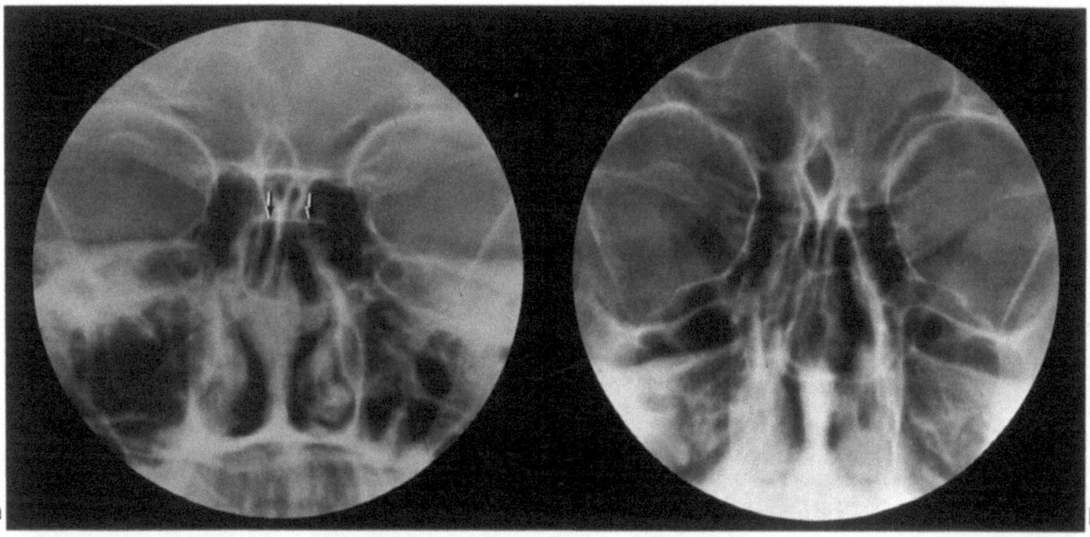

a b

Fig. 5.72a, b. Variations in the visibility of the sellar floor *(arrows)* with angulation of the incident beam

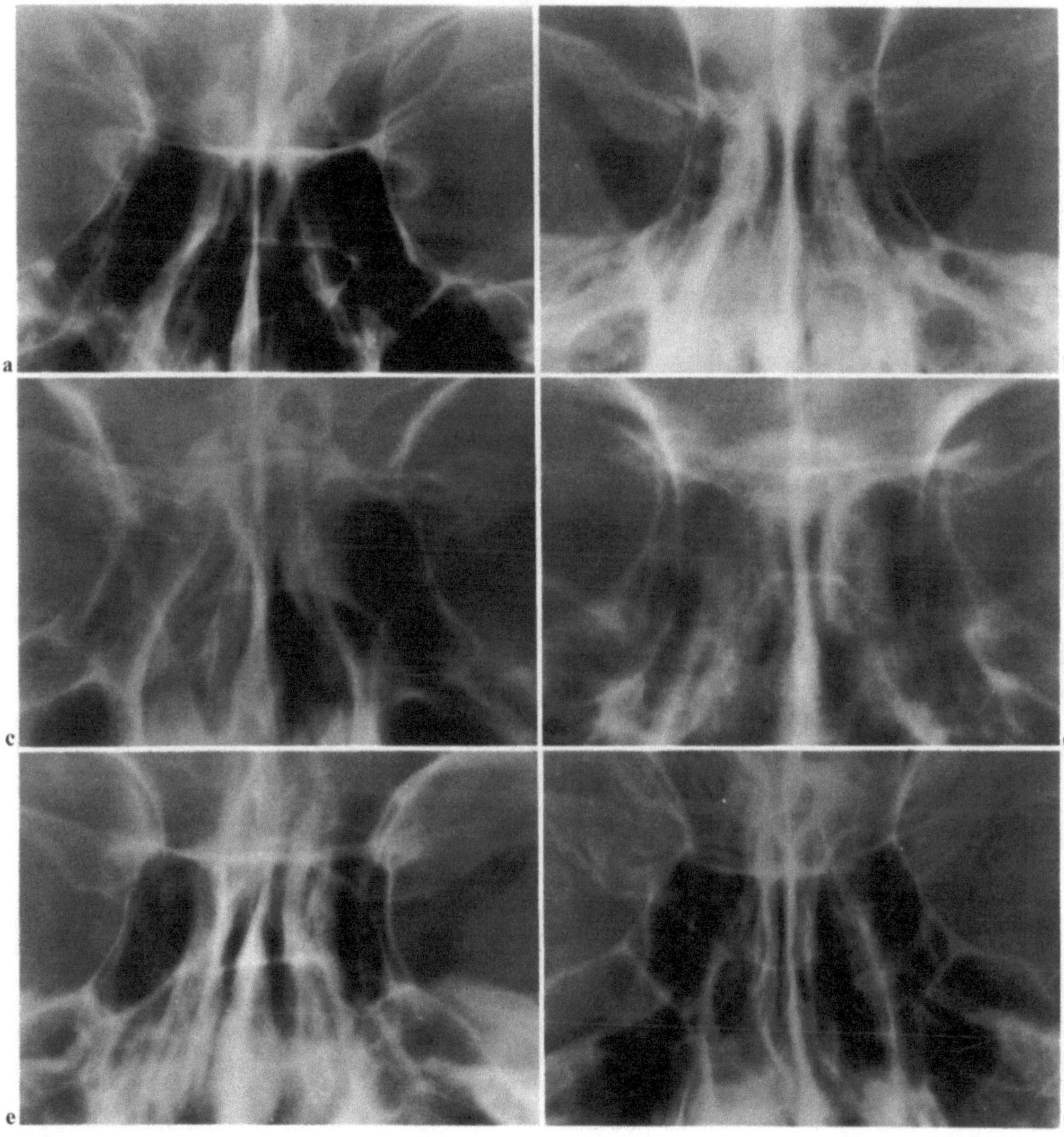

a b c d e f

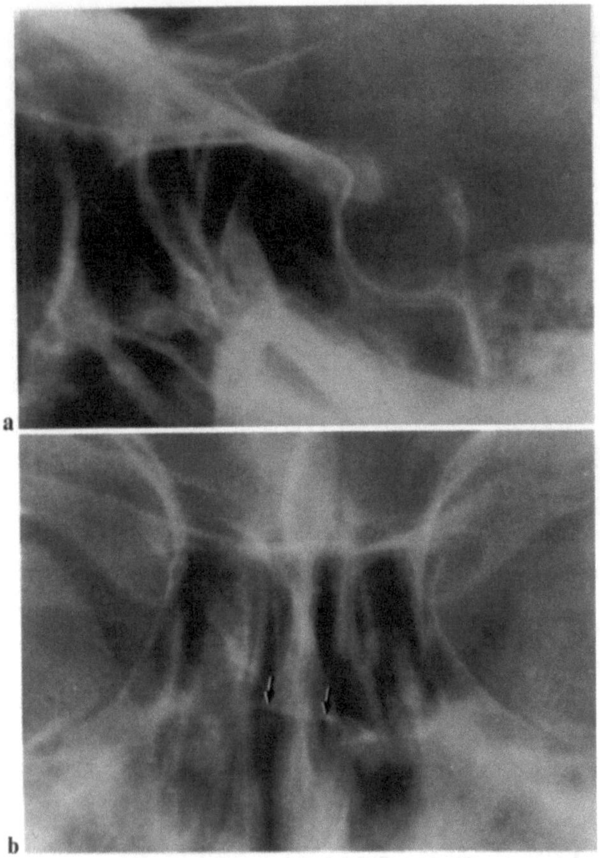

a

b

Fig. 5.74 a, b. Normal double floor: sloping of the floor toward the left, **b** without changes in the cortical bone *(arrows)*

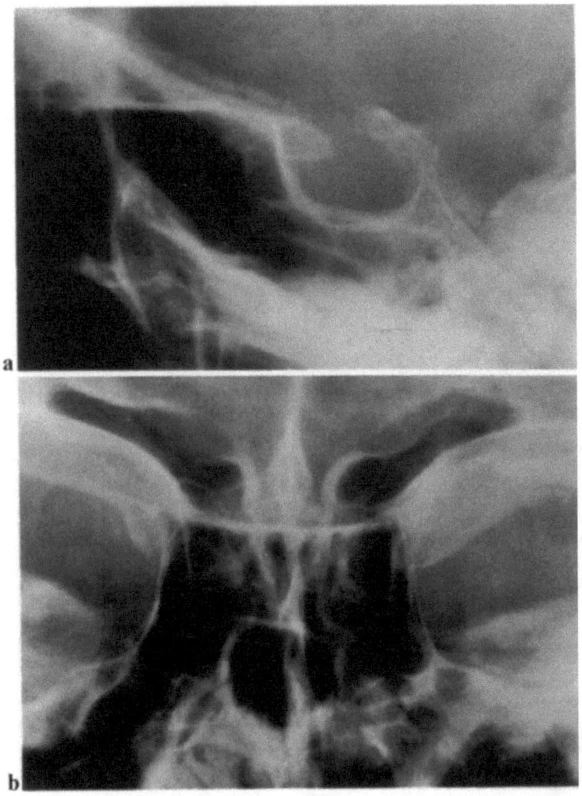

a

b

Fig. 5.77 ▷

a False double floor: the projection is not correct, the two orbital roofs are not superimposed *(arrows)*.

b True lateral projection shows no double floor

Fig. 5.78 a, b. False double floor: on both lateral and frontal projections the floor is doubled by the cortical bone of the sphenoid sinus, which is seen here as strictly parallel to the floor *(arrows)*. There remains a small plate of cancellous bone between the floor of the sella turcica and the sphenoid sinus ▷

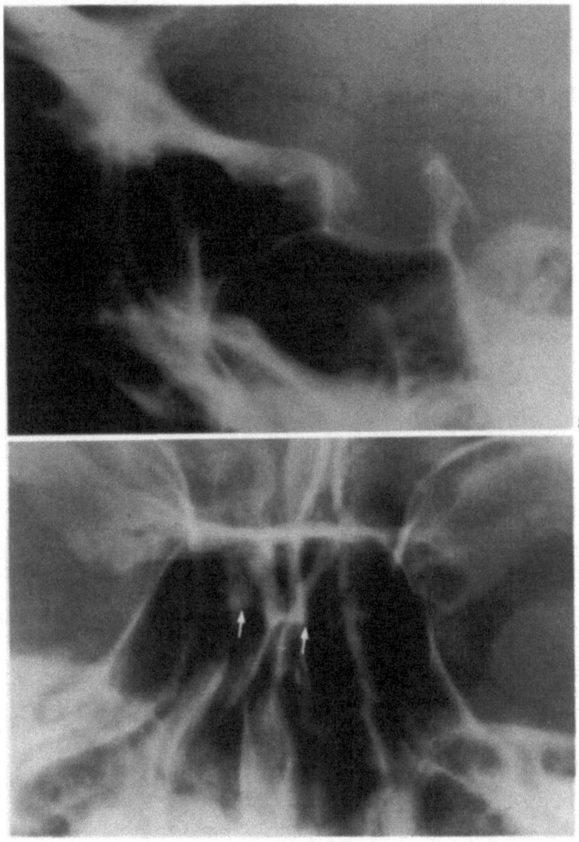

a

b

Fig. 5.75 a, b. Normal double floor: the lower line of the double floor is as thick as the upper; **b,** sloping of the sellar floor toward the left and asymmetry of sphenoid sinus pneumatization *(arrows)*

◁ **Fig. 5.76 a, b.** Very slight doubling of the middle part of the floor of the sella; the lower line of the double floor is thinner than the upper. However, the lower line presents a more harmonious and rounded outline than the upper, which appears slightly broken. Actually, this is simply slight asymmetry of the two halves of the sella. On the frontal projection the floor slopes slightly to the left

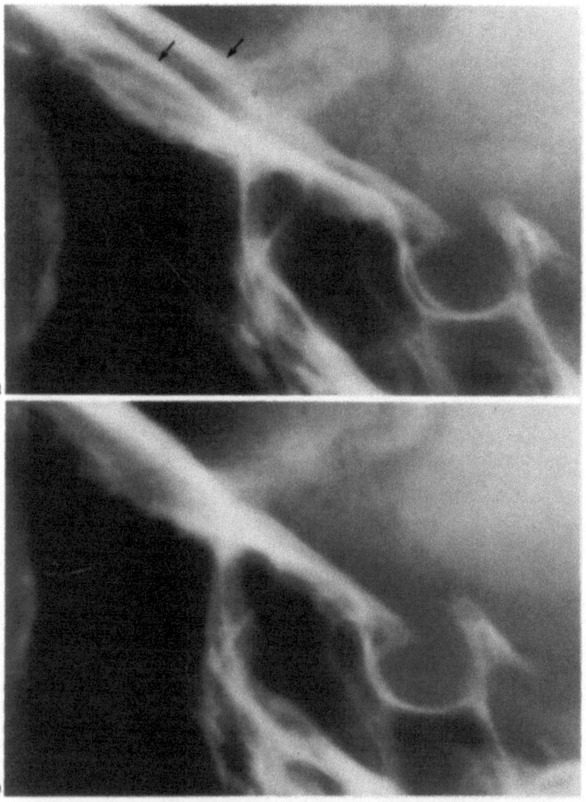

a

b

Fig. 5.77a, b

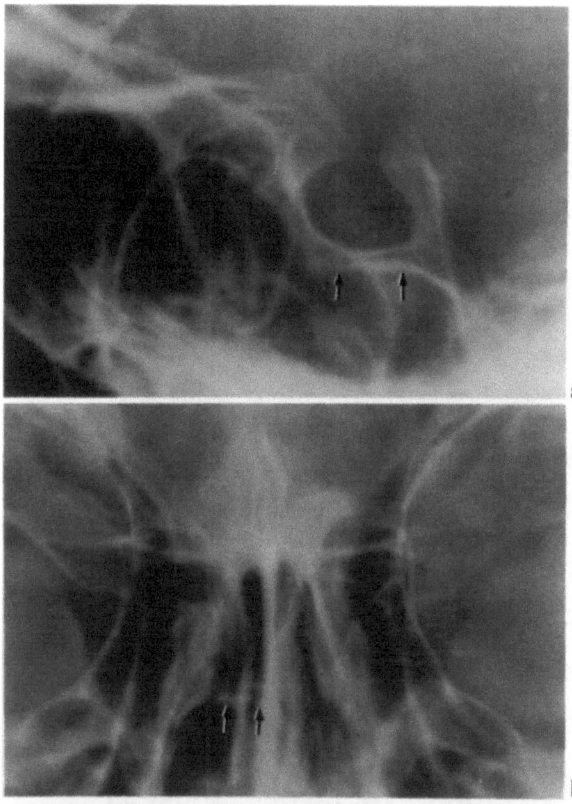

a

b

Fig. 5.78a, b

a

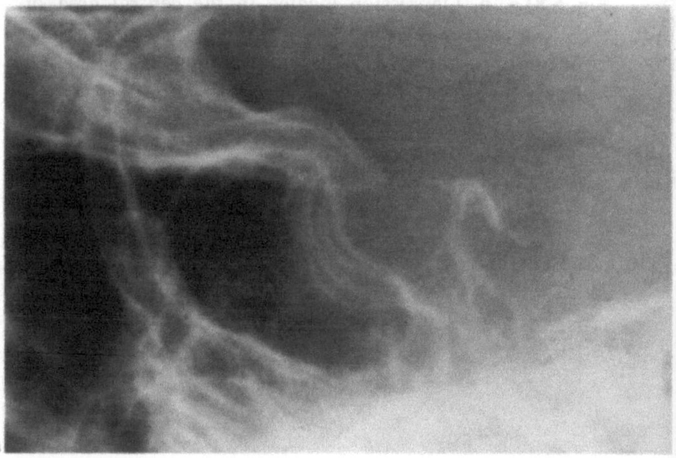

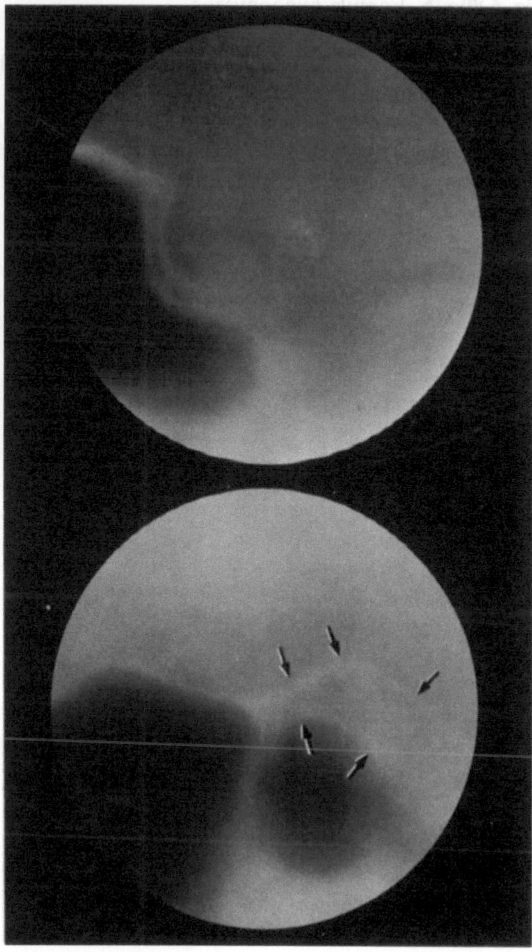

b

c

Fig. 5.79a–c. Quadruple floor to the sella.

a Lateral projection.

b Sagittal tomography to the left of the midline: a plate of cancellous bone persists between the anterior part of the floor of the sella and the sphenoid sinus, giving a double line.

c Frontal tomogram passing through the sellar floor and explaining the constitution of a quadruple floor: the sellar floor presents a clear upward concavity, broken at the point of insertion of the sagittal septum of the sphenoid sinus; the right margin of the floor is slightly lower than the left one. Finally, one finds once again incomplete pneumatization of the left part of the sphenoid sinus *(arrows)*

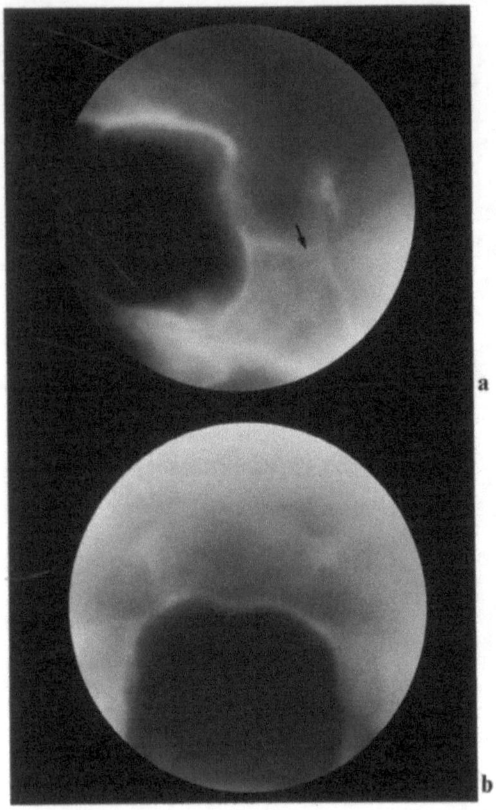

Fig. 5.80 a, b. Fossula hypophyseos.

a Lateral tomogram; note a depression limited to the posterior part of the base of the sella, rising slightly on the dorsum sellae *(arrow)*.

b Frontal tomogram immediately in front of the dorsum sellae: small depression of the most posterior part of the floor in the midline without changes in the cortical bone

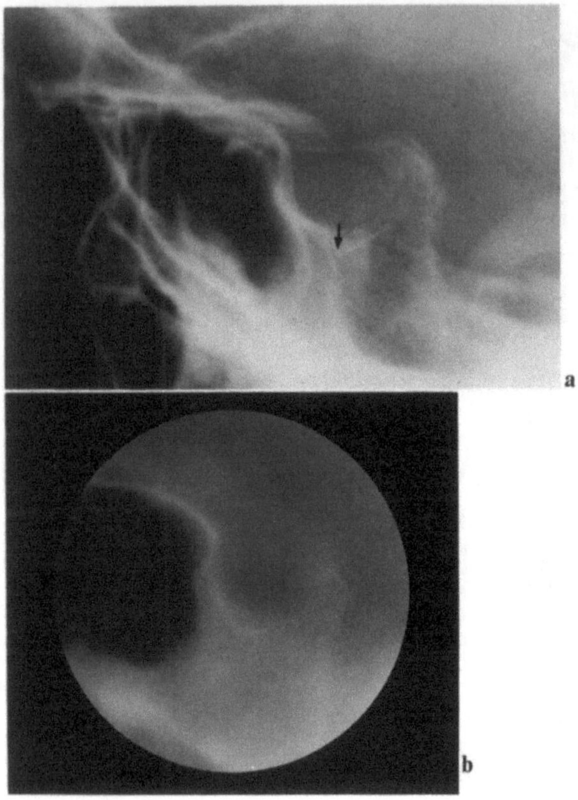

Fig. 5.81 a, b. Depression located in the central area of the floor *(arrow)* without any thinning of the cortical bone on tomography. This appearance must be considered normal

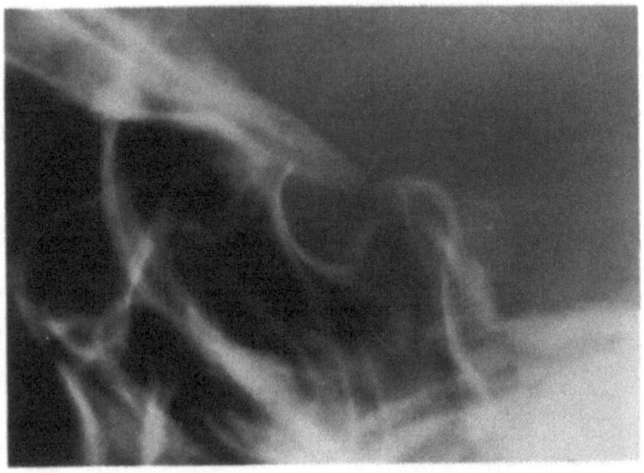

◁ **Fig. 5.82.** Localized blistering of the posterior part of the floor and the base of the dorsum sellae, due to hyperpneumatization of the sphenoid sinus

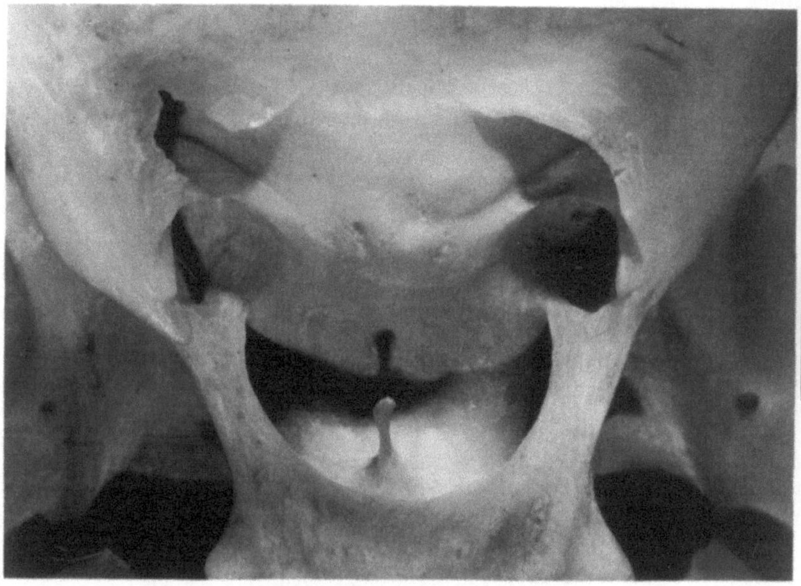
a

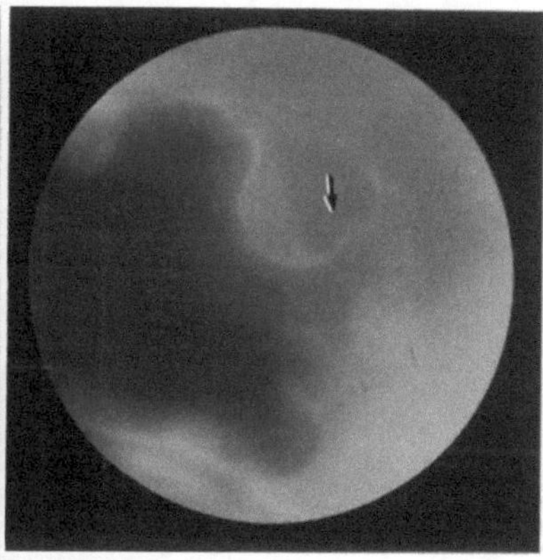
b

Fig. 5.83
a Anatomic specimen of a sellar spine (Courtesy J. LANG)
b Sellar spine inserted at the junction of the floor and the dorsum sellae, in the midline *(arrow)*.

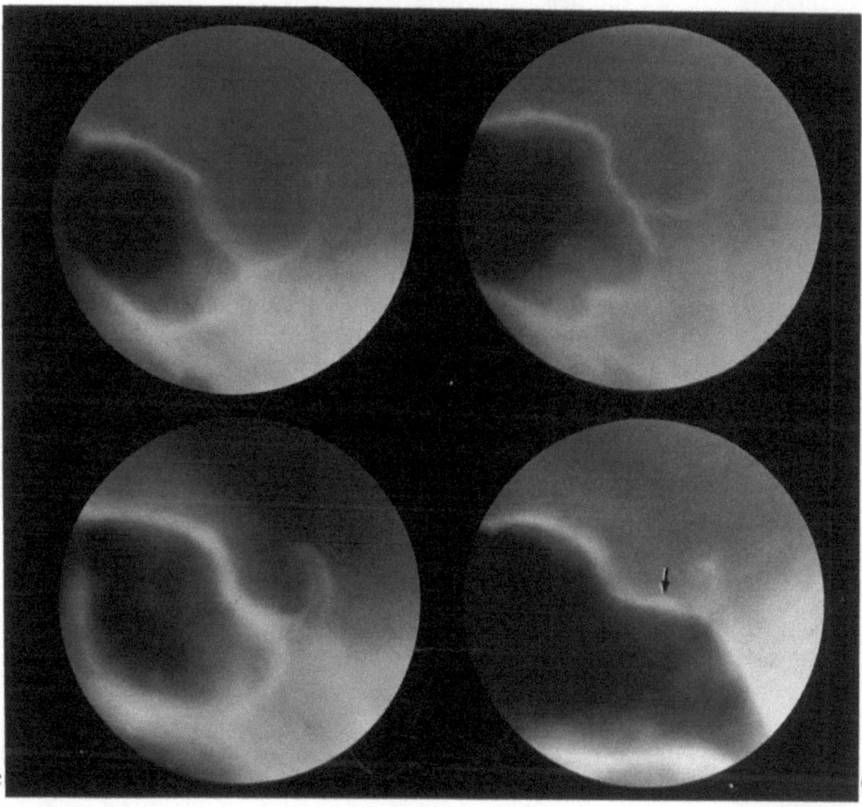
a b c d

Fig. 5.84 a–d. Variation of the floor of the sella in contact with the sphenoid sinus: normal patterns.
a Generally thin cortical bone opposite the sinus.
b Thicker cortical bone opposite the sinus.
c Generally thick cortical bone; cortical bone of the sphenoid sinus actually joins that of the sella turcica.
d Localized nonpathologic cortical thickening *(arrow)*

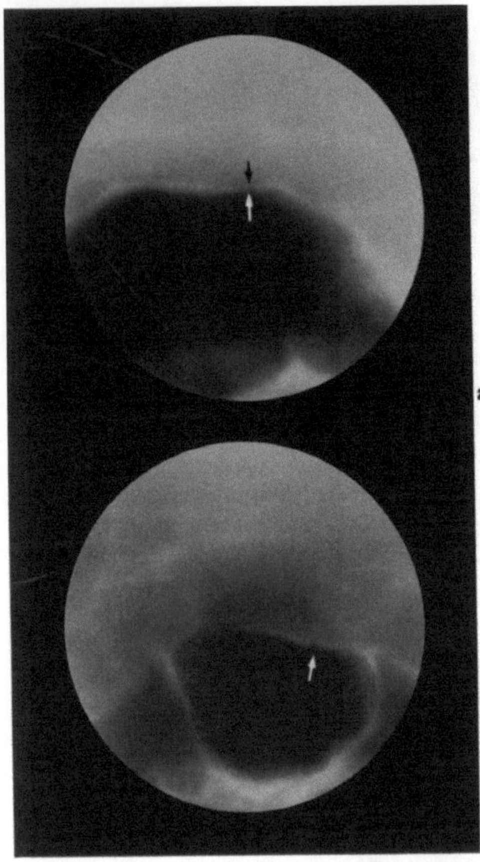

a

b

Fig. 5.85
a Nonpathologic, localized thinning of the floor of the sella: the cortical bone of the floor is thinned on both surfaces *(arrows)*; obviously this cannot be the result of a purely intrasellar process.
b Nonpathologic thinning of the inferior surface of the floor of the sella *(arrow)*

Fig. 5.86. False thinning of the floor of the sella: tomographic artifacts, i.e., blurring of the air contained in a well-developed sphenoid sinus *(arrows)*

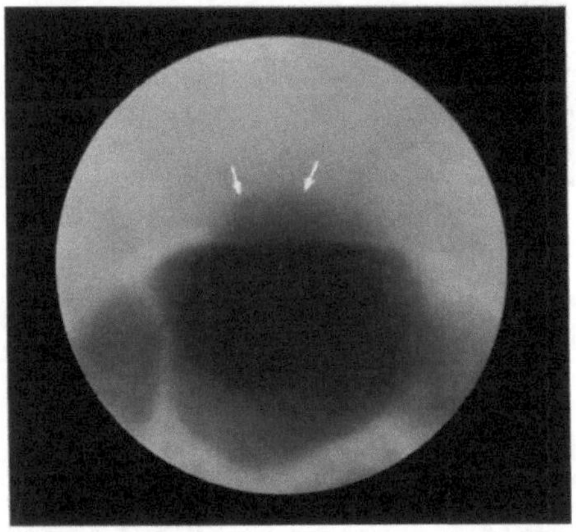

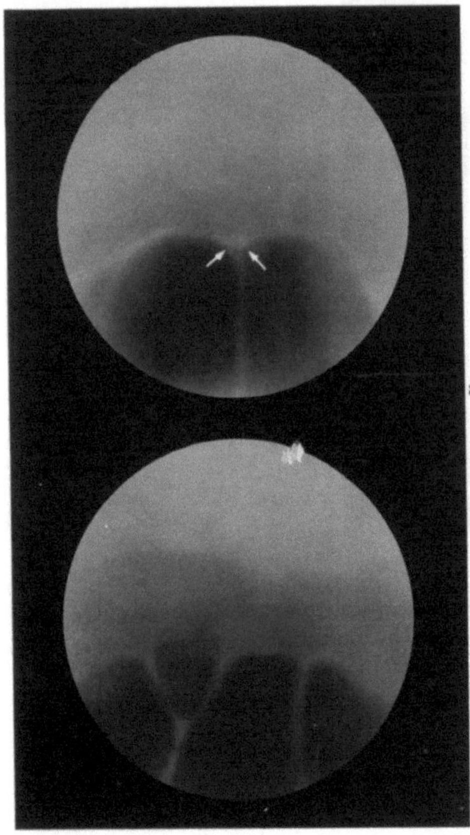

a

b

Fig. 5.87
a Localized triangular thickening of the floor of the sella at the point of insertion of a sagittal septum of the sinus *(arrows)*.
b Localized triangular depression of the floor of the sella opposite the insertion of a sinus septum; the cortical bone remains unchanged. This appearance must be considered physiologic

Fig. 5.89
a On the lateral projection: anteriorly predominant double floor; the lower line is very thin.
b Lateral tomography: regular thinning of the anterior wall and the floor between the insertion of a small crest of the sphenoid sinus and the most posterior area of the sinus *(arrows)*, the evenly thin character of the cortical bone and the sudden changes at its front and back boundaries, indicate physiologic "thinness" rather than pathologic "thinning."
c, d Frontal and sagittal pneumotomograms; no suggestion of an intrasellar microadenoma; strictly normal diaphragma sellae; the pituitary stalk is not displaced *(arrows)*

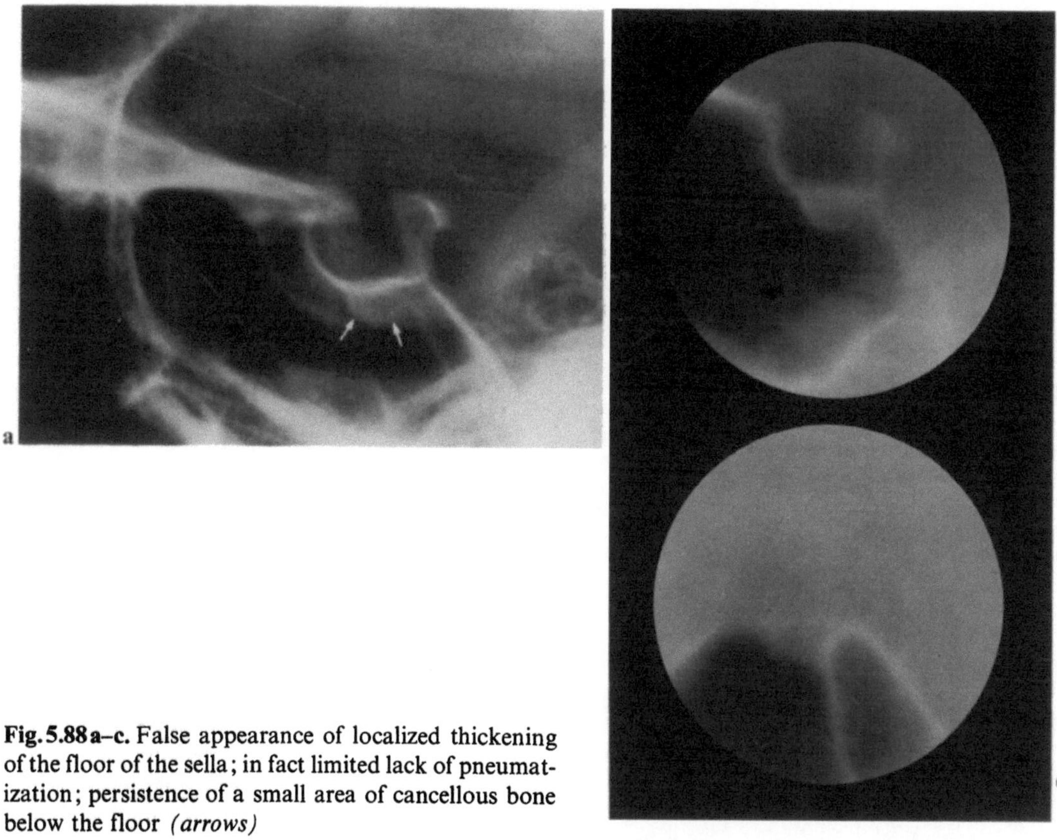

Fig. 5.88 a–c. False appearance of localized thickening of the floor of the sella; in fact limited lack of pneumatization; persistence of a small area of cancellous bone below the floor *(arrows)*

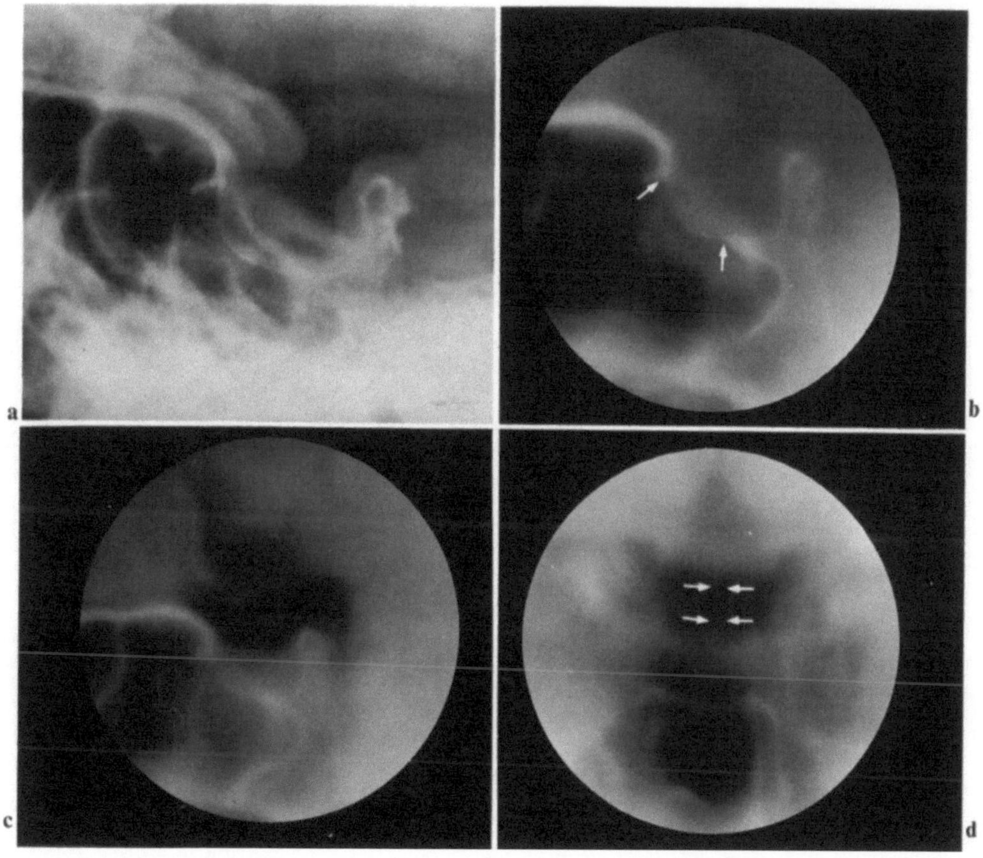

Fig. 5.89 a–d (caption see opposite page)

Chapter 6 Intrasellar Pathology

A. The Empty Sella Turcica[12]

The empty sella turcica is defined as an intrasellar herniation of the suprasellar subarachnoid space resulting from an incomplete diaphragma sellae, with compression of the pituitary gland (Fig. 6.1) producing in many cases a gradual change in the radiologic appearance of the sella turcica.

It is important to distinguish the primary empty sella turcica related to a congenital deficiency of the diaphragma sellae from the secondary empty sella turcica, which may follow treatment (surgery, radiotherapy) or spontaneous necrosis of pituitary adenoma. Secondary empty sella turcica has also been reported in cases of increased intracranial pressure.

I. Primary Empty Sella Turcica

1) History

During the past 10 years, many works have been published on the clinical signs, laboratory investigations, plain films, CT, pneumography, and surgical treatment of the empty sella. The term empty sella turcica has been commonly used since Busch's anatomic work (1951). Busch was the first to correlate the apparently partially empty sella turcica with congenital deficiency of the diaphragma sellae. Other authors had, however, already noted the abnormalities. As early as 1875, Key and Retzius observed extension of the chiasmatic cistern along the infundibular stalk toward the intrasellar space. In 1924, Schaeffer pointed out the considerable variability of the diaphragma sellae. The first radioanatomic correlation was reported in 1937 by Haas, who showed, by pneumoencephalography, air within the intrasellar cavity in a patient whose autopsy revealed deficiency of the diaphragma sellae and intrasellar herniation of the subarachnoid space, which compressed the pituitary gland. Busch's anatomic observations have been confirmed by more recent anatomic works such as those of Bergland et al. (1968), McLachlan et al. (1968a), Kaufman and Chamberlain (1972), and Renn and Rhoton (1975). Radiologic confirmation of the intrasellar subarachnoid herniation was obtained by Robertson (1957) using pneumoencephalography and by Engels (1958) with Pantopaque cisternography.

Over the past 10 years, a number of papers have been devoted to clinical aspects of the empty sella (Neelon et al. 1973; Messimy et al. 1974; Ommaya et al. 1968; Brisman et al. 1972; Schaison and Metzger 1976; Metzger et al. 1970; Leclerq et al. 1974; Jordan et al. 1977). It should be noted that several papers have dealt with the coexistence of a primary empty sella turcica and a secreting microadenoma (Fig. 6.8) (Zatz et al. 1969; Sutton and Vezina 1974; Ganguly et al. 1976; Domingue et al. 1978).

2) Pathogenesis

Dehiscence of the diaphragma sellae is indispensable for intrasellar subarachnoid space

12 Refs. [22, 30, 33, 36, 46, 56, 65, 68, 69, 86, 105, 112, 113, 114, 118, 126, 132, 142, 153, 190, 196, 237, 267, 280, 289, 297, 298, 319, 321, 323, 349, 355, 367, 423, 441, 470, 477, 478, 479, 480, 481, 482, 492, 494, 528, 546, 552, 556, 563, 577, 608, 628, 630, 631, 637, 649, 658, 660, 665, 687, 691, 694, 695, 696, 702, 723, 764, 782, 784, 798, 804, 834, 882, 921, 934, 972, 1006, 1007, 1010, 1044, 1047]

herniation. All other factors are only predisposing, whether causing increased pressure in the suprasellar subarachnoid space or reduction in size of the pituitary gland.

a) Congenital Deficiency of the Diaphragma Sellae

Variations of the diaphragma sellae were described in Chap. 2 (Figs. 2.6, 2.7). Both the high incidence of total absence of the diaphragma (20.5% of normal subjects) and the frequency of the empty sella turcica in a normal population (5.5% according to Busch 1951) are noteworthy.

b) Suprasellar Promoting Factors

Most authors believe that the flattening of the pituitary toward the posteroinferior part of the pituitary fossa is related to pulsations of the cerebrospinal fluid in both the chiasmatic cistern and the third ventricle. When the diaphragma sellae is incomplete, cerebrospinal fluid pulsations act directly on the upper aspect of the pituitary gland. This suprasellar cistern pressure is, of course, accentuated in cases in which intracranial pressure is raised. A posteriorly placed optic chiasm (see Chap. 2) may also expose the upper surface of the pituitary gland and thereby increase cerebrospinal fluid pressure on it (Zatz et al. 1969).

c) Pituitary Promoting Factors

Any reduction in size of the pituitary gland favors intrasellar extension of suprasellar subarachnoid spaces. This reduction may be due to:

1. Physiologic involution. These involutional processes occur essentially in women. Pregnancies bring about large variations in the size of the pituitary gland. After the menopause, there is a reduction in pituitary volume, changes which explain a very clear predominance of the primary empty sella turcica in women, especially after 40 years of age. Such involution may also be seen in end organ failure leading to adenohypophyseal hyperplasia; commencement of hormone replacement therapy induces a secondary involution of the pituitary gland (Neelon et al. 1973).

2. Pathologic processes. Shrinkage may occur after postpartum necrosis (Sheehan's syndrome), or hypophyseal infarction in patients with vascular diseases, notably diabetics (Neelon et al. 1973), those with raised intracranial pressure (Wolman 1956), with head injury (Kornblum and Fisher 1969), or with meningitis or cavernous sinus thrombosis.

d) Rupture of an Intrasellar Cyst

Such rupture probably occurs only rarely. Raiti et al. (1976) reported the most convincing observation, demonstrating the existence of a membrane formed by pseudostratified cuboid epithelium, which itself was covered by the arachnoid mater. Before erupting into the suprasellar subarachnoid space, the cyst distends and deforms the diaphragma sellae. Rupture of the cyst allows intrasellar extension of the subarachnoid space.

e) Pathogenesis of the Radiologic Appearances

Pressure of the arachnoidocele is maximal in the midline (diaphragmatic deficiency often being midline). This pressure will affect the floor of the sella through the pituitary gland, causing symmetric depression of the floor.

Retrospective study of more than 150 personal cases of the primary empty sella turcica shows that this depression is more distinct when the sphenoid sinus does not extend below the floor of the sella, which would thus be less resistant (Mahmoud 1958). Despite decreased resistance, the floor of the sella appears generally thick and dense; this change could be related to a sinking of the cancellous bone adjacent to the cortex of the floor.

3) Clinical Symptomatology

The primary empty sella turcica is found essentially in women (80% of cases) especially after the menopause. It is rare during childhood. Often, there are no clinical manifestations and therefore the empty sella turcica is discovered by chance on plain skull films, pneumoencephalography, metrizamide cisternography (Fig. 6.9), or CT scans. Clinical manifestations are very varied.

Headache and obesity are the most frequently encountered abnormalities. Headache could be related to a stretching of the dura mater of the floor of the sella (Mortara and Norrell 1970).

Visual signs. There may be diminished visual acuity or even uni- or bilateral optic atrophy. The visual fields may be affected in a variable fashion (Messimy et al. 1974). Papilledema may occasionally be present (Lepage et al. 1975; Szecsi 1978).

Rhinorrhea. A rare occurrence, rhinorrhea can be explained by rupture of the cortical bone of the floor of the sella due to the cerebrospinal fluid pulsations. Communication between the intrasellar subarachnoid space and the sphenoid sinus obviously raises a risk of meningitis (Brisman et al. 1969; Uria et al. 1979). Rhinorrhea requires extradural surgical packing of the sella turcica by the transsphenoidal approach (Welch and Stears 1971; Olson et al. 1972; Keravel 1973).

Endocrine symptoms are rarely encountered. However, secondary amenorrhea and hypothyroidism may be seen. These symptoms reflect insufficiency of the adenohypophysis (Caplan and Dobben 1969; Brisman et al. 1972; Neelon et al. 1973; Schaison and Metzger 1976; Jordan et al. 1977). It is important to note that neurohypophysis is almost never involved (Matisonn and Pimstone 1973). Some authors have emphasized the frequent association of diabetes and the primary empty sella turcica (Neelon et al. 1973).

Some cases of primary empty sella turcica are associated with hyperpituitarism (hypersecretion of prolactin, growth hormone, ACTH). Such hypersecretion may be due to the existence of an empty sella turcica but it is mandatory to look for an associated secreting microadenoma.

4) Radiology

It is possible to observe intrasellar extension of the subarachnoid spaces with cisternography in a radiologically normal sella. In general, diagnosis of the empty sella turcica is nearly always easy on plain films, with frontal and sagittal tomograms when necessary. A distinction can be made between the empty sella in the early stage and established empty sella.

a) Empty Sella in the Early Stage (Chap. 5)

On the lateral projection the radiologic appearance is quite characteristic (Figs. 5.12, 5.13).

Above a poorly aerated sphenoid the lateral film shows the appearance of a double floor of the sella, with two parallel lines corresponding to the floor and occasionally also to the dorsum sellae and the anterior wall; the lower line of this double floor is very well defined and thick, but the upper line is blurred. On the frontal projection the floor shows a symmetric downward convexity.

b) Established Empty Sella

The radiologic appearance of the established empty sella is traditionally that of a sella which is enlarged mainly in the vertical direction without erosion of the cortical bone. The dorsum, which may be thinned, is almost never pushed backward (Fig. 6.4). Actually, three radiologic patterns are fundamental to the diagnosis of the empty sella turcica (Figs. 6.2, 6.3).

Radiologic changes are located below the level of the diaphragma sellae (Kaufman 1968).

The downward convexity of the floor of the sella shown on frontal radiographs is very regular (Leclercq et al. 1974). This appearance is totally characteristic when it seems to be *en cuvette* and when there is no sinus below the floor.

The appearances of a primary empty sella turcica above a well-aerated sphenoid are different, and may be confused with those of a pituitary adenoma (Figs. 6.5, 6.6). In fact, the radiologic abnormalities are also quite characteristic. In general the cortical bone in contact with the cancellous bone is preserved, or even dense. This pattern is particularly clear on sagittal tomograms and also on frontal tomograms, when pneumatization of the sphenoid is asymmetric (Fig. 6.7). In cases where sphenoid pneumatization extends below the floor of the sella turcica one almost never sees deformation *en cuvette* similar to that seen above a poorly aerated sphenoid. Here the cortical bone of the floor of the sella is clearly and regularly thinned, showing on frontal tomography a well-defined downward discontinuity from one end; in this type of sinus, the appearance is generally that of a depression *en plateau* with thinning of the cortical bone.

Attachment of an intersinus septum to the floor of the sella may considerably change the configuration of the floor in a primary empty sella turcica. It is in these cases that distinguishing an empty sella from a pituitary adenoma is most difficult (Fig. 6.7 f).

II. Special Types of Empty Sella Turcica

Secondary empty sellae turcicae after surgical or radiotherapy, or after spontaneous necrosis of a pituitary adenoma are described in Chap. 6.B.

Fig. 6.2a–d. Typical empty sella turcica.
a Lateral view: deep sella and regularly doubled floor. The two cortical lines are parallel: the inferior one, indicating the deepest part of the floor in the midline, is well-defined and dense *(black arrow)*; the upper line, corresponding to the lateral limits of the sella, is less well-defined *(black* and *white arrow)*.
b Frontal view: pronounced concavity of the floor *(arrows)*.
c Midsagittal tomogram: thinning of the anterior portion of the floor, adjacent to the sphenoid sinus *(white arrows)*. The cortical bone is preserved opposite the nonpneumatized sphenoid bone *(black arrows)*.
d Frontal tomogram: there is no pneumatization below the floor of the sella, which shows a very regular, symmetric concavity *en cuvette*. The lamina dura is dense

Anterior pituitary necrosis occurring in the postpartum (Sheehan's syndrome) and pituitary gland agenesis may also be related to the development of an empty sella turcica. In marked hydrocephalus, dilated optic and infundibular recesses of the third ventricle may give rise to an empty sella turcica by penetrating the sellar cavity. Raised intracranial pressure, regardless of its origin, may also give rise to an empty sella turcica, as a result of increased cerebrospinal fluid pressure in the suprasellar cistern.

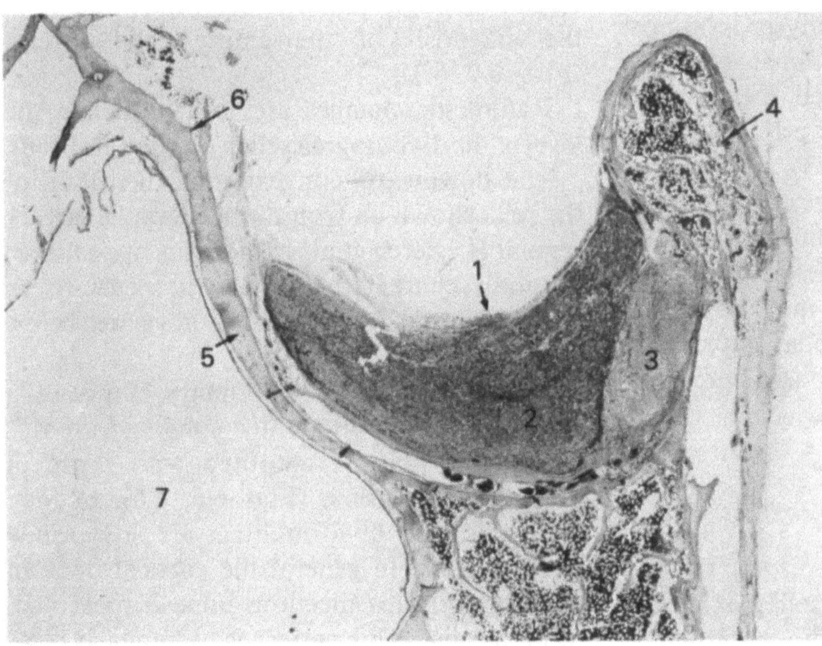

Fig. 6.1. Midsagittal section of specimen showing an empty sella turcica. The diaphragma sellae is absent and the pituitary gland is flattened. There is intrasellar herniation of the subarachnoid space.
1 Upper surface of pituitary gland
2 Adenohypophysis
3 Neurohypophysis
4 Dorsum sellae
5 Anterior wall of sella turcica
6 Tuberculum sellae
7 Sphenoid sinus
(Courtesy, N. Heldt, Strasbourg)

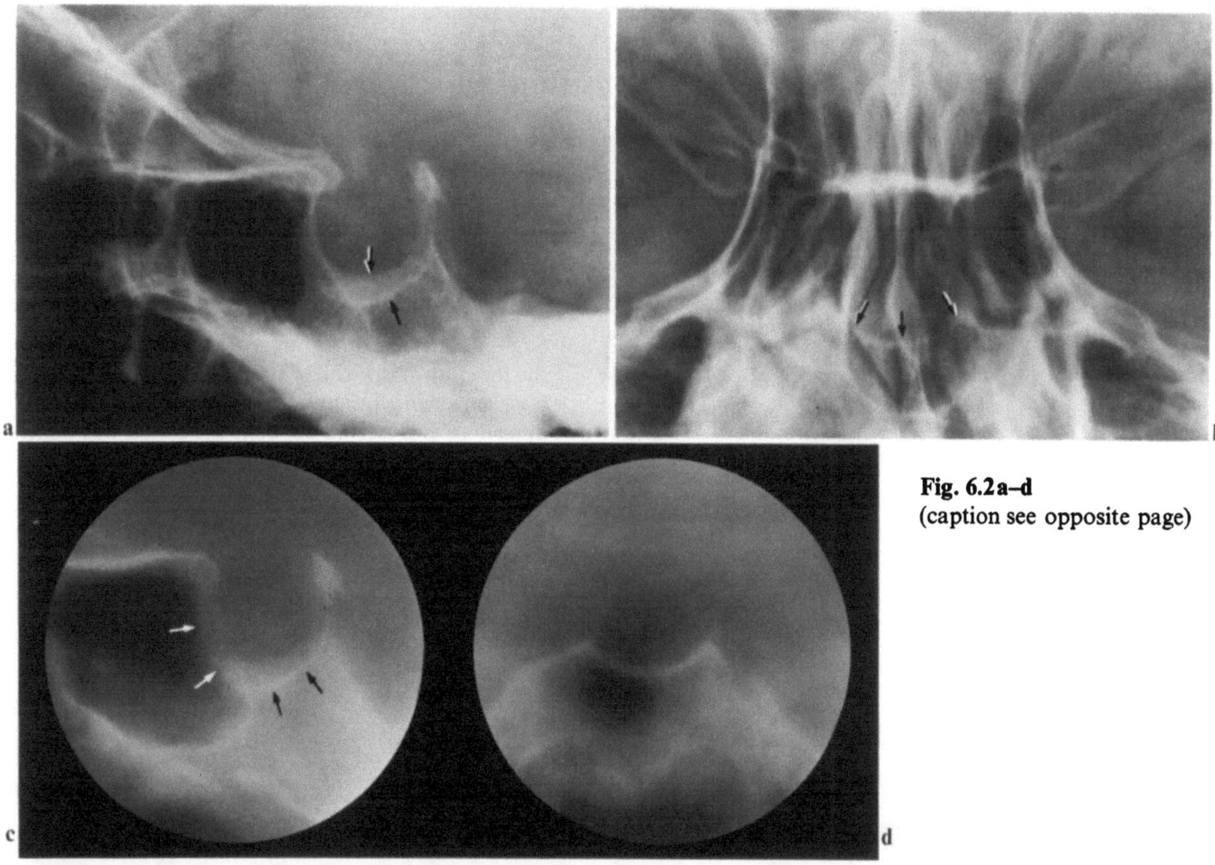

a b c d

Fig. 6.2a–d
(caption see opposite page)

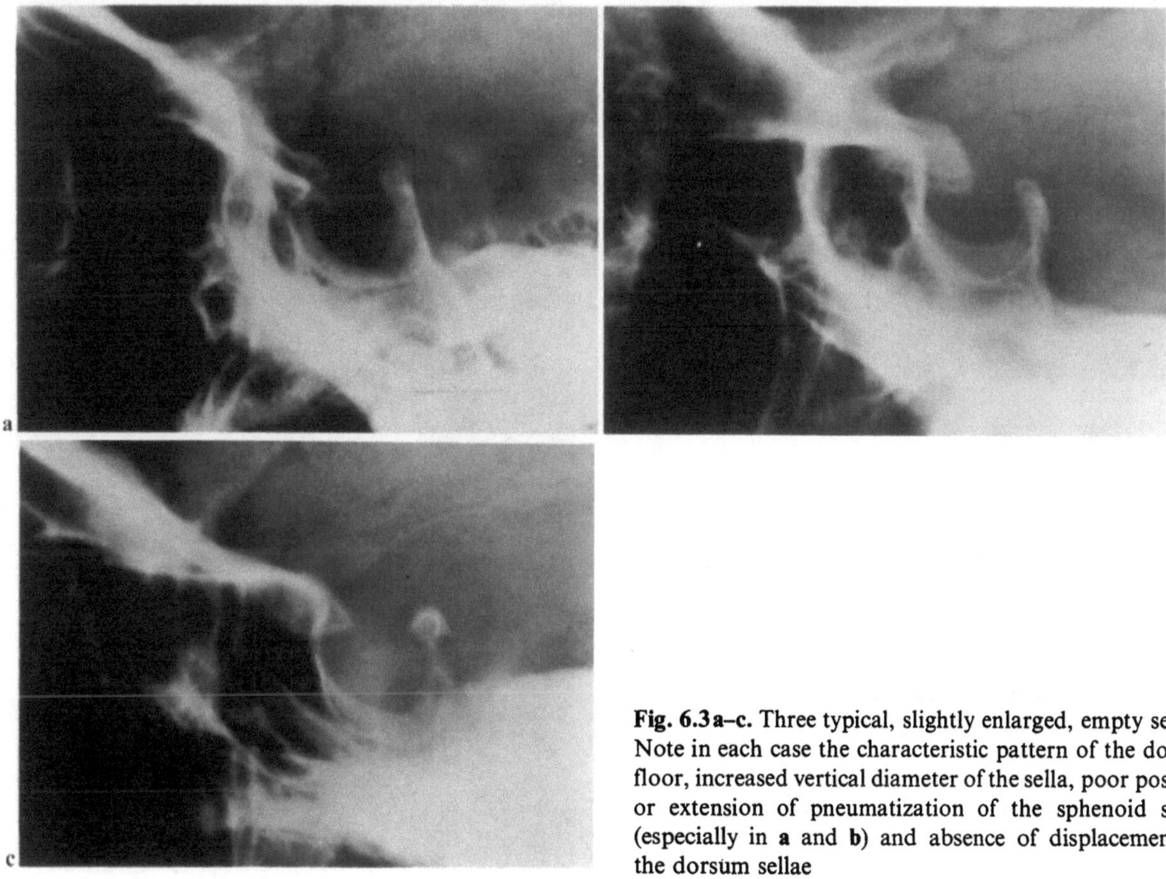

a b c

Fig. 6.3a–c. Three typical, slightly enlarged, empty sellae. Note in each case the characteristic pattern of the double floor, increased vertical diameter of the sella, poor posterior extension of pneumatization of the sphenoid sinus (especially in **a** and **b**) and absence of displacement of the dorsum sellae

Fig. 6.5a–d. Empty sella. ▷
a Lateral projection: bulging of the anterior wall of the sella, which appears demineralized *(arrows)*.
b Frontal projection: the floor of the sella is slightly tilted and is blurred *(arrows)*. This appearance may suggest a pituitary adenoma.
c Frontal tomogram: regular thinning of the floor of the sella opposite the sphenoid sinus. The floor is depressed on the right.
d Sagittal pneumotomogram; demonstrating clearly the thinning of the wall of the sella anterior to the intrasellar subarachnoid herniation *(arrows)*; the lamina dura is preserved opposite the cancellous bone. Note the posterior position of the optic and infundibular recesses of the third ventricle corresponding to a postfixed chiasm

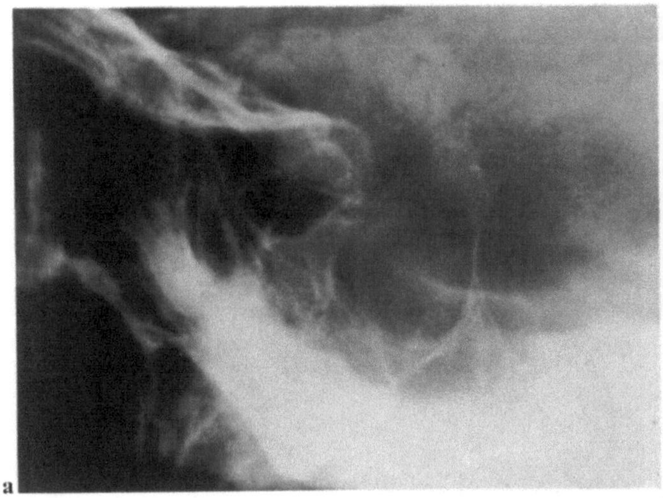
a

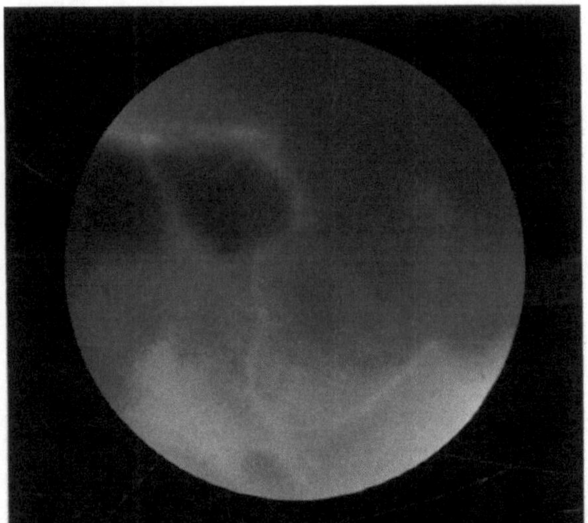
b

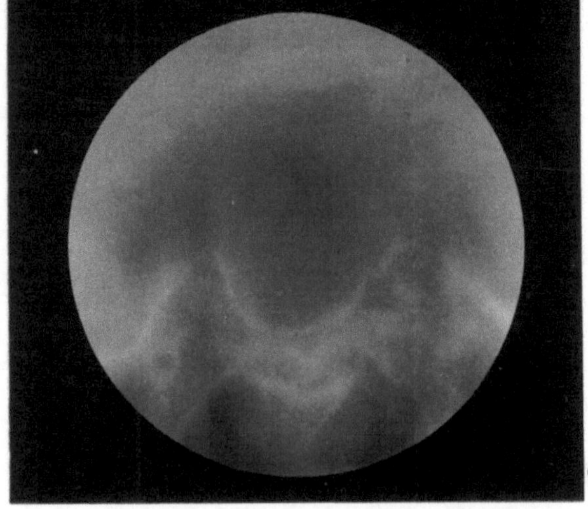
c

Fig. 6.4a–c. Very deep empty sella.
a Lateral view: marked deepening of the sella turcica, and thinning of the dorsum sellae.
b Sagittal tomogram: pronounced hollowing of the floor, whose lamina dura is well-preserved.
c Frontal tomogram: symmetric, markedly depressed floor. Note absence of the sphenoid sinus below the floor of the sella

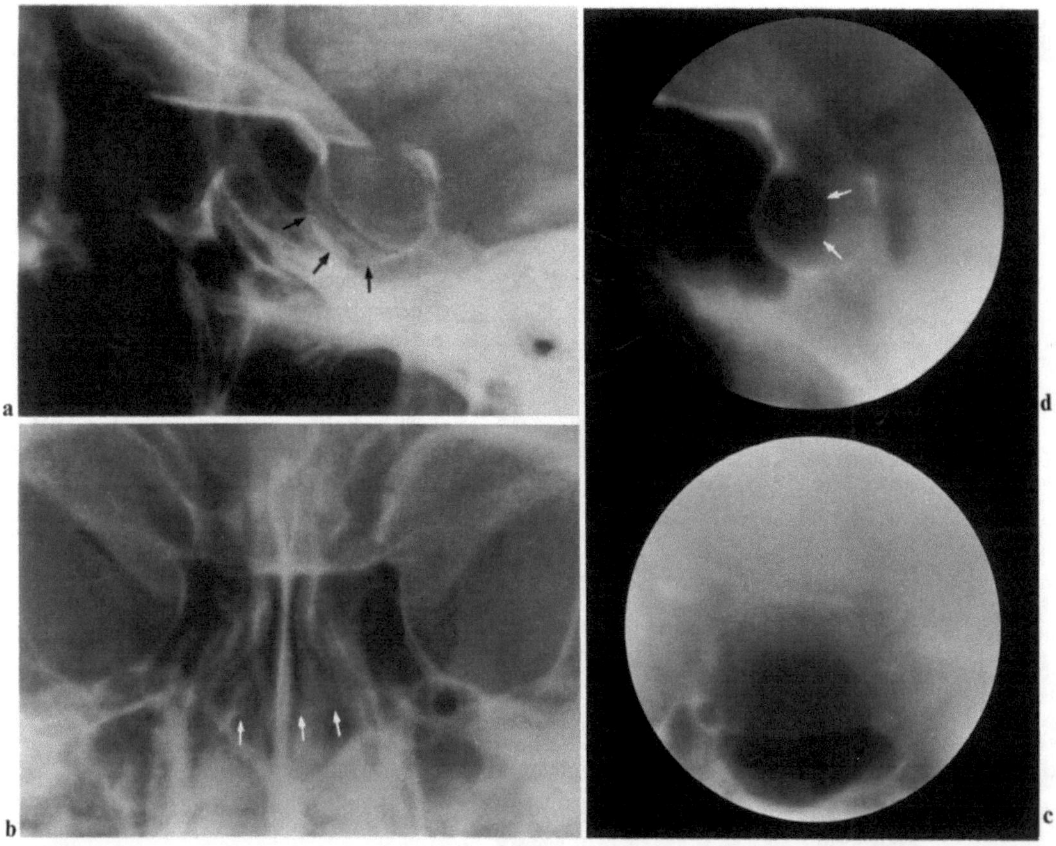

Fig. 6.5a–d (caption see opposite page)

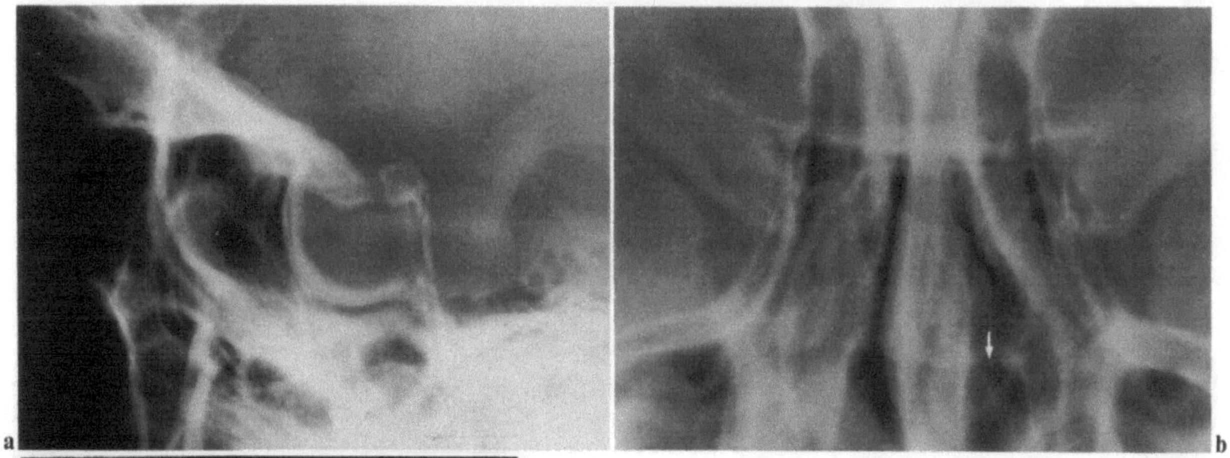

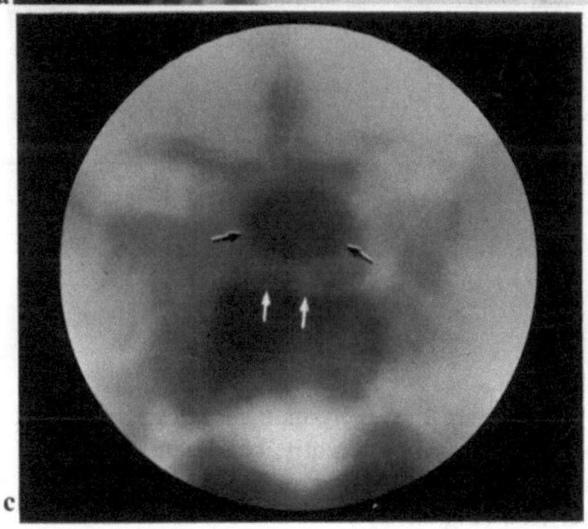

Fig. 6.6a–c. Empty sella.

a Lateral projection: slight double floor with minor demineralization.

b Frontal projection: the floor of the sella is slightly depressed on the left *(arrow)*.

c Frontal pneumotomogram: large intrasellar extension of the subarachnoid space *(black and white arrows)*: regular thinning of the floor, which is slightly depressed *(en plateau) (white arrows)*

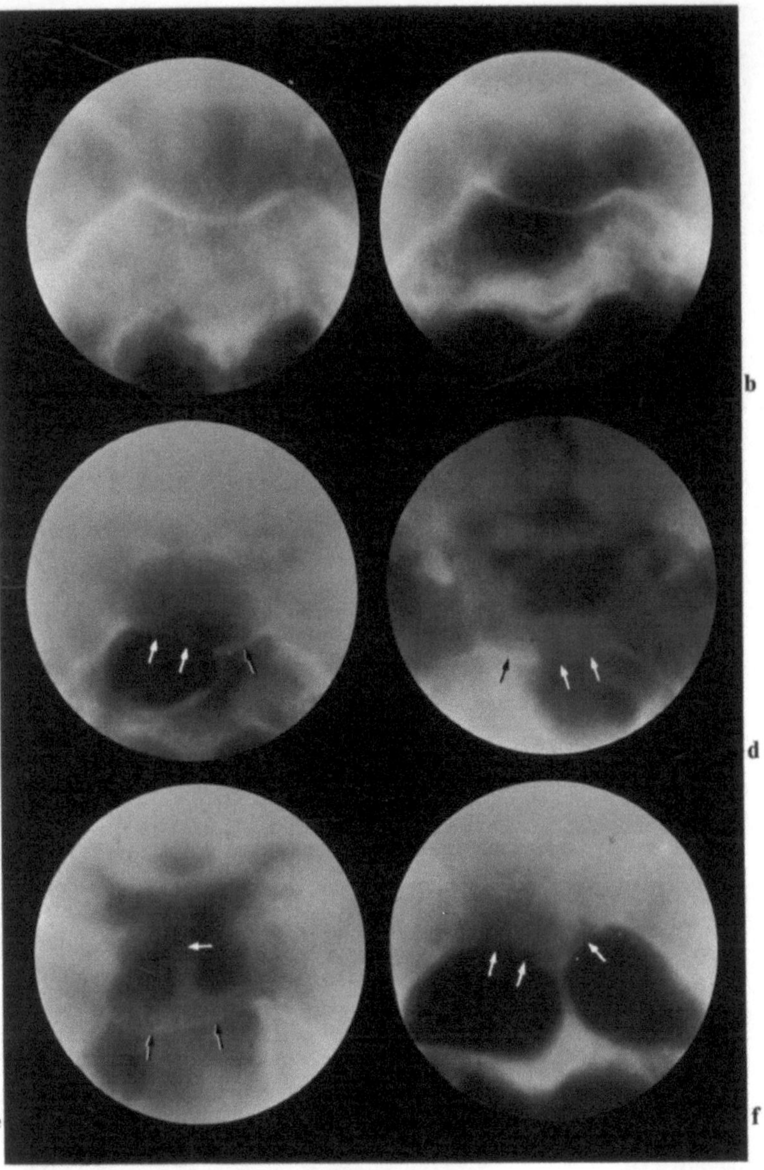

Fig. 6.7 a–f. Different appearances of the floor in empty sellae.

a, b Symmetric upward concavity with dense cortical bone in nonpneumatized sphenoid bone *(en cuvette)*.

c, d Asymmetric aeration of the sphenoid sinus. The cortical bone is thinned opposite the well-aerated part *(white arrows)* and preserved opposite the cancellous bone *(black and white arrows)*.

e Note thinning of the floor *(black and white arrows)* opposite the wellpneumatized sinus (clouded by sinusitis). The pituitary stalk appears elongated *(white arrow)*.

f Thinning of the floor of the sella, *(white arrows)* and angulation centered on the attachment of the sagital septum of the sinus. In such cases diagnosis between empty sella and pituitary adenoma may be difficult

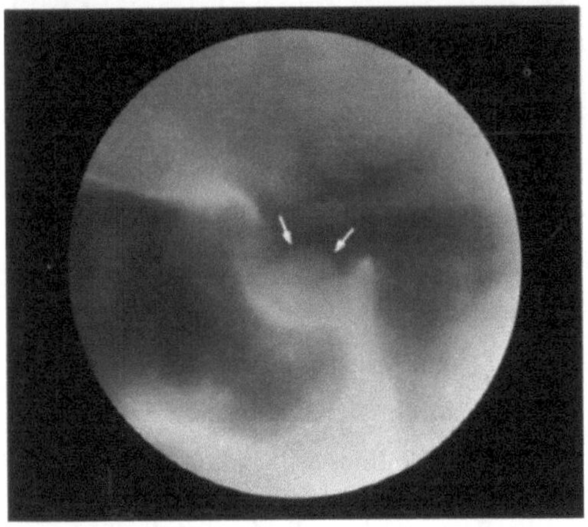

Fig. 6.8. Microadenoma coexisting with an empty sella (midsagittal pneumotomogram)

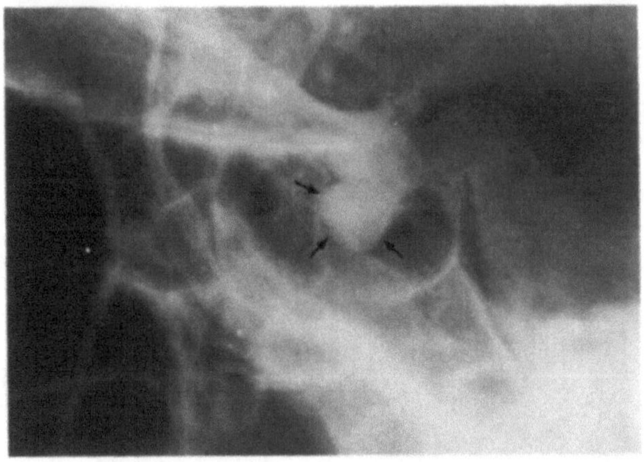

Fig. 6.9. Empty sella shown by metrizamide cisternography. The sella turcica is slightly ballooned. Note the typical appearance of the floor of the same patient on the frontal view (Fig. 6.7 a)

B. Pituitary Adenomas[13]

I. General Considerations

Pituitary adenomas develop from the cells of the adenohypophysis. The normal hypophysis includes six different types of cells: secreting growth hormone (HGH), prolactin, gonadotropic hormone FSH-LH, adrenocorticotropic hormone (ACTH), thyroid stimulating hormone (TSH) and melanocyte stimulating hormone (MSH). It would seem that these cells have a definite topographical distribution within the adenohypophysis. Based on his surgical findings, Hardy (1973) established the topography of the main types of secreting adenomas, localization of these adenomas corresponding to islands of secreting cells present in the normal adenohypophysis. Thus growth hormone secreting cells are in the anterolateral part of the adenohypophysis, whereas prolactin-secreting cells, though also lateral, are deeper and more posterior, and both ACTH and TSH secreting cells are midline.

Adenomas were previously classified as eosinophilic, basophilic, or chromophobe according to the staining properties of the tumor cells. Lack of correlation between histology and clinical observations was frequent; progress in pituitary cytology (in light and electron microscopy) and in immunohistochemistry has enabled a precise cytofunctional classification (Racadot 1966; Racadot et al. 1974). This classification distinguishes the following:

13 Refs. [4, 11, 16, 34, 47, 73, 87, 91, 93, 111, 119, 152, 158, 160, 162, 165, 175, 180, 183, 185, 186, 200, 205, 207, 209, 213, 214, 220, 229, 237, 250, 270, 277, 294, 299, 314, 325, 328, 333, 347, 348, 354, 356, 358, 359, 378, 380, 383, 384, 385, 388, 390, 405, 406, 409, 412, 413, 435, 464, 469, 473, 483, 505, 523, 525, 526, 535, 544, 560, 564, 573, 583, 585, 592, 599, 605, 609, 615, 620, 632, 633, 635, 639, 647, 648, 650, 653, 656, 662, 668, 669, 670, 689, 693, 700, 708, 756, 757, 758, 759, 761, 776, 781, 788, 793, 795, 796, 799, 800, 801, 805, 806, 807, 817, 820, 821, 838, 844, 868, 869, 887, 898, 904, 905, 919, 935, 945, 952, 956, 957, 960, 963, 968, 978, 984, 985, 986, 991, 995, 1003, 1005, 1009, 1012, 1013, 1025, 1033, 1042]

Nonsecreting (chromophobe or nonfunctioning) adenomas. These present as tumors, causing hypopituitarism, by compression of the normal hypophysis. The cells of these adenomas are agranular, thus "chromophobe."

Secreting (or functioning) adenomas. Five in number, these consist of growth hormone secreting adenomas, causing acromegaly, prolactin-secreting giving amenorrhea and galactorrhea (Forbes-Albright syndrome), ACTH-secreting adenomas giving Cushing's disease, TSH-secreting adenomas giving hyperthyroidism and FSH- and LH-secreting adenomas (Trouillas et al. 1979).

Cytofunctional classification is of special value for determining with precision the type of adenoma. Radiologic appearances depend on this typing, but also on the macroscopic pattern of the adenoma. Surgical prognosis is also related to the macroscopic pattern.

Guiot (1958) and Hardy et al. (1973) divided pituitary tumors into four macroscopic categories (Fig. 6.10):

Microadenoma with a diameter of less than 10 mm. This tumor, which is intrasellar, causes localized erosion of the floor of the sella.

Enclosed adenomas. These have a diameter greater than 10 mm. Although they have intrasphenoid and/or suprasellar extensions of variable size, they remain well enclosed by the dura mater.

Locally invasive adenomas. The dura mater and the floor of the sella are destroyed and the tumor extends into the sphenoid sinus.

Diffuse invasive adenomas. The floor of the sella, sphenoid sinus, cavernous sinus, dorsum sellae, and diaphragma sellae are invaded.

The two latter types have a worse prognosis than the others.

II. Clinical Findings

1) Nonsecreting Adenomas

Generally quite large when first discovered, these adenomas present with symptoms of chiasmatic compression (bitemporal hemianopia). Hypopituitarism rarely draws attention to nonsecreting adenomas. Hypopituitarism has different clinical manifestations in adults and in children. In the former, clinical findings include weakness (with arterial hypotension and a tendency to hypoglycemia), sexual disorders (amenorrhea in women and impotence, with decreased libido, in men), cutaneous signs (pallor and skin atrophy). In children, growth hormone deficiency causes dwarfism and sexual immaturity. In practice, adenohypophyseal deficiency is due to a pituitary adenoma only rarely in children; it is, however, found especially with craniopharyngiomas and suprasellar germinomas.

2) Secreting Adenomas

These adenomas have predominant endocrine manifestations; visual symptoms appear later and are less frequent.

a) Growth Hormone Secreting Adenomas

These are seen particularly between the ages of 18 and 70, with maximal frequency at about 40. Prepubertal types, giving rise to gigantism, are infrequent. In adults, the clinical features include a marked dysmorphic syndrome with prognathism, widening and lengthening of hands and feet, and thoracic kyphosis. In addition to this dysmorphism, there is hypertrophy of numerous tissues; hypogonadism, hypothyroidism, and adrenal insufficiency are rare. The most characteristic laboratory abnormalities are diabetes mellitus and hyperlipidemia. Radioimmunoassay of plasma growth hormone confirms the diagnosis.

b) Prolactin-Secreting Pituitary Adenomas

This type of adenoma is seen essentially in young women between the ages of 20 and 30. The first symptom is often spontaneous or induced galactorrhea, either uni- or bilateral, and usually intermittent. However, this sign is often misinterpreted or neglected. Hypogonadism with amenorrhea preceded by oligomenorrhea and infertility is frequent. These symptoms indicate hypersecretion of prolactin; the possibility of induced hyperprolactinemia (oral contraceptives, phenothiazines, alphamethyldopa, reserpine, benzodiazepines, and isoniazid) must be eliminated. In men prolactin-secreting adenomas are rare; the clinical manifestations are sexual impotence and decreased libido, sometimes associated with gynecomastia. In children (David et al. 1962; Richmond and Wilson 1978; Grisoli et al. 1978), prolactin-secreting adenomas seem to be rather frequent, appearing during puberty, predominantly in girls. Before puberty, there is growth failure with primary amenorrhea, while after puberty, amenorrhea is seen, as in adult women. Visual field defects related to suprasellar expansion of the adenoma are frequent. In every case, the plasma prolactin levels serve to confirm the diagnosis. Very high levels indicate an adenoma. Tests of stimulation and suppression of prolactin secretion are the determining factors in confirmation of the presence of a prolactin-secreting adenoma.

c) ACTH-Secreting Adenomas

Hypersecretion of ACTH causes bilateral adrenal hyperplasia, the clinical manifestations of which include central obesity, skin changes (purpura, acne, hypertrichosis, pink striae, and hyperpigmentation), cardiovascular disorders (arterial hypertension), and bone problems (backache, pathologic fractures) related to osteoporosis.

Laboratory tests show disordered blood sugar regulation, with hyperglycemia. The levels of plasma cortisol and of urinary 17-hydroxy-corticosteroids and 17-ketosteroids are increased. Estimation of plasma ACTH concentration and adrenocortical catabolites confirms the diagnosis.

d) Pituitary Adenomas in Nelson's Syndrome

A pituitary adenoma secreting ACTH frequently appears some years after bilateral adrenalectomy for Cushing's syndrome. The clinical manifestation is increased skin pigmentation. It is not clear at present whether such adenomas have simply hypertrophied after adrenalectomy, or have been induced by adrenalectomy.

A review by Moore et al. (1976) suggests that Nelson's syndrome is relatively frequent, i.e., 8% of patients treated by bilateral adrenalectomy for bilateral adrenocortical hyperplasia developed an ACTH-secreting pituitary adenoma. In this study, pituitary tumors appeared between 6 months and 16 years after adrenalectomy. Prior pituitary irradiation does not reduce the frequency of these tumors.

e) TSH-Secreting Adenomas

Secondary hyperthyroidism with a TSH-secreting adenoma is relatively rare.

f) Polyadenomatosis

In polyadenomatosis (Werner's syndrome), adenomas of the adenohypophysis are associated with other endocrine adenomas.

III. Radiology

1) Nonsecreting (Chromophobe) Adenomas

Chromophobe adenomas are often discovered late, when they have already attained a large size. Obvious changes in the sella turcica are therefore almost invariably present (Ross and Greitz 1966; Pribram and du Boulay 1971; Taveras and Wood 1976) (Fig. 6.11). The sella turcica is generally enlarged in all diameters, i.e., ballooned. The cortical bone is completely thinned and the dorsum shows extensive demineralization and backward displacement (Figs. 6.14, 6.15). The posterior clinoid processes may be "rubbed out." Anterior expansion of the tumor tends to flatten the tuberculum sellae. The anterior clinoid processes may be elevated and tapered, indicating lateral expansion of the adenoma. On the frontal projection the floor of the sella appears asymmetric, tilted to one side, or is invisible. The frontal projection may also demonstrate erosion of a superior orbital fissure by anterolateral extension (Fig. 6.12).

Discovery of a small nonsecreting adenoma is less common. The appearance of the sella turcica on the lateral projection suggests regular expansion of its contents (Fig. 6.13).

Nonsecreting adenomas generally have a definite suprasellar expansion which may be large, and can be shown on sophisticated investigations such as CT, pneumography, or angiotomography (Figs. 6.14, 6.15).

Calcification is found within adenomas in about 4% of cases (Ozonoff and Burrows 1971). There is curvilinear capsular calcification, which often resembles that seen in craniopharyngiomas. Less commonly, calcification is floccular or granular. Calcification developing after radiotherapy may represent the result of hemorrhage within the adenoma. In other cases, the appearance of calcification may be an indication that the tumor has ceased to grow. Calcification has been noted in all types of adenomas and does not therefore constitute a fundamental element in diagnosis.

2) Prolactin-Secreting Pituitary Adenomas

A large number of adenomas, previously considered to be chromophobe, were actually prolactin-secreting pituitary adenomas. The appearances of these tumors when large do not therefore differ from those of chromophobe adenomas, i.e., demineralization of the cortical bone and changes in the dorsum sellae (Fig. 6.16). The appearance of the sella turcica never permits confident prediction of the presence or absence of a suprasellar extension, which may threaten the optic chiasm (Figs. 6.17, 6.22).

At an earlier stage, unilateral enlargement of the adenoma gives the appearance of a double floor. The volume of one-half of the fossa is normal (Fig. 6.18). The floor of the sella is clearly tilted on the frontal projection (Fig. 6.19; see also Fig. 6.20).

When the prolactin-secreting pituitary adenoma is still smaller, its appearance is usually quite characteristic (Vezina and Sutton 1974, 1978; Metzger et al. 1975). Typically, it produces a lateral indentation of the anteroinferior part of the floor (Fig. 6.21) or, less commonly, the anterior wall of the sella (Fig. 6.22). In 20% of cases, these changes are located in the posteroinferior part of the pituitary fossa (Fig. 6.26); absence of lateralization of the adenoma is exceptional. The change in the floor may be influenced by the septa of the sphenoid sinus (Fig. 6.23). Although the anteroinferior depression of the floor is very small, even questionable, frontal and sagittal millimetric polytomograms often show, only on one or two contiguous sections, a thinning of the cortical bone, confirming the diagnosis (Figs. 6.24, 6.26).

In a few cases, plain films may appear strictly normal, without a double floor in the lateral projection. Direct magnification radiographs and tomograms constitute, therefore, the only means of showing a microadenoma at a very early stage, when it is causing only a localized demineralization of the floor, which is not yet depressed. Thus, it is sometimes difficult to make an absolute differentiation between pathologic thinning and physiologic thinness; correlation with clinical and laboratory data is, of course, of fundamental importance (Fig. 6.25).

3) Growth Hormone Secreting (Eosinophilic) Adenomas

Growth hormone hypersecretion usually produces a characteristic change in the general appearance of the skull (Steinbach et al. 1959): the vault is thick and this is accentuated, particularly in women, by hyperostosis frontalis interna. Hyperostosis of the external occipital protuberance, extensive pneumatization of the sinuses (less commonly of the mastoid), and elongation of the mandible, with prognathism and opening of the mandibular angle, are frequently present (Fig. 6.27). Changes in the region of the sella itself are generally characteristic of the presence of a growth hormone secreting adenoma (McLachlan et al. 1970). The size of the adenoma may vary from a few millimeters to several centimeters. The volume of the sella turcica is highly variable. Generally, however, the size of eosinophilic adenomas is less than that of chromophobe adenomas. Moreover, eosinophilic adenomas tend to extend downward and rarely have large suprasellar extension (Fig. 6.28). This preferential downward extension occurs even when there is extensive pneumatization of the sphenoid sinus below the floor of the sella.

Contrary to certain conventional opinions (Wiggli and Oberson 1975), hyperpneumatization of the sphenoid does not cause blistering of the planum sphenoidale during acromegaly.

Certain morphological characteristics of the sella turcica are particular to eosinophilic adenomas: the configuration of the enlarged sella is usually angular, square rather than actually ballooned. The tuberculum sellae may be very tapered whereas the anterior clinoid processes are often large and dumpy. The dorsum sellae is displaced backward to a variable extent, and appears straighted, but not demineralized. The cortical bone of the pituitary fossa may be demineralized, as with chromophobe adenomas (Fig. 6.29). However, with large eosinophilic adenomas, the cortical bone is sometimes increased in density (Taveras and Wood 1976) (Fig. 6.29). There would appear to be a balance between the cortical erosion due to extension of the tumor and the action of growth hormone.

Calcification is rarer than with chromophobe adenomas (Metzger et al. 1975). Calcium deposits, of granular of floccular type, have been compared to orange pips (Ozonoff and Burrows 1971). It is quite possible that calcifications appear or increase in number after radiotherapy (Fig. 6.30).

Nowadays, eosinophilic adenomas are often discovered when they are still small. There is no correlation between tumor size and the degree of the manifestations of acromegaly (Hardy et al. 1973). Eosinophilic microadenomas are located laterally in the anteroinferior part of the gland. They give rise to doubling of the floor of the sella, identical to that produced by the prolactin-secreting microadenoma when the cortical bone is thinned (Fig. 6.31).

Certain adenomas, even when small, tend to be invasive (Vezina and Maltais 1973). Thus, focal destruction of the floor of the sella, with herniation of the tumor into the sphenoid sinus (Figs. 6.32, 6.33) may be seen. An empty sella turcica may be found during radiologic investigation of acromegaly. This may be explained by a chance association between a primary empty sella turcica and a microadenoma (Zatz et al. 1969; Domingue et al. 1978), or more probably by an adenoma becoming partially or totally necrotic (Login and Santen 1975). In the latter case, conservation of the cortical bone of the sella above an extensively pneumatized sinus may suggest hypersecretion of growth hormone (Fig. 6.41).

4) Other Hypersecreting Adenomas

Basophilic adenomas found in Cushing's disease are often small enough not to deform the sella, and polytomograms may be strictly normal (Bricaire et al. 1974; McErlan and Doyle 1976; Bigos et al. 1977; Hardy 1979); in any case, changes are always minimal. According to Hardy, ACTH-secreting adenomas develop preferentially in the middle part of the gland (Fig. 6.34). Diagnosis of these adenomas is especially difficult since the thinning of the cortical bone is not easy to evaluate in the presence of the osteoporosis of Cushing's disease. Adenomas occurring after bilateral adrenalectomy (Nelson's syndrome) can grow very rapidly (Fig. 6.35) (Nelson et al. 1958).

In the rare cases of TSH and FSH-LH adenomas, which are usually quite small, the sella turcica does not show specific radiologic changes (Guinet and Tourniaire 1974; Trouillas et al. 1979).

IV. Development of Pituitary Adenomas

1) Spontaneous Progression of Adenomas

It is widely recognized that chromophobe adenomas develop more rapidly than eosinophilic adenomas, while prolactin-secreting microadenomas usually grow slowly unless their development is accelerated by exogenous or endogenous estrogens. Rapid growth of tumors accompanied by severe visual loss and even hemorrhage into the adenoma has been reported during pregnancy (Nelson et al. 1978), or after treatment by bromocriptine (Linquette et al. 1977). These serious complications may justify routine radiography of the sella turcica before prescribing ovulatory agents. Eosinophilic adenomas generally grow slowly and certain small adenomas of this type may regress spontaneously, giving a "pituitary calculus" (Taveras and Wood 1976) (Fig. 6.36).

The adenomas of Nelson's syndrome are known to grow rapidly. Whatever their cellular type, certain slow-growing pituitary adenomas may develop in a sella whose cortical bone is preserved.

2) Expansion

Expansion of the adenoma in individual cases depends on the opening of the sella and the pneumatization of the sphenoid sinus (Mahmoud 1958). In a sella having a wide upper opening, with pronounced sphenoid pneumatization, the adenoma would tend to develop upward into the suprasellar area. Conversely, in a closed sella with a poorly pneumatized sphenoid, the adenoma would expand downward. In fact, it appears likely that the extension of adenomas depends essentially on their enclosed or invasive character (Fig. 6.10) and on their histologic type. Eosinophilic and chromophobe adenomas are the most markedly invasive, while nonsecreting adenomas produce the largest tumors (Fig. 6.37). Cushing's observation that "the largest frontal tumors were chromophobe adenomas" is worthy of note (Fig. 6.38). It is not uncommon for chromophobe adenomas to extend into structures as far apart as the ethmoidal cells and orbit, the sphenoid sinus and nasopharynx, the cavernous sinus and middle cerebral fossa, and the upper part of the clivus and the petrous apex (Krueger and Unger 1966; Stargardter and Margolis 1972). In such cases, axial tomograms are very useful in defining the extent of the lesion precisely (Fig. 6.39). These extensions must be distinguished from the rare ectopic pituitary adenomas which may arise in the suprasellar region (Rothman et al. 1976), within the sphenoid (Erdheim 1903, and a personal case), or on the posterior surface of the nasopharynx (Russel and Rubinstein 1977).

3) Rapid Growth of Pituitary Adenomas

Rapid growth of the adenoma, extension to the neural structures of the cavernous sinus, and reactive hyperostosis of the adjacent skull base indicating osseous infiltration (Fig. 8.13) would constitute strong arguments in favor of malignant histology (Scott et al. 1974; Lee 1976). In addition, certain rare histologically malignant adenomas can cause intracranial dissemination (Ogilvy and Jakubowski 1973). Contrary to the opinion of certain authors, it does not appear that adenomas found in children or in adolescents are particularly agressive (Richmond and Wilson 1978).

4) Abscess Formation

In the development of some adenomas, abscess formation may occur (Rudwan 1977; Domingue and Wilson 1977), either through hematogenous dissemination or secondary to sinusitis. In the latter case the radiologic appearances of pituitary adenoma, with destruction of the cortical bone of the floor of the sella and homogeneous cloudiness of the sphenoid sinus, must suggest the diagnosis of a pituitary abscess (Fig. 6.40). Clinical findings of a severe infectious syndrome must be taken into account.

5) Spontaneous Necrosis of Pituitary Adenomas

Spontaneous necrosis of certain adenomas may occur, particularly in acromegaly (Fig. 6.41) (Login and Santen 1975). As a result of this necrosis, the sella turcica appears empty, but the radiologic picture is different from that of a primary empty sella. The appearances of a secondary empty sella turcica with necrosis of a pituitary adenoma obviously depend on the underlying adenoma (Fig. 6.42). In these cases, complementary investigations or surgical exploration are indispensable to the diagnosis (Fig. 6.43). A secondary empty sella turcica must be distinguished from the chance association of a microadenoma and a primary empty sella turcica (Fig. 6.8) (Sutton and Vezina 1974). Another cause of a secondary empty sella turcica is treatment, surgical or radiotherapeutic (Rådberg 1963; Unger and Roswit 1959). The area left vacant by removal or destruction of the adenoma may be filled by cerebrospinal fluid. In some cases, the anterior recesses of the third ventricle are pulled downward by the cavity in the sella (Fig. 6.44). The radiologic appearance of the secondary empty sella turcica again depends on its appearance before treatment.

6) Remodeling of the Sella Turcica After Treatment

After surgery or radiotherapy, remodeling, recalcification, or even condensation of the margins of the sella (Fig. 6.45) are frequent. Several authors have reported reduction in size of large prolactin-secreting adenomas with correction of visual field defects following treatment with bromocriptine; in these cases slight recalcification of the floor of the sella may occur (Corenblum 1978; Matsumura et al. 1979).

Transsphenoidal removal of pituitary tumors leads to radiologic changes; an operative bone defect may be present, sometimes filled with bone fragments (Fig. 6.46) (Rothman S.L.G. et al. 1976); clouding of the sphenoid sinus is usual.

Radiologic diagnosis of recurrence of a pituitary adenoma is difficult (Sheline 1971). It is based on demineralization of the margins of the sella, or sometimes on the shifting of surgical clips placed on the diaphragma sellae.

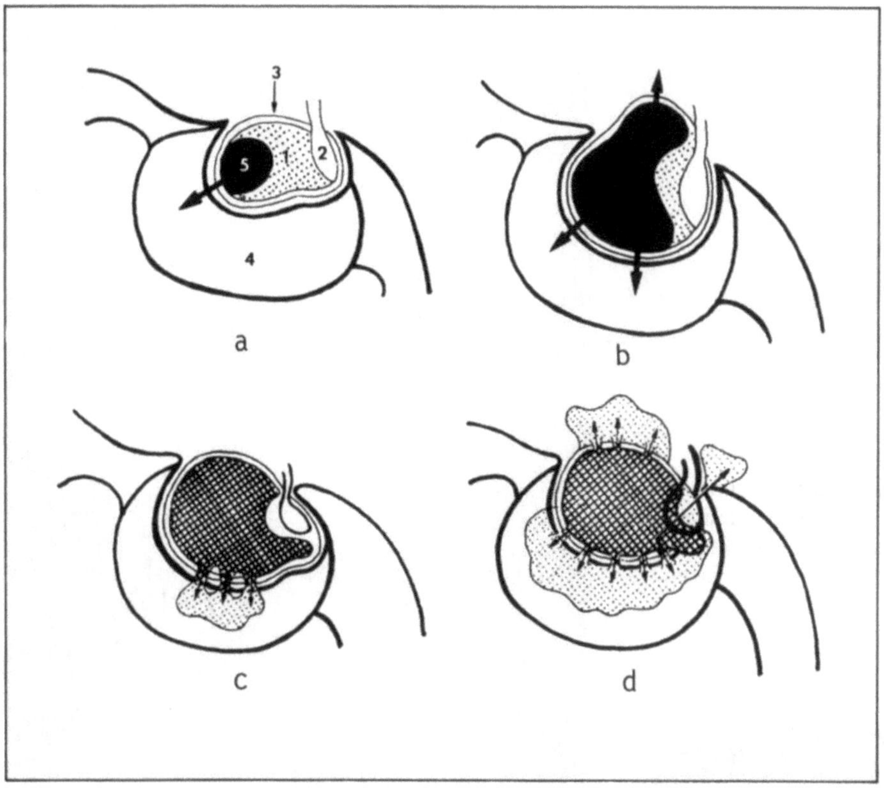

1 Adenohypophysis
2 Neurohypophysis
3 Diaphragma sellae
4 Sphenoid sinus
5 Adenoma.

Fig. 6.10a–d. Diagram of the macroscopic types of pituitary adenomas (modified from Hardy et al. 1973):

a Microadenoma: the diameter of the tumor is less than 10 mm. A fibrous capsule separates the tumor from the normal pituitary tissue. The diaphragma sellae is not displaced. The pituitary fossa is not enlarged. Opposite the microadenoma, the floor of the sella turcica bulges into the sphenoid sinus *(arrow)*.

b Circumscribed adenoma with intrasphenoid and suprasellar extension. The diaphragma sellae is displaced upward. The pituitary fossa is enlarged and the floor of the sella bulges into the sphenoid sinus *(arrows)*.

c Locally invasive adenoma: the dura mater and the floor of the sella turcica are invaded and the tumor extends into the sphenoid sinus *(arrows)*.

d Diffuse invasive adenoma: the tumor invades the dural and bony walls of the pituitary fossa *(arrows)*

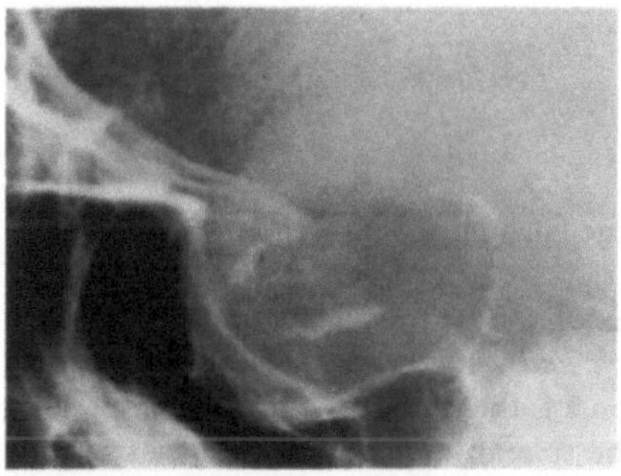

Fig. 6.11. Chromophobe adenoma. Enlarged sella turcica with marked thinning of its walls and posterior displacement of the dorsum. Note the overlying calcification of the carotid siphons

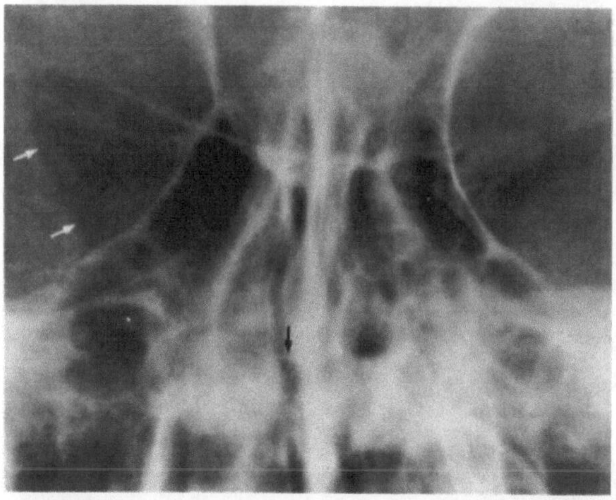

Fig. 6.12. Chromophobe adenoma. On the frontal projection the floor is depressed on the right and markedly demineralized *(black arrow)*. The enlarged superior orbital fissure reflects the anterolateral extension of the tumor *(white arrows)*

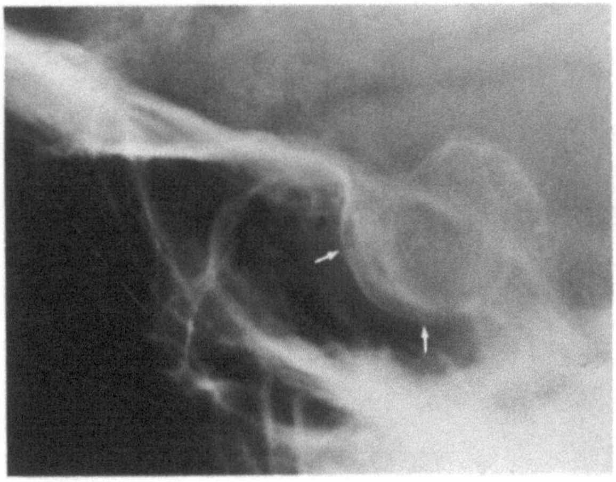

Fig. 6.13. Small intrasellar chromophobe adenoma. The regularly ballooned sella with a thinned lamina dura *(white arrows)* indicates an increase in the size of the contents of the pituitary fossa

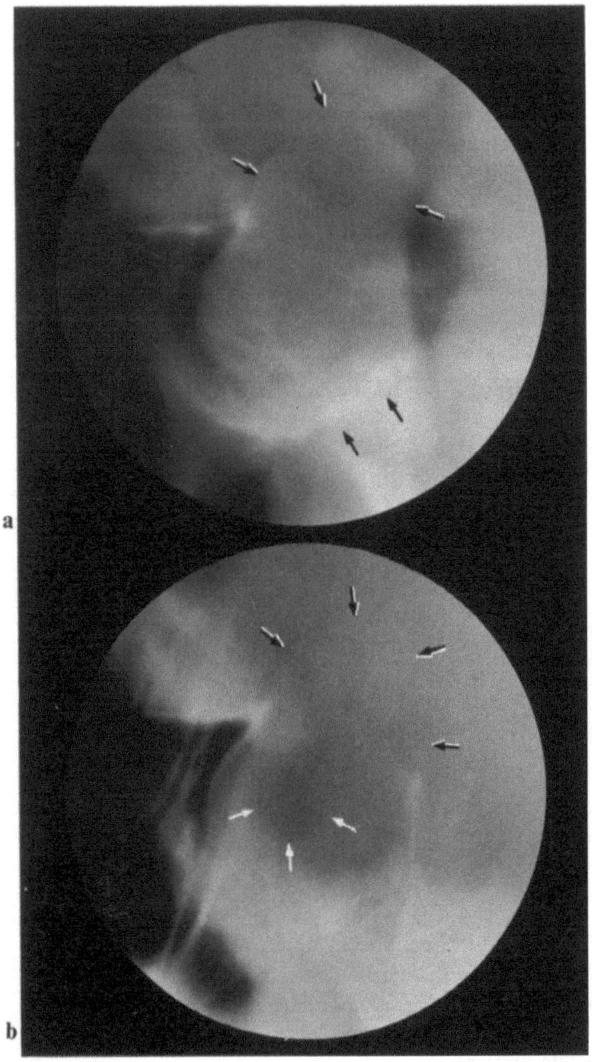

a

b

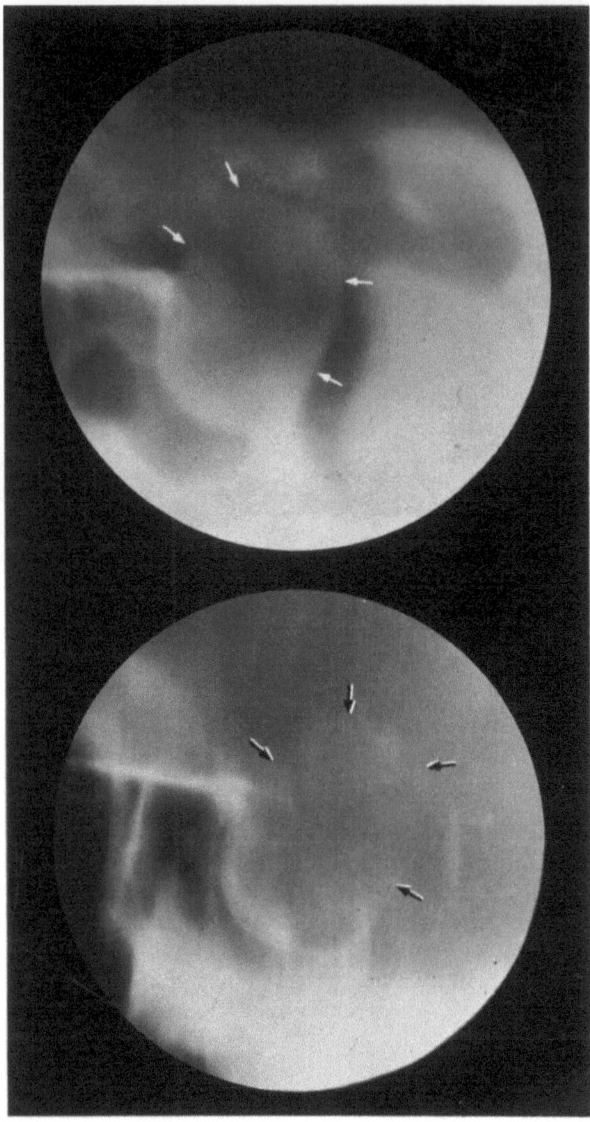

a

b

Fig. 6.14 a, b. Chromophobe adenoma.

a Pneumoencephalogram: suprasellar extension of the tumor *(arrows)*. Note destruction of the dorsum sellae.

b Sagittal angiotomogram: the suprasellar extension of the tumor is well-defined *(arrows)*

◁ **Fig. 6.15 a, b.** Large chromophobe adenoma in a 15-year-old boy.

a Pneumoencephalogram: intrasphenoid and suprasellar extension *(black and white arrows)*. Note the extreme demineralization of the walls of the sella, especially the dorsum. The spheno-occipital synchondrosis limits posteroinferior extension of the adenoma *(black arrows)*.

b Sagittal angiotomogram: faint suprasellar blush *(black and white arrows)* and intrasellar translucency corresponding to a cystic part of the tumor, confirmed at surgery *(white arrows)*

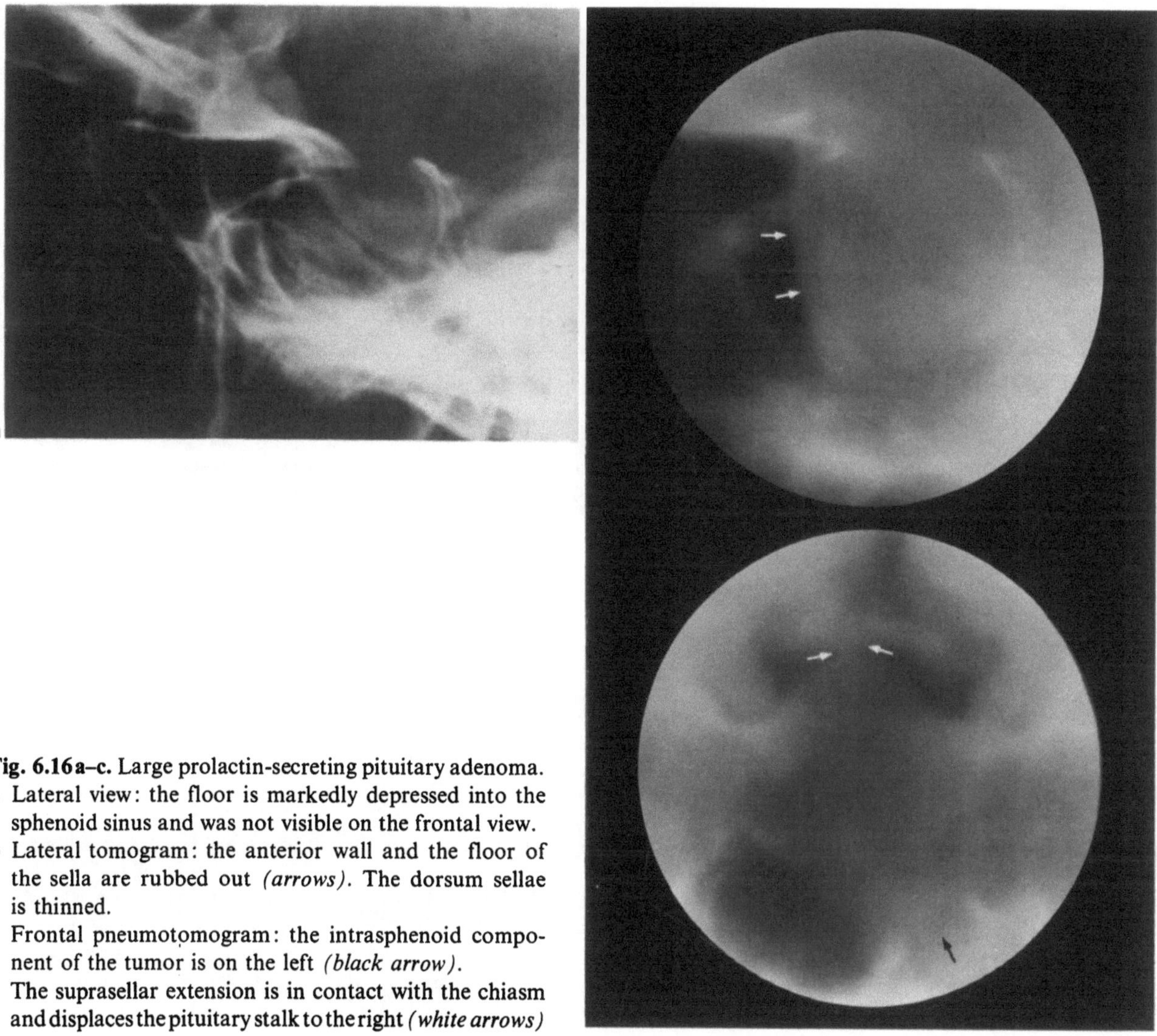

Fig. 6.16a–c. Large prolactin-secreting pituitary adenoma.

a Lateral view: the floor is markedly depressed into the sphenoid sinus and was not visible on the frontal view.

b Lateral tomogram: the anterior wall and the floor of the sella are rubbed out *(arrows)*. The dorsum sellae is thinned.

c Frontal pneumotomogram: the intrasphenoid component of the tumor is on the left *(black arrow)*. The suprasellar extension is in contact with the chiasm and displaces the pituitary stalk to the right *(white arrows)*

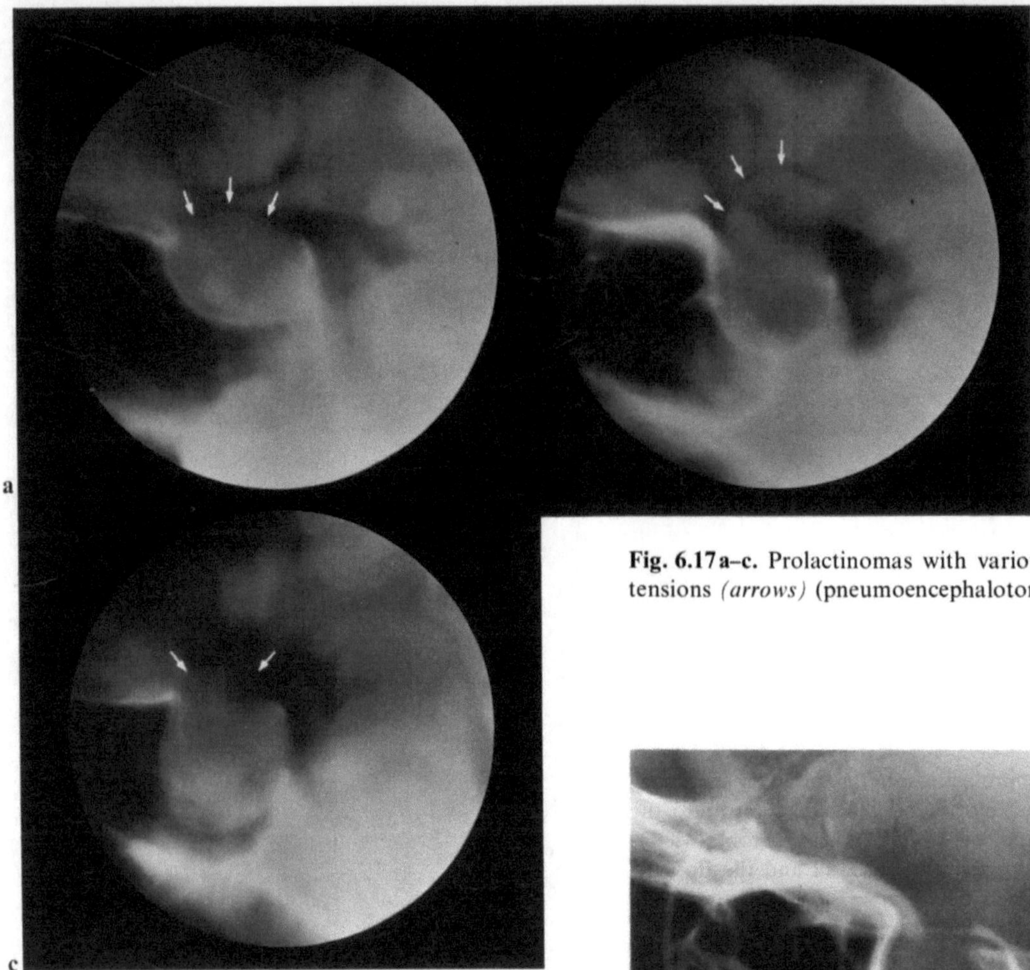

Fig. 6.17 a–c. Prolactinomas with various suprasellar extensions *(arrows)* (pneumoencephalotomogram)

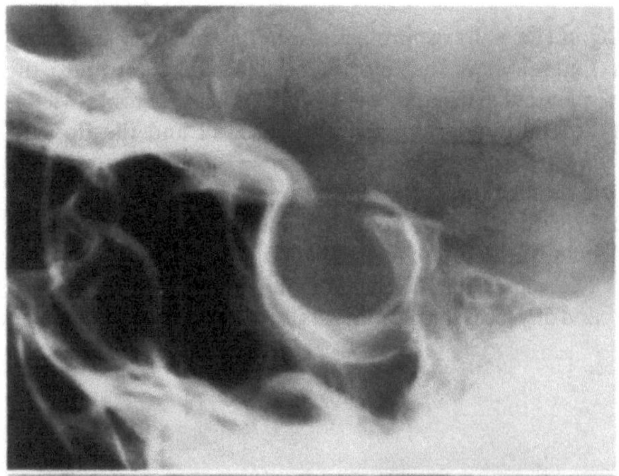

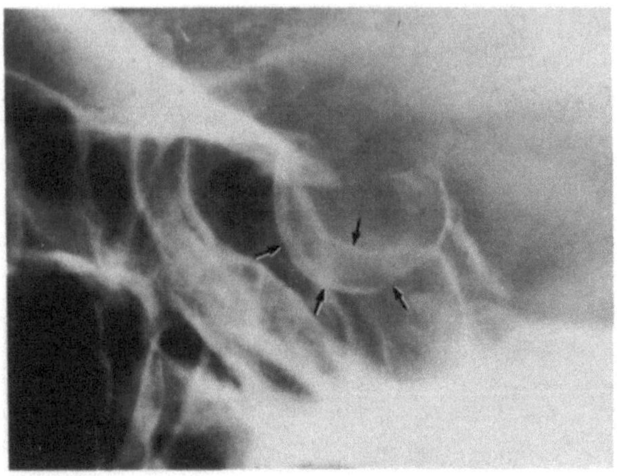

Fig. 6.18. Prolactinoma. Unilateral development. Half of the pituitary fossa is normal in size *(black arrow)*, the other half is concentrically enlarged and its lamina dura is demineralized *(black and white arrows)*

Fig. 6.19 a, b. Prolactinoma. Unilateral development.
a Lateral view: half of the pituitary fossa is normal in size.
b Frontal view: the floor is depressed on the left *(arrows)*

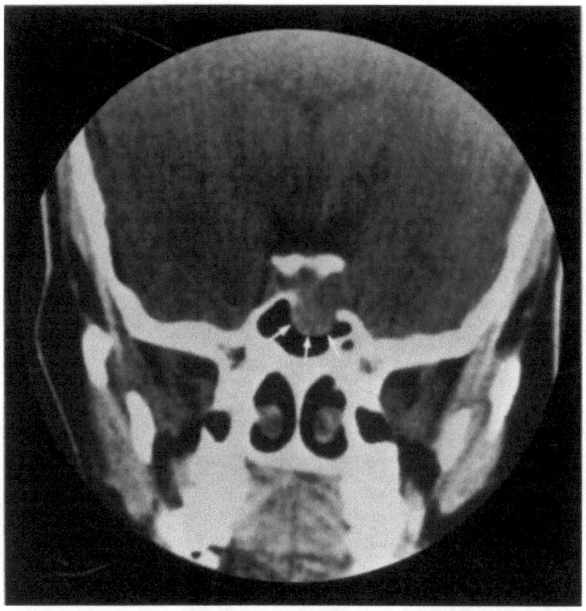

Fig. 6.20. Prolactinoma with extension to sphenoid sinus shown by CT with coronal section. Depression and thinning of the left side of the floor of the sella *(arrows)* (Courtesy, V. Haughton)

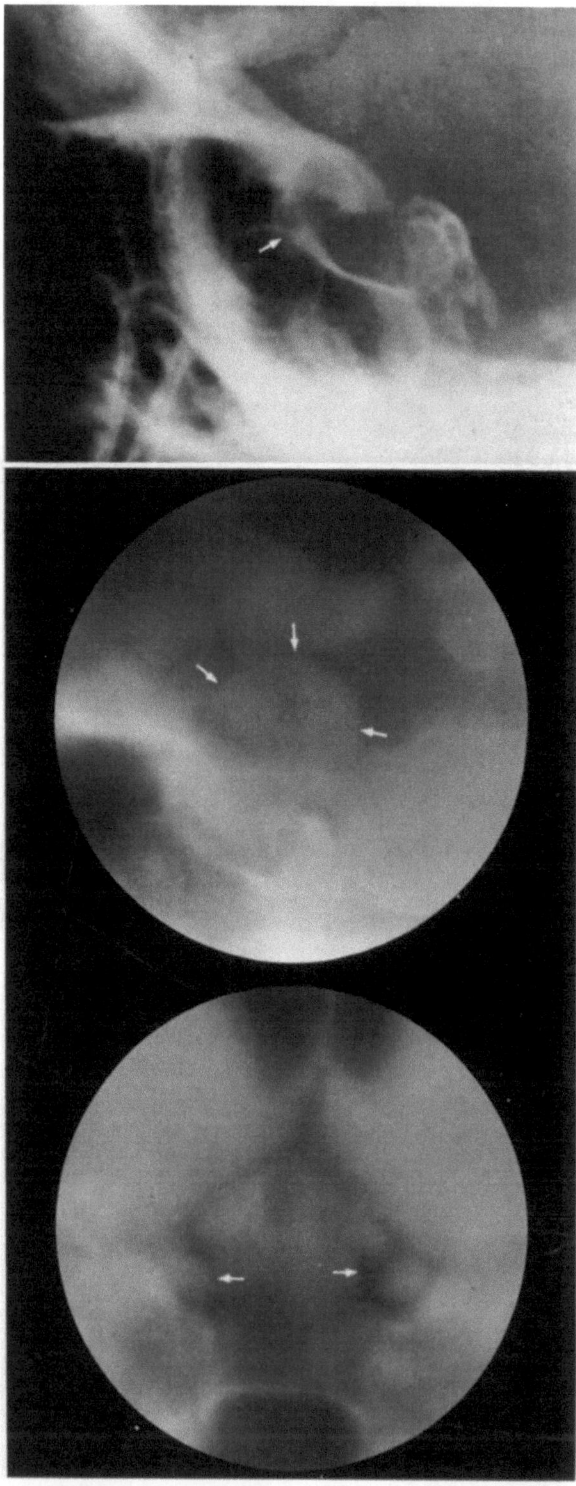

a

b

c

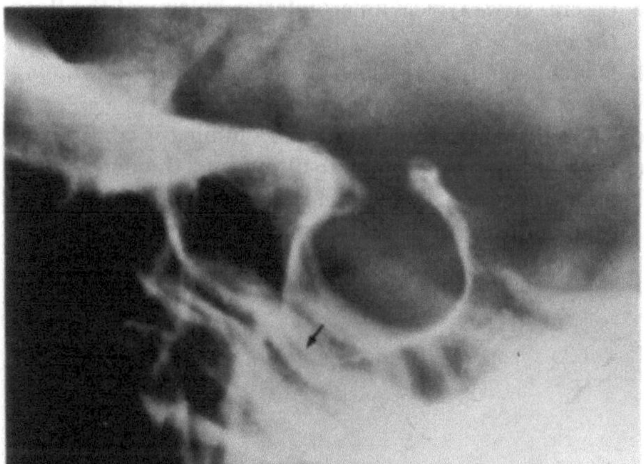

Fig. 6.21. Intrasellar prolactinoma. Enlarged pituitary fossa with bulging and thinning of the anteroinferior part of the floor *(arrow)*

Fig. 6.22 a–c. Prolactinoma.

a Lateral view: radiologic changes are confined to the anterior wall of the pituitary fossa *(arrow)* without erosion of the floor itself.

b, c Sagittal and frontal tomogram: large suprasellar extension *(arrows)* with compression of the chiasm. It is impossible to predict the degree of suprasellar extension from the bony changes

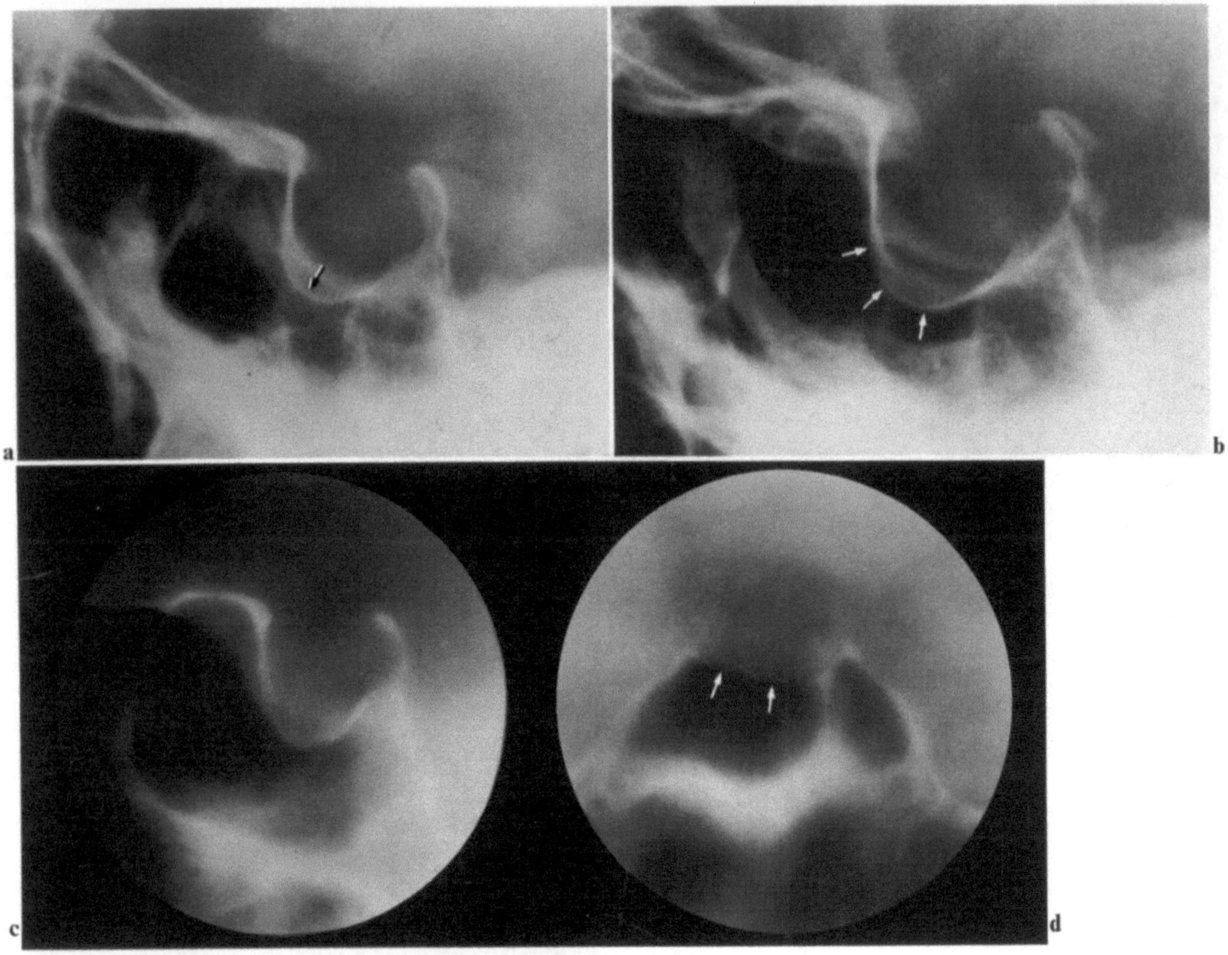

△
Fig. 6.23a–d. Intrasellar prolactinoma.

a Lateral view: typical anteroinferior bulging of the floor *(arrow)*.

b Direct magnification (microfocus): the cortical thinning is clearly demonstrated *(arrows)*.

c Sagittal tomogram: the cortical thinning is confirmed.

d Frontal tomogram: lateral tilting and thinning of the floor *(arrows)*. The intersinusal septum of the sphenoid sinus seems to limit extension of the adenoma

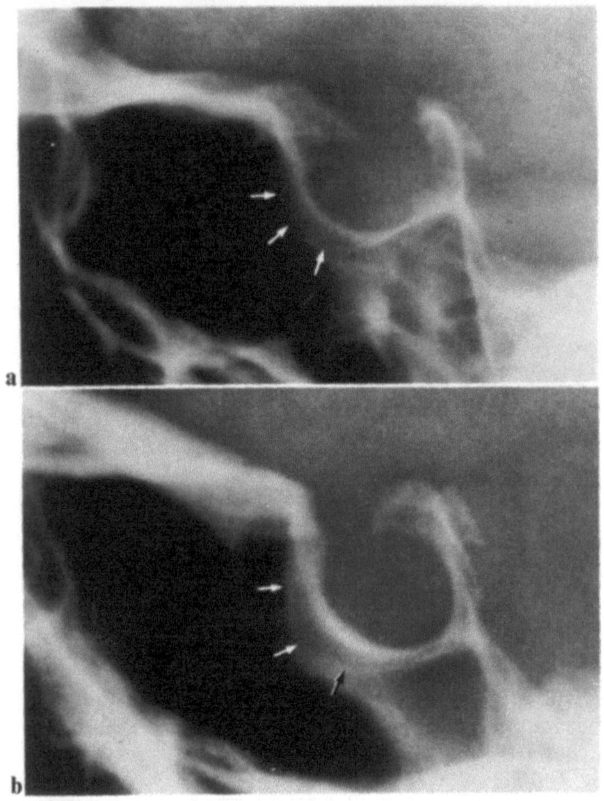

Fig. 6.24a, b. Prolactin-secreting microadenomas. Localized doubling of the anteroinferior part of the floor, with thinning of the lower portion *(arrows)*

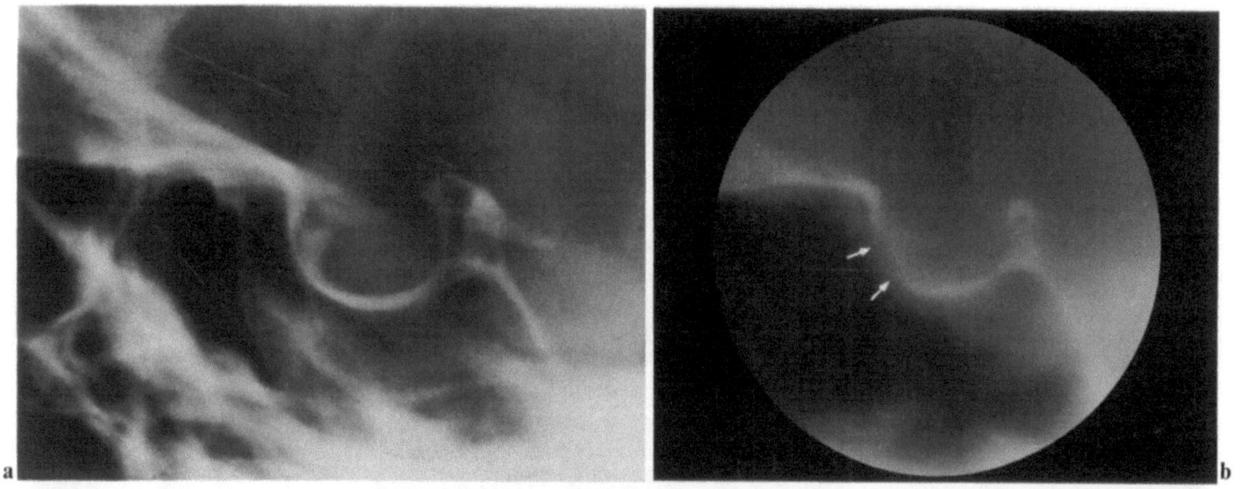

△

Fig. 6.25a, b. Prolactin-secreting microadenoma.

a Lateral view: the pituitary fossa seems normal.

b Lateral tomogram: hypocycloidal tomogram reveals minor thinning of the lamina dura *(arrows)*

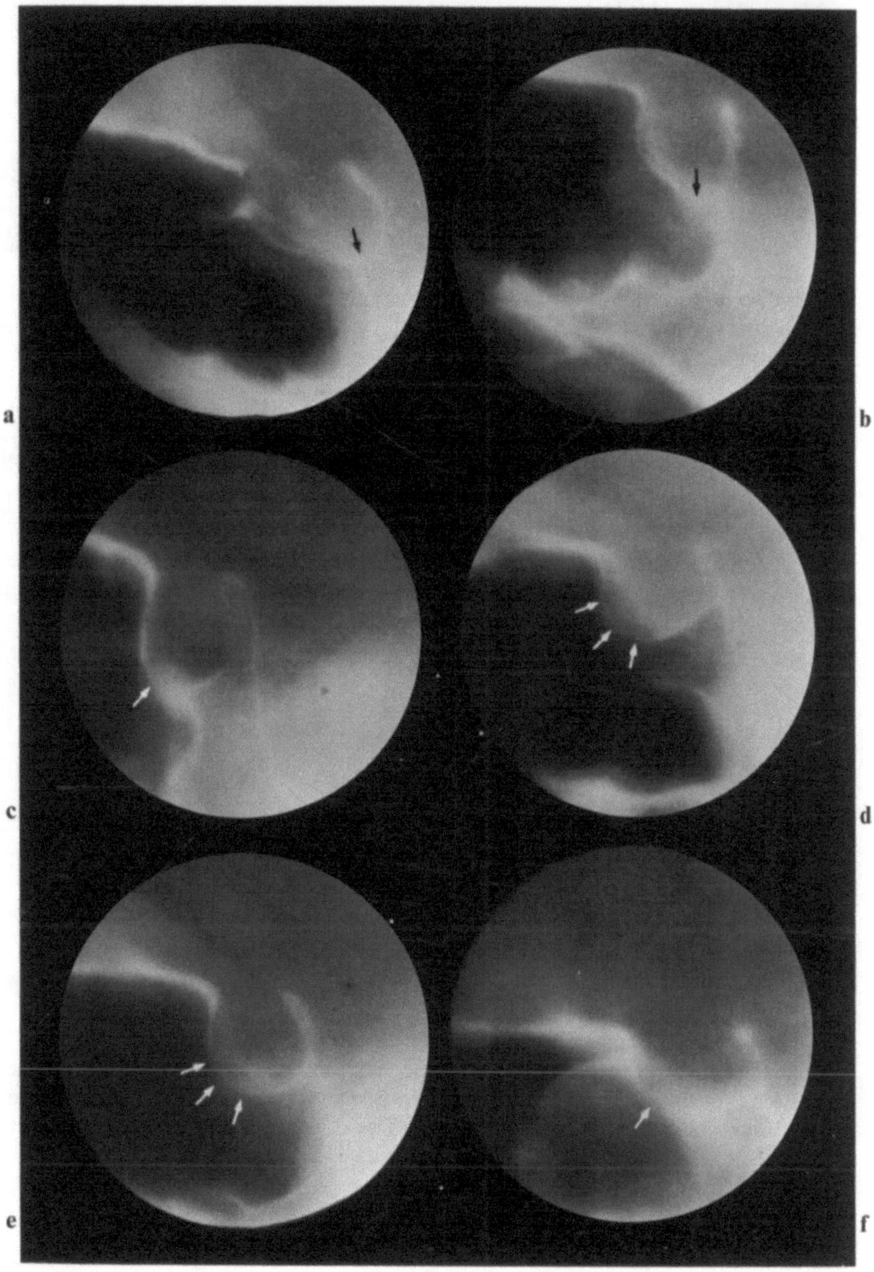

Fig. 6.26a–f. Changes in the sella turcica with prolactin-secreting microadenomas; polytomograms.

a, b Note the posteroinferior bone changes *(arrows)*.

c, d, e Anteroinferior cortical thinning *(arrows)* associated, in **f**, with sphenoid sinusitis

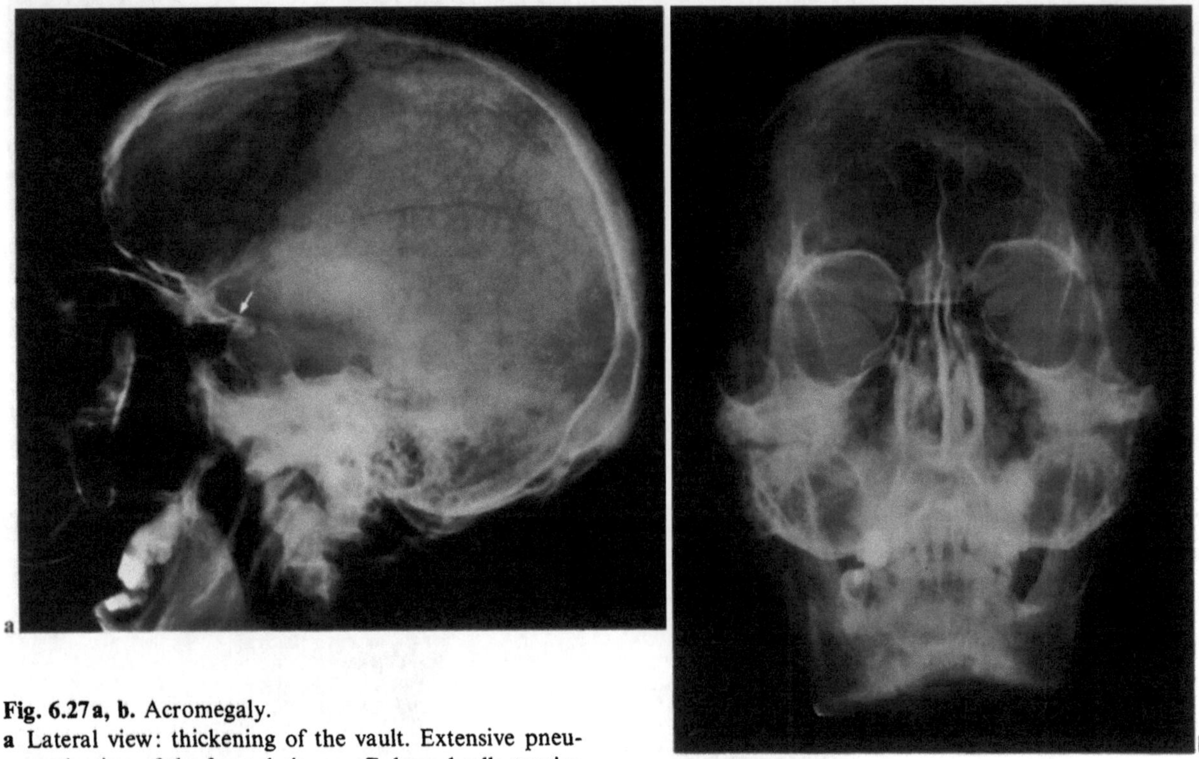

Fig. 6.27a, b. Acromegaly.
a Lateral view: thickening of the vault. Extensive pneumatization of the frontal sinuses. Enlarged sella turcica with demineralization of its walls. The tuberculum sellae is pointed *(arrow)*.
b Frontal view: the demineralized floor is invisible. Note the elongated mandible

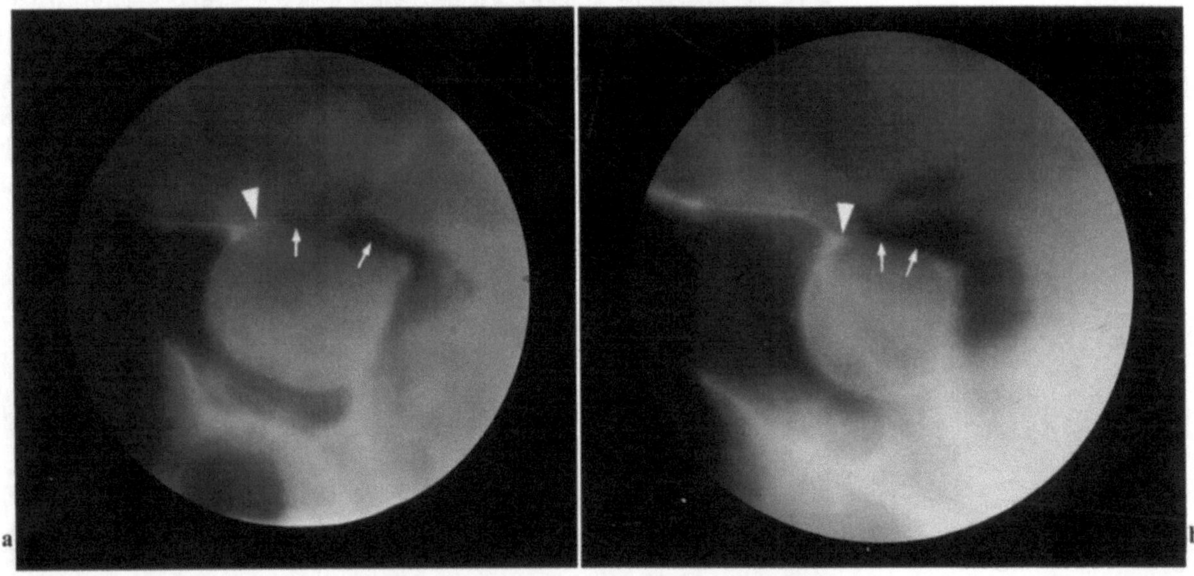

Fig. 6.28a, b. Eosinophilic adenomas shown by midsagittal pneumotomograms. Note in each case the tapered appearance of the tuberculum sellae *(arrowhead)*; the cortical bone of the floor is well preserved. The dorsum sellae is straightened. The expansion is mainly intrasphenoid. Diaphragma sellae *(arrows)*. Compare with the pneumographic appearance of chromophobe adenomas (Fig. 6.14) and prolactinoma (Fig. 6.17)

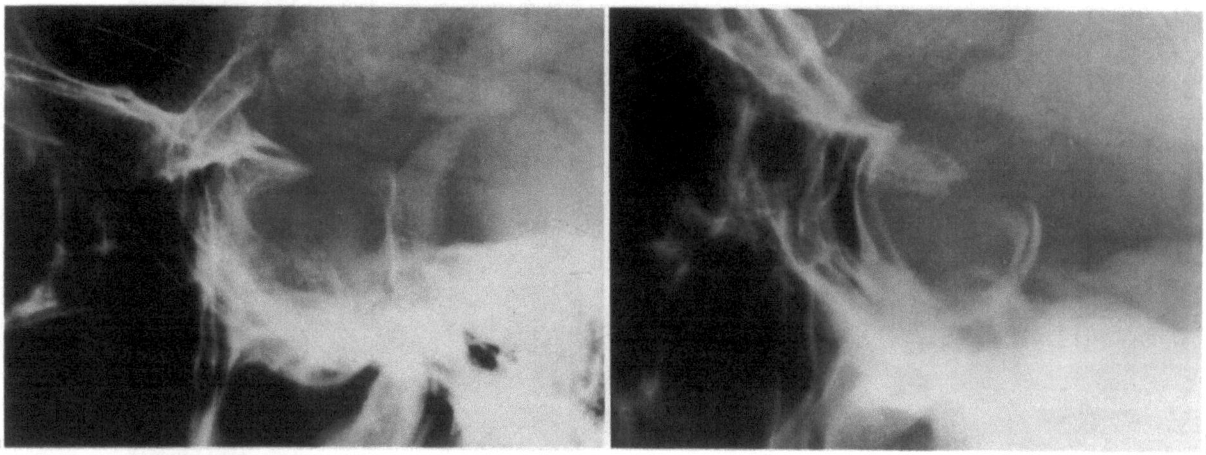

a b

Fig. 6.29 a, b. Large eosinophilic adenomas.
a The sphenoid sinus is invaded. Note the straight appearance of the dorsum sellae.
b Massive anterior clinoid processes. Note the dense appearance of the cortical bone of the pituitary fossa

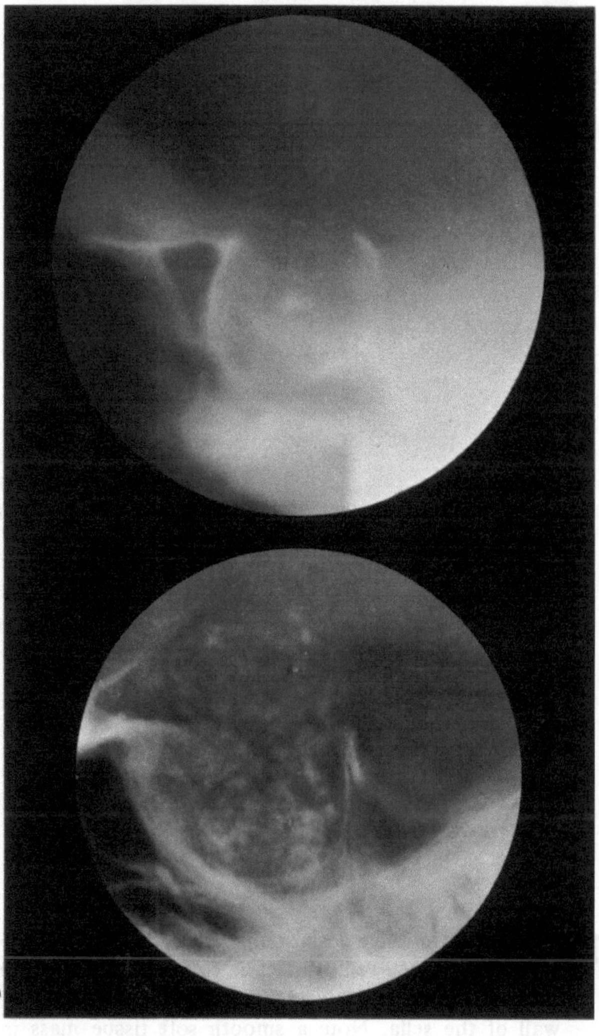

a b

Fig. 6.30 a, b. Calcified eosinophilic adenomas.
a Nodular tumor calcification.
b Numerous floccular intra- and suprasellar calcifications following radiotherapy
(Courtesy, J. Metzger)

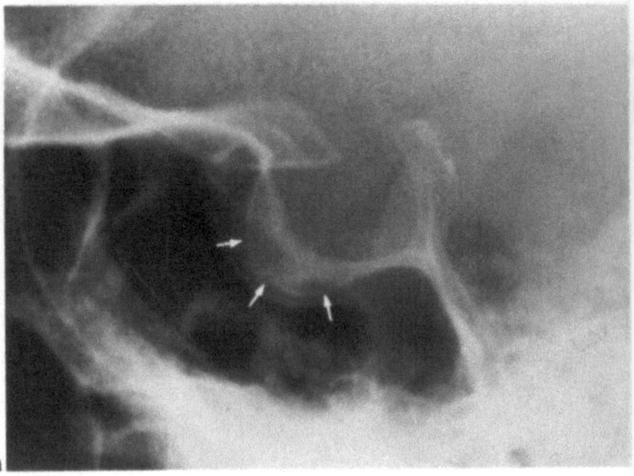

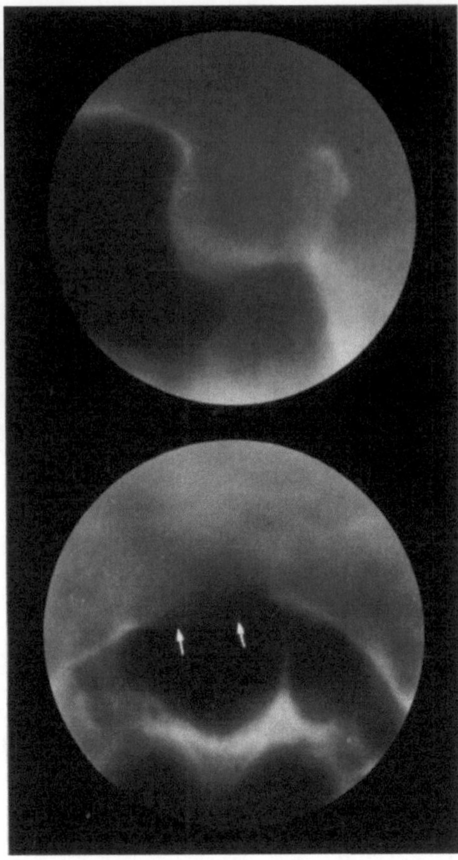

Fig. 6.31 a–c. Eosinophilic microadenoma.
a Lateral direct magnified view: anterior double floor with bulging of the anterior wall *(arrows)*.
b Lateral tomogram: anteroinferior depression and thinning of the floor of the sella. Note the straightened dorsum sellae.
c Frontal tomogram: the markedly thinned floor is slanted toward the right *(arrows)*

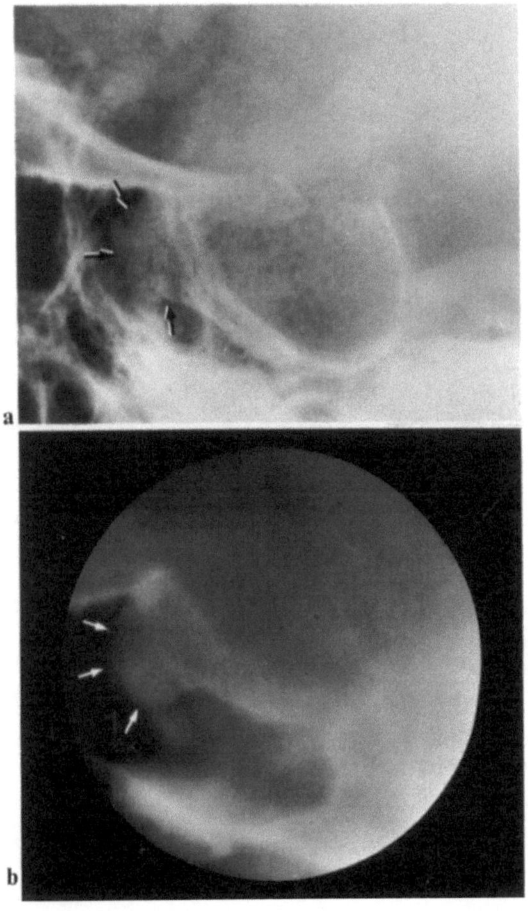

◁ **Fig. 6.32 a, b.** Invasive eosinophilic adenoma.
a Lateral view; the sella turcica is enlarged: the dorsum sellae is thinned; localized destruction of the anterior wall of the sella. Note a smooth soft tissue mass in the sphenoid sinus opposite the bone erosion *(arrows)*.
b Midsagittal tomogram; the soft tissue mass within the sphenoid sinus, representing the intrasphenoid expansion of the adenoma is well demonstrated *(arrows)*

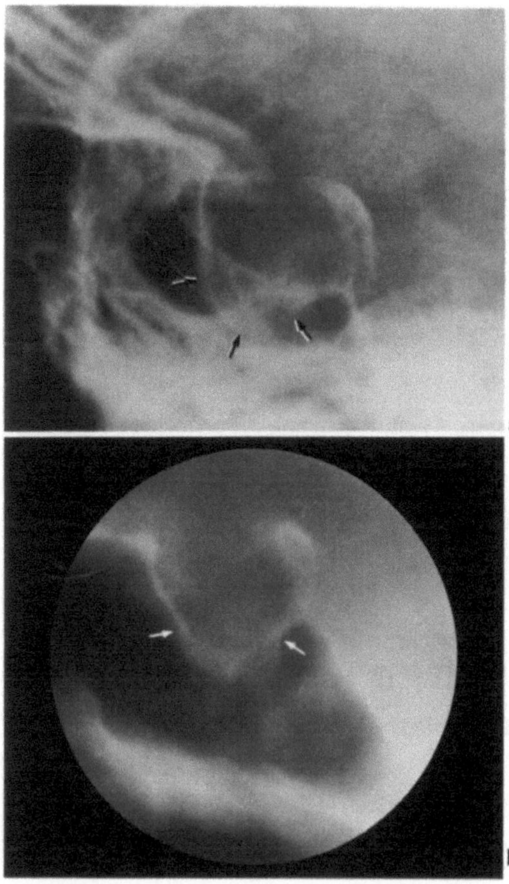

a

b

Fig. 6.33a, b. Small eosinophilic adenoma.
a Lateral view: anteroinferior double floor *(arrows)*.
b Midsagittal tomogram demonstrates a small invasive adenoma. Note the localized abrupt break in the lamina dura and the soft tissue mass within the sphenoid sinus opposite the cortical bone *(arrows)*

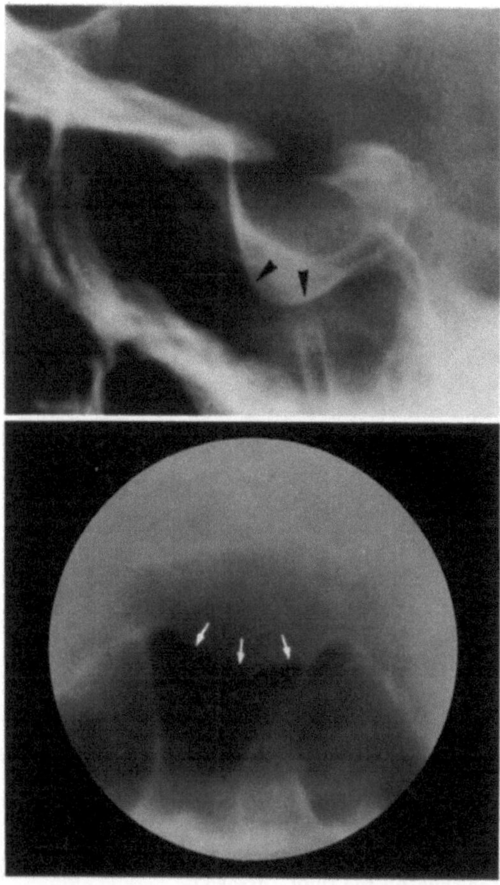

a

b

Fig. 6.34a, b. ACTH-secreting adenoma (basophilic adenoma).
a Lateral view: bulging and thinning of the anteroinferior part of the floor of the sella *(arrowheads)*.
b Frontal tomogram: symmetric depression and thinning of the floor *(arrows)*

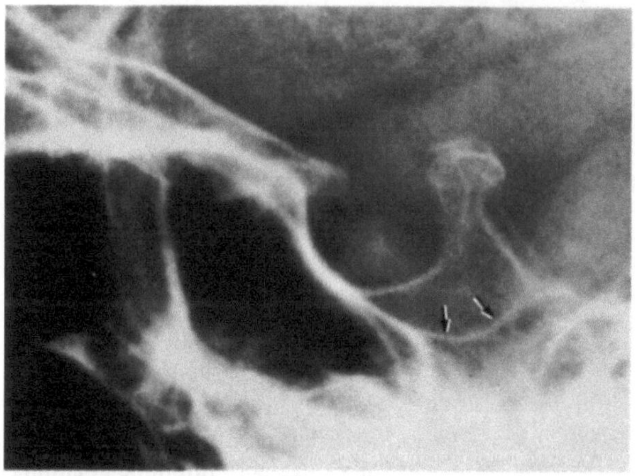

Fig. 6.35. Nelson's syndrome (ACTH-secreting adenoma). The lateral radiograph demonstrates a marked double floor posteroinferiorly *(arrows)* (the sella turcica was normal before adrenalectomy)

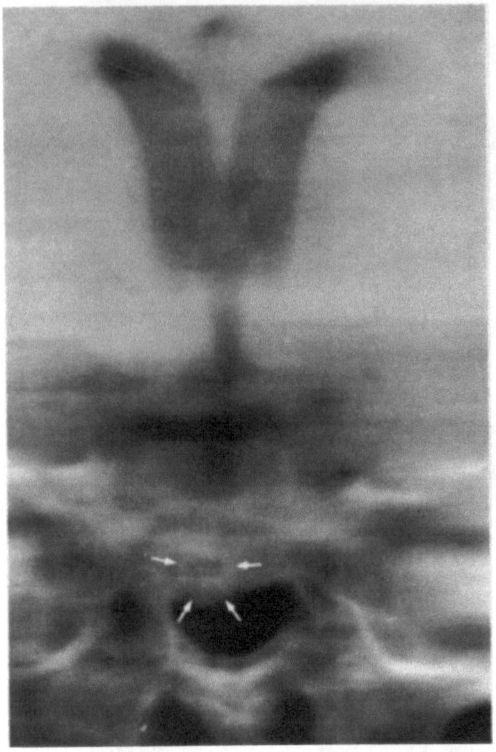

Fig. 6.36. Pituitary calculus. In a spontaneously healed acromegalic patient, the frontal linear pneumotomogram shows extensive intrasellar calcification depressing the floor of the sella on the right *(arrows)*

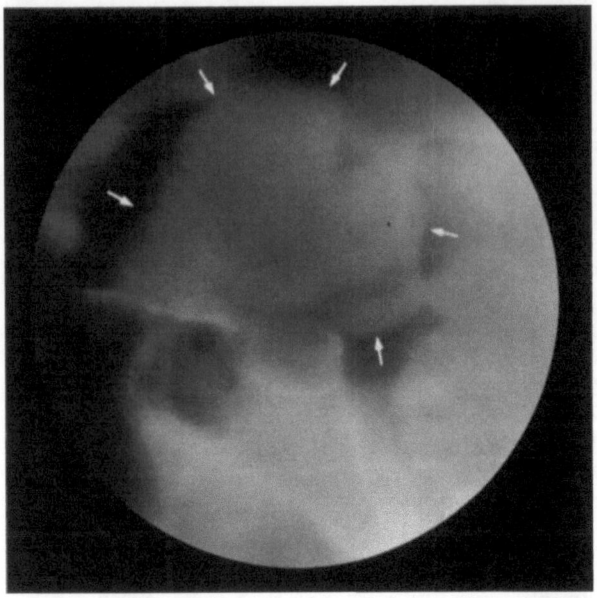

Fig. 6.37. Huge chromophobe adenoma with suprasellar and subfrontal extension *(arrows)*. The lateral tomogram shows an open sella turcica. Note the absence of intrasphenoid extension

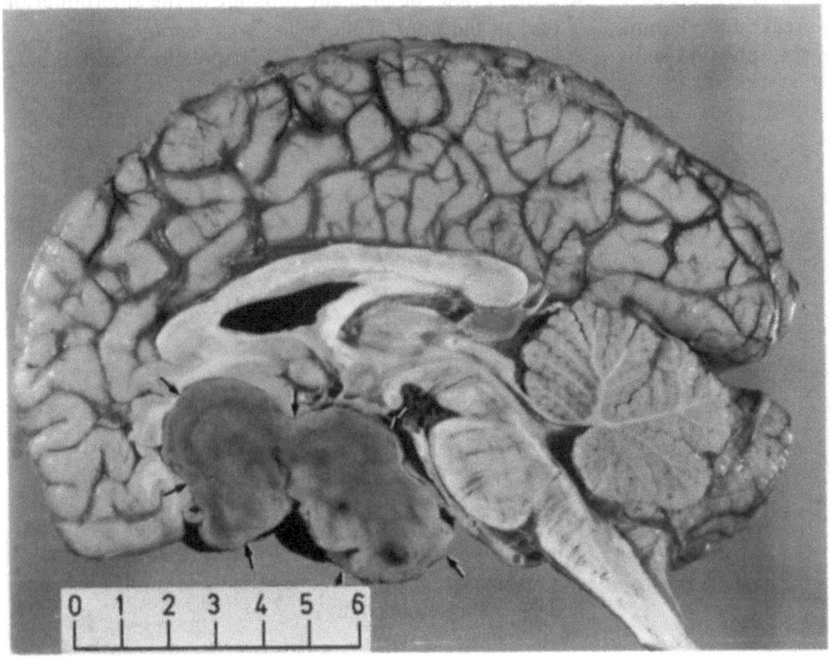

Fig. 6.38. Midsagittal section of brain; autopsy specimen. Suprasellar and subfrontal extension of a huge chromophobe adenoma *(arrows)* (Courtesy, N. Heldt)

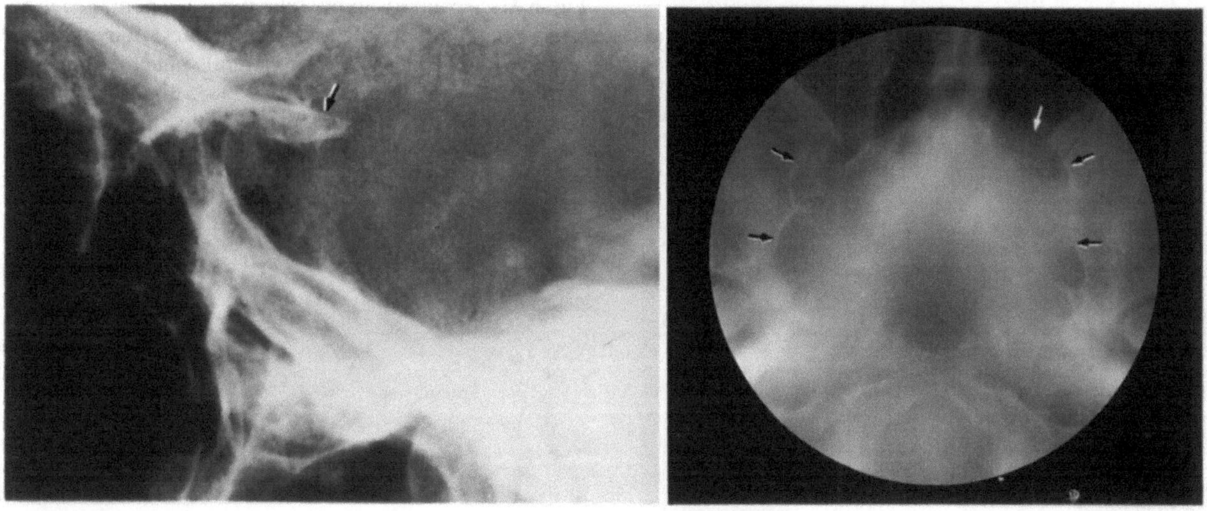

Fig. 6.39 a, b. Giant invasive chromophobe adenoma.
a Lateral view: ghost sella turcica; one of the anterior clinoid processes remains visible *(arrow)*.
b Axial tomography: Marked anterior and lateral extension of the tumor invading the anterior ethmoid cells, widening the superior orbital fissure *(white arrow)*, and destroying the medial part of both middle cranial fossae *(black and white arrows)*

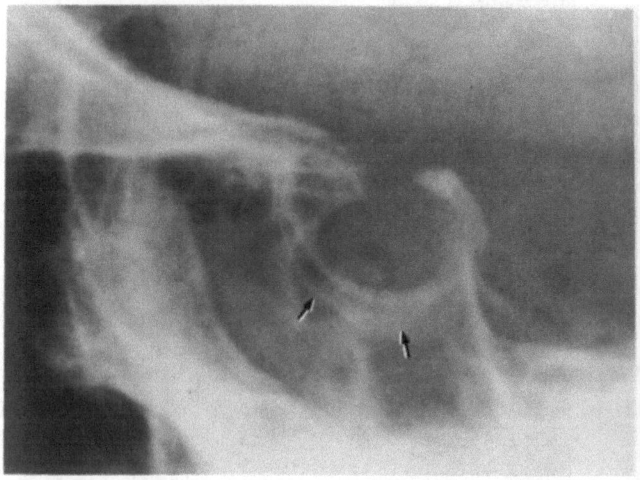

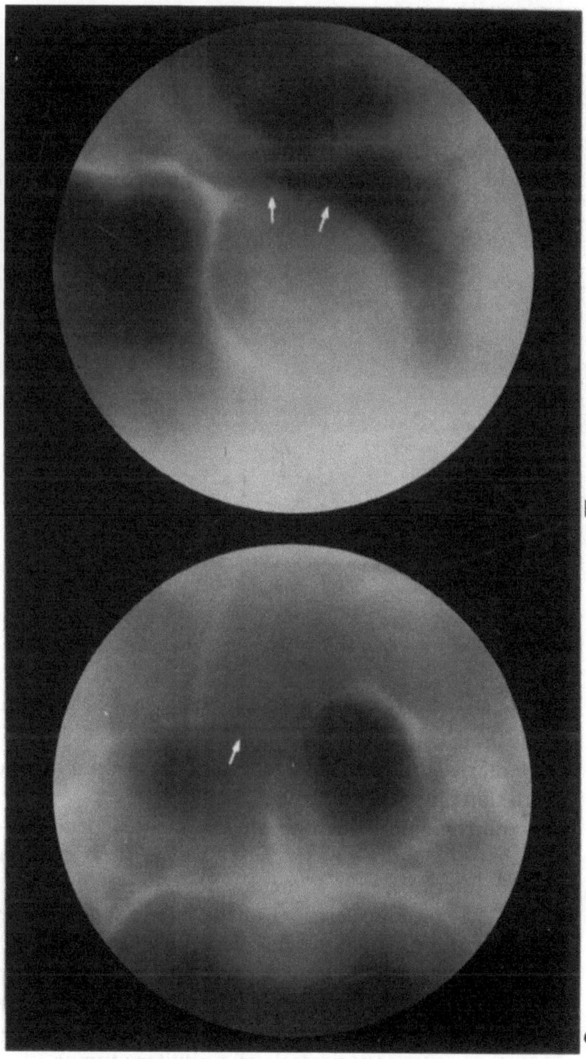

Fig. 6.40 a–c. Pituitary abscess developing within a chromophobe adenoma (meningitislike symptoms and panhypopituitarism).
a Lateral view: slight double floor with thinning of the cortical bone *(arrows)*; clouded sphenoid sinus.
b Lateral pneumotomogram: the dorsum sellae is demineralized; clouding of the sphenoid sinus; upward bulging of the diaphragma sellae *(arrows)*.
c Frontal tomogram: cloudiness of the sphenoid sinus and defect in the posterior part of the floor of the sella *(arrow)*

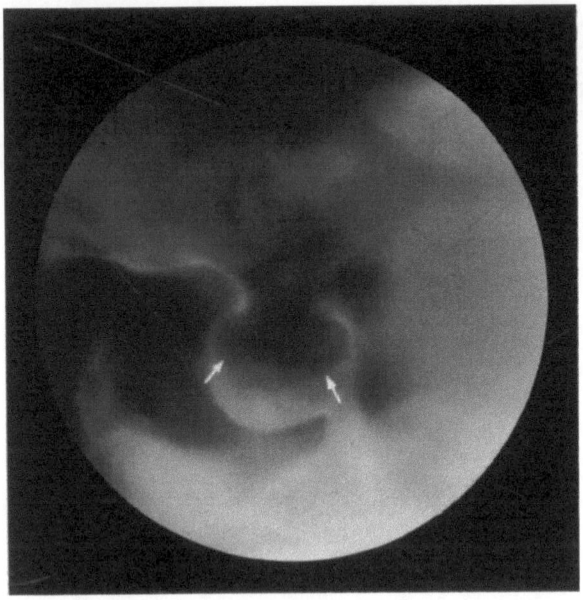

Fig. 6.41. Secondary empty sella turcica. Spontaneous necrosis of an eosinophilic adenoma; clinical and endocrine picture of acromegaly. The lateral pneumotomogram demonstrates an empty sella *(arrows)*. Slight regular enlargement of the pituitary fossa. Note the well-aerated sphenoid sinus and the preserved lamina dura of the floor

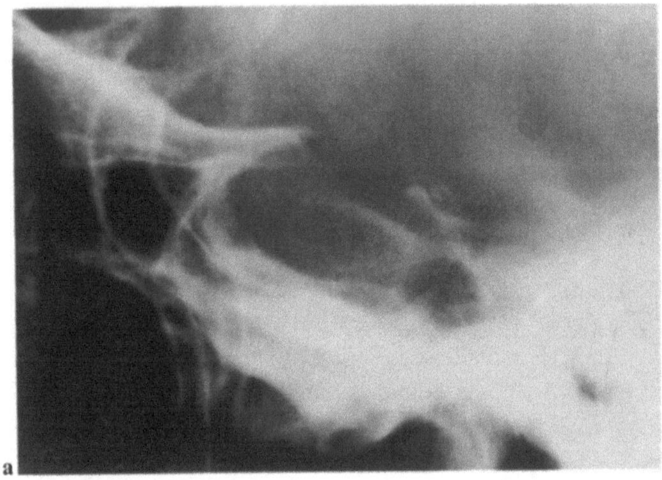

a

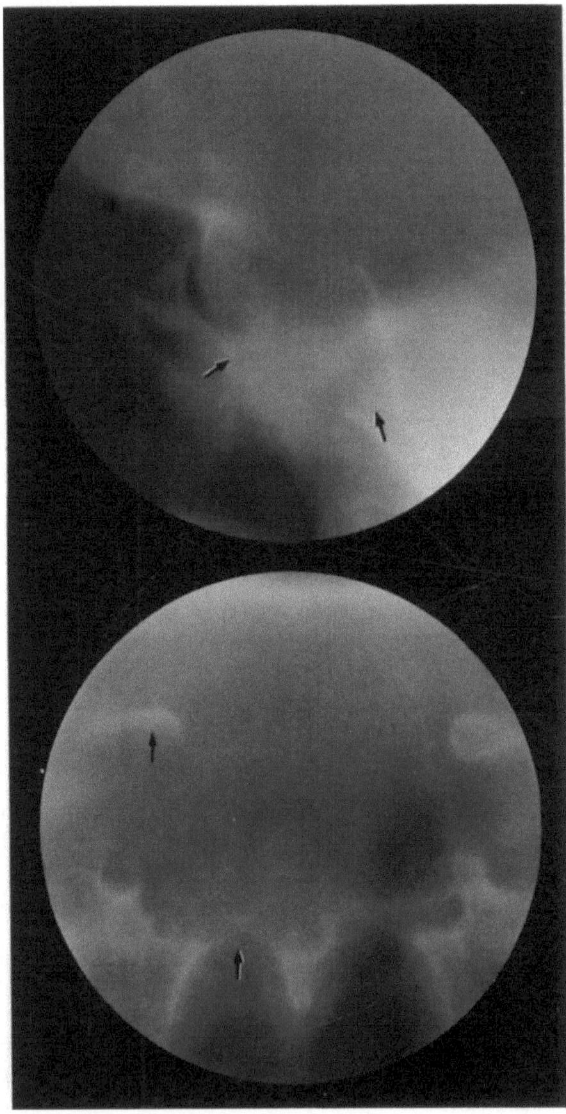

b

c

Fig. 6.42a–c. Secondary empty sella. Necrosis of a pituitary adenoma. Recurrent meningitis.

a Lateral view: marked enlargement of the pituitary fossa with thinning of the dorsum.

b Lateral tomogram: depression and marked demineralization of the floor of the sella *(arrows)* and erosion of the anterior aspect of the dorsum sellae.

c Frontal tomogram: erosion of the inferior wall of the sphenoid sinus on the right *(black and white arrow)* and erosion of the inferior aspect of the right anterior clinoid process *(black arrow)*. This appearance strongly suggests that the empty sella is not primary, but secondary, to spontaneous necrosis of a pituitary tumor

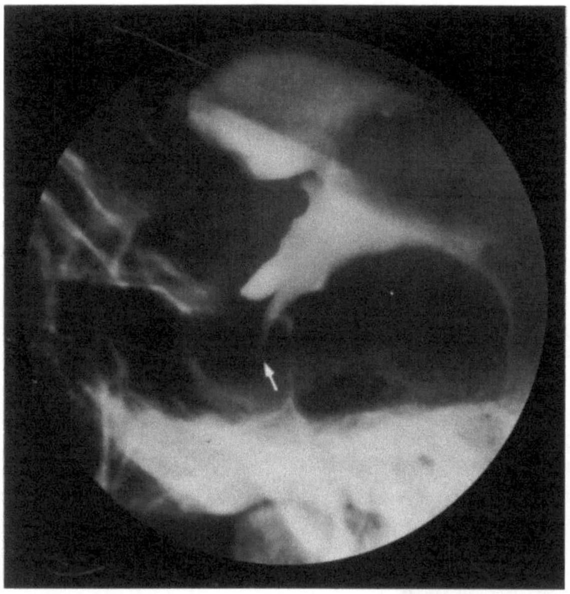

Fig. 6.43. Probable necrosis of a pituitary tumor. Clinical picture of hypopituitarism. Enlargement of the pituitary fossa and marked thinning of its walls, simulating chromophobe adenoma. Metrizamide ventriculography demonstrates an empty sella turcica; note the downward intrasellar displacement of the infundibular recess of the third ventricle *(arrow)*

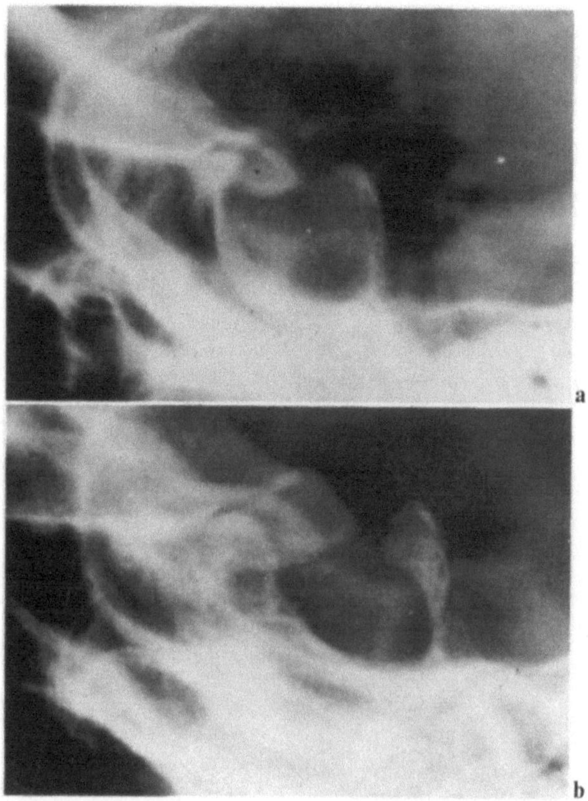

a

b

Fig. 6.45 a, b. Prolactinoma.

a Preoperative lateral projection: slight enlargement of the pituitary fossa and demineralization of its margins.

b Postoperative lateral projection (1 year after surgery): remodeling of the sella turcica, which decreased in size: remineralization of the floor and dorsum: note the uneven anterior wall and the cloudiness of the sphenoid sinus, related to transsphenoidal surgery

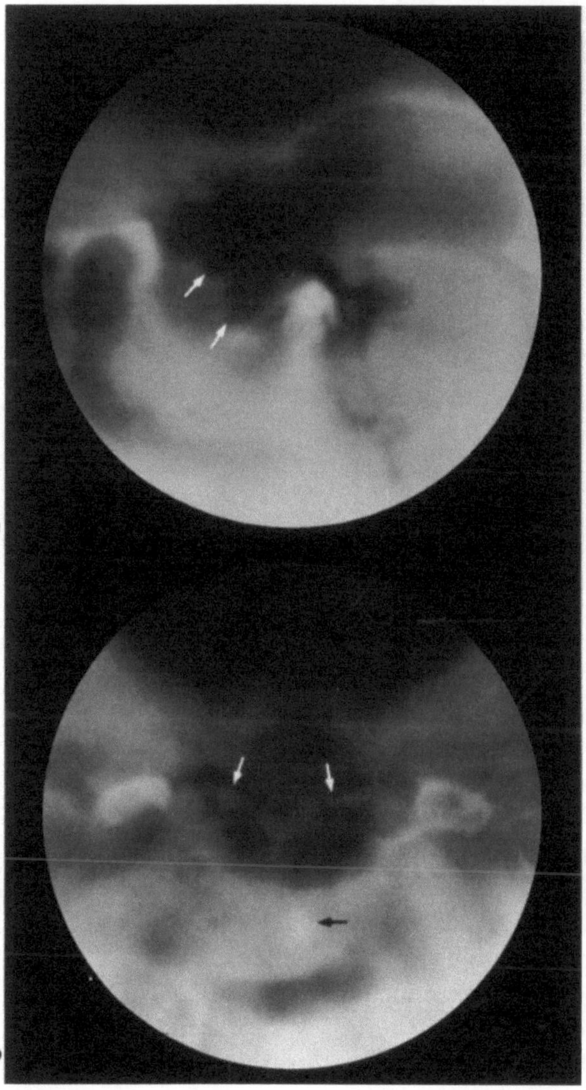

a

b

Fig. 6.44 a, b. Empty sella turcica after transsphenoidal removal of a pituitary adenoma.

a Lateral pneumotomogram: the dilated anterior recesses of the third ventricle lie within the sella turcica *(arrows)*.

b Frontal tomogram: the optic chiasm is deformed and displaced downward *(white arrows)*. Note amorphous calcification within the sella turcica, related to surgery *(black arrow)*

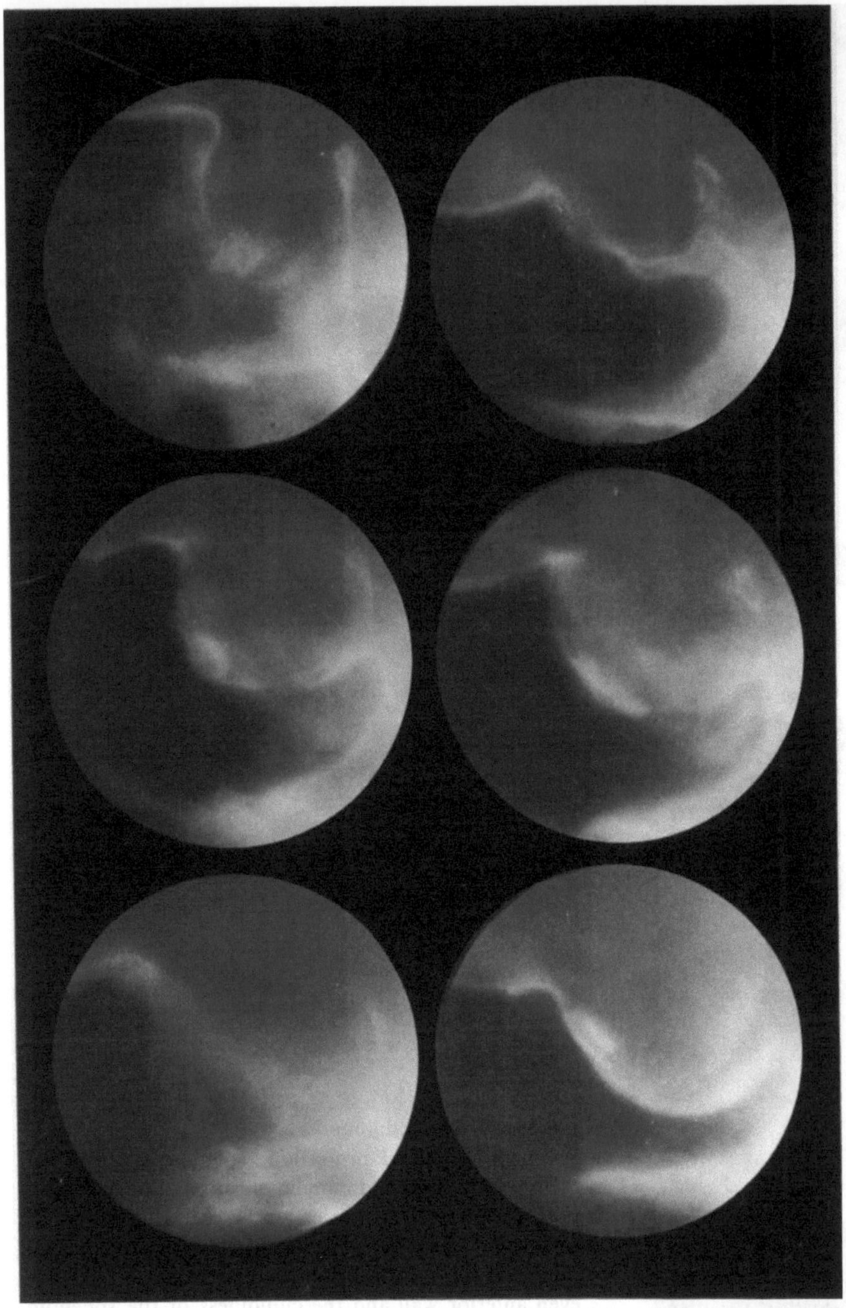

Fig. 6.46. Various tomographic appearances of postoperative sellae: bone defect, localized or diffuse thickening of the floor, intrasellar opacities, and surgical clips

C. Intrasellar Craniopharyngiomas

The embryologic origin, pathologic features, and clinical aspects of craniopharyngiomas are described in Chap. 7. Craniopharyngiomas which are initially intrasellar occur much less frequently than suprasellar craniopharyngiomas. The first clinical symptom is adenohypophyseal deficiency.

From a radiologic standpoint, these tumors are very frequently calcified (Figs. 6.47, 6.49). The calcification is almost always nodular, reflecting the fact that these intrasellar craniopharyngiomas are only rarely cystic. They usually enlarge the sella turcica in a regular manner. The dorsum sellae is thinned, with an anterior concavity, and usually becomes vertical when there is a suprasellar extension (Fig. 6.48). As with suprasellar craniopharyngiomas, in the frontal projection, the floor of the sella often shows a midline depression *en cuvette*. The cortical bone of the sella is usually well preserved and only in rare cases does the tumor extend toward the sphenoid sinus (Majlessi et al. 1978) (Fig. 6.49). Taveras and Wood (1976) believed that an enlarged sella turcica, without intrasellar calcification but with a persistent enlarged craniopharyngeal canal, suggests a craniopharyngioma.

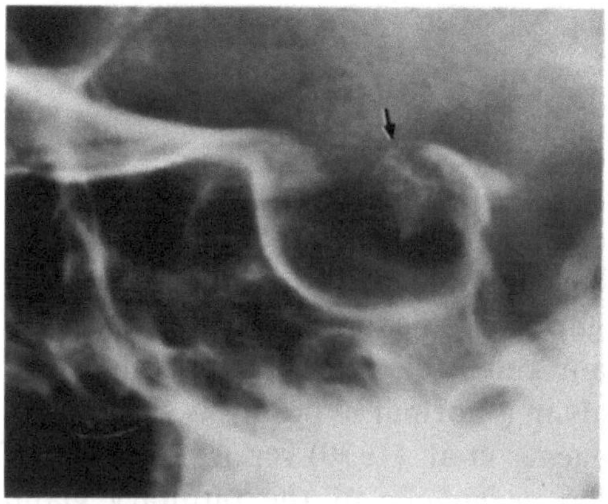

Fig. 6.47. Intrasellar craniopharyngioma. The lateral radiograph shows intrasellar calcification within a fairly normal sella *(arrow)*

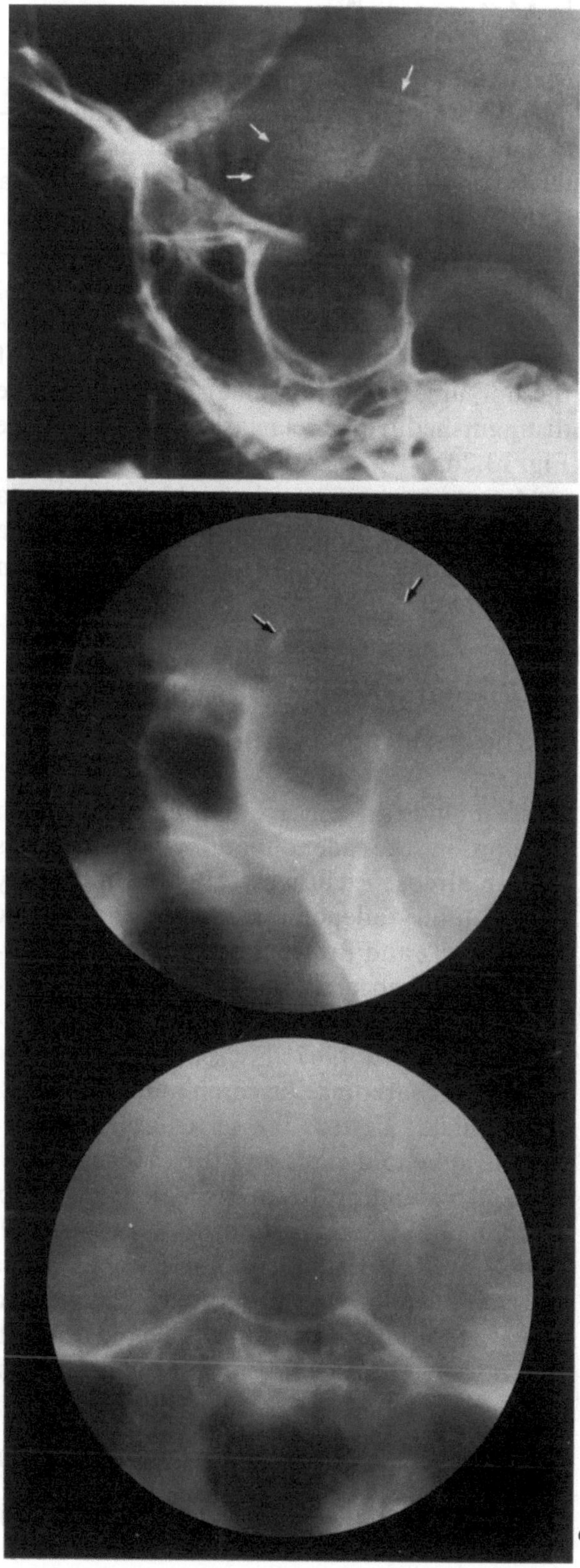

a

b

c

Fig. 6.48 a–c. Intrasellar craniopharyngioma with large suprasellar extension. ▷

a Lateral view

b Lateral tomogram: the sella turcica is enlarged and has a double floor while the dorsum sellae is thinned and straightened. Note curvilinear suprasellar calcification *(arrows)*, representing the upper limit of the tumor. Small nodular calcification is present between the anterior and posterior clinoid processes.

c Frontal tomogram: the floor of the sella turcica presents a symmetric midline depression *(en cuvette)*. The cortical bone of the floor is well preserved

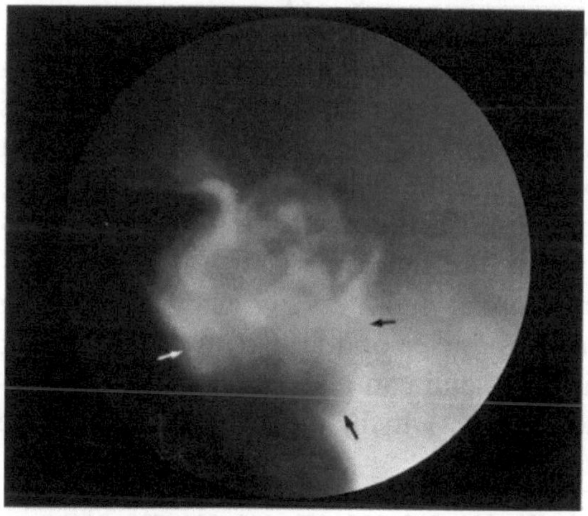

Fig. 6.49. Intrasellar craniopharyngioma with infrasellar extension: lateral tomogram. Intrasellar calcified mass invading sphenoid sinus and clivus *(arrows)*

D. Miscellaneous Disorders

Rare intrasellar space-occupying lesions do not have specific radiologic appearances.

I. Metastases [14]

Metastases to the pituitary gland are very frequent, but usually without clinical or radiologic signs. They almost always occur in the context of multiple metastases. Breast carcinoma is the most frequent primary tumor.

Metastasis to the neurohypophysis is common (Grisoli et al. 1974; Delarue et al. 1964; Duchen 1966) and may be associated with invasion and destruction of the inferior part of the dorsum sellae (Fig. 6.50), which must be distinguished from metastasis of the bone (see Fig. 13.20).

Metastases may sometimes enlarge the sella turcica and even give the appearance of a chromophobe adenoma (Kistler and Pribram 1975).

II. Primary Malignant Tumors of the Pituitary Gland [15]

These include sarcomas and carcinomas which develop in the adenohypophysis. Sarcomas appear almost exclusively after radiotherapy of a pituitary adenoma (Solitaire and Jatlow 1967; Waltz and Brownell 1966). Pituitary carcinomas (adenocarcinomas) are degenerate pituitary adenomas (Newton et al. 1962; Solitaire and Jatlow 1967; Fleischer et al. 1972).

Whether sarcomas or carcinomas, these tumors quickly destroy the sella turcica, invading the sphenoid sinus and the cavernous sinus, and extending toward the petrous apex and the Gasserian ganglion. As Newton et al. (1962) indicated, the radiologic appearances of adenocarcinomas are often identical to those of pituitary adenomas. Rapid growth and invasiveness should suggest the diagnosis.

III. Sarcoidosis [16]

Pituitary gland involvement is rare (Delaney 1977). Anterior pituitary insufficiency and diabetes insipidus reveal involvement of the hypophysis. The sella turcica is rarely enlarged.

IV. Abscesses [17]

Abscesses develop within a pituitary adenoma, craniopharyngioma or Rathke's cleft cyst (Domingue and Wilson 1977; Rudwan 1977; Obenchain and Becker 1972), and much less commonly in a normal hypophysis (Fig. 6.40). Wollschlaeger et al. (1970) reported a case of *Aspergillus granuloma* of the pituitary gland with infiltration of the walls of the internal carotid artery. The sella turcica was demineralized.

V. Pituitary "Calculus" [18]

Several authors (Taveras and Wood 1976; Glasser and Earll 1968; Ozonoff and Burrows 1971; Destee et al. 1977) have reported amorphous intrasellar calcification which is usually rounded, dense, the size of a pea, and lying in the anterior two-thirds of the sella turcica. The calcification may be midline (Destee et al. 1977) or paramedian (Ozonoff and Burrows 1971). Contradicting hypotheses have been suggested as to the pathogenesis of these calcific deposits. Ozonoff and Burrows (1971) believe that they represent degenerative deposits (Fig. 6.51); in such cases the floor of the sella is normal and the sella turcica is not enlarged. Calcification in the posterior third of a normal-size sella turcica may suggest an intrasellar craniopharyngioma (Fig. 6.47). According to Taveras and Wood (1964), we believe that pituitary calculi can also occur in eosinophilic adenomas which have become inactive

14 Refs. [216, 233, 253, 346, 371, 389, 493, 503, 520, 830].

15 Refs. [5, 339, 344, 396, 522, 626, 676, 809, 853, 890, 999]

16 Refs. [79, 212, 215, 335, 452, 543, 736, 780, 864]

17 Refs. [26, 221, 236, 572, 652, 686, 688, 813, 925, 989, 1016]

18 Refs. [219, 236].

(Fig. 6.36). Finally, pituitary gland calcification may be observed in infants without pituitary dysfunction (Harwood-Nash and Fitz 1976) (Fig. 6.52).

VI. Rathke's Cleft Cysts and Other Intrasellar "Cysts"[19]

Macroscopic study of the hypophysis often reveals the existence of small cysts, several millimeters in diameter, which have developed either within Rathke's cleft or within the adenohypophysis. These cysts are lined by cuboidal epithelium, often ciliated, with mucus-secreting cells. Epithelial cysts contain a white mucoid fluid (Rasmussen 1928; Shanklin 1949, 1951 a, b; Fager and Carter 1966; Shuangshoti et al. 1970). These cysts therefore have a common embryologic origin with craniopharyngiomas since both types of tumor are derived from remnants of Rathke's pouch. However the epithelial lining of epithelial cysts is distinct from that of craniopharyngiomas. As Yoshida et al. (1977) suggest, this is due to different cytologic differentiation. Although frequent, epithelial cysts only rarely give rise to clinical or radiologic signs (Frazier and Alpers 1934; Messimy et al. 1955; Berry and Schlesinger 1959; Naiken et al. 1961; Fairburn and Larkin 1964; Fager and Carter 1966; Ringel and Bailey 1972; Concha et al. 1975; Yoshida et al. 1977). When these cysts increase in size they cause pituitary insufficiency and occasionally visual disturbance. Radiologically, they enlarge the sella turcica, thinning and straightening the dorsum sellae; the floor of the sella is not usually destroyed (Fig. 6.53).

Epithelial cysts must be distinguished from intrasellar cysts which do not possess an epithelial layer. These may be intrasellar arachnoid cysts which do not communicate with the suprasellar subarachnoid space (Guiot et al. 1971) or secondary cysts after necrosis of a pituitary adenoma (Weber et al. 1970).

19 Refs. [56, 70, 157, 169, 254, 279, 281, 311, 360, 491, 627, 664, 764, 767, 785, 808, 860, 861, 862, 873, 874, 884, 959, 1001, 1041]

VII. Granular Cell Tumors[20]

These develop in the neurohypophysis from granular cells normally present in this area (Satyamurti and Huntington 1972; Janny et al. 1974; Gardeur et al. 1978). Synonyms for this type of tumor are: choristoma (a term applied by Sternberg, who thought that granular cells were foreign to the hypophysis), pituicytoma, myoblastoma, or Abrikossof's tumor. Radiologic findings are not specific and reflect an intrasellar mass. On the lateral projection, the sella turcica appears ballooned; the cortical bone is thinned. In the frontal projection, depression of the floor is midline and the floor appears enlarged, representing a symmetric, midline, space-occupying lesion (Fig. 6.54). Other neuroradiologic investigations usually show a tumor with a lobulated suprasellar extension. In exceptional cases, these tumors may have a suprasellar origin (Sholkoff et al. 1977).

VIII. Vascular Disease

1. Aneurysms and ectatic carotid siphons with intrasellar extensions (see Chap. 9).
2. Carotid siphon calcification (see Chap. 13 and Fig. 13.26).
3. Transsellar or transsphenoid carotid anastomosis asssociated with agenesis of the internal carotid artery (see Chap. 13 and Fig. 13.9).
4. Persistent trigeminal artery (see Chap. 13 and Fig. 13.10).

IX. Rare Intrasellar Disorders

1. Germinomas[21]. These tumors usually develop in the pineal and suprasellar region (see Chap. 7 and Fig. 7.10). Banna et al. (1976) described a case of intrasellar germinoma in the adenohypophysis. The sella turcica was enlarged, the floor destroyed, the dorsum sellae thinned, and the sphenoid sinus invaded by the tumor.

20 Refs. [7, 122, 238, 324, 337, 392, 460, 829, 871, 933]
21 Refs. [40, 330]

2. Epidermoid tumors. More frequently suprasellar than intrasellar, an epidermoid may enlarge the sella turcica, eroding the dorsum and the floor (Steimle et al. 1977) (Fig. 6.55).

3. Chordomas. Intrasellar chordomas are exceptional. This unusual site can be explained by ectopic intrasellar remnants of the notochord (Mathews and Wilson 1974); the cranial end of the notochord is initially in contact with the infundibular recess. Toward the end of the 4th week of fetal life, the infundibular recess moves towards the sella turcica to form the neurohypophysis. At this time, the cranial end of the notochord may be drawn toward the sella. Intrasellar chordomas produce the same bony changes as other intrasellar tumors. The sella turcica is at first regularly ballooned without destruction of the floor or dorsum (Mathews and Wilson 1974); later, the margins of the sella may be destroyed, but the dorsum is always the last to be destroyed (Belza 1966). When intrasellar and suprasellar expansion are present, it is impossible to distinguish suprasellar calcifications from that of an adenoma or a craniopharyngioma.

4. Chondromas. Intrasellar chondromas are rare (Asenjo and Chiorino 1956; David and Constans 1957; Falconer et al. 1968; Paillas and Alliez 1974). In the case described by Paillas and Alliez, the sella turcica was enlarged but its floor was not destroyed. Calcification was present within the sella. There was moderate suprasellar extension of the tumor.

5. Meningioma. In 1968, Hardy and Robert described a case of intrasellar meningioma, causing pituitary insufficiency. The sella turcica was enlarged with an asymmetrically depressed floor. There was neither hyperostosis nor tumor calcification.

6. Gangliocytomas. These tumors arise in the neurohypophysis, from residual nerve cells. Radiologically, the sella turcica is ballooned; the floor and the dorsum sellae are demineralized (Jakumeit et al. 1974).

7. Hemangioblastoma. Dan and Smith (1975) described a case of pituitary hemangioblastoma in a patient with von Hippel-Lindau's disease.

8. Melanoma of the pituitary gland. Neilson and Moffat (1963) described a case of intrasellar melanoma of the pituitary gland and advance the hypothesis of an origin from pigmented cells in the pars nervosa. Scholtz and Siu (1976) report another case, with an enlarged sella turcica and an eroded dorsum sellae.

9. Leiomyoma. This very rare tumor (Kroe et al. 1968) develops from smooth muscles of the vessels supplying the hypophysis. Sellar changes are identical to those of a chromophobe adenoma.

10. Cysticercosis. Prosser et al. (1978) reported a case of intrasellar cysticercosis.

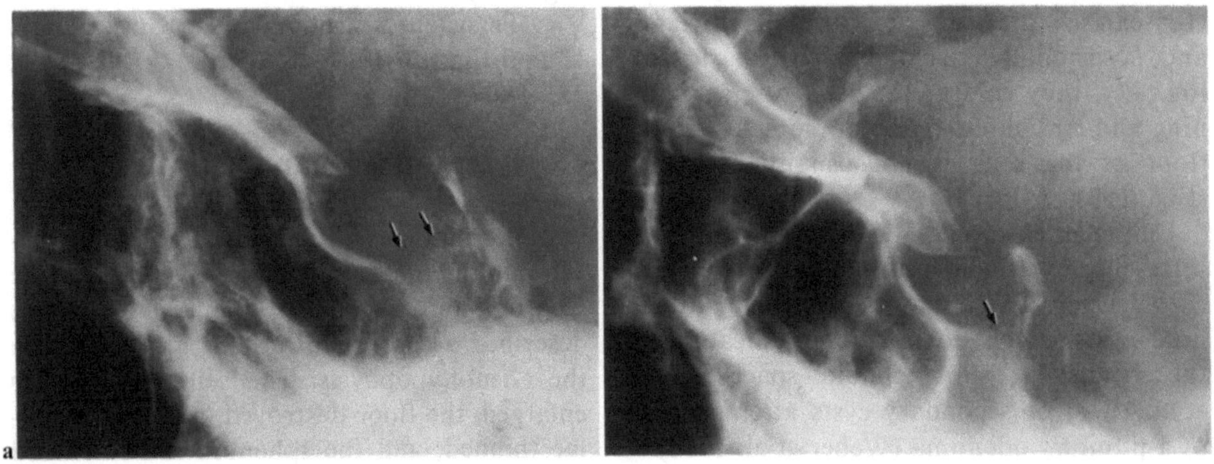

Fig. 6.50 a, b. Pituitary metastasis (breast carcinoma).
a The metastasis, in the neurohypophysis, has destroyed the posteroinferior part of the floor of the sella and the dorsum sellae *(arrows)*.
b Limited erosion of the anterior aspect of the base of the dorsum sellae *(arrow)* simulating raised intracranial pressure. Note also skull vault metastasis

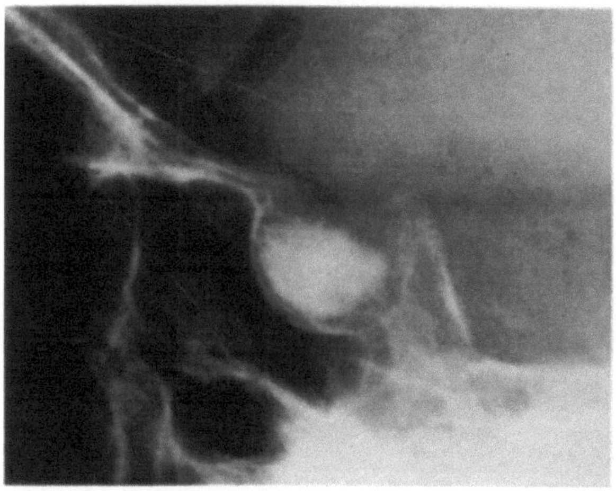

Fig. 6.51. Pituitary calculus in a 70-year-old male without endocrine disturbance. The lateral film shows a massive calcareous deposit in the adenohypophysis. The sella turcica is not enlarged (Courtesy H. Deramond, Amiens)

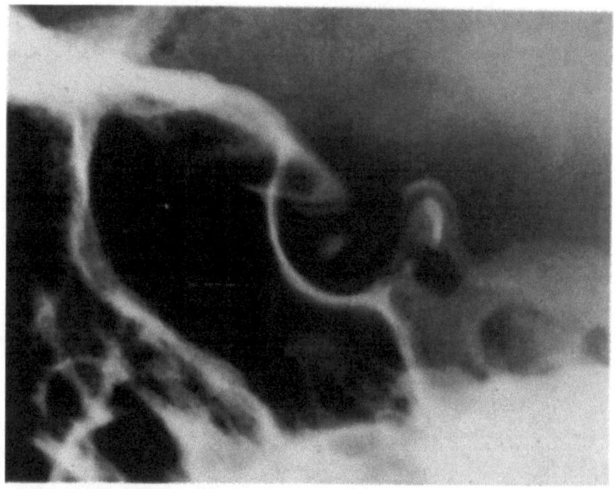

Fig. 6.52. Pituitary calculus: note a small calcareous deposit. The sella turcica is normal in size

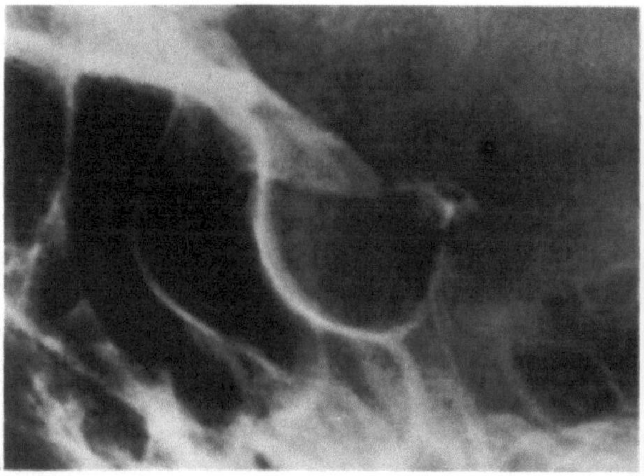

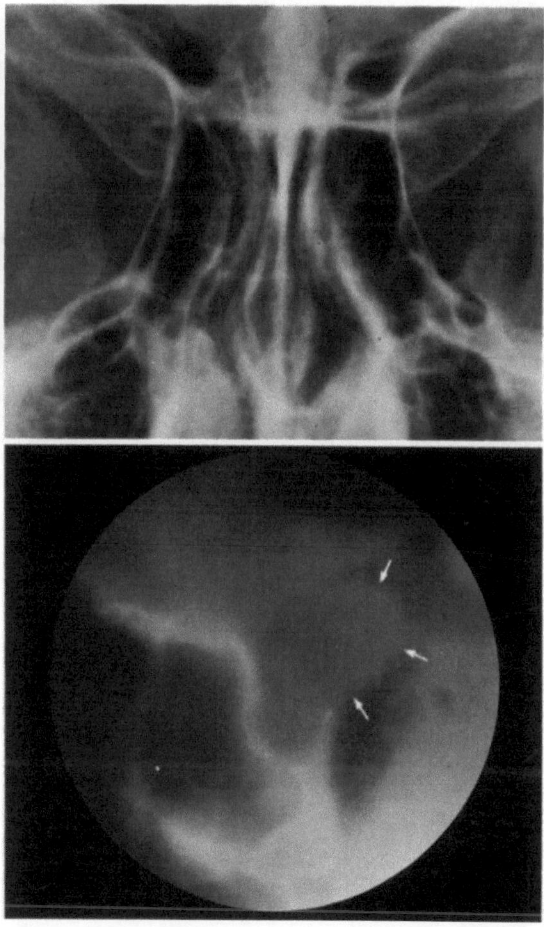

Fig. 6.53a–c. Intrasellar Rathke's cleft cyst in a 18-year-old female with bitemporal hemianopia.

a Magnified lateral view: the sella turcica is not enlarged, but the dorsum sellae is thinned. The floor seems normal.

b Frontal view: the floor of the sella turcica appears normally flat; however, it is not normal: a blurred appearance from right to left indicates slight demineralization.

c Midsagittal pneumotomogram: the intrasellar cyst has a large suprasellar extension *(arrows)*

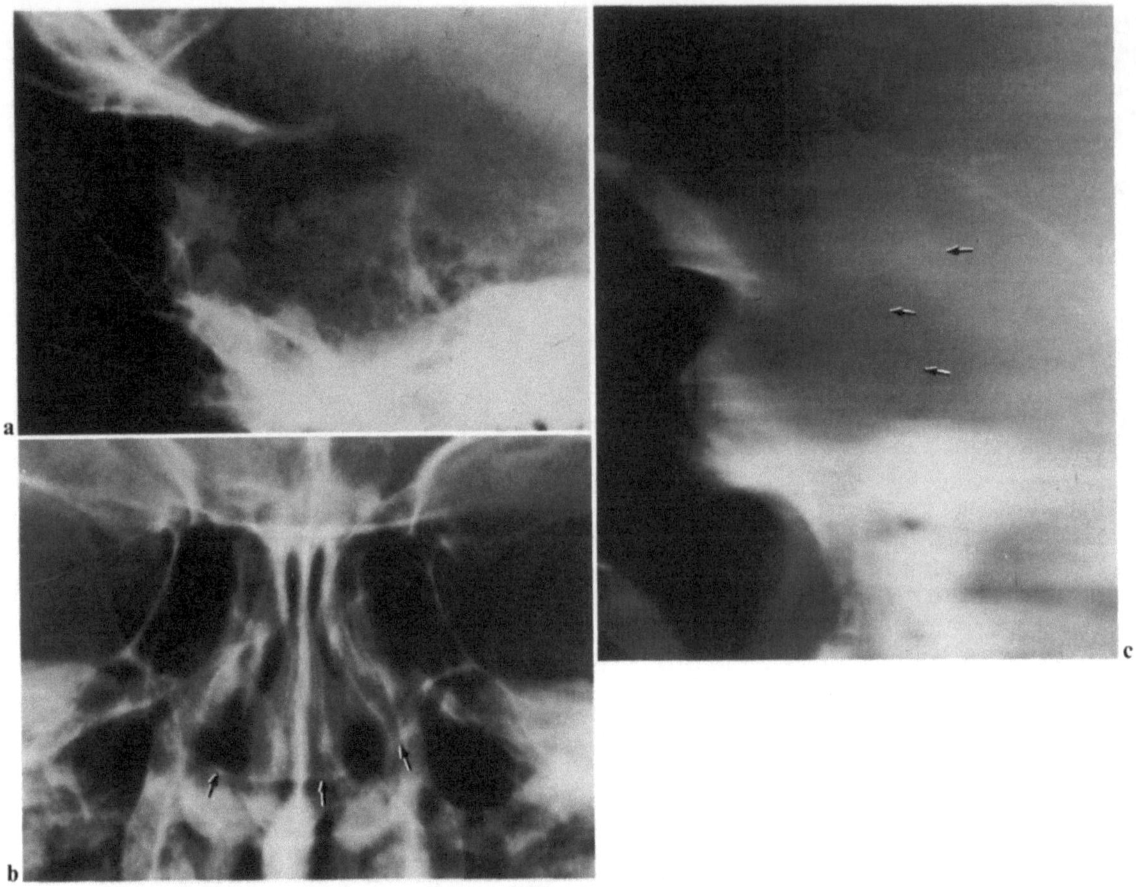

Fig. 6.54a–c. Intrasellar granular cell tumor in a 75-year-old female with bitemporal hemianopia.

a Lateral view: destruction of the floor of the sella turcica and the dorsum sellae. The tumor extends into the sphenoid sinus.

b Frontal view: the destroyed floor is invisible. *Arrows* indicate the upper limit of the preserved part of the clivus.

c Sagittal angiotomogram: the tumor appears bilobular *(arrows)*

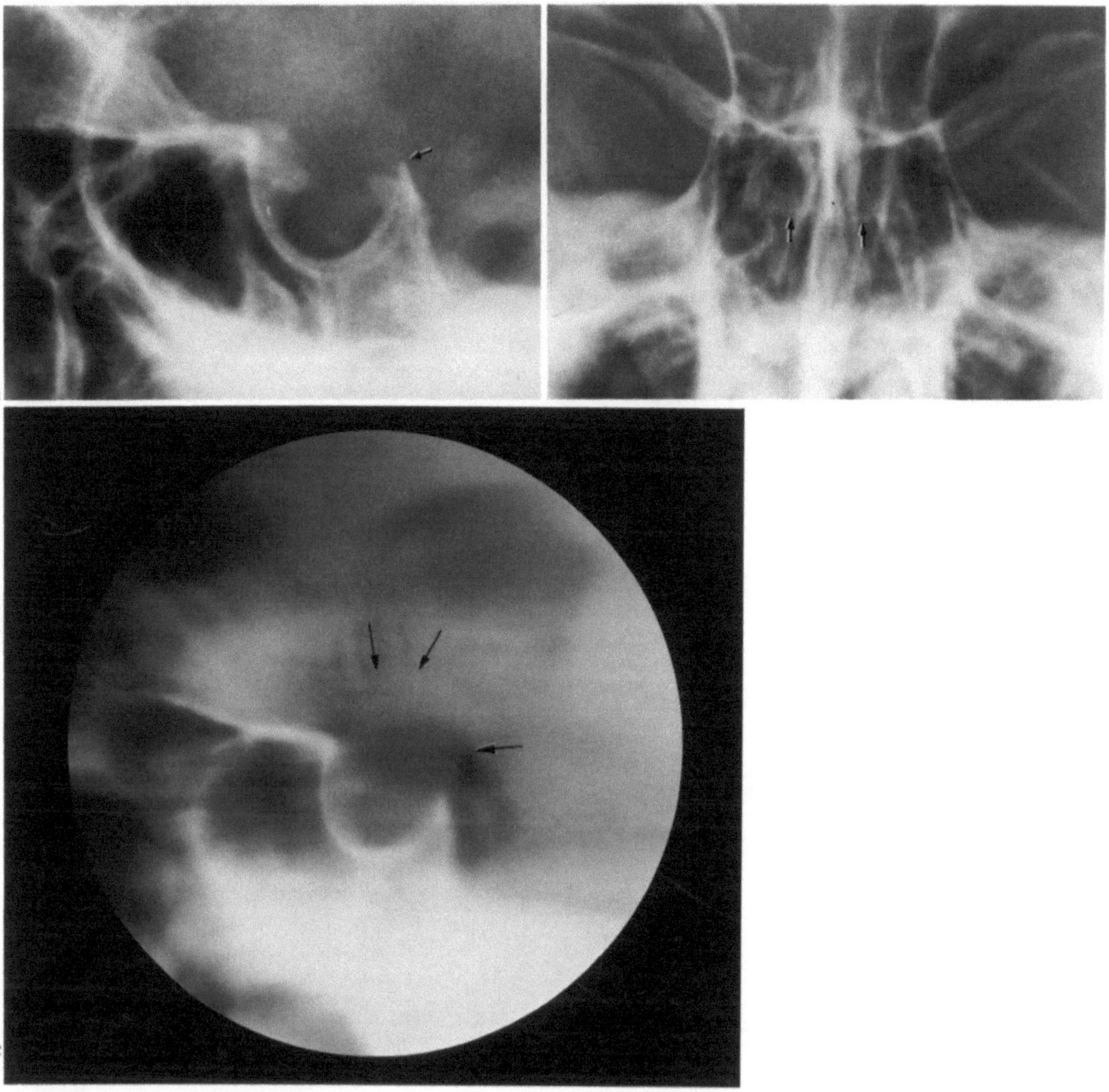

a b c

Fig. 6.55 a–c. Intrasellar epidermoid in a 20-year-old female with amenorrhea and polydipsia.

a Lateral view: the pituitary fossa is not enlarged. The top of the dorsum sellae is thinned *(arrow)*.

b Frontal view: the floor of the sella turcica appears markedly demineralized *(arrows)*.

c Midsagittal pneumotomogram: note the suprasellar extension *(arrows)*

Chapter 7 Suprasellar Pathology

A. Craniopharyngiomas[22]

I. General Considerations

Intrasellar craniopharyngiomas are benign tumors arising from epithelial remnants of Rathke's pouch. They are, therefore, still referred to as Rathke's pouch tumors, tumors of the craniopharyngeal pouch, or Erdheim's tumor (1903).

1) Pathology

Craniopharyngiomas are encapsulated tumors, very often either totally or partially cystic. Calcium deposits are almost always found on histologic examination.

2) Topography

Craniopharyngiomas arise along the course of the craniopharyngeal duct. However, most craniopharyngiomas develop in the pars tuberalis. These suprasellar craniopharyngiomas are classified according to their position with regard to the optic chiasm in prechiasmatic, subchiasmatic, and retrochiasmatic tumors (Pertuiset 1975). Intrasellar craniopharyngiomas are more rarely encountered. Their suprasellar expansion is usually backward (Bailey 1933). These tumors are rarely cystic. In fact, craniopharyngiomas are often both supra- and intrasellar, a primarily intra- or suprasellar tumor extending to or arising in both regions. Tumors in other sites, e.g., pharyngeal (Podoshin et al. 1970), intrasphenoid (Cooper and Ransohoff 1972), or within the third ventricle (Cashion and Young 1971; Bollatti et al. 1973) are exceptional.

3) Age, Sex, and Incidence

Intrasellar craniopharyngiomas constitute 5% of all intracranial tumors (Barnett 1959), but 13% in children (Banna et al. 1973). These tumors are usually diagnosed before the age of 25 and may even be discovered in the newborn (Majd et al. 1971; Azarkia et al. 1975). Intrasellar craniopharyngiomas are less common in adults and are found essentially after the age of 55. Craniopharyngiomas are more frequent in males during childhood, whereas in the adult the incidence is the same in both sexes.

II. Symptoms

The following symptoms are typical:

1. Pituitary deficiency and hypothalamic dysfunction with retarded growth and hypogonadism, obesity related to hypothyroidism, adrenocortical insufficiency, hypothalamic dysfunction, somnolence, hyper- or hypohidrosis, hypothermia, and diabetes insipidus in children.

2. Visual disturbances consisting of a uni- or bilateral decrease in visual acuity and visual field defect. Papilledema is often present.

3. Headaches and vomiting related to increased intracranial pressure.

22 Refs. [9, 23, 31, 32, 35, 37, 39, 42, 45, 55, 82, 89, 138, 140, 146, 160, 173, 269, 271, 310, 329, 338, 351, 380, 408, 424, 447, 571, 596, 597, 601, 616, 634, 688, 703, 728, 797, 802, 924, 926, 948]

III. Radiology

The radiologic features of suprasellar craniopharyngiomas differ in children and adults. The picture depends on the site of the tumor. Craniopharyngiomas in atypical sites, e.g., intrasellar, intrasphenoid, and pharyngeal, are discussed in the corresponding chapters.

Craniopharyngiomas have two radiologic characteristics: they often calcify and they cause changes in the sella turcica.

1) Calcification

Calcification is frequent, but the incidence decreases with age – about 85% in children and 50% in adults (Ozonoff and Burrows 1971). Visualization of the calcification is facilitated by high contrast film and by tomography. The calcification is midline, suprasellar, occasionally extending toward the upper part of the sella turcica. There are two types of calcification: nodular or floccular calcification lie in the solid part of the tumor (Fig. 7.1); capsular calcification is in the cyst wall, appearing curvilinear, and often very thin (Figs. 7.2, 7.3). Unlike nodular calcification, which does not indicate the size of the lesion, capsular calcification often delimits the tumor perfectly.

2) Changes in the Sella Turcica

In certain cases, the sella turcica appears normal (Fig. 14.60), while in others there are marked changes, related to the tumor itself or to raised intracranial pressure. Hoff and Patterson (1972) suggested that bony changes in the sella turcica are more frequent in adults than in children.

Some craniopharyngiomas, which are both supra- and intrasellar, produce changes which do not indicate their origin as supra- or intrasellar lesions. In such cases, the sella turcica is ballooned, the dorsum sellae thinned and usually vertical, and the posterior clinoid processes are eroded. The floor of the sella shows a midline depression, generally with cortical bone which is better preserved than with adenomas, or even dense (Fig. 7.2).

When seen in association with the type of calcification previously discussed, the changes produced by the majority of suprasellar craniopharyngiomas, be they pre-, sub-, or retrochiasmatic, are characteristic. The most important radiologic changes are located mainly above the diaphragma sellae. The upper margin of the dorsum sellae is eroded, depressed, and extends outward into lengthened tapered posterior clinoid processes (Figs. 7.4, 7.5). This erosion of the dorsum sellae may be asymmetric, creating on the lateral projection, the appearance of a truncated hemidorsum. As an isolated observation, this truncated hemidorsum may be difficult to differentiate from a morphological variation (Figs. 5.61, 14.37). Retrochiasmatic craniopharyngiomas may also tilt the dorsum sellae forward (Fig. 7.4).

It must be emphasized that the dorsum sellae is frequently not demineralized, even when bony changes are extensive. Associated increased intracranial pressure, or rapid growth of the tumor is likely to be accompanied by demineralization of the dorsum sellae (Figs. 7.5, 7.6). The tuberculum sellae is flattened; the chiasmatic sulcus becomes vertical, forming a straight line with the anterior wall of the sella turcica (Fig. 7.1). At a later stage, the chiasmatic sulcus is deepened, giving an omega appearance to the sellar region (Fig. 7.5). At the same time, the anterior clinoid processes, which have been reshaped by the tumor, are separated and tilted; their upper surface becomes supermedial. On the lateral projection, the height of the anterior clinoid processes therefore appears increased (Fig. 7.6). The degree of demineralization of the cortex of the chiasmatic sulcus depends on the rapidity of growth of the tumor. Changes of the floor of the sella are less characteristic with suprasellar craniopharyngiomas. Sometimes the floor appears normal; often it presents a midline depression with a *cuvette* appearence. The cortex of the floor is usually preserved (Fig. 6.48). Careful appraisal of these changes may be a valuable contribution when the craniopharyngioma is not calcified.

B. Hypothalamic Gliomas

The sella turcica is often normal (Fig. 14.59), but sometimes these tumors erode the dorsum sellae. Hypothalamic gliomas often become very large and are occasionally cystic. They therefore destroy the tuberculum sellae, the chiasmatic sulcus, and even the planum sphenoidale; the region of the sella therefore appears much enlarged (Fig. 7.8). These gliomas rarely calcify.

C. Gliomas of the Optic Chiasm

Anatomically and neuroradiologically, differentiation between gliomas of the optic chiasm and those of the hypothalamus is often impossible (see Chap. 8).

D. Miscellaneous Disorders

1) Histiocytosis X of the Hypothalamus [23]

Primary, isolated lesions of the hypothalamus are rare. Changes in the region of the sella, when present, are nonspecific (Pressman et al. 1975; Tibbs et al. 1978).

2) Hypothalamic Sarcoidosis [24]

Hypothalamic involvement in sarcoidosis is also rare (Wiederholt and Siekert 1965; Vesely et al. 1977). In general, the sella turcica appears normal.

3) Colloid Cysts of the Third Ventricle

The appearance of the sella turcica is normal when intracranial pressure is not raised and when the tumor is small. In all other cases there are changes of raised intracranial pressure associated with shortening of the dorsum sellae (Batnitzky et al. 1974). Colloid cysts do not calcify (Figs. 7.9 and 12.3).

4) Suprasellar Germinomas [25]

Germinomas or atypical teratomas are rare tumors, histologically identical to seminomas and dysgerminomas of the gonads. These tumors are often wrongly referred to as ectopic pinealomas, because their cells have an apparent histologic resemblance to cells of the fetal pineal gland (Ghatak et al. 1969). Sometimes these tumors affect both the pineal and suprasellar regions (Steimle et al. 1979). Frequent in males, these tumors develop particularly before the age of 20. The clinical findings are visual disorders, diabetes insipidus, and pituitary insufficiency. In 50% of cases, the sella turcica is quite normal but it is enlarged in the remainder (Fig. 7.10), with erosion of the dorsum sellae (Takeuchi et al. 1979). These germinomas are not calcified. Certain authors, e.g., Pribram and du Boulay (1971) and Takeuchi et al. (1979), have pointed out the presence of a dense floor to the sella.

5) Suprasellar Arachnoid Cysts [26]

These cysts represent unusual suprasellar masses whose pathogenesis remains uncertain: they may be inflammatory (Sansregret et al. 1969), post-traumatic (Contreras et al. 1977), or congenital (Murali and Epstein 1979). Clinical manifestations are slight and are related to raised intracranial pressure, due to obstruction of the foramen of Monro (Fig. 7.11). Plain films of the sella turcica are often normal (Danziger and Bloch 1974; Sansregret et al. 1969). When radiologic changes are present, they involve the dorsum sellae, with direct erosion by the suprasellar mass and demineralization secondary to raised intracranial pressure (Sansregret et al. 1969). These arachnoid cysts often fill with air at pneumoencephalography.

23 Refs. [67, 751, 947]
24 Refs. [212, 215, 490, 543, 709, 736, 864, 983, 1018]
25 Refs. [107, 133, 164, 472, 812, 879, 902, 930, 938]
26 Refs. [172, 197, 663, 824, 856, 1029]

6) Dermoid and Epidermoid Tumors

These tumors occur typically in the suprasellar region, as well as in other sites; they develop very slowly from embryonic remnants of cutaneous ectoderm. The sella turcica shows shortening of the dorsum (Fig. 7.12). There may be calcification of the wall of the tumor.

7) Hamartomas of the Tuber Cinereum[27]

Precocious puberty is a common presenting symptom in cases of hamartoma. These tumors rarely cause changes in the sella (Harwood-Nash and Fitz 1976). Erosion of the upper part of the dorsum sellae or suprasellar calcification may occasionally be present (Lin et al. 1978a, b).

8) Meningiomas of the Diaphragma Sellae

These are rare tumors which may be calcified. Suprasellar meningiomas are discussed in Chap. 8.

27 Refs. [569, 570, 575, 683, 847]

9) Esthesioneuroblastomas[28]

Olfactory esthesioneuroblastoma is a rare tumor, particularly within the cranium (David et al. 1960; Seaman 1951; Jakumeit 1971; Hamilton et al. 1973; Sarwar 1979). These tumors develop from Locy's ganglion (Jakumeit 1971). The case reported by Sarwar (1979) showed nodular calcification above and alongside the sella turcica.

10) Suprasellar Aneurysms

Only giant aneurysms, arising at the bifurcation of the internal carotid artery or basilar artery, give rise to bony changes. They erode the dorsum sellae and sometimes the upper surface of the anterior clinoid processes. Their wall often shows thin curvilinear calcification. When these aneurysms are huge, the floor of the sella may be eroded by an intrasellar extension.

11) Suprasellar Arachnoiditis

Tuberculous meningitis often heals with suprasellar calcification of floccular and irregular appearance (Ozonoff and Burrows 1971). The sella turcica is normal (Fig. 7.13).

28 Refs. [148, 204, 375, 456, 825, 854]

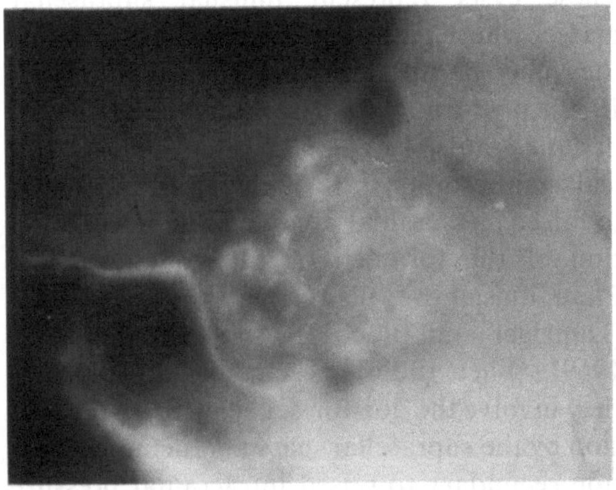

Fig. 7.1. Craniopharyngioma. The lateral pneumotomogram reveals floccular calcification in the sellar area, and destruction of the top of the dorsum sellae. The tuberculum sellae is flattened. The chiasmatic sulcus is vertical, forming a straight line with the anterior wall of the sella turcica (Courtesy, J. Metzger)

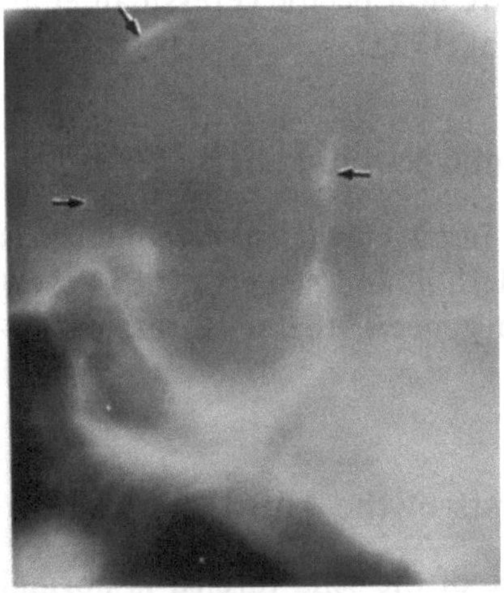

Fig. 7.2. Craniopharyngioma in a 14-year-old boy. The pituitary fossa is enlarged: the dorsum sellae is thinned and straightened. Note the curvilinear calcification in the posterior and superior parts of the tumor *(arrows)*. The chiasmatic sulcus and the anterior wall of the sella turcica are in line

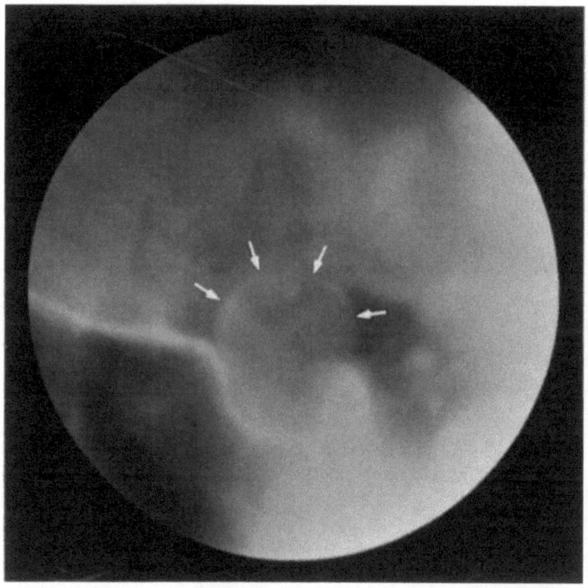

Fig. 7.3. Craniopharyngioma in a 25-year-old female with amenorrhea. The lateral pneumotomogram shows a suprasellar mass with eggshell calcifications *(arrows)*. The sella turcica is normal

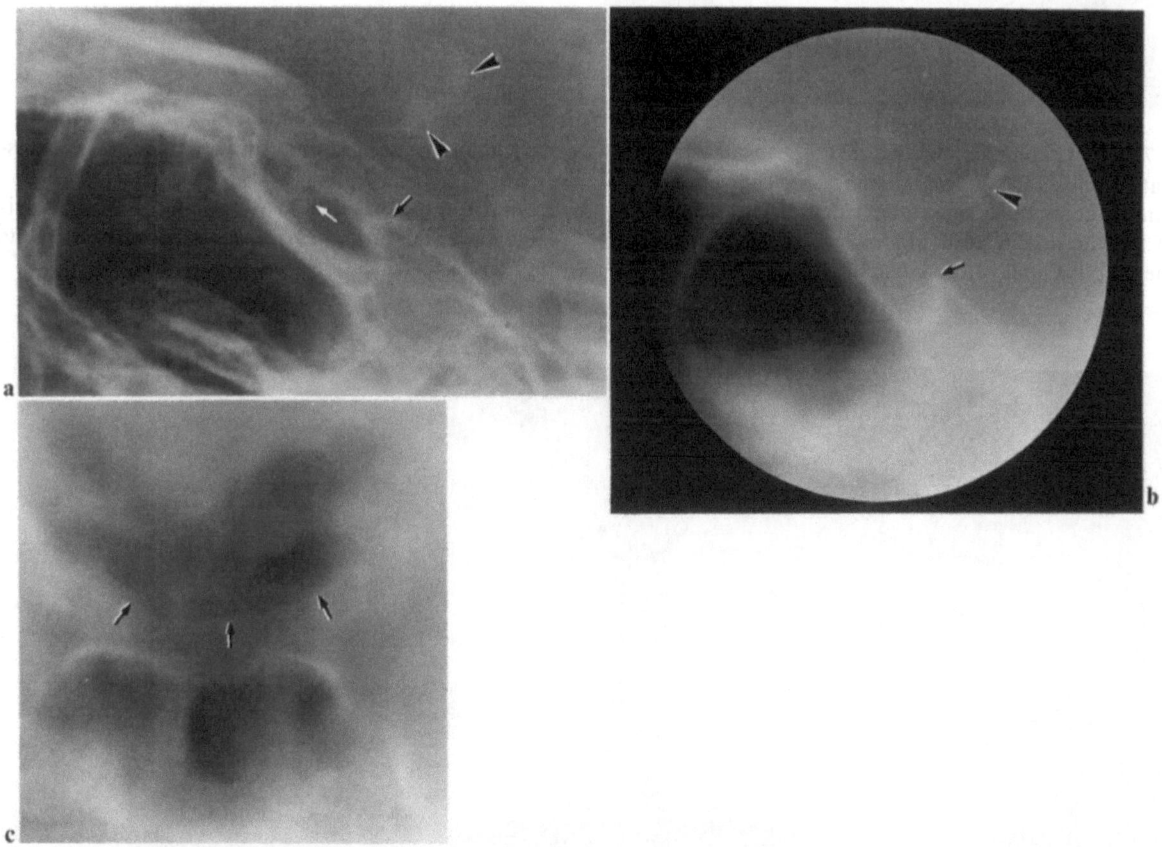

Fig. 7.4a–c. Craniopharyngioma in a 16-year-old boy with retarded growth.

a Lateral view, and **b** lateral tomogram: the dorsum sellae is tilted forward and shortened *(black and white arrows)*. Note nodular suprasellar calcification *(arrowheads)* and calcified interclinoid ligaments *(white arrow)*.

c Frontal pneumotomogram: the tip of the dorsum sellae shows a midline depression *(arrows)*, beneath the tumor

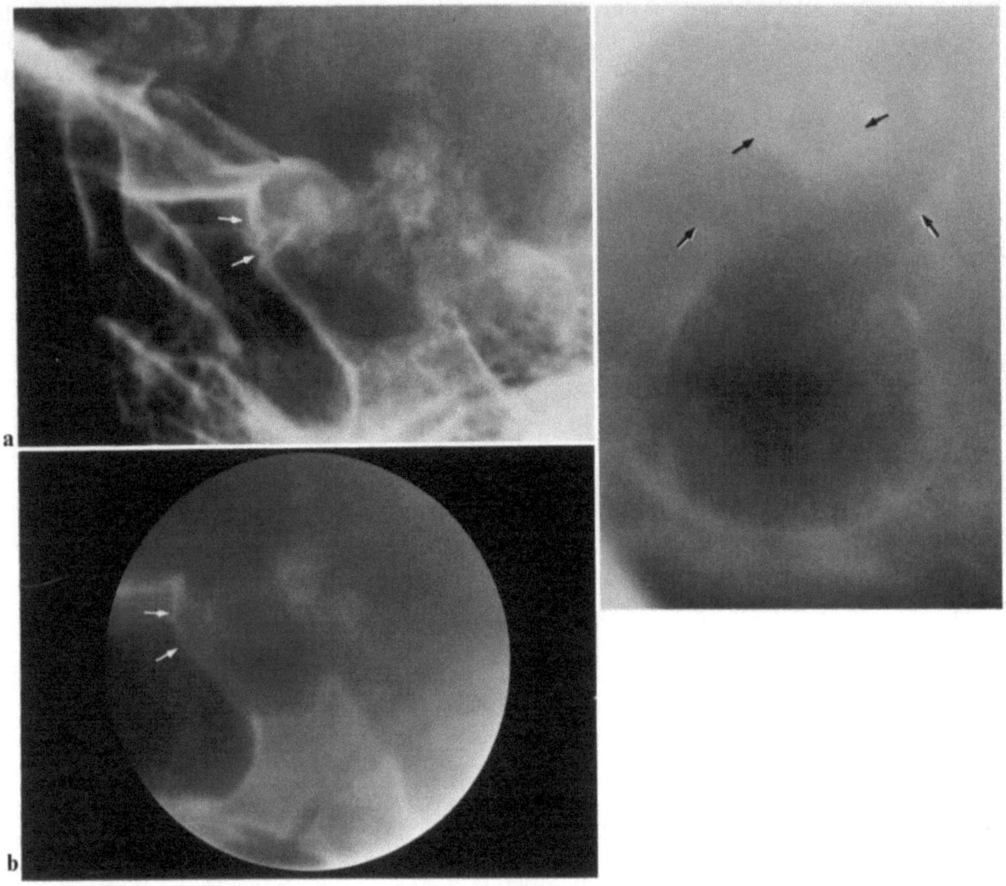

Fig. 7.5a–c. Craniopharyngioma in a 9-year-old girl.
a Lateral view, and **b** lateral tomogram: shortened dorsum sellae; nodular suprasellar calcification; deepening of the chiasmatic sulcus *(arrow)* which is in line with the anterior wall of the sella. Demineralization of the margins of the sella is due to raised intracranial pressure.
c Frontal tomogram: the posterior clinoid processes is lengthened and tapered *(black and white arrows)*. Note the nodular calcification *(black arrows)*

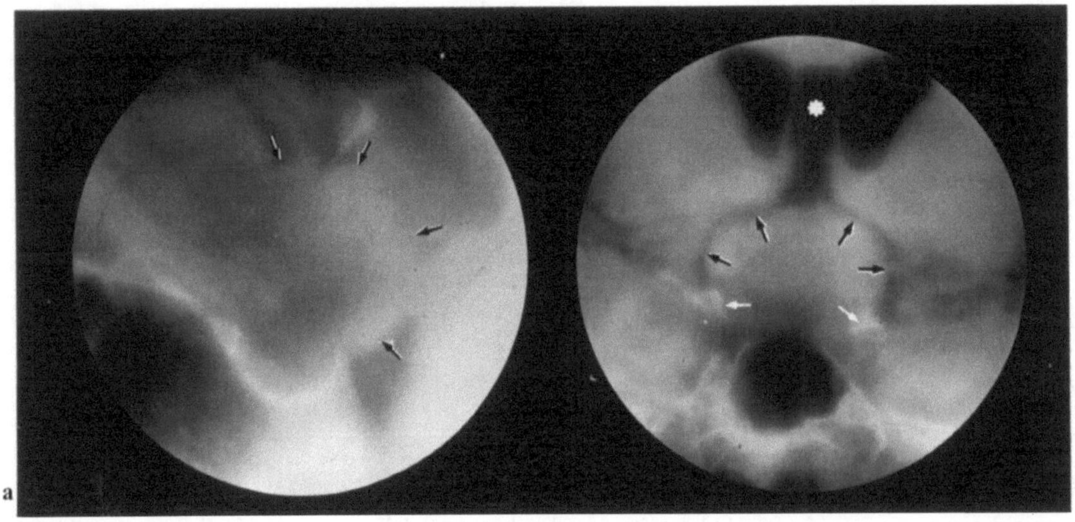

Fig. 7.6a, b. Craniopharyngioma in a 35-year-old male. **a** Lateral and **b** frontal pneumotomograms show a large suprasellar tumor with anterior extension *(black and white arrows)*. No tumor calcification is demonstrated. Note the erosion of the upper part of the dorsum sellae. The frontal tomogram **b** demonstrates separated anterior clinoid processes which are eroded on the medial surfaces *(white arrows)*. Note a cavum septi pellucidi *(star)*

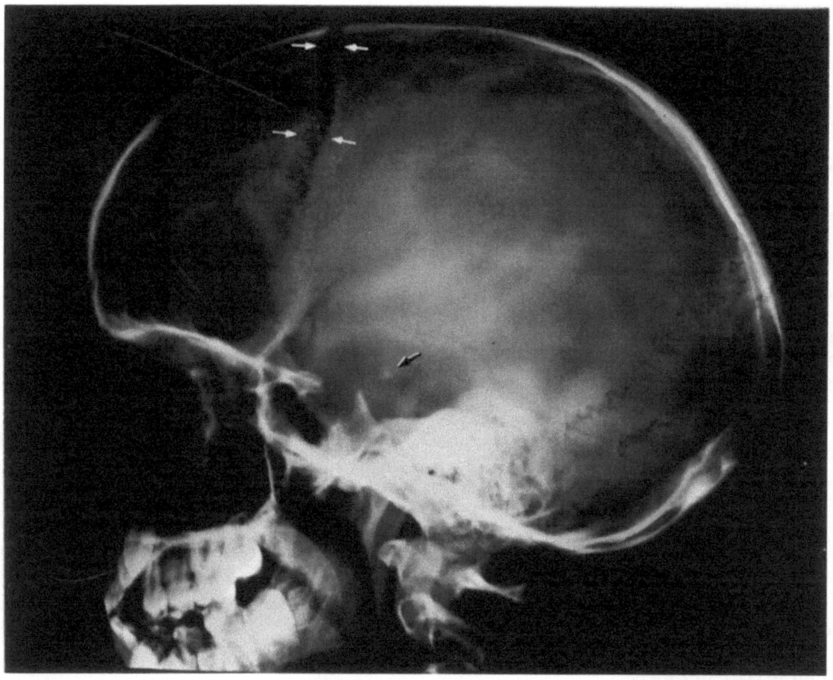

Fig. 7.7. Craniopharyngioma in a 10-year-old boy with bilateral loss of visual acuity. The lateral radiograph of the skull shows an enlarged and demineralized sella turcica with pointed dorsum and retrosellar calcification *(black and white arrow)*. Note widening of the coronal suture reflecting raised intracranial pressure *(white arrows)*

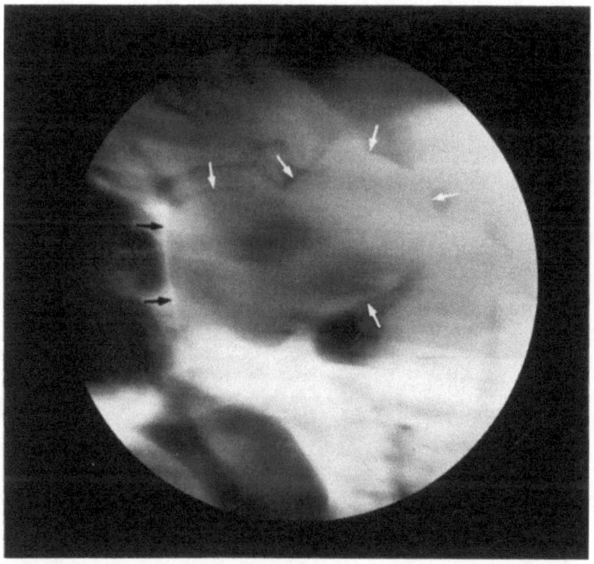

Fig. 7.8. Huge cystic hypothalamic astrocytoma *(white arrows)*. Enlargement of the sella turcica; the dorsum sella is destroyed; the chiasmatic sulcus is vertical and displaced anteriorly *(black arrows)*. The lamina dura of the sella is preserved

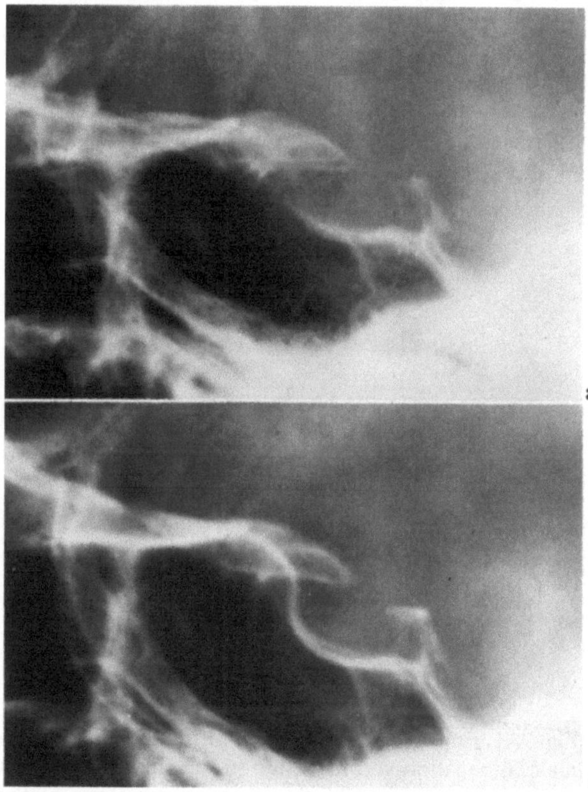

a

b

Fig. 7.9a, b. Colloid cyst of the third ventricle.

a Preoperative lateral radiograph: the tip of the dorsum sellae is eroded. The whole sellar region appears demineralized because of raised intracranial pressure.

b Lateral radiograph 6 months after operation: the dorsum sellae is still truncated with relief of the raised intracranial pressure, there is remineralization of the cortical bone

a 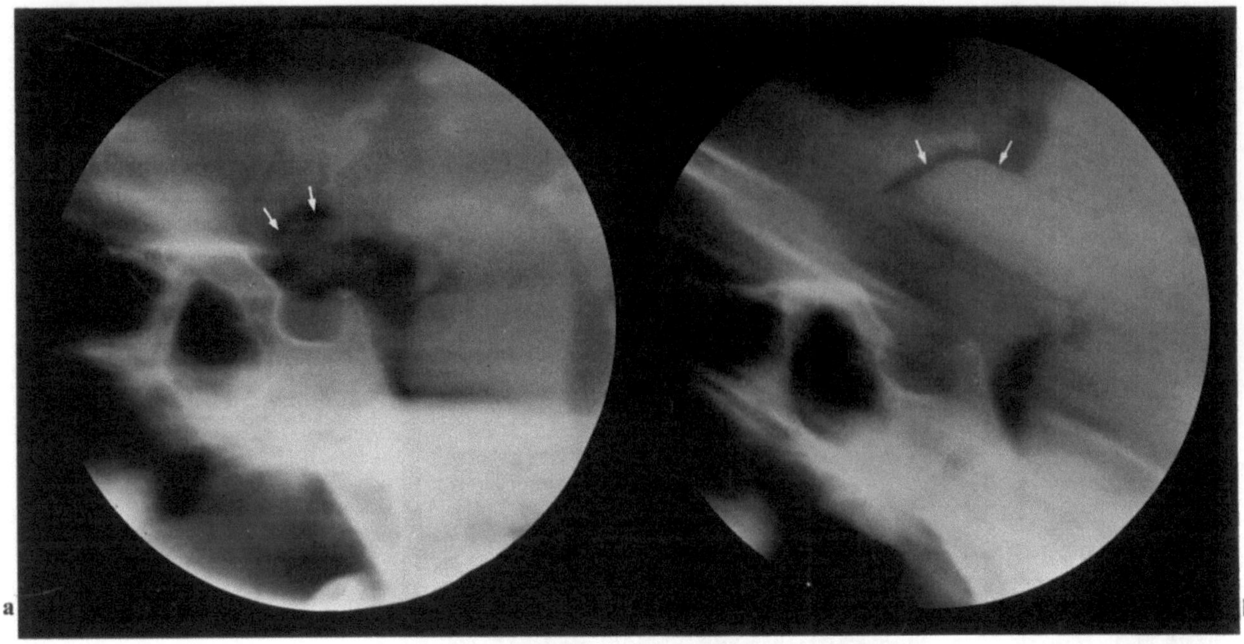 b

Fig. 7.10a, b. Suprasellar germinoma in an 8-year-old girl; lateral pneumoencephalograms.

a Diabetes insipidus (18.05.76). Normal sella turcica. Filling defect in the chiasmatic cistern *(arrows)*.

b Persistent diabetes insipidus and raised intracranial pressure (22.09.77). Large suprasellar tumor *(arrows)*. The dorsum sellae is slightly demineralized. The patient has a second tumor within the fourth ventricle

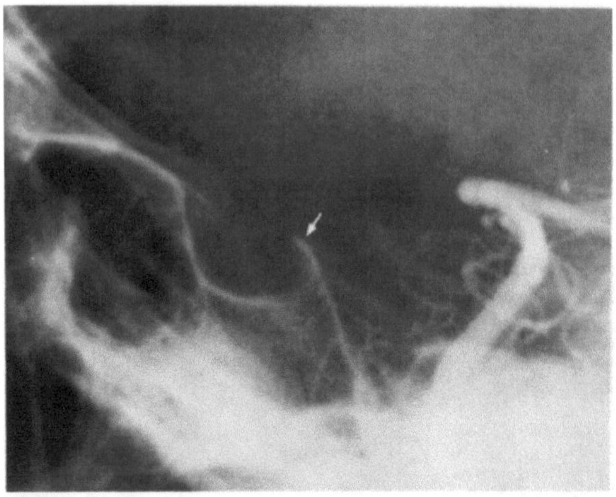

Fig. 7.12. Suprasellar epidermoid in a 50-year-old male with headache. Shortened, pointed dorsum sellae *(arrow)*. Note posterior displacement of the basilar artery

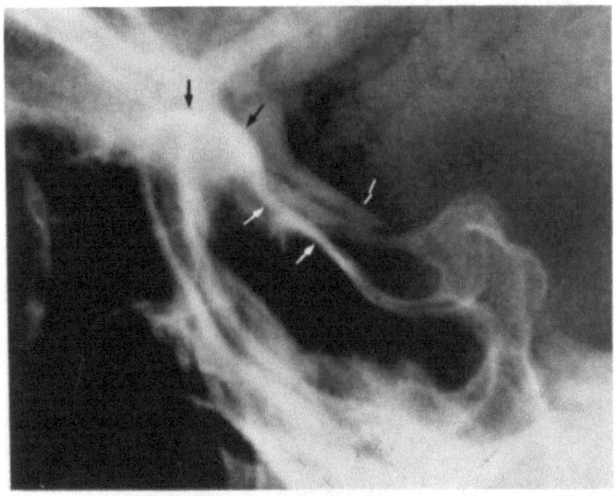

Fig. 7.11. Suprasellar arachnoid cyst in a 30-year-old female. Elongation and flattening of the chiasmatic sulcus *(white arrows)*. Slight blistering and hyperostosis of the planum sphenoidale *(black arrows)*. Downward displacement of the anterior clinoid processes, which appear tapered *(black and white arrow)* (Courtesy, D. Gardeur)

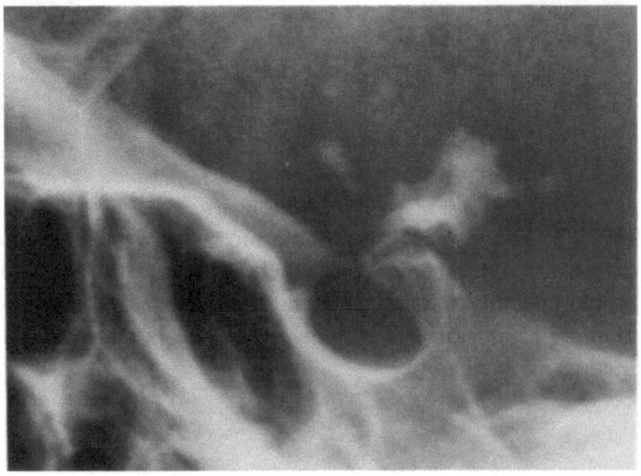

Fig. 7.13. Calcified optochiasmatic arachnoiditis following tuberculous meningitis. Note the cluster of suprasellar calcific deposits. The sella turcica is normal

Chapter 8 Presellar Pathology

Pathology of the presellar region has traditionally been typified by optic glioma in the child and presellar meningioma in the adult. In addition to these two particular diseases, routine examination of the presellar region may be very useful in everyday X-ray diagnosis.

A. Gliomas of the Optic Pathways (Nerve and Chiasm)[29]

I. General Considerations

Gliomas of the optic pathways occur particularly in children, where they represent 7% of all gliomas (Fowler and Matson 1957). Neurofibromatosis is present in about one-quarter of affected children, and approximately one-half of affected individuals if both children and adults are considered (Harwood-Nash and Fitz 1976). Optic gliomas usually occur between 2 and 9 years of age with the highest frequency around the age of 4. Gliomas of the optic nerves generally develop slowly and are sometimes present at birth (Holman 1959). Therefore, the developing sphenoid bone will grow around the tumor; this could explain the absence of decalcification of the walls of the optic canal in some cases. In addition, the form of certain parts of the sphenoid, especially the chiasmatic sulcus and the optic canals, would be determined by the tumor, which would stop growing during adolescence when the skull itself stops growing. This hypothesis is challenged by observations of size reduction of the optic canals after surgical treatment or radiotherapy of optic gliomas (Taveras and Wood 1964). In patients with von Recklinghausen's disease, the presence of a large optic canal does not necessarily indicate an optic glioma; it can also be caused either by hyperplasia of the optic nerve (Spencer and Borit 1967), or by an arachnoid cyst surrounding the nerve (Hoyt and Baghdassarian 1969).

Tumors affecting the hypothalamus and the chiasm grow more rapidly than gliomas of the optic nerves, hence resembling astrocytomas of the cerebral hemispheres; it is often impossible to know whether large tumors arise from the chiasm or the hypothalamus. Apart from involving the hypothalamus, these tumors involve an optic nerve in about one-quarter of cases; one or both optic nerves and the chiasm in half, and the chiasm alone in one-quarter of cases (Harwood-Nash and Fitz 1976).

Clinical and radiologic appearances of optic gliomas differ according to whether the neoplasm involves the optic nerve, and/or the chiasm or hypothalamus. In gliomas of the optic nerves, visual acuity remains normal in one-third of the cases. However, there is often gradual visual failure and optic atrophy or papilledema may occur. Proptosis is constant when there is extension into the orbital cavity. Reduction of visual acuity is more constant with gliomas affecting the chiasm and visual field defects may occur. Symptoms of raised intracranial pressure can be seen in large tumors, which may cause hydrocephalus by blocking the foramina of Monro. Very large tumors involving the hypothalamus may be accompanied by precocious puberty (particularly when there is related neurofibromatosis),

29 Refs. [159, 163, 278, 304, 305, 331, 425, 438, 439, 603, 612, 851, 895, 964, 1046]

panhypopituitarism, obesity, or diabetes insipidus.

II. Radiology

In gliomas of the optic nerve, an enlarged optic canal is a practically constant finding. On plain films of the optic canal, it is especially the thicker orbital segment which is apparent; limited enlargement of the cranial part of the canal may not be visible on conventional films (Potter and Trokel 1971). Axial tomograms are therefore of great importance. Initially, the optic canal loses its normal oval shape and becomes circular (Fig. 8.1). The cranial end of the optic canal appears deeper below the anterior clinoid processes. Since these tumors grow very slowly, there is rarely obvious erosion of the walls of the optic canal. In gliomas of the chiasm and optic nerves, radiographs are very seldom completely normal (about 5%) but the optic canals themselves are normal in one-third of cases.

Separation, undercutting, and tilting of the anterior clinoid processes may be seen; the clinoid processes appear abnormally "dumpy" on the lateral projection (Fig. 8.2). In addition, enlargement and excavation of the chiasmatic sulcus and erosion of the dorsum sellae and the posterior clinoid processes may be visible. The degree of these radiologic changes obviously depends on the size of the tumor and on its exact location. Pronounced excavation of the chiasmatic sulcus occurs particularly when an anterior, prefixed chiasm is affected by the glioma (Figs. 8.3, 8.4). Conversely, a glioma in a postfixed chiasm may give only very slight changes in the sella, with erosion of either the dorsum or a posterior clinoid process (Fig. 8.5).

The presence within the tumor of nodular calcification identical to those seen in craniopharyngiomas or hypothalamic gliomas is rare.

Finally, nonspecific signs of raised intracranial pressure, in particular with diastasis of the sutures, are present in about one-quarter of optic gliomas.

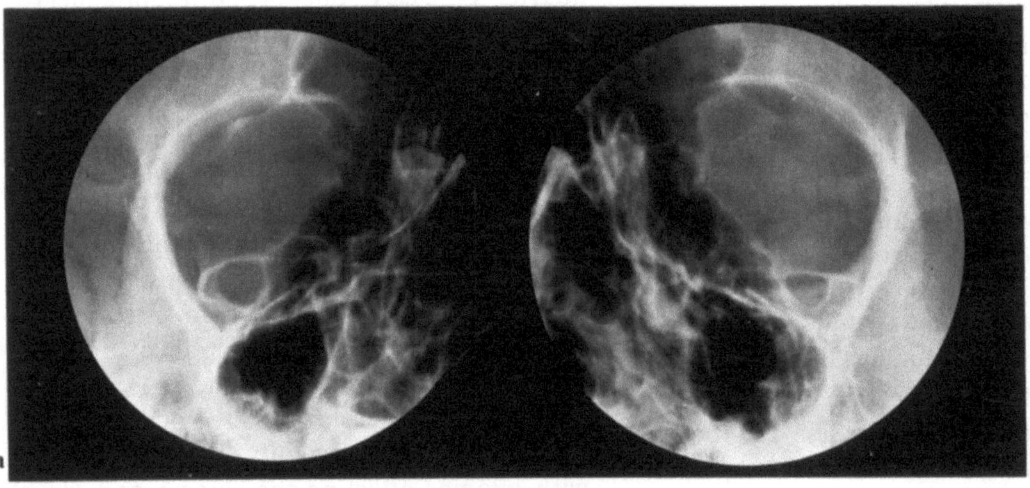
a b

Fig. 8.1 a, b. Optic nerve glioma; conventional views of optic canals. The right canal is enlarged and rounded; the cortical margin is intact

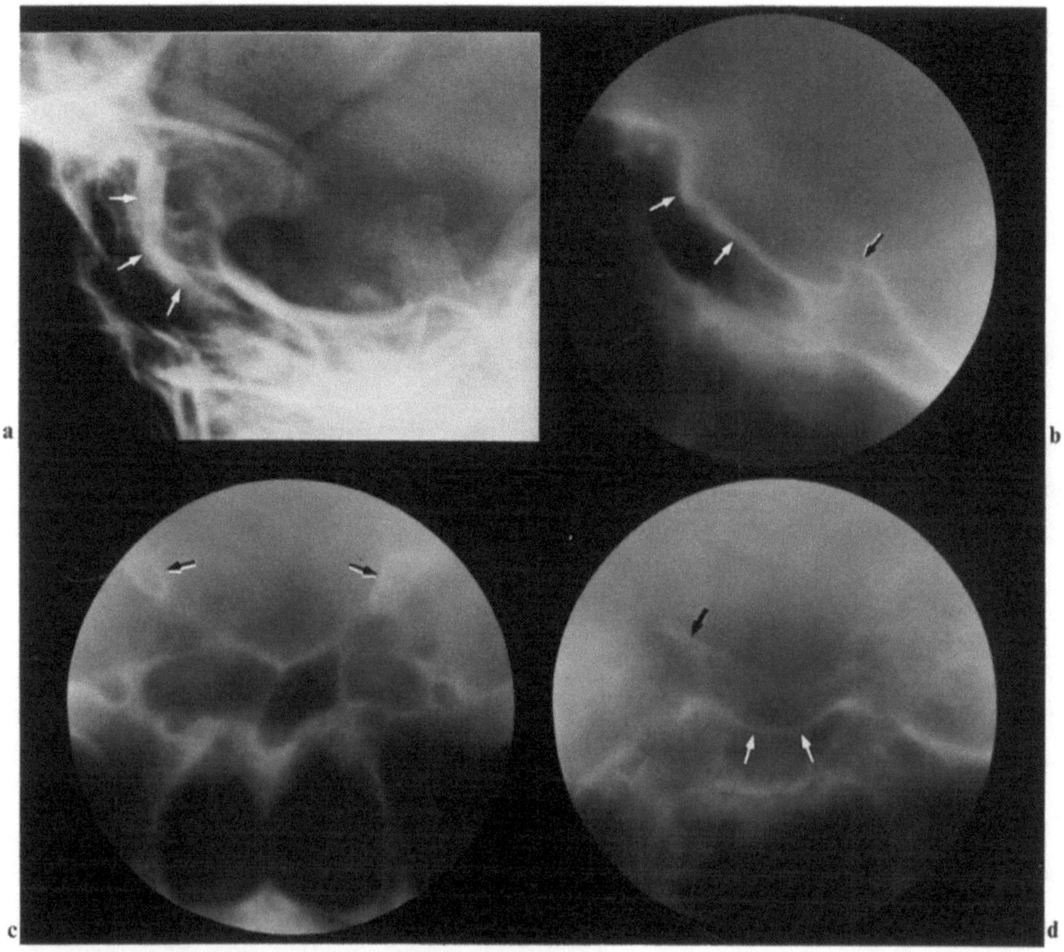

Fig. 8.2 a–d. Optic chiasm glioma in a 16-year-old girl.
a Lateral magnified projection of the sella turcica. Lengthening of the sellar region; the deepened chiasmatic sulcus *(arrows)* is nearly in line with the anterior margin of the sella. Undercutting of massive anterior clinoid processes, which appear increased in height.
b Midsagittal polytomogram: excavation and lengthening of the chiasmatic sulcus *(white arrows)*. Truncation of the middle part of the dorsum *(black and white arrow)*. The lamina dura is intact.
c, d Frontal polytomograms: tilting of the anterior clinoid processes whose superior aspect has become supermedial *(black and white arrows)*; regular depression of the posterior part of the floor of the sella *(white arrows)*; the posterior clinoid processes are lengthened and tapered, especially on the right *(black arrow)*

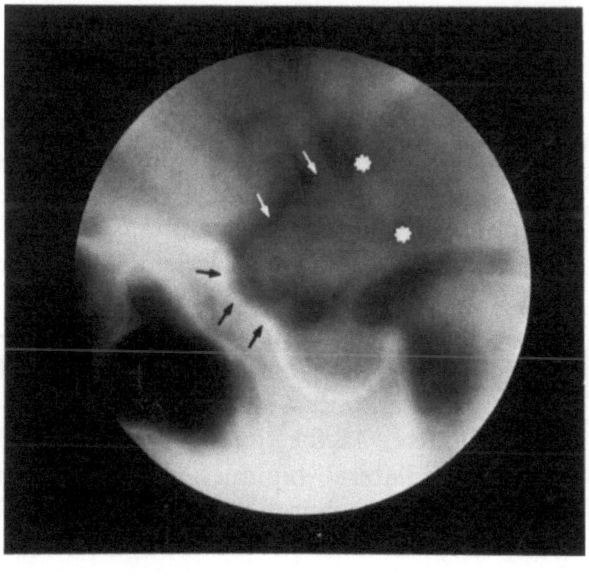

Fig. 8.3. Optic nerve and chiasm glioma with an anterior prefixed chiasm shown by a midsagittal pneumopolytomogram. Lengthening and deepening of the chiasmatic sulcus, without erosion of the cortical bone *(black arrows)*. The anterior recesses of the third ventricle *(stars)* are separated by the tumor *(white arrows)*. The dorsum sellae is almost normal

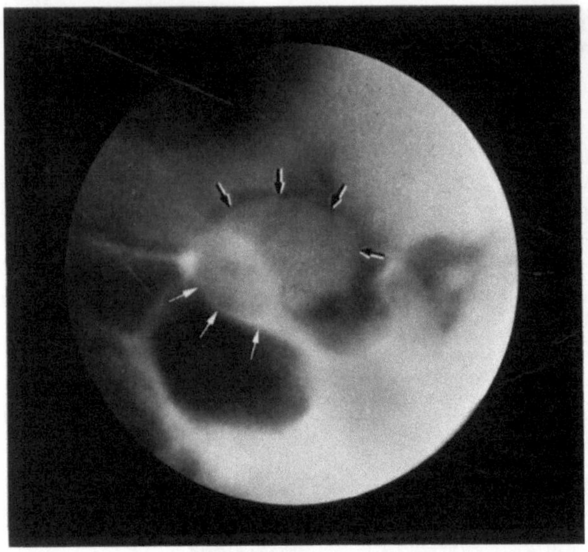

Fig. 8.4. Optic nerve and chiasm glioma with an anterior prefixed chiasm *(black and white arrows)*. Sagittal pneumopolytomogram, 5 mm to the right of midline. Marked excavation of the chiasmatic sulcus with thinning of the lamina dura *(white arrows)*. The dorsum sellae is intact

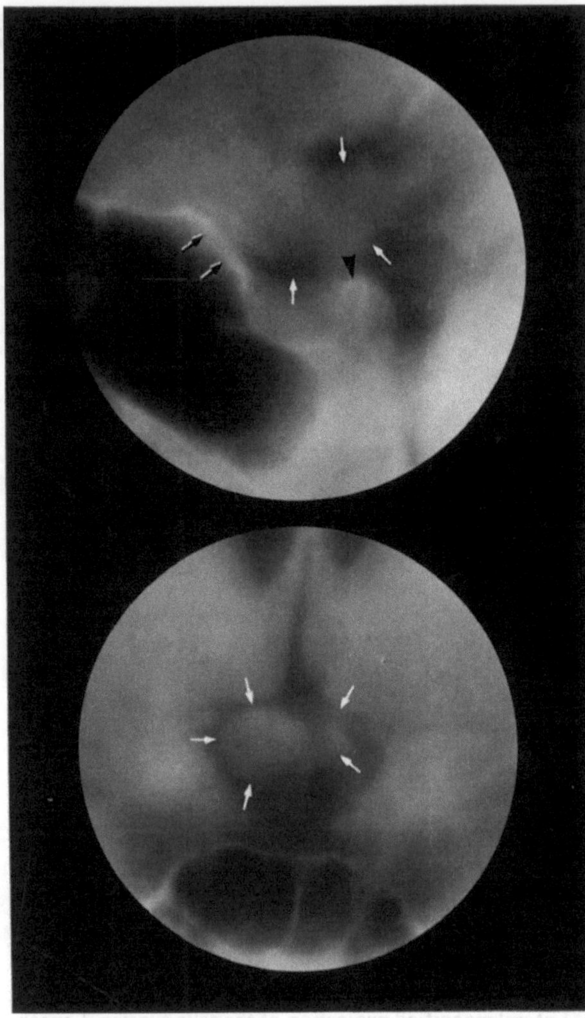

a

b

Fig. 8.5a, b. Optic nerve and chiasm glioma with a postfixed chiasm.
a Midsagittal pneumopolytomogram: slight lengthening of the chiasmatic sulcus, which is not excavated *(black and white arrows)*; the tip of the dorsum sellae is barely truncated *(arrowhead)*; air outlines the tumor *(white arrows)*.
b Frontal pneumopolytomogram: the floor of the sella is normal; the tumor is well outlined by air *(white arrows)* (Courtesy J. Metzger)

B. Presellar Meningiomas[30]

I. General Considerations

Presellar meningiomas arise on the anterior tuberculum sellae, the chiasmatic sulcus, or planum sphenoidale. Meningiomas of the diaphragma sellae are rare.

Presellar meningiomas occur particularly in women about 50 years of age. These tumors are slow growing; on average, 4 years separate the appearance of the first symptoms and surgical treatment. Visual disorders constitute the most important symptoms. Most commonly, there is unilateral deterioration of visual acuity, slowly progressive with slight pallor of the optic disk and occasionally a small temporal field defect. At the time of diagnosis and treatment, bilateral blindness is present in 10% of cases, and in the remainder visual acuity is lower than one-tenth on the worst-

30 Refs. [136, 187, 188, 227, 262, 264, 291, 300, 445, 454, 466, 548, 549, 551, 562, 579, 636, 651, 657, 962, 1015, 1021, 1023]

affected side. There is optic atrophy in 80% of cases.

Careful examination of the visual fields indicates unilateral impairment in only one-third of cases; in another third, total blindness is seen on one side and a temporal defect on the other. In the remaining cases impairment of the visual fields is more or less symmetric, leading to bitemporal hemianopia. Three out of four patients undergo surgery when visual acuity is lower than one-tenth on the worst-affected side (Metzger et al. 1976).

Other symptoms of presellar meningiomas are much less common: headaches, frontal signs, raised intracranial pressure, or convulsions are only found in very large tumors and more often with meningiomas of the olfactory groove. It should be noted once again that the presence of multiple meningiomas suggests the possibility of neurofibromatosis.

II. Radiology

Radiologic signs of presellar meningiomas include hyperostosis, calcification, blistering, and distant bony changes.

Hyperostosis (Di Chiro and Lindgren 1952; Lee 1976) is present in nine out of ten cases, but is not always obvious without tomography. An actual osteoma may be situated on the planum sphenoidale (Fig. 8.6), obscuring the angle formed by superimposed orbital roofs and the planum on the lateral projection (Fig. 8.7). More often the planum simply appears thick; this observation is particularly important if the thickness is irregular from front to back. In other cases, the planum sphenoidale loses its smooth appearance, its upper or lower surface appearing mamillated (Fig. 8.8). Finally, reactive hyperostosis may simply give the planum sphenoidale a laminated appearance. The following points should be remembered: the form of the hyperostosis does not reliably indicate the site of attachment of the meningioma, whose size is not related to the degree of hyperostosis. Hyperostosis in this region is often quite difficult to detect on plain films, especially since one is not used to analyzing this area. A strictly lateral projection must be examined very carefully both for any obscuring of the angle separating the orbital roofs from the planum sphenoidale and, especially, for any bony abnormality other than hyperostosis (Fig. 8.7).

Calcification above the presellar region may indicate a meningioma at the base (Fig. 8.10); however, it occurs in only 15% of cases and is very faint, sometimes apparent only on tomography (Fig. 8.11).

Blistering (Dyke 1941) is a fundamental sign which is present to a variable extent in more than half of the cases (Figs. 8.9, 8.11). The planum sphenoidale loses its flat appearance, showing an upward convexity. This may be the only indication of the presence of a meningioma, unassociated with any hyperostosis. Since this essential sign is simple to assess, it must be certain that the planum sphenoidale is flat before a lateral film of the skull is judged to be normal. Blistering, which represents a reaction of the planum sphenoidale to the presence of a meningioma is not, however, a pathognomonic sign. Other causes are discussed below.

In everyday practice, particularly when tomography is not available, blistering is always much easier to detect than an isolated hyperostotic reaction. These two features are frequently combined in presellar meningiomas.

Nearly two-thirds of cases show *bony changes at some distance* from the attachment of the meningioma, consisting of demineralization of the floor and dorsum of the sella, sometimes resulting in amputation of the upper part of the dorsum (Figs. 8.7, 8.8, 8.11).

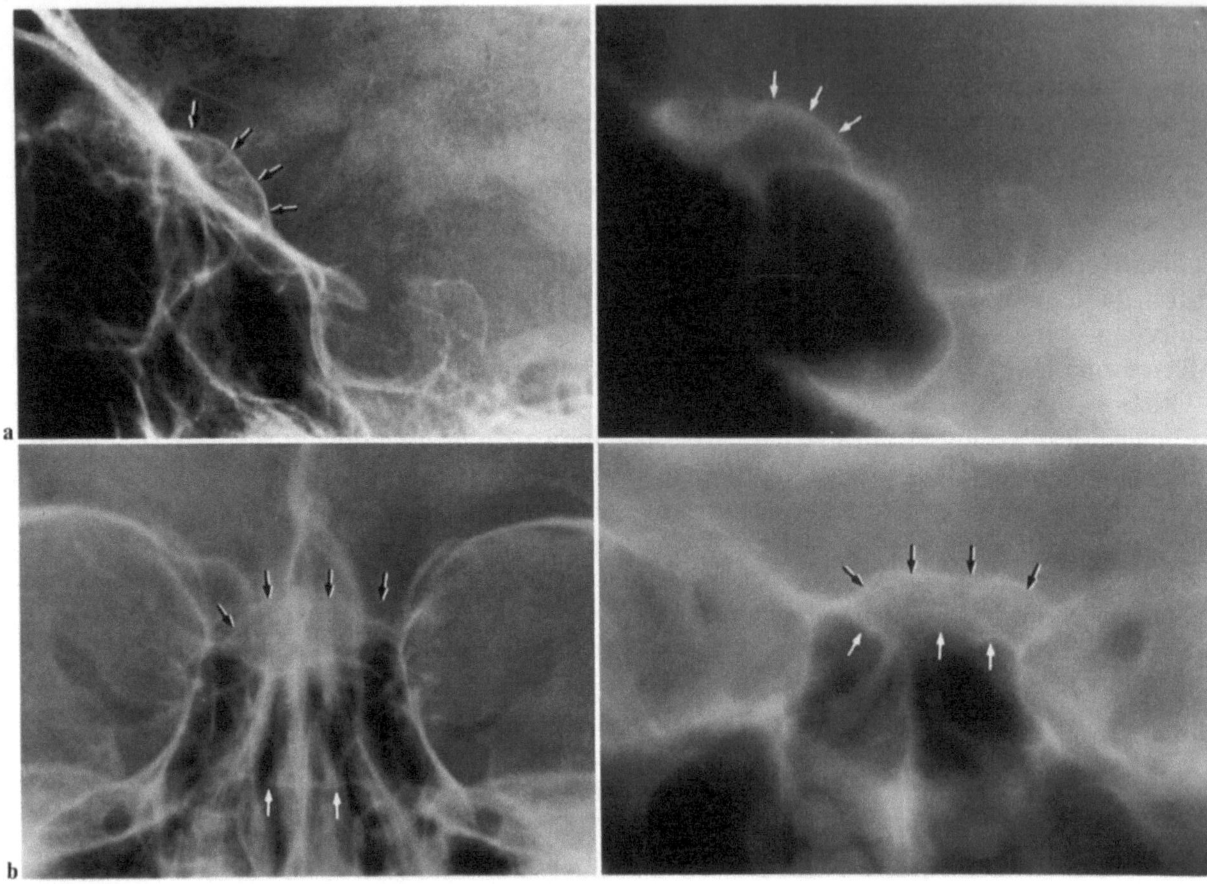

a d b c

△

Fig. 8.6a–d. Meningioma of the planum sphenoidale.

a Lateral projection: hyperostosis of planum *(arrows)*.

b Frontal projection. Hyperostosis of the planum *(black and white arrows)*. The floor of the sella is normal *(white arrows)*.

c Frontal tomogram: thickening of the planum *(black and white arrows)*.

d Sagittal tomogram: the hyperostosis is well shown; note also upward convexity of the sphenoid sinus *(white arrows)*

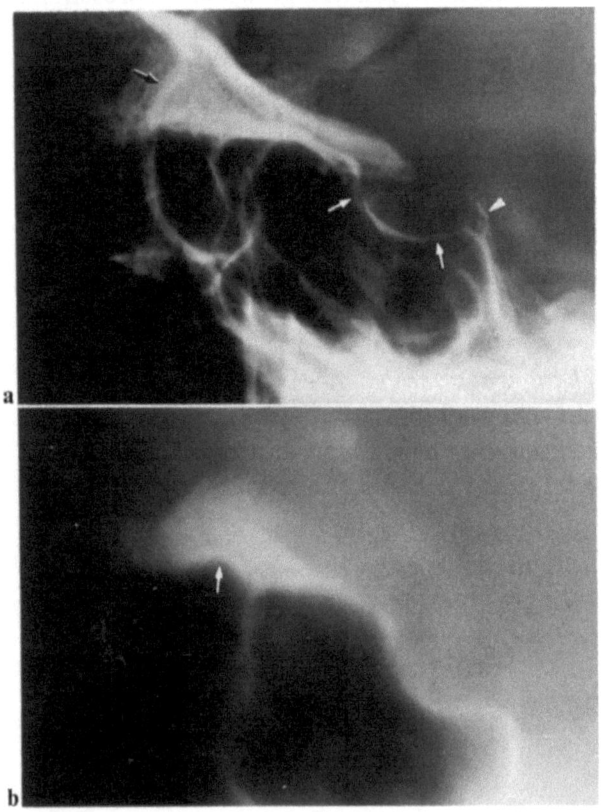

a b

Fig. 8.7a, b. Meningioma of the planum sphenoidale.

a Lateral projection: the angle between the roof of the orbit and the planum is obscured *(black and white arrow)*. Erosion of the floor of the sella *(white arrows)* and the dorsum sellae *(arrowhead)* is evident.

b Midsagittal polytomogram: dense hyperostosis of the planum; localized blistering of the anterior part of the planum *(arrow)*; erosive changes in the sella

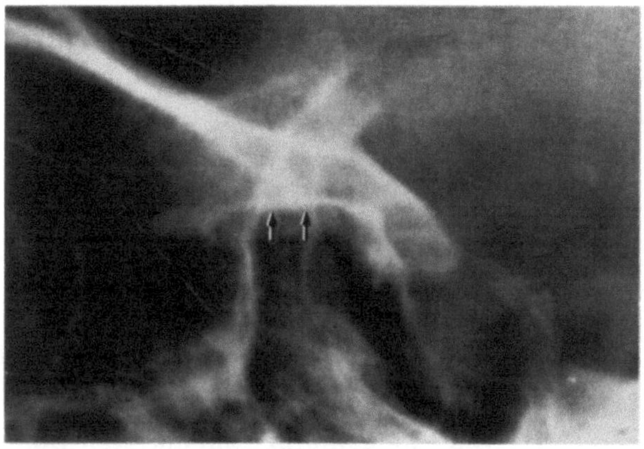

Fig. 8.8. Meningioma of the planum sphenoidale. Irregular appearance of the inferior surface of the planum *(arrows)*, without hyperostosis. Marked demineralization of the cortex of the sella turcica

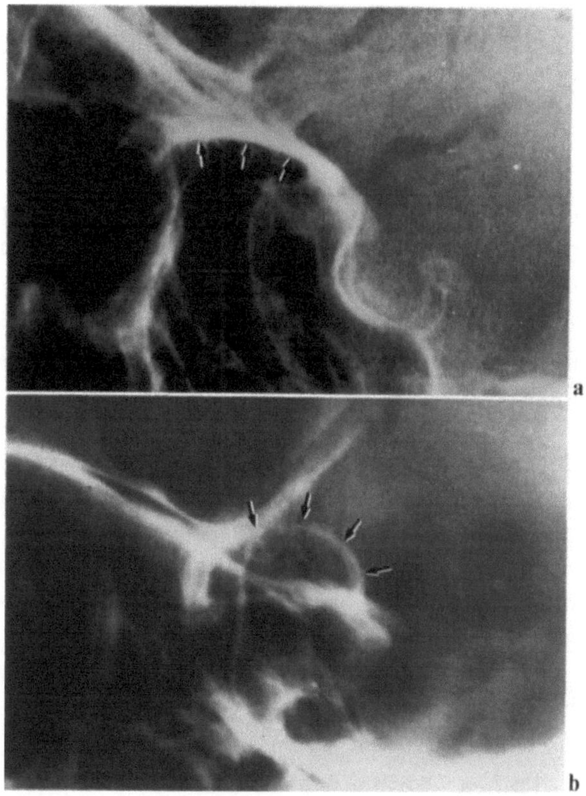

Fig. 8.9 a, b. Meningiomas of the planum sphenoidale; lateral projection. ▷

a Regular thickening of the planum, which has an abnormal upward convexity *(arrows)*. The sella turcica is normal.

b Marked blistering of the planum *(arrows)*. Ghostlike sella turcica

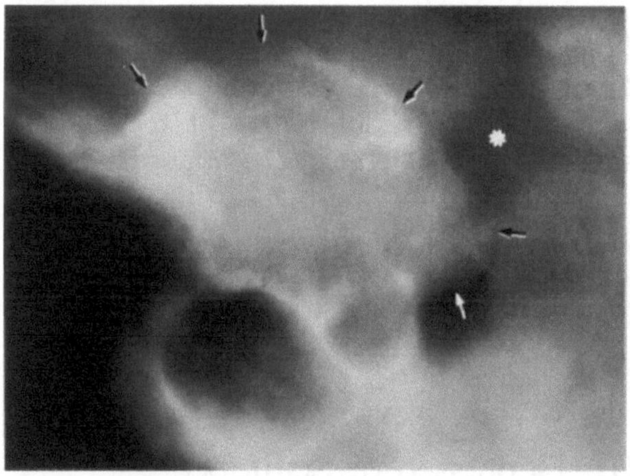

Fig. 8.10. Calcified meningioma of the planum sphenoidale in neurofibromatosis. Midsagittal pneumotomogram: extensive calcification above the sella and presellar region *(black and white arrows)*. Note that the anterior part of the third ventricle *(star)* is pushed backward by the tumor, which also occupies the interpeduncular fossa *(white arrow)*

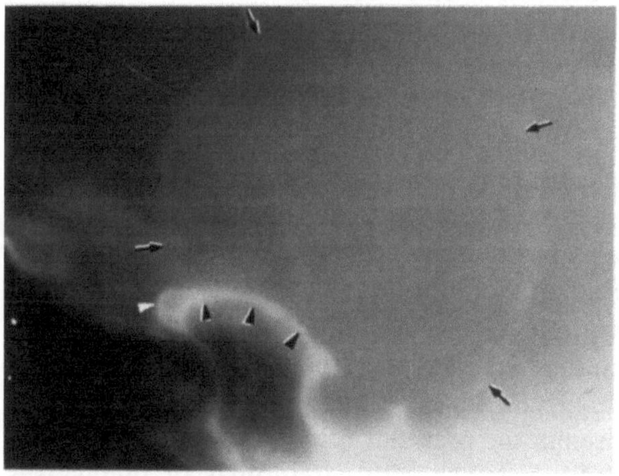

Fig. 8.11. Huge calcified meningioma of the planum sphenoidale; midsagittal polytomogram: the limits of the tumor are clearly outlined by faint calcification *(black and white arrows)*. Hyperostosis of the anterior part of the planum *(white arrowhead)*, with obvious blistering *(black and white arrowheads)*. The dorsum sellae is pointed

C. Diagnosis of an Abnormal Presellar Region[31]

Gliomas of the optic chiasm and presellar meningiomas are not the only diseases which modify the radiologic appearances of the presellar region. Differential diagnosis requires precise analysis of the presellar region, giving particular attention to the following:

Any abnormality of the planum sphenoidale, which may be too dense, too thick, too short, demineralized, depressed, scalloped, or blistered.

Changes in the chiasmatic sulcus, which may be hollowed or deepened, giving an omega sella.

Precise study of the presellar region may be extremely useful in everyday X-ray diagnosis.

I. Excessively Dense or Thick Planum Sphenoidale

A planum sphenoidale which is too dense, thick, uneven, or has a laminated appearance raises first the possibility of meningioma. Simple osteomas of the planum, the chiasmatic sulcus (Fig. 8.12), and the tuberculum may also be encountered. These osteomas, without related meningioma, may in some cases cause compression of the optic pathways simply due to their size. Hyperostosis (Belanger and Dyke 1943; Potter 1970; Lee et al. 1976) may be seen in the presellar region not only with meningiomas, but with carcinomas or metastases of either the sphenoid sinus or the nasopharynx, in contact with thrombosed aneurysms or dural arteriovenous malformations, with chronic sphenoid sinusitis, with a craniopharyngioma or even with an adenoma (Fig. 8.13). Such hyperostoses are never accompanied by blistering. However, hyperostosis, often present following fractures of the presellar region, may then be accompanied by blistering similar to that caused by meningiomas (Bonneville et al. 1978) (Figs. 8.18–8.20).

II. Excessively Short or Demineralized Planum Sphenoidale

The planum sphenoidale may appear to be short if the chiasmatic sulcus is long (Figs. 5.39, 5.40) or in the case of craniostenosis especially anterior, or hypotelorism. It may appear demineralized or eroded when in contact with a tumor (Fig. 8.25) or if local radiotherapy has been given. A planum having a shortened, demineralized appearance and which does not extend anteriorly beyond the greater wings of the sphenoid must raise the distinct possibility of chronically raised intracranial pressure. These changes always predominate in the anterior part of the planum and, at least initially, spare the chiasmatic sulcus and tuberculum sellae. Erosion related to raised intracranial pressure is very frequently seen with slowly growing tumors (Figs. 12.8, 12.5, 12.10), as well as in large cerebral and dural arteriovenous malformations (Bonneville et al. 1979). Shortening of the planum sphenoidale is all the more interesting since with certain slowly growing tumors it may appear before the increased intracranial pressure has begun to affect the sella turcica (Fig. 8.24).

III. Depressed or Scalloped Planum Sphenoidale

Routine analysis of the presellar region on all skull films sometimes reveals depression of the planum, indicating a former fracture, which is very often unrecognized. In addition, a good knowledge of this region makes immediate diagnosis of these fractures much easier (Fig. 8.21). Such fractures often cause rhinorrhea (Fig. 13.24). Contrary to classical opinion, one may observe in planum sphenoidale fractures a fragment displacement which is identi-

31 Refs. [92, 94, 95, 421 bis 501, 553, 610, 639, 671, 746, 833, 863, 961, 1028]

cal to the anomaly described as "unfused planum" (Smith and Kier 1971), (Fig. 8.22) (see also Fig. 5.47). Scalloping of the presellar area may be produced by advanced hydrocephalus.

IV. Blistering of the Planum Sphenoidale

Blistering of the planum sphenoidale has always been considered a sign strongly indicative of a presellar meningioma. It may also appear in contact with tumors originating within the sphenoid sinus. In fact, blistering of the planum sphenoidale may be seen in extremely varied conditions.

Blistering

In association with hyper-pneumatization	Constitutional Pneumosinus dilatans Acromegaly (?) Psychomotor retardation
"Malformative"	Phakomatoses
Reactive	Fibrous dysplasia Mucocoele Meningioma and other tumors Prolonged CSF shunting procedures Post-traumatic

Hyperpneumatization of the sphenoid sinus may be accompanied by an upward convexity of the planum without pathologic significance (Fig. 5.44). The term pneumosinus dilatans seems to be reserved for very rare cases in which expansion of the sinus is such that it may, like a tumor, compress the brain tissue or the optic pathways (Lombardi et al. 1968). In fact, pneumosinus dilatans may be associated with occult intracanalicular optic nerve meningioma (Hirst et al. 1979). Contrary to traditional notions (Wiggli and Oberson 1975) we have never seen blistering with acromegaly.

Severely mentally retarded patients often show an upward convexity of the planum sphenoidale (Figs. 8.16, 13.6). This malformation, greater than that which may accompany a highly pneumatized sphenoid, does not appear totally related to hyperpneumatization; we have observed this in young severely retarded children in whom pneumatization has not begun (Fig. 8.17). Discovery of such an anomaly always implies a very unfavourable mental prognosis (Bonneville and Dietmann to be published).

Blistering of the planum may be present in phakomatoses. We have observed this in the basal cell nevus syndrome and in von Recklinghausen's neurofibromatosis (Fig. 8.15). Such a malformation in the latter disease does not necessarily indicate the existence of a meningioma. Upward bowing and thickening of the planum sphenoidale may be observed after prolonged cerebrospinal fluid shunting procedures (Kaufman et al. 1970). Expansion of the planum sphenoidale is common in fibrous dysplasia (Figs. 8.14, 13.27, 13.28), however, the general appearance of the bone makes its diagnosis quite simple (Leeds and Seaman 1962). If they are situated in the anterior part of the sphenoid, mucocoeles may expand the planum sphenoidale; diagnosis is simplified both by the opacity of the sinus and the bulging of its walls (Nugent et al. 1970).

Fractures of the planum sphenoidale often heal, creating an appearance analogous to that of meningiomas, combining blistering and hyperostosis (Figs. 8.18–8.20) (Bonneville et al. 1978). These changes, which occur during the 1st year following trauma, appear only if there is a certain air pressure within the adjacent sinus. Blistering does not occur as long as the sinus remains opaque. These fractures of the skull base may subsequently cause visual disorders. In such cases, possible blistering of the planum sphenoidale may be attributed to a known injury but not to a meningioma.

V. Abnormal Chiasmatic Sulcus

When the chiasmatic sulcus appears depressed or deepened, creating eventually the appearance of an omega-shaped sella turcica, a glioma of the chiasm must be suspected. This appearance may also be seen in suprasellar tumors, particularly craniopharyngiomas (Fig. 7.5) and rarely in pituitary adenomas. This pattern has been described in the Hurler syndrome (Fig. 13.11); it may definitely be related to the presence of an arachnoid cyst. Lengthening, verticalization, and deepening of the chiasmatic sulcus occur very frequently in certain types of hydrocephalus, especially aqueduct stenosis (Fig. 8.26). Osteolysis of metastatic origin, limited to the sulcus, may also change its appearance (Fig. 8.27).

It is also possible to observe post-traumatic depression of the sulcus, a diagnosis which may be difficult if preceding trauma is unrecognized (Fig. 8.23).

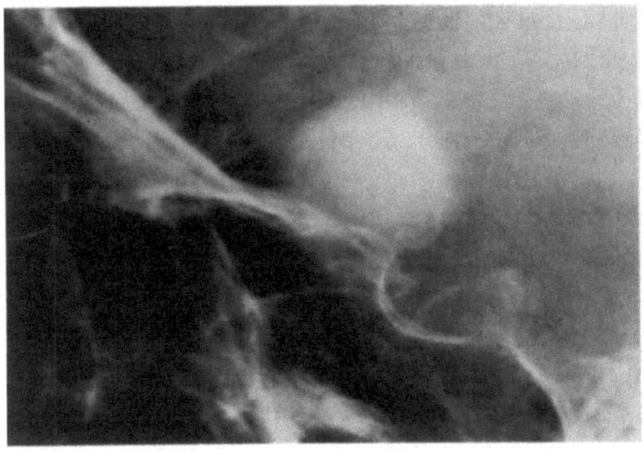

Fig. 8.12. Osteoma of the chiasmatic sulcus without meningioma

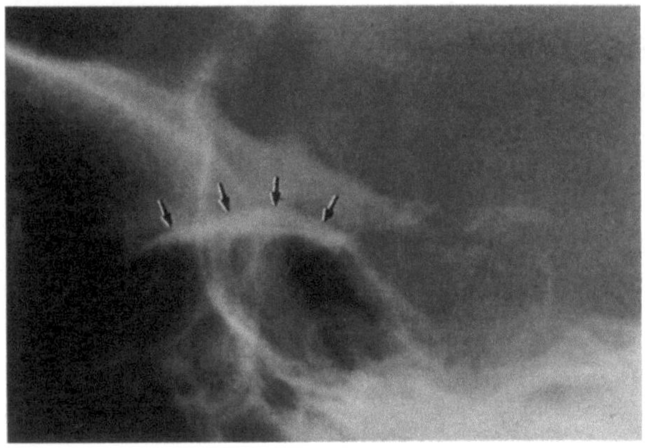

Fig. 8.13. Presellar hyperostosis unrelated to meningioma *(arrows)*. Chromophobe adenoma; the sella is destroyed

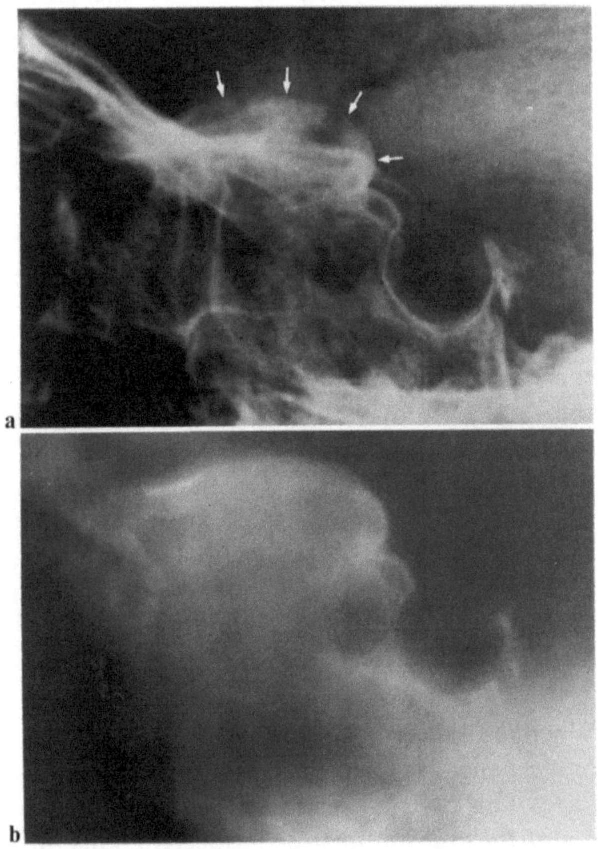

Fig. 8.14a, b. Fibrous dysplasia.
a Lateral projection and **b** midsagittal polytomogram: marked, regular blistering of the planum sphenoidale *(arrows)*. A predominantly soft tissue mass obliterates the sphenoid sinus

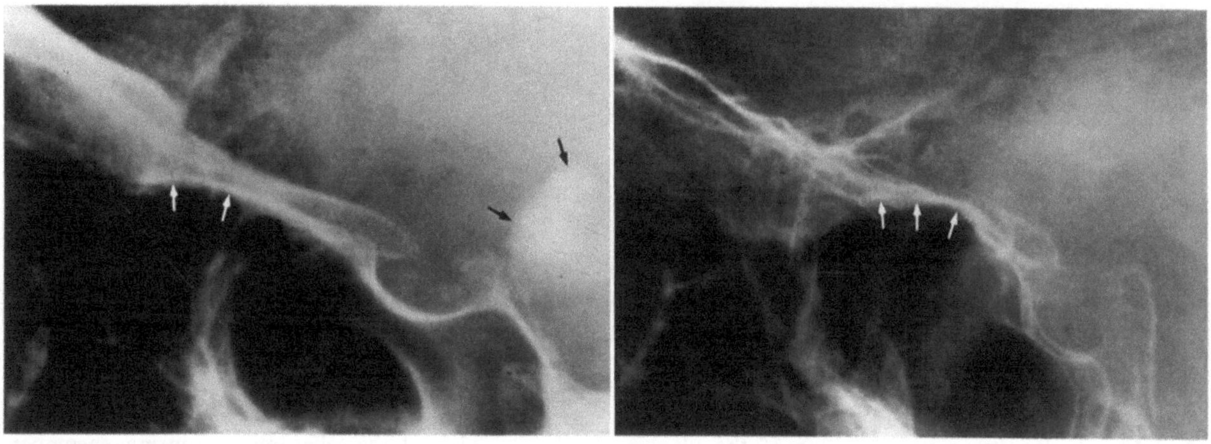

a b

Fig. 8.15a, b. Slight blistering of the planum sphenoidale *(white arrows)* in phakomatoses.
a Neurofibromatosis; note a calcified temporal tumor *(black arrows)*.
b Basal cell nevus syndrome

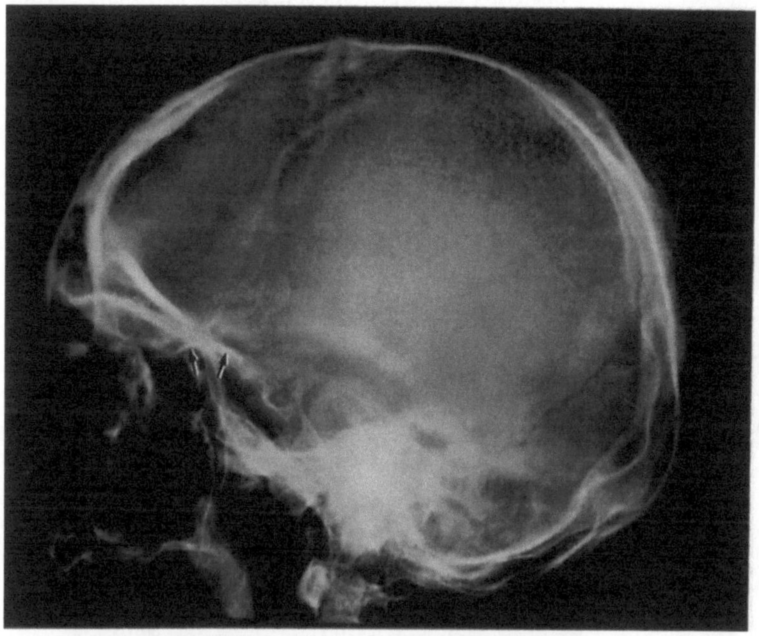

Fig. 8.16. Blistering of the planum sphenoidale in cerebral hypoplasia *(arrows)*. Note the abnormally thick cranial vault

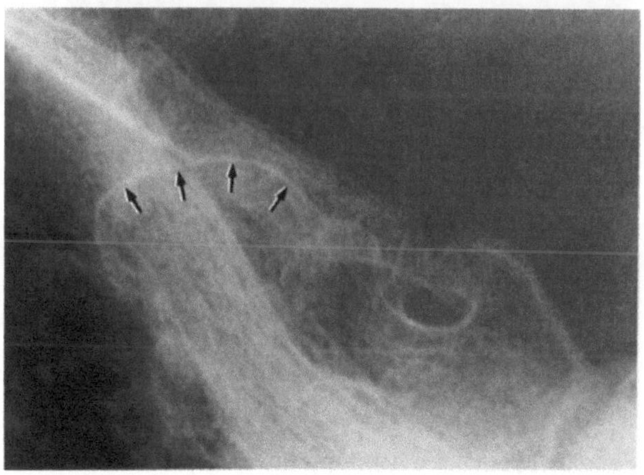

Fig. 8.17. Blistering of the planum sphenoidale *(arrows)* in a 2-year-old boy with severe psychomotor retardation. The sphenoid bone is not pneumatized

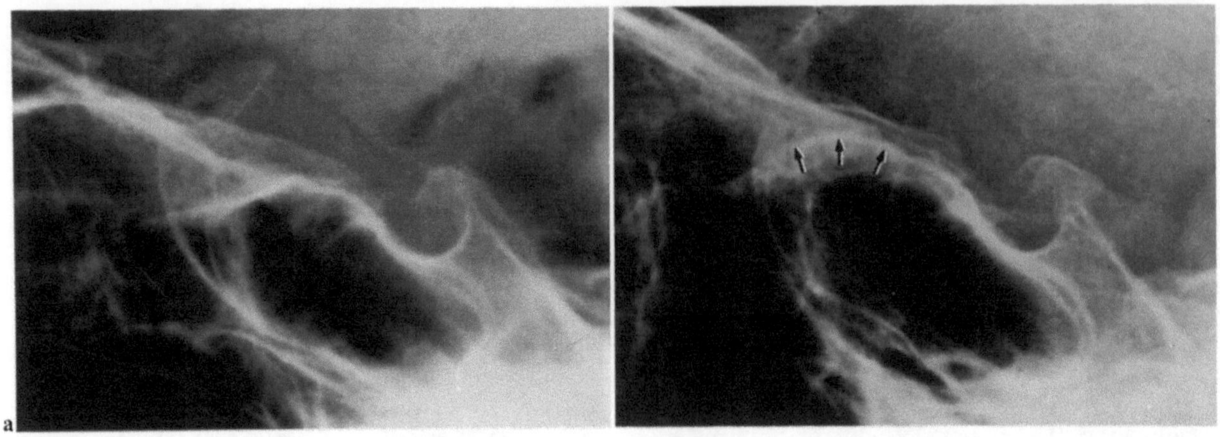

Fig. 8.18 a, b. Post-traumatic blistering of the planum sphenoidale.

a Lateral view of the sellar region (during pneumoencephalography) in an 18-year-old patient. A bifrontal fracture involving the floor of the anterior cranial fossa was demonstrated on frontal skull view (not shown). The lateral view appears normal.

b Lateral view of the same patient, now complaining of visual loss, 7 years later. There is hyperostosis and blistering of the planum sphenoidale *(arrows)*. Carotid angiogram and pneumoencephalogram normal (reproduced from Bonneville et al. 1978 by permission of the publisher)

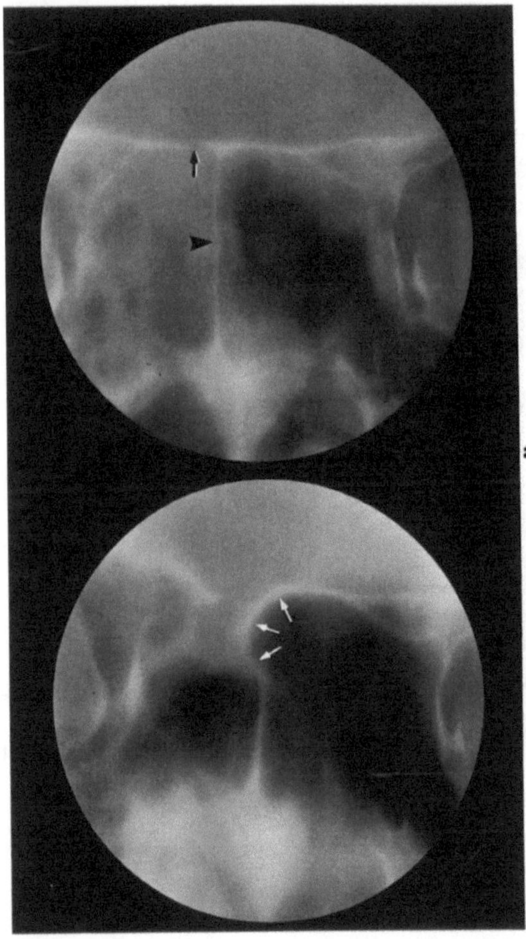

Fig. 8.19 a, b. Post-traumatic blistering of the planum sphenoidale.

a Frontal tomogram in a 29-year-old patient with severe head injury. Undisplaced fracture of the right side of the planum *(arrow)*; undisplaced fracture of the intersinus septum of the sphenoid sinus *(arrowhead)*. There is blood within the sinus, particularly on the right.

b Follow-up frontal tomogram 2 years later. Definite bulging to the right of the intersinus septum *(white arrows)*; slight elevation of the planum

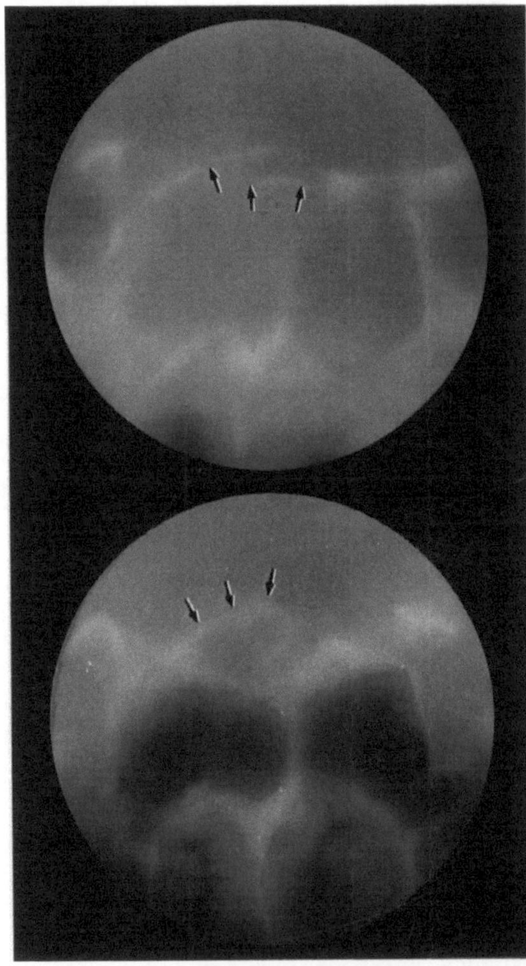

Fig. 8.20 a, b (caption see opposite page)

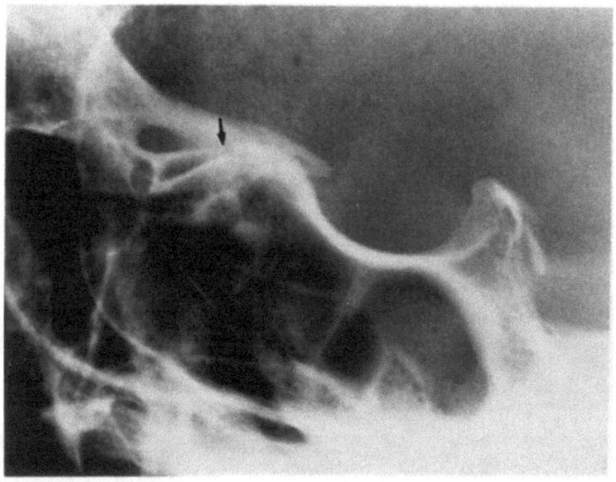

Fig. 8.21. Fracture of the planum sphenoidale (lateral magnified projection) *(arrow)*

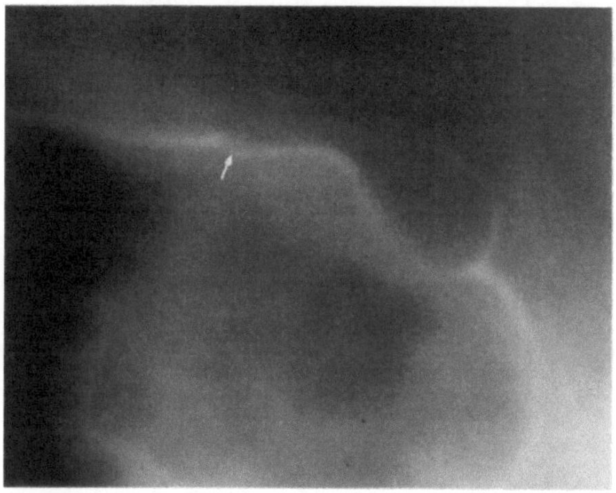

Fig. 8.22. Fracture of the planum sphenoidale *(arrow)*, shown by midsagittal tomogram. The posterior fragment of the planum is depressed, resembling an unfused planum. Haziness of the sphenoid sinus

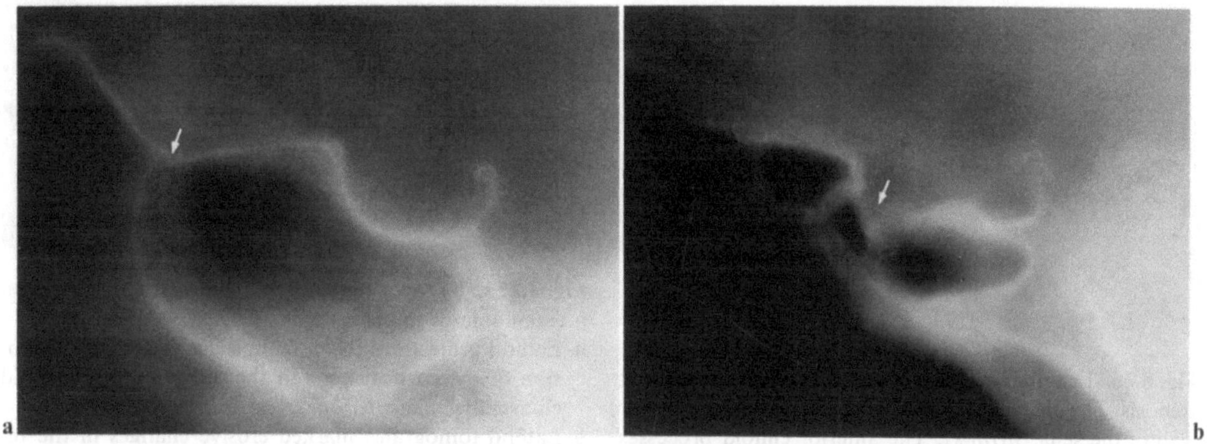

Fig. 8.23a, b. Old fractures of the presellar region on midsagittal polytomograms *(arrows)*

Fig. 8.24. Longstanding raised intracranial pressure due to temporal astrocytoma. Demineralization and shortening of the planum sphenoidale *(arrow)*. The sella itself is almost normal

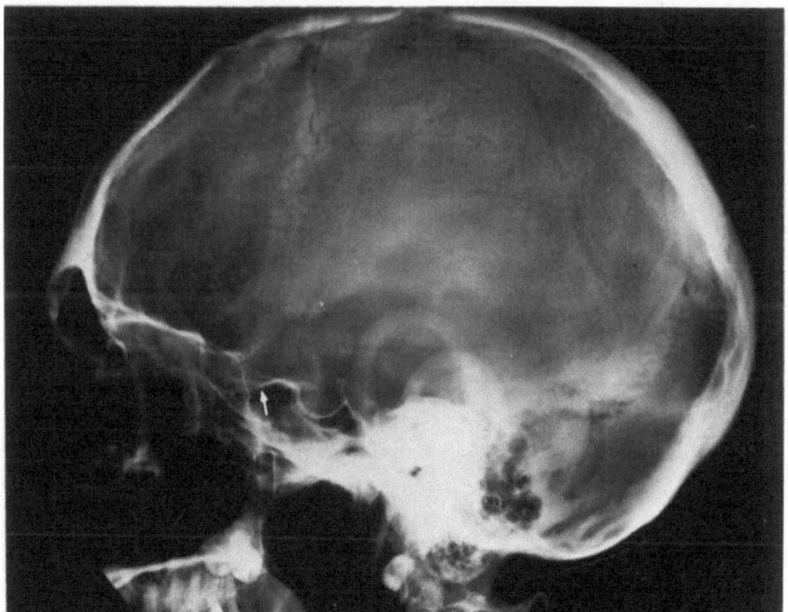

Fig. 8.20a, b. Post-traumatic blistering of the planum sphenoidale.
- **a** Frontal tomogram in a 29-year-old patient with head injury. Fracture of the planum *(arrows)* with blood in the sphenoid sinus, which appears cloudy.
- **b** Frontal tomogram 2 years later: clearing of the sinus, hyperostosis and blistering of the planum *(arrows)*

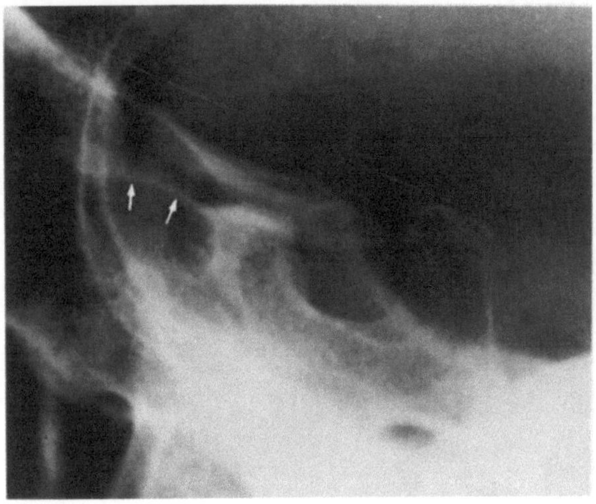

Fig. 8.25. Carcinoma of the ethmoid sinus. Erosion and thinning of the planum *(arrows)*

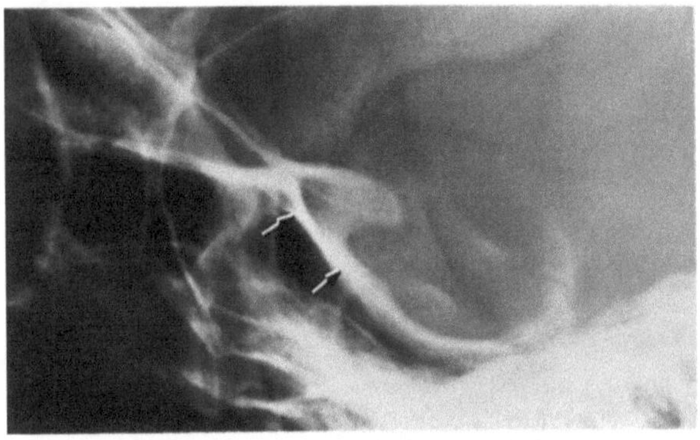

Fig. 8.26. Lengthening of the sellar region in aqueduct stenosis. The chiasmatic sulcus is in line with the anterior wall of the sella *(arrows)*. The anterior clinoid processes are dumpy

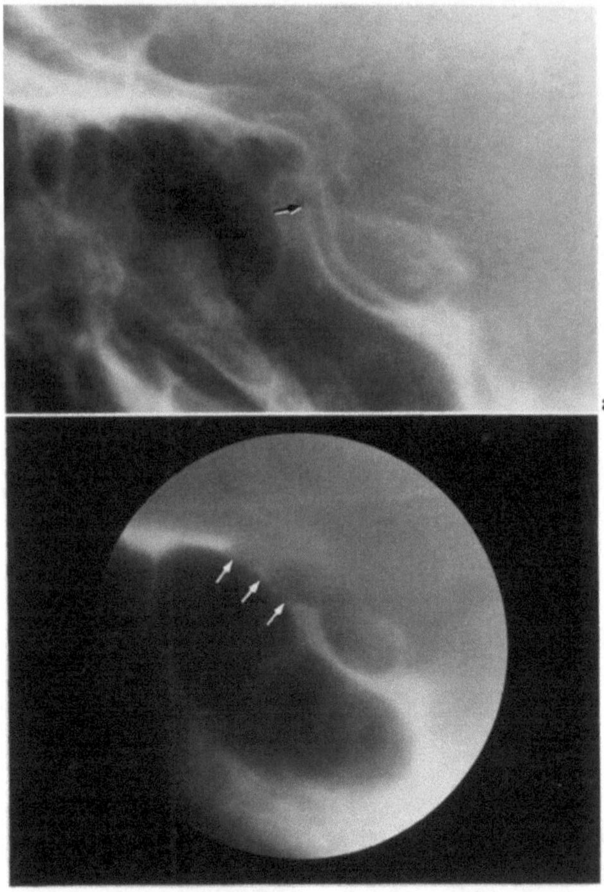

a

b

Fig. 8.27a, b. Metastasis to the chiasmatic sulcus from a carcinoma of the breast.

a Lateral projection: the dorsum sellae is destroyed; erosion of the cortical bone of the tuberculum sellae and chiasmatic sulcus *(arrow)*.

b Lateral tomogram: marked erosive changes in the region of the chiasmatic sulcus *(arrows)*

Chapter 9 Parasellar Pathology

The changes produced in the sella turcica by parasellar lesions are variable. Those lesions closest to the sella turcica bring about the greatest bony changes. These changes are unilateral, and frontal projections are essential to show the asymmetry.

Clinically, parasellar lesions cause compression of the vessels and nerves in the cavernous sinus, resulting in paralysis of the third (oculomotor), fourth (trochlear), and sixth (abducens) cranial nerves, and the ophthalmic division of the fifth (trigeminal). Compression with stenosis, or even obstruction, of the internal carotid artery may also occur. Moreover, extension of these parasellar lesions may give rise to compression of the medial aspect of the temporal lobe (provoking temporal lobe epilepsy), or of the pituitary gland (with panhypopituitarism), and compression, often unilateral, of the visual pathways, or of the vessels and nerves traversing the superior orbital fissure. A carotid-cavernous fistula (following trauma, or rupture of an intracavernous aneurysm) causes uni- or bilateral pulsatile exophthalmos.

A. Vascular Disease*[32]

I. Aneurysms of the Internal Carotid Artery

Intracavernous aneurysms constitute the main source of bony changes in the sellar region. However, aneurysms of the supracavernous internal carotid artery may also cause changes in the sella turcica, as may certain large aneurysms of the horizontal segment of the anterior cerebral artery or the anterior communicating artery.

Intracavernous aneurysms of the internal carotid artery may extend into the pituitary fossa, compress the pituitary gland, and cause panhypopituitarism, simulating a pituitary tumor (Mitchell 1968; Cushing 1912; Lodge et al. 1927; White and Ballantine 1961; Dussault et al. 1969; Houdart et al. 1975). Detailed analysis of bone changes in the sellar region is therefore of some importance.

The changes most characteristic of intracavernous aneurysms are:

1) Bony Changes

In the floor of the sella; on the frontal projection, there is depression of the floor in relation to the aneurysm (Fig. 9.2). The cortical bone is thinned (Fig. 9.3) or destroyed, or more rarely dense. On the lateral projection the sella turcica has a ballooned appearance with a double floor (Fig. 9.3). The lateral wall of the sphenoid sinus may be destroyed (Fig. 9.1).

In the anterior clinoid processes; on the side of the aneurysm, the anterior clinoid process is elevated, its lower surface being eroded (Fig. 9.1). In the cases of large aneurysms, it may be destroyed (Lombardi et al. 1963; Houdart et al. 1975).

In the dorsum sellae and posterior clinoid processes; unilateral erosion of the lateral portion of the dorsum sellae is a frequent finding.

In the superior orbital fissure and optic canal; widening of the superior orbital fissure is fre-

---

* For tortuosity of the carotid siphons see Chap. 5; for carotid siphon calcification see Chap. 13; for persistent trigeminal artery see Chap. 13.

32 Refs. [7, 23, 43, 48, 182, 241, 258, 286, 341, 342, 436, 576, 578, 623, 645, 701, 787, 841, 857, 858, 894, 980, 1017, 1037, 1050].

quent (Jefferson et al. 1937). The optic strut which unites the optic canal and the superior orbital fissure is often destroyed (Alajouanine et al. 1955; White and Ballantine 1961).

In the middle cranial fossa; Lombardi et al. (1963) commented on the frequency of erosion of the floor of the middle cranial fossa. This may be very extensive (Fig. 9.1). Frontal tomograms and axial views are essential for demonstration of these abnormalities.

In the petrous apex; very large aneurysms may erode the tip of the petrous pyramid (Lombardi et al. 1963).

2) Calcification

Calcification is frequent (30% of cases); it may be juxta-, supra-, and/or intrasellar (Fig. 9.1). Eggshell calcification may outline the aneurysm completely (Lombardi et al. 1963).

3) Presence of Soft Tissue Mass Within the Sphenoid Sinus (Figs. 9.1, 9.4)

Certain workers have noted the frequent occurrence of bilateral intracavernous aneurysms (Seltzer and Hurteau 1957; Gouaze et al. 1967).

Supracavernous aneurysms of the internal carotid artery cause bony changes less often, as they are usually smaller. When they are large, as is the case with certain aneurysms of the horizontal segment of the anterior communicating artery, they may erode or destroy the anterior clinoid process on one or both sides. Deepening of the chiasmatic sulcus or excavation of the planum sphenoidale may be seen (Wren 1969; Raymond and Tew 1978).

II. Carotid-Cavernous Fistulas

Plain films are of little value, since the history (e.g., trauma) and clinical findings (pulsating exophthalmos) often indicate the diagnosis. Long-standing carotid-cavernous fistulas erode the lateral portion of the floor of the sella, and of the lower part of the dorsum sellae (due to a dilated basilar sinus). Widening of the superior orbital fissure by a dilated superior ophthalmic vein is common. Radiologic investigation of the region of the sella may reveal the cause of a carotid-cavernous fistula: it is often possible to observe a fracture of the floor of the sella in post-traumatic fistulas.

B. Meningiomas[33]

Meningiomas, arising in the dural walls of the cavernous sinus, will invade the sinus and destroy the more lateral bony structures of the sellar region. Meningiomas which arise from the anterior clinoid process and the medial part of the sphenoid ridge or from the floor of the middle cranial fossa (greater sphenoid wing) provoke sellar changes at a later stage.

1. Meningiomas arising from the lateral wall of the cavernous sinus erode the walls of both the sella turcica and the sphenoid sinus on one side (Figs. 9.6, 9.7). They may extend to erode the anterior clinoid process and widen the superior orbital fissure (Fig. 9.7).

2. Meningiomas arising from the medial part of the greater sphenoid wing are often accompanied by hyperostosis, and lead at a later stage to unilateral erosion of the lateral walls of the sella turcica.

3. Meningiomas of the anterior clinoid process and the medial surface of the sphenoidal ridge produce hyperostosis at their insertion (Fig. 9.8); the roof of the optic canal may show hyperostosis (du Boulay 1965). These tumors cause changes in the sella turcica as a late phenomenon.

C. Gasserian Neurinomas and Meningiomas[34]

Neurinomas of the gasserian ganglion are rare. The sella turcica is directly affected only

33 Refs. [332, 908, 965]

34 Refs. [13, 302, 334, 340, 403, 422, 427, 465, 624, 706, 810, 842]

when the tumor is large. The tip of the petrous bone is first eroded. When the tumor extends toward the middle cranial fossa, the lateral margin of the dorsum sellae, the floor of the sella and the anterior clinoid process may be eroded (Holman et al. 1961). The foramina ovale and rotundum may be widened. Large tumors sometimes cause considerable destruction of the floor of the middle cranial fossa; the superior orbital fissure may be widened (Mellow and Tänzer 1972).

Meningiomas of Meckel's cave destroy the petrous apex and may erode the lateral margin of the dorsum sellae.

D. Diseases of Temporal Lobe

I. Gliomas

Certain large high-grade gliomas may invade the skull base and erode the dorsum sellae, the floor of the sella and, laterally, the anterior clinoid process.

II. Epidermoids

Large slow-growing tumors may erode the lateral border of the sella turcica and the sphenoid sinus (Fig. 9.9).

III. Lipoid Proteinosis[35]

Described by Urbach and Wiethe in 1929, lipoid proteinosis (Hyalinosis cutis, lipoglycoproteinosis, lipoidosis cutis and mucosae, Urbach-Wiethe disease) is a rare disease characterized by nodular deposits of glycoprotein and lipid in the dermis, in numerous mucous membranes (lips, oral and pharyngeal mucosa, larynx), and in the central nervous system. The disease causes characteristic intracranial calcifications, first described by Ramos and Silva (1943), situated symmetrically in the anteromedial part of the temporal lobe (located around capillaries of the hippocampal gyri).

On the lateral projection, the calcification overlying is projected above the dorsum sellae (McCusker and Caplan 1962; Ozonoff and Burrows 1971). They are bean, comma, or sickle-shaped. In Caldwell's view, the calcification is projected symmetrically on both sides of the sella turcica in the lateral part of the superior orbital fissures (Francis 1975). Williams et al. (1972) reported a case of hippocampal calcification without cutaneous manifestations of lipoid proteinosis.

35 Refs. [307, 614, 765, 971, 1004, 1022]

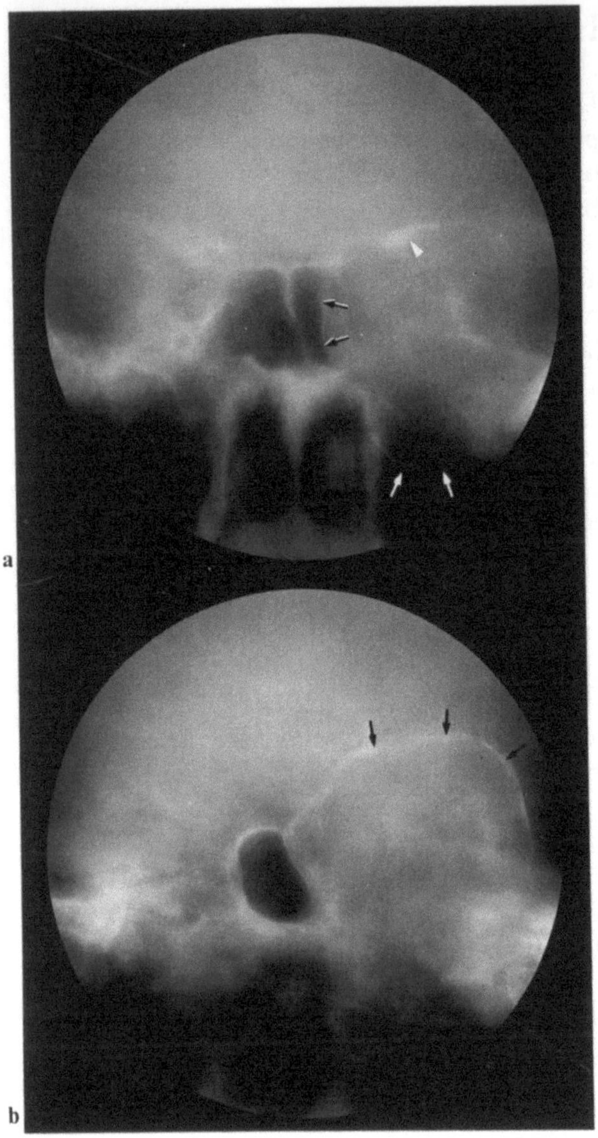

a

b

Fig. 9.1 a, b. Intracavernous aneurysm of the left internal carotid artery.

- **a** Frontal tomogram: anterior section taken through the superior orbital fissure. The left lateral wall of the sphenoid sinus and the medial part of the floor of the middle cranial fossa are destroyed. A soft tissue mass bulges into the left sphenoid sinus *(black and white arrows)*. The medial part of the lesser wing is displaced upward and eroded *(arrowhead)*. Note fine eggshell calcification outlining the lateral, medial, and inferior portion of the wall of the aneurysm *(white arrows)*.
- **b** Frontal tomogram: posterior section taken through the floor of the sella; thin eggshell calcification outlines the superior and lateral margins of the aneurysm *(black arrows)*

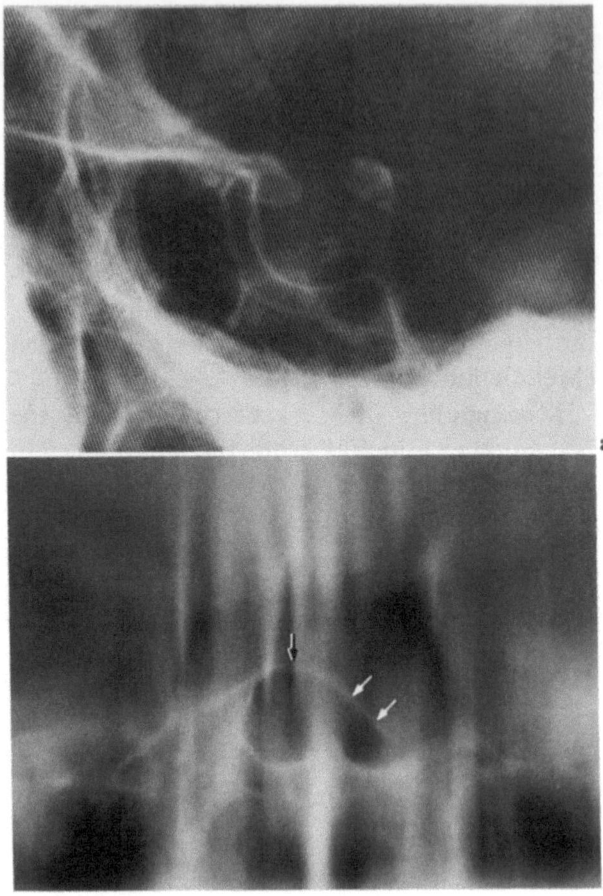

a

b

Fig. 9.2 a, b. Intracavernous aneurysm of the left carotid artery. 50-year-old female with left third and fifth cranial nerve palsies.

- **a** Lateral projection: the sella turcica seems normal, without doubling of its floor.
- **b** Frontal linear tomogram: the floor of the sella is narrowed *(black and white arrow)* and the lateral wall of the sphenoid sinus is displaced medially *(white arrows)*

a
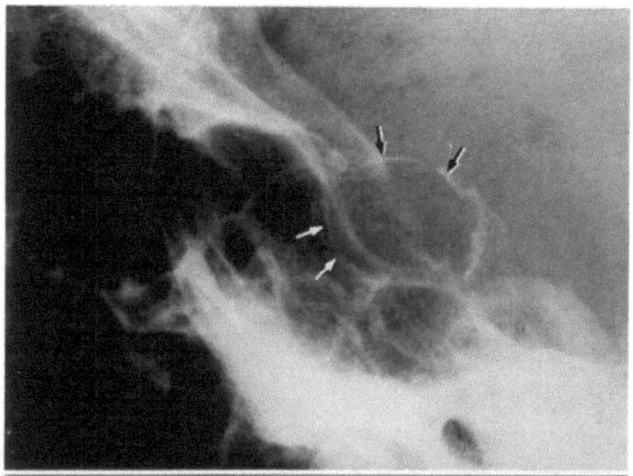

b
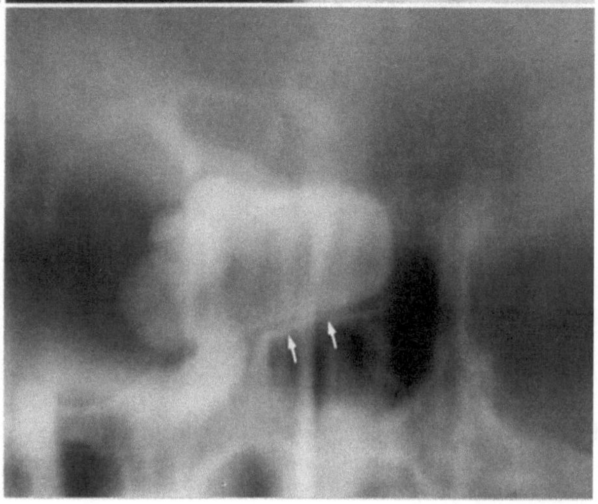

Fig. 9.3a, b. Intracavernous aneurysm of the right internal carotid artery extending into the pituitary fossa. Hypopituitarism.

a Magnified lateral projection: the anteroinferior part of the floor *(white arrows)* and the dorsum sellae are demineralized. Note curvilinear calcification with a downward concavity corresponding to the upper border of the aneurysm *(black and white arrows)*.

b Anteroposterior angiotomogram: large intracavernous aneurysm extending into the sella turcica, depressing its floor on the right *(white arrows)*

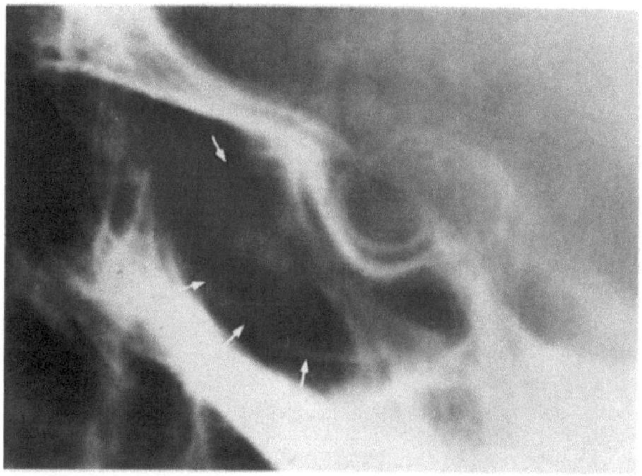

Fig. 9.4. Intracavernous aneurysm of the internal carotid artery. Enlargement of the carotid sulcus *(white arrows)*. Soft tissue mass overlying the sphenoid sinus

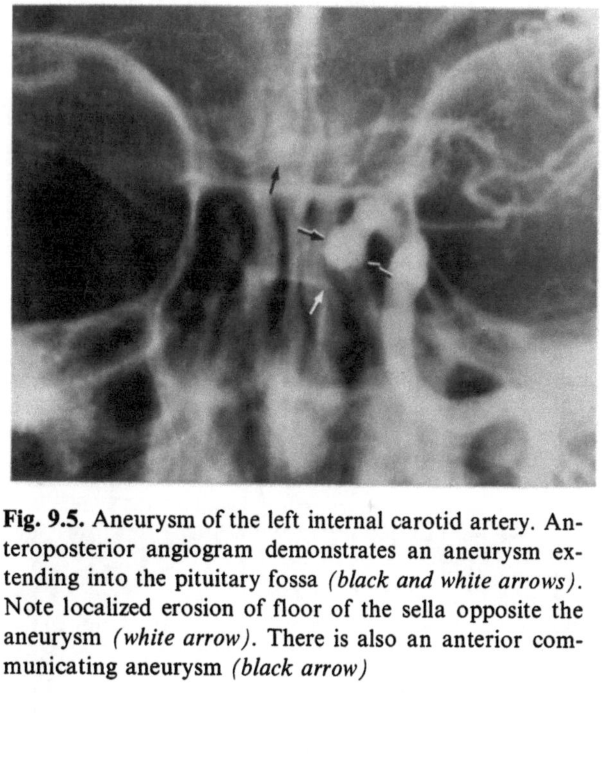

Fig. 9.5. Aneurysm of the left internal carotid artery. Anteroposterior angiogram demonstrates an aneurysm extending into the pituitary fossa *(black and white arrows)*. Note localized erosion of floor of the sella opposite the aneurysm *(white arrow)*. There is also an anterior communicating aneurysm *(black arrow)*

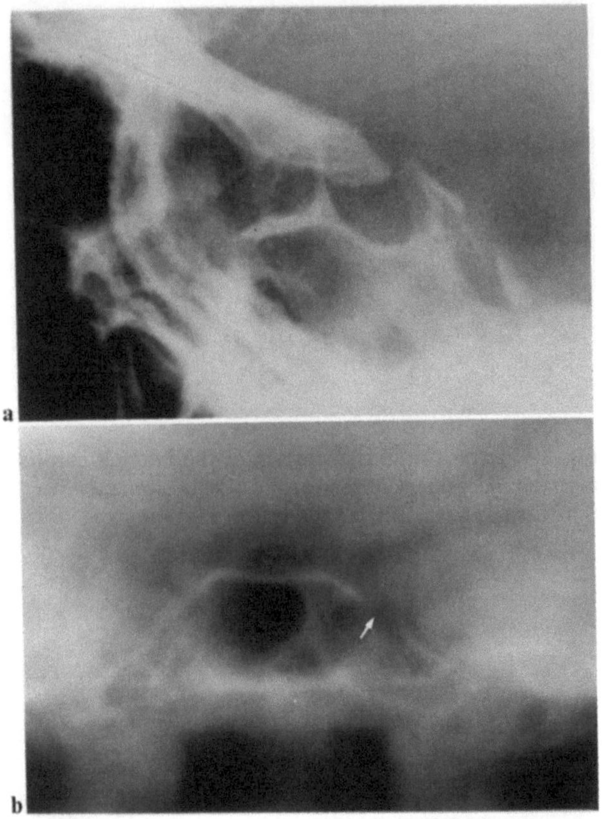

Fig. 9.6a, b. Left cavernous sinus meningioma in a 35-year-old female with third nerve palsy during pregnancy.
- **a** Lateral projection: sella turcica of normal size; clouded sphenoid sinus.
- **b** Frontal tomogram: erosion of the left lateral wall of the sphenoid sinus *(arrow)*

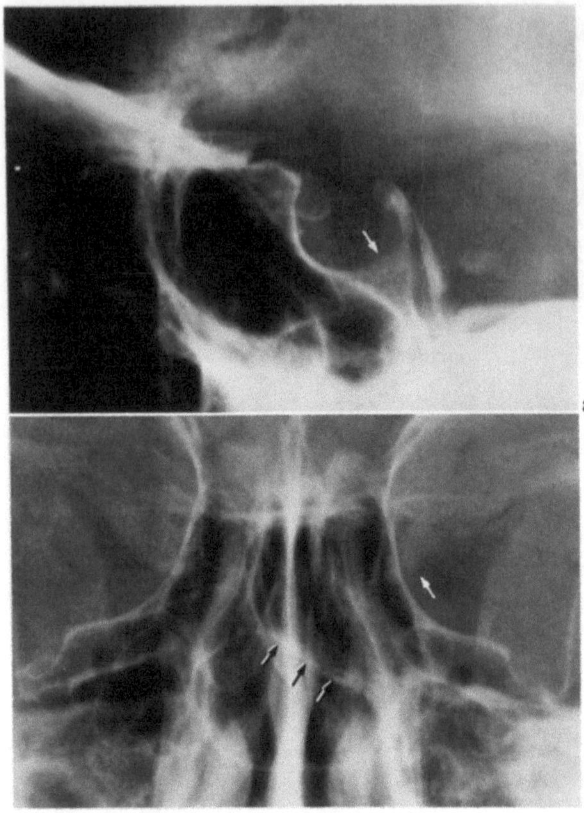

Fig. 9.7a, b. Left cavernous sinus meningioma in an 18-year-old pregnant patient with antepartum involvement of left third, fifth, and sixth cranial nerves.
- **a** Lateral projection: sella turcica is not enlarged: demineralization of the lamina dura of dorsum sellae *(arrow)*.
- **b** Frontal projection: tilting of floor of sella to left *(black and white arrows)*; widening of the left superior orbital fissure *(white arrow)*

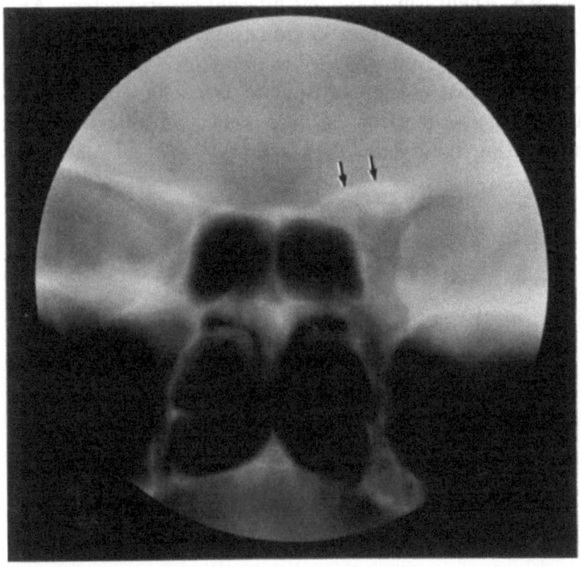

Fig. 9.8. Meningioma of left anterior clinoid process. Frontal tomogram shows thickening and sclerosis of the left anterior clinoid process *(arrows)*

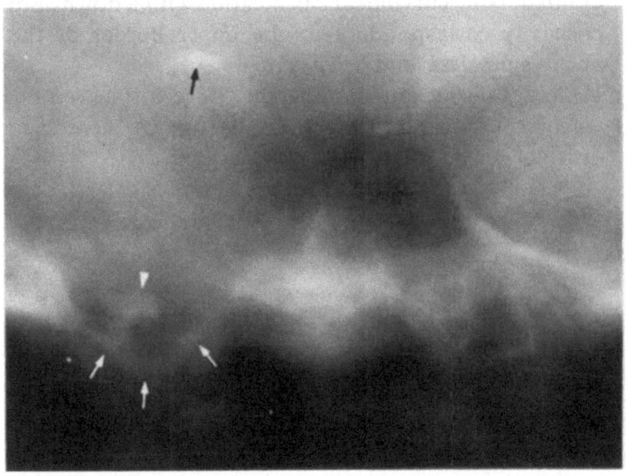

Fig. 9.9. Epidermoid tumor of the right middle cranial fossa. Frontal tomogram demonstrates erosion of the right side of the sphenoid sinus and depression of the floor of the middle cranial fossa *(white arrows)* with calcification *(arrowhead)*. Erosion of inferior aspect of right anterior clinoid process *(black arrow)*

Chapter 10 Retrosellar Pathology

Retrosellar lesions usually affect the dorsum sellae, but the other bony structures of the sellar region often become involved. From a clinical standpoint, retrosellar space-occupying lesions usually remain latent for a considerable length of time, which explains why these tumors are often quite large when they are discovered. These tumors compress cranial nerves (oculomotor, trochlear, abducens, trigeminal, facial, acoustic, and even the glossopharyngeal, vagus, and hypoglossal nerves in very advanced cases); compression of the optic chiasm and sometimes of the pituitary gland and brain stem may occur.

Generalized bone diseases (Paget's disease, fibrous dysplasia, and osteopetrosis) are discussed in Chap. 13, as are other primary and secondary bone diseases involving the clivus.

A. Chordomas[36]

I. General Considerations

Chordomas are rare, forming only 0.1%–0.2% of intracranial tumors (Mouren et al. 1970). They develop from remnants of the notochord, which may be found from the sellar region to the sacrococcygeal region. Chordomas are intracranial in 40% of cases, vertebral in 10%, and sacrococcygeal in 50%. Intracranial remnants of the notochord are located in three areas: the posterior aspect of the dorsum sellae, the spheno-occipital synchondrosis region, and the retropharyngeal tissues (Petit-Dutaillis et al. 1951; Arey 1965; Patten 1968). It is for this reason that one can distinguish three varieties among intracranial chordomas: sellar region chordomas, which arise in the dorsum sellae and extend intracranially; clivus chordomas, the most frequent, which develop from intraosseous remnants in the regions of the spheno-occipital synchondrosis; and nasopharyngeal chordomas, which arise from prebasilar remnants. It is noteworthy, however, that many nasopharyngeal chordomas which arise from prebasilar chordomas are secondary and related to invasion of the nasopharynx by a chordoma of either the clivus or the sellar region (Mouren et al. 1970).

Chordomas occur at any age, but are rare in children (Becker et al. 1975; Harwood-Nash and Fitz 1976; Nolte 1979). They are seen generally between the ages of 30 and 60.

Chordomas are malignant tumors and must be distinguished from so-called benign chordomas which are merely notochordal remnants (Bonneville et al. 1980). These benign chordomas were described for the first time by Virchow (1856) who called them "ecchondrosis physaliphora spheno-occipitalis" because of their histologic resemblance to cartilaginous cells. In 1858, Muller first suggested that the tumors arose from the notochord, and Stewart and Morin (1926) accordingly suggested the term "ecchordosis physaliphora spheno-occipitalis." These benign tumors are not usually progressive, but they may give clinical symptoms. In exceptional cases, they give rise to malignant chordomas. The first symptomatic or malignant chordoma was described by Klebs in 1864. These so-called malignant chordomas rarely metastasize (Stough et al. 1970).

36 Refs. [6, 50, 55, 88, 96, 167, 170, 189, 198, 222, 224, 283, 292, 381, 398, 402, 415, 418, 434, 458, 474, 489, 506, 529, 566, 568, 591, 634, 659, 661, 681, 703, 730, 734, 739, 745, 749, 828, 837, 843, 855, 859, 911, 915, 973, 981, 1024, 1036, 1048]

II. Radiology

Chordomas may be manifest in three different ways (Fig. 10.1): by bony changes they produce in the clivus and sellar region, by calcification within the tumor, and finally by the presence of a soft tissue mass within the sphenoid sinus or nasopharynx.

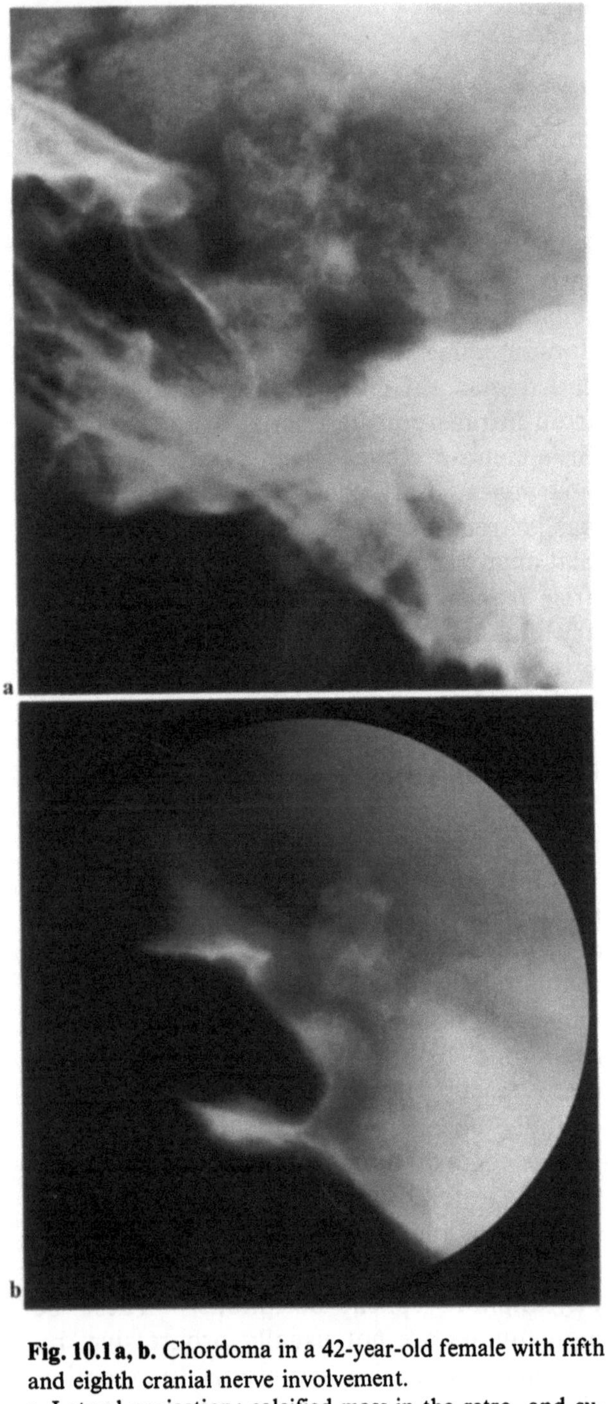

Fig. 10.1 a, b. Chordoma in a 42-year-old female with fifth and eighth cranial nerve involvement.

a Lateral projection: calcified mass in the retro- and suprasellar area; destruction of the posterior aspect of the clivus.

b Lateral tomogram: extensive destruction of the clivus and dorsum sellae. Tumor calcification of floccular type is seen in the midline

1) Bone Changes

The midline portion of the clivus is usually eroded, with destruction of the dorsum sellae. Later the tumor extends forward, destroying the floor of the sella and even the anterior clinoid processes and the margin of the superior orbital fissure (Rischbieth and Bull 1958), and invading the sphenoid sinus. Laterally, the tumor destroys the petrous apex. Large tumors may invade the posterior wall of the nasopharynx. Although classically midline, erosion of the clivus may be asymmetric (Kendall and Lee 1977). In the adult, bone changes are not always lytic; hyperostosis may be seen (Kendall and Lee 1977).

Although bone erosion is frequent in adults, it occurs much more rarely in children (Zingesser and Schechter 1964; Di Chiro and Anderson 1965; Becker et al. 1975; Nolte 1979). In children, chordomas characteristically demineralize the clivus and dorsum sellae (Harwood-Nash and Fitz 1976).

2) Tumor Calcification

Calcification is frequent, appearing in more than 70% of cases (Ozonoff and Burrows 1971). Such calcification is usually midline, in the retrosellar region. Extension of the tumor may produce intra- and suprasellar, as well as nasopharyngeal calcification.

According to Petit-Dutaillis (1951), calcification in chordomas is characteristic, giving the appearance of small quadrangular, triangular, or linear spots. In some cases, the tumor is clearly outlined. Calcification is sometimes of floccular type, resembling that seen in craniopharyngiomas.

3) Soft Tissue Mass

A soft tissue mass is found in 43% of cases, in either the sphenoid sinus or the nasopharynx (Kendall and Lee 1977).

B. Chondromas[37]

I. General Considerations

Chondromatous tumors are even more rare than chordomas. Intracranial chondromas may develop in the cranial vault or in the intracranial structures, but arise most often from the skull base. Such tumors develop from embryonic cartilaginous remnants; only the bones of the base of the skull are of cartilaginous origin. Primary intracranial chondromas develop from the spheno-occipital synchondrosis. These tumors, which are initially extradural, displace the dura and the brain without invasion. Chondromas extend upward and forward toward the sella turcica, and laterally toward the petrous apex. Some originate in the nasal cavities and invade the sellar region.

Chondromas are histologically benign tumors but malignant transformation is possible (Sarwar et al. 1976). They develop between the second and fifth decades (Acquaviva et al. 1965).

II. Radiology

Tumors arising from the spheno-occipital synchondrosis destroy the clivus in 50% of cases

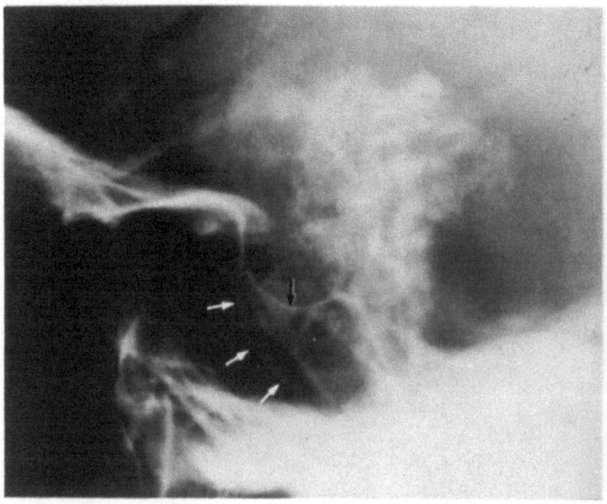

a

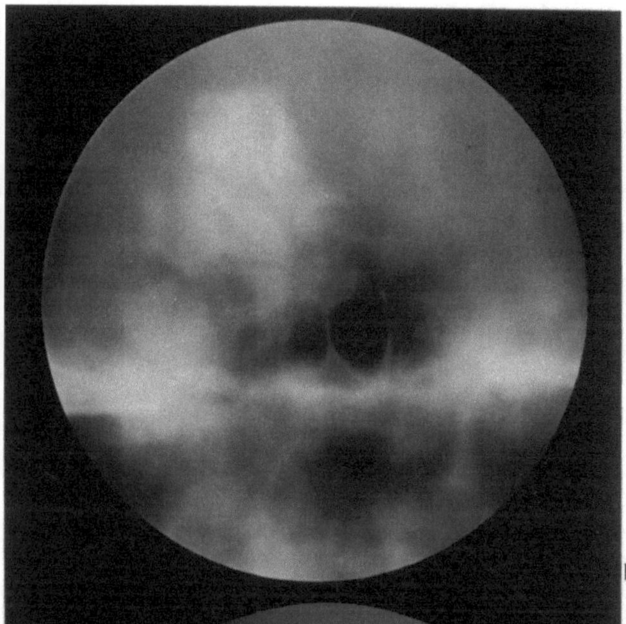

b

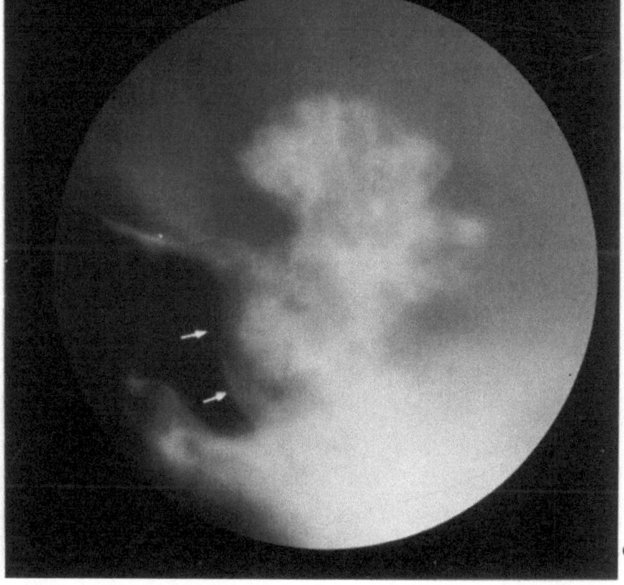

c

Fig. 10.2a–c. Chondroma in a 27-year-old male with headache and third nerve palsy.
a Lateral view: calcified supra- and retrosellar mass. Marked depression of the right side of the floor of the sella *(white arrows)*; the left side is not depressed *(black and white arrow)*.
b Frontal tomogram: unilateral destruction of the dorsum sellae, clinoid process and floor of the sella on the right. Tumor calcification is seen to the right of the midline.
c Right lateral tomogram: the floor of the sella bulges downward *(arrows)*. Note the floccular appearance of the tumor calcification

37 Refs. [3, 25, 66, 167, 202, 274, 283, 285, 320, 343, 387, 421, 468, 498, 504, 507, 574, 589, 634, 642, 705, 826, 901, 1048]

(Gabrielsen and Kingman 1964; Acquaviva et al. 1965; Minagi and Newton 1969). Unlike that seen in chordomas, this erosion of the clivus is usually unilateral. Anterior and upward extension of the tumor destroys the margins of the sella on one side (Ozonoff and Burrows 1971) (Fig. 10.2). Calcification is present in 60% of the cases according to Acquaviva (1965). When it gives the tumor a "beehive" appearance, the calcification is characteristic, but nodular or massive calcification is less typical.

C. Clivus Meningiomas[38]

Clivus meningiomas are rare (Castellano and Ruggiero 1953). They develop in the spheno-occipital synchondrosis and, much less frequently, in the dorsum sellae, There may be increased bone density opposite their dural insertion (Coin and Malkasian 1971; Pribram and du Boulay 1971) (Fig. 10.3); in rare cases, there is osteolysis (Dechaume and Wertheimer 1936; Dany et al. 1963). It is, however, common to see large tumors which do not cause bony changes in the clivus but merely changes of raised intracranial pressure. Absence of bony changes facilitates the differentiation between meningiomas and chordomas or chondromas; the latter often destroy the clivus (Castellano and Ruggiero 1953). Like any meningiomas, those arising on the clivus may calcify.

38 Refs. [147, 155, 167, 195, 210, 224, 532, 729, 814, 1048]

D. Aneurysms of the Basilar Artery

When basilar aneurysms extend anteriorly, they may destroy the dorsum sellae and posterior clinoid processes (Pribram and du Boulay 1971). Very large aneurysms of the basilar artery may even erode the supermedial surface of the anterior clinoid processes and deform the floor of the sella. They sometimes have a calcified wall.

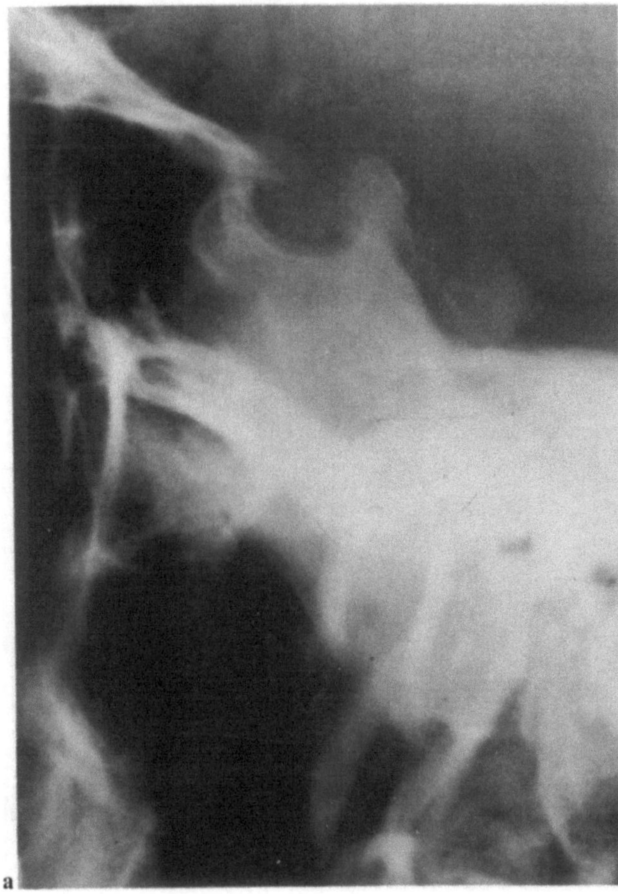

a

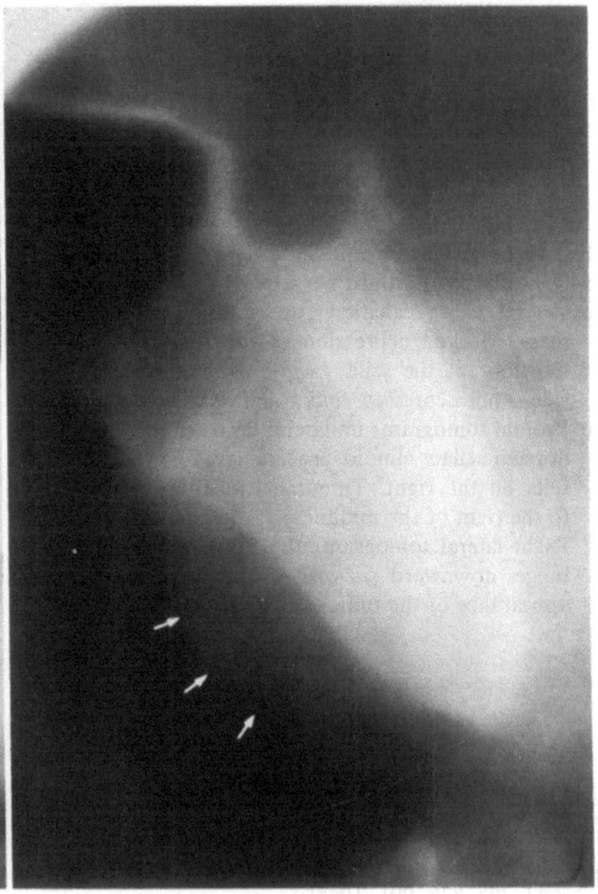

b

Chapter 11 Infrasellar Pathology

Diseases of infrasellar origin are essentially inflammatory or infectious diseases of the sphenoid sinus, or they are neoplastic diseases, usually malignant, which develop within the sphenoid sinus or, more frequently, propagate from the nasopharynx or the sinuses. Benign neoplastic diseases, parasitoses, and mycoses are exceptional. Primary bone diseases and trauma will be discussed in Chap. 13. In general, infrasellar expansive processes are most often latent until they have reached a volume large enough to cause signs indicating extension of the tumor into the sella itself or into the cavernous sinus or nasopharynx.

A. Inflammatory, Infectious, and Mycotic Lesions of the Sphenoid Sinus

I. Acute Sinusitis

Sphenoid sinusitis appears initially as a thickening of the mucous membrane lining; accumulation of secretion then gives the sinus a cloudy appearance. Sphenoid sinusitis rarely produces a fluid level, which indicates the presence of serous fluid, pus, or mucus. Demonstration of a fluid level is quite simple on the lateral projection taken in the vertical position with a horizontal beam. Tilting the head leads to the formation of a different fluid level; however, it should be remembered that, due to the viscous character of this fluid within the sinus a new fluid level may only appear after a short delay.

In general, the bony walls of the sphenoid sinus do not change in acute sinusitis. When there is cloudiness of the sinus, contrast modification may bring less calcified bony walls into view. However, during severe sinusitis, hyperemia may lead to demineralization of the bony walls of the sphenoid sinus (Fig. 11.1). The resulting appearance may be disturbing and wrongly suggest the diagnosis of malignant disease of the sinus. In fact, such severe sinusitis never brings about actual destruction of the bony walls, and the walls may be seen to remineralize very quickly several weeks after treatment.

Fluid levels in the sphenoid sinus must be distinguished from the cloudiness caused in certain subjects by superimposition of the zygomatic arch (see Fig. 14.14).

II. Chronic Sinusitis

When the infection has spread beyond the mucoperiosteal line, changes in the bony walls of the sinus occur; the walls of the sella turcica and the planum sphenoidale may therefore show a sclerotic appearance and in most cases, a very regular thickening (Figs. 11.2, 11.3), which must be distinguished from bony changes found with meningioma or in fibrous dysplasia.

◁ **Fig. 10.3a, b.** Clivus meningioma.
a Lateral projection.
b Lateral tomogram: marked hyperostosis of the clivus. The anterior and posterior surfaces of the thickened clivus appear convex. The floor and the anterior wall of the sella turcica appear thick and dense. Note forward displacement of the posterior wall of the nasopharynx *(white arrows)*

III. Mucoceles [39]

Mucoceles are caused by accumulation of secretions in a sinus whose ostium is occluded. Mucoceles are relatively rare in the sphenoid sinus. They may rapidly become destructive and cause severe lesions, particularly in the chiasm and optic nerves, if they are not discovered in time. Headaches are common. The radiologic appearance of sphenoid mucocele is generally quite characteristic (Nugent et al. 1970). This expansive lesion displaces very evenly the sinus walls it has thinned and then destroyed. Cloudiness of the sinus itself is usually very homogeneous. The floor of the sella may be completely destroyed by expansion of the mucocele; however, the sella turcica is not enlarged, at least at the beginning; at a later stage, destruction may become more widespread, leading to erosion of the optic canal, enlargement of the superior orbital fissure, and elevation of the anterior clinoid process. Differential diagnosis between mucocele, pituitary adenoma, and malignant tumor of the sphenoid sinus region may then be more difficult. Tomography almost always confirms the diagnosis: the erosion of the floor of the sella is from below rather than from above (Fig. 11.4). Absence of calcification eliminates the possibility of a craniopharyngioma or of a chordoma, which is also generally calcified. Malignant tumors of the sphenoid sinus have a more destructive radiologic appearance, quite different from the expansive appearance of a mucocele. If the mucocele arises from the anterior part of the sphenoid sinus, blistering of the planum may result. Finally, those mucoceles which have reached a large size may cause extensive destruction, affecting the cavernous sinus and optic nerves, sometimes even leading to hypopituitarism.

IV. Fungal Infections [40]

The frequency of fungal infections of the paranasal sinuses has clearly increased during the past few years, due to massive use of antibiotics and immunosuppressive drugs. Mycoses of the sphenoid sinus occur more often therefore in patients with cancer and especially in those with malignant lymphoma. *Aspergillus* is the most commonly encountered organism (Becker et al. 1968; Warder et al. 1975). Radiologic changes are not specific and are identical to those changes found in bacterial sinusitis, i.e., mucosal thickening and opacification of the sinus. Cases of osteolysis of the walls of the sphenoid sinus have also been reported (Le Clech et al. 1978). In addition to aspergillosis, mucormycosis may affect the sphenoid sinus.

V. Sarcoidosis

Sarcoid granuloma may involve the sphenoid sinus, without bone destruction. An opaque sphenoid sinus with well-preserved margins is the most common finding.

B. Infrasellar Neoplastic Diseases

I. Malignant Tumors [41]

Primary tumors of the sphenoid sinus are not frequent. Those encountered include carcinoma, sarcoma, and in rare cases, melanoma or plasmocytoma. On the other hand, the sphenoid sinus is often invaded by neighboring tumors, usually from the nasopharynx. Secondary deposits may also involve the sphenoid sinus, particularly from lymphosarcoma or in rare cases, metastatic carcinoma, the origin of which may be kidney, lung, prostate, or breast.

In general, diagnosis of a malignant tumor in the sphenoid sinus is only considered when the lesion has become very large. The symptoms therefore depend on the direction of extension of the tumor: decline in visual acuity, oculomotor palsies, and proptosis if the extension is anterolateral and anterosuperior; paralysis of the sixth nerve and trigeminal neu-

39 Refs. [38, 83, 372, 596, 644, 685, 707, 731, 733, 773, 792, 877, 878, 944, 976]

40 Refs. [51, 318, 432, 545, 1000, 1034]

41 Refs. [53, 98, 398, 418, 684, 719, 746, 786, 961, 1038]

ralgia if the tumor grows laterally toward the cavernous sinus; nasopharyngeal signs if the tumor extends downward or if it has arisen in the nasopharynx. On the whole, one cannot predict an intra- or extrasphenoidal source of the tumor from the clinical findings. In some cases, a malignant tumor of the sphenoid sinus may be discovered before there are any clinical findings, e.g., during a radiologic examination for sinusitis.

From a radiologic standpoint, the appearance is almost always stereotyped: more or less homogeneous cloudiness of the sphenoid sinus, decalcification, thinning, and destruction usually extending from the sinus walls. In anterosuperiorly extending tumors, the walls of the sella turcica may be completely destroyed, thus creating a ghostlike picture (Fig. 11.6). The general appearance may sometimes cause confusion between a malignant tumor of the sphenoid sinus and an invasive pituitary adenoma. Axial tomograms are often very valuable in such cases in order to evaluate the lateral extent of the lesion.

II. Nonmalignant Infrasellar Tumors [42]

Primary benign tumors of the sphenoid sinus are exceptional and remain clinically silent for a long time. They appear radiologically as well-defined intrasphenoid soft tissue masses. When they become large, erosion and destruction of the walls, simulating malignant disease, may take place. The tumor may be a papilloma, giant cell tumor, or aneurysmal bone cyst; moreover, parasellar meningiomas, craniopharyngiomas, adenomas, germinomas, epidermoid tumors, and chordomas may either develop within the sinus or invade it.

Osteomas of the sphenoid sinus are also exceptional. Their radiologic appearance is totally characteristic and therefore, presents no diagnostic problems (Fig. 11.5).

Juvenile angiofibroma, generally arising in the nasopharynx, is a noninfiltrating, highly vascular, benign tumor. It usually appears in the young male, presenting clinically as nasal obstruction with episodes of epistaxis. These tumors tend to grow toward the cavities adjacent to the nasopharynx, particularly the sphenoid sinus (Fig. 11.7). Thus, both cranial nerve palsies and proptosis may occur.

From a radiologic viewpoint, these tumors appear as soft tissue masses arising in the roof of the nasopharynx without destroying the bone. When angiofibromas extend to the pterygomaxillary space, bowing of the posterior wall of the maxillary sinus is entirely characteristic. Upward extension of the angiofibroma may cause erosion of the walls of the sella and enlargement of a superior orbital fissure.

42 Refs. [72, 313, 425, 555, 598, 640, 741, 750, 770, 899, 1026]

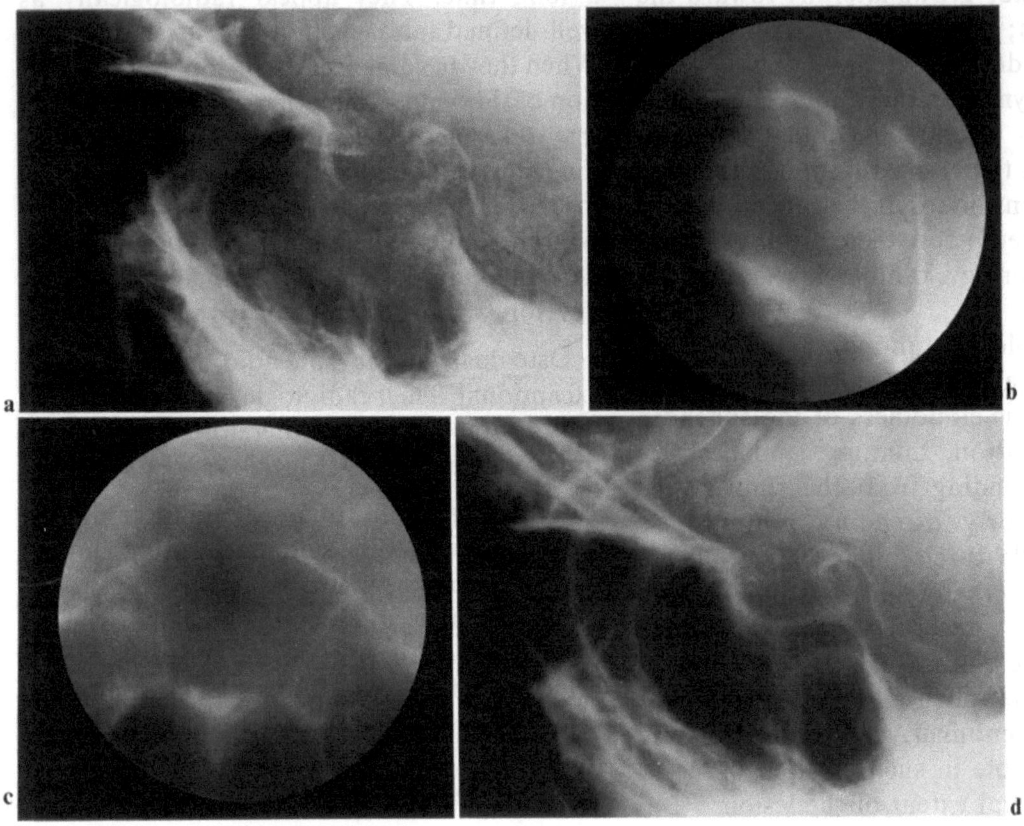

Fig. 11.1a–d (caption see opposite page)

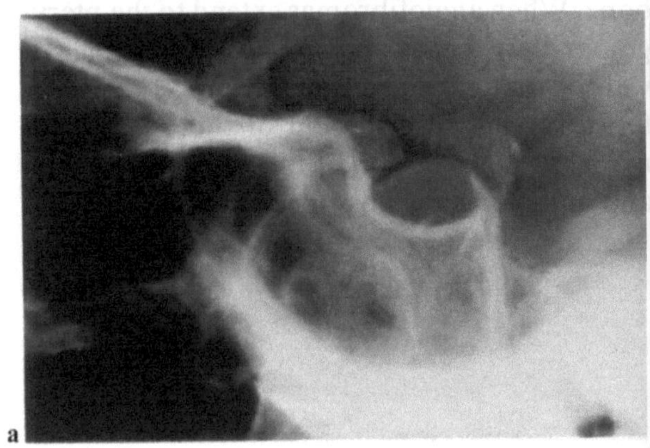

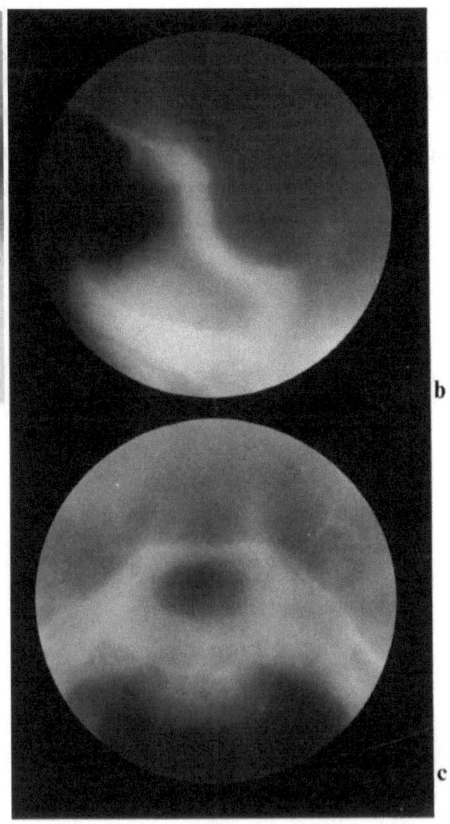

Fig. 11.2a–c. Chronic sinusitis.

a Lateral view: dense opacity of the sphenoid sinus; thickening and sclerosis of the lamina dura of the floor of the sella. Note also calcified carotid siphons.

b Sagittal tomogram.

c Frontal tomogram: thickening and sclerosis of the wall of the sphenoid sinus are well demonstrated; calcification of the left internal carotid artery

◁ **Fig. 11.1a–d.** Acute sinusitis.
a Lateral view: generalized erosion of the walls of the sella turcica and the dorsum sellae. Haziness of the sphenoid sinus.
b Sagittal tomogram.
c Frontal tomogram: homogeneous cloudiness of the sphenoid sinus. The cortical bone of the floor of the sella and of the base of the dorsum is very blurred.
d Lateral view 6 weeks later, after antibiotic therapy. Sphenoid sinus is now clear. Obvious remineralization of the floor of the sella

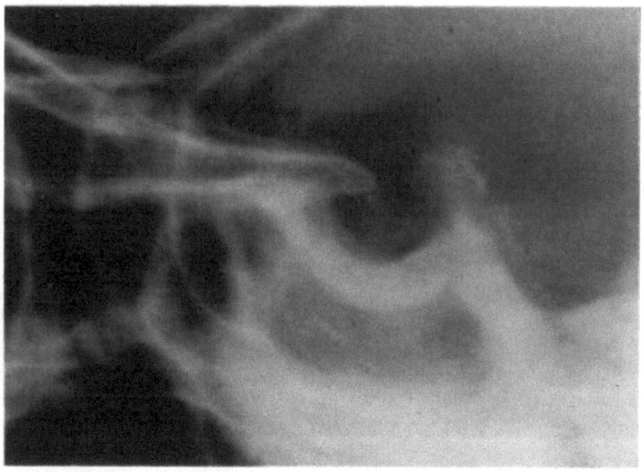

Fig. 11.3. Chronic sinusitis. Homogeneous clouding of the sphenoid sinus; very regular thickening and sclerosis of the superior and posterior walls of the sphenoid sinus

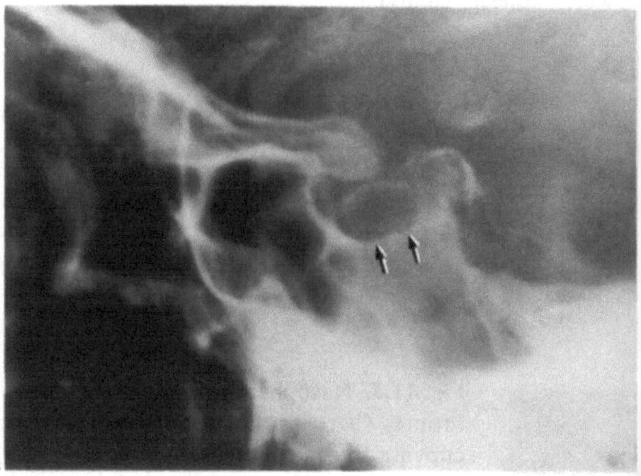

a

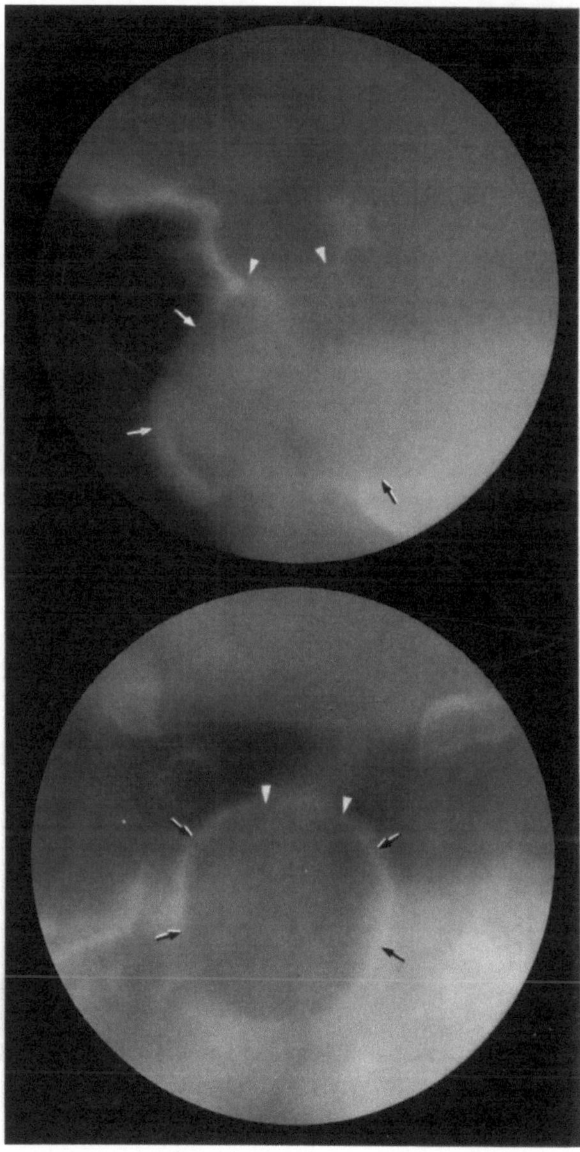

b

c

Fig. 11.4a–c. Mucocele.
a Lateral view: opacity of the posterior part of the sphenoid sinus. Thinning of the posterior part of the floor of the sella *(arrows)*.
b Sagittal tomogram and **c** frontal tomogram: a large soft tissue mass occupies the sphenoid sinus. Thinning and expansion of the lateral walls of the sphenoid sinus *(arrows)* with ballooned appearance. There is a large defect in the floor of the sella turcica *(arrowheads)*. Note the regular anterior *(white arrows)* and posterior limit *(black and white arrow)* of the mucocele in **b**

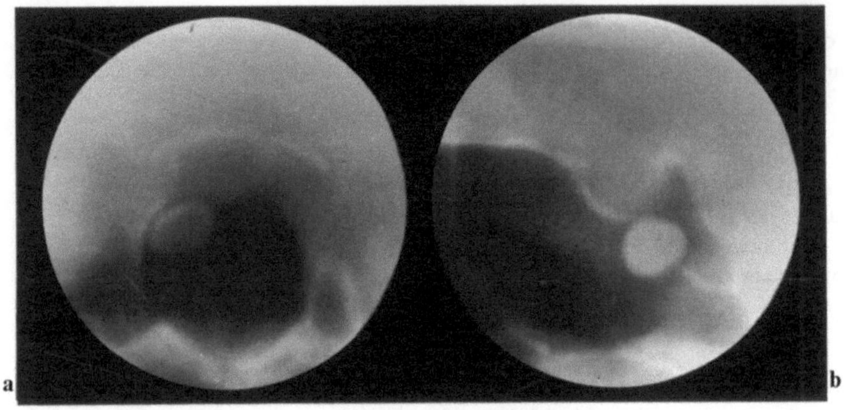

Fig. 11.5a, b. Osteoma of the sphenoid sinus. **a** Frontal polytomogram and **b** sagittal polytomogram: A dense bony opacity occupies the right part of the sphenoid sinus, below the floor of the sella

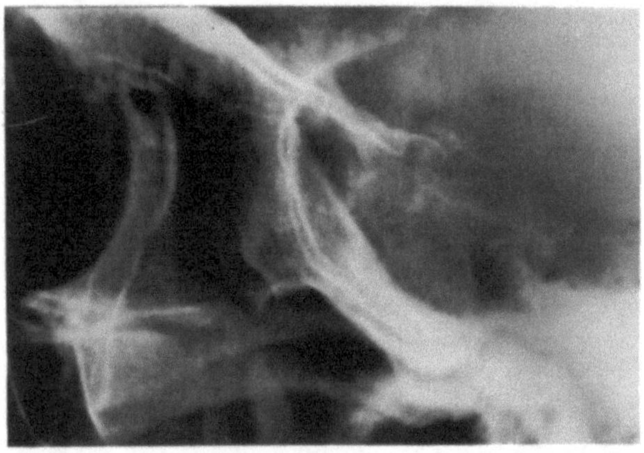

Fig. 11.6. Advanced squamous cell carcinoma of the ethmoid and maxillary region with extension to the sella turcica. Opacity of the ethmoid and sphenoid sinuses. Complete destruction of the walls of the pituitary fossa, which appears enlarged

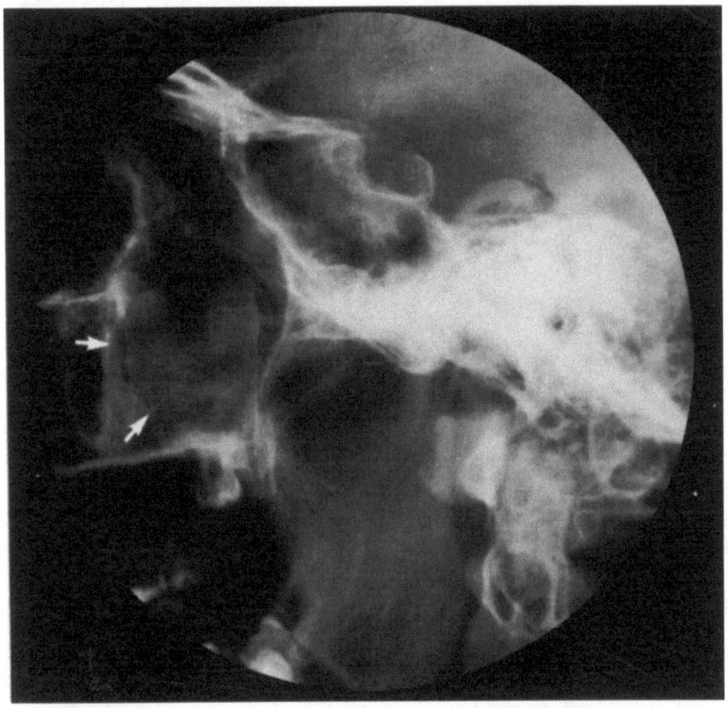

Fig. 11.7. Nasopharyngeal juvenile angiofibroma. There is a large soft tissue mass occupying the maxillary, ethmoid, and sphenoid sinuses. Note anterior bulging of the posterior wall of the maxillary sinus *(arrows)* (Courtesy, J. Théron)

Chapter 12 Sella Turcica in Raised Intracranial Pressure and Hydrocephalus[43]

Changes in the sella may occur very rapidly, i.e., in several weeks or several months, in the presence of raised intracranial pressure (Fig. 12.1). In the adult, changes in the sella turcica vary according to whether or not this raised intracranial pressure is accompanied by chronic hydrocephalus.

A. Raised Intracranial Pressure

Erosion of the cortical bone lining the sella turcica begins on the anterior surface of the base of the dorsum sellae, because it is thinner here than on the other walls, being composed of a fine, compact layer overlying cancellous bone (du Boulay and El Gammal 1966; Kornblum and Osmond 1935) (Fig. 12.2). The floor itself is also often demineralized in the early stages. Initially, however, this change is only apparent on the frontal projection (Fig. 12.3), which displays the cortical bone of the floor better than the lateral projection. In our experience, a floor which appears blurred and lacking in contrast on the frontal view could indicate the possibility of raised intracranial pressure. During the development of certain frontal or temporal tumors, raised intracranial pressure may cause erosion of the planum sphenoidale before the cortical bone of the sella turcica itself is changed (Fig. 12.4). Erosion of the planum sphenoidale usually begins in the most anterior portion and, at least initially, spares the chiasmatic sulcus and the tuberculum sellae (Fig. 8.24). Isolated erosion of the chiasmatic sulcus is much less common (Fig. 12.5). A shortened planum sphenoidale which no longer extends forward beyond the greater wings of the sphenoid must, indeed, suggest the possibility of raised intracranial pressure.

Erosive changes in the planum sphenoidale, without concomitant involvement of the sella turcica, may also be seen in cases of cerebral or dural arteriovenous malformation (Fig. 12.6).

In some cases, raised intracranial pressure may be manifest only by shortening, demineralization, or a pointed appearance of the dorsum sellae (Fig. 12.7). When intracranial pressure has been raised for a longer period of time, erosive changes eventually affect all margins of the sella, including the anterior wall. The planum sphenoidale often undergoes changes (Fig. 12.8) and the anterior clinoid processes may be tapered.

At a later stage, the outline of the sella turcica, the planum sphenoidale, and the chiasmatic sulcus may seem completely rubbed out; a false appearance of widening of the sellar region may also be seen. Widening may occur when chronically raised intracranial pressure has created a secondary empty sella turcica.

B. Sella Turcica in Chronic Obstructive Hydrocephalus

Like other causes of raised intracranial pressure, obstructive hydrocephalus may produce erosive changes in the sellar region (Fig. 12.9). Tumors of the anterior part of the third ventricle, causing dilatation of only the lateral ventricles, may bring about nonspecific erosive

43 Refs. [247, 248, 249, 252, 269, 322, 411, 444, 517, 682, 835, 836, 840, 850, 953, 954, 955, 1007, 1035]

changes in the sella turcica; considerable dilatation of the frontal horns may cause a scalloping of the planum (Fig. 12.10).

Downward extension of a dilated third ventricle generally produces erosion of the tip of the dorsum sellae, which therefore appears truncated (Figs. 12.11, 12.13). Infratentorial herniation of a greatly enlarged infundibular recess may also erode the posterior wall of the dorsum, giving a pointed rather than truncated appearance or sometimes forward displacement, due to pressure of the third ventricle.

Lesions causing chronic dilatation of the third ventricle change the sella turcica in a particular way. This is the case of aqueduct stenosis, during the long evolution of which there is considerable remodeling of the sellar region (du Boulay and Trickey 1970; Schechter and Zingesser 1967) (Fig. 12.12). The longest axis of the sella turcica seems nearly in line with the clivus. The sella turcica appears enlarged from front to back because the chiasmatic sulcus elongated and vertical is in line with the anterior wall of the sella. The tuberculum sellae is hardly visible, whereas the limbus sphenoidale is quite prominent.

The anterior clinoid processes are separated by the anterior part of the dilated third ventricle and their medial surface is widened; they therefore appear abnormally dumpy on the lateral projection. On the frontal projection, the floor of the sella seems widened, showing a slight upward concavity.

These characteristic abnormalities make the diagnosis of aqueduct stenosis simple, especially since they are almost always accompanied by an increase in convolutional markings and slight suture diastasis (Fig. 12.13). In most adults with aqueduct stenosis, the lamina dura does not show any erosion.

After prolonged cerebrospinal fluid shunt procedures, changes in the sella turcica may occur, giving sometimes the appearance of a "small sella turcica" (Kaufmann et al. 1970).

C. Changes in the Sella Turcica in Childhood

In infancy, communicating hydrocephalus does not produce very pronounced changes in the sella turcica. A simple elongation of the sella turcica may accompany increased head size. For the most part, increased intracranial pressure is manifested by widening of sutures and does not affect the region of the sella.

In older children, raised intracranial pressure is often related to intracranial mass lesions; radiologic signs are the result of obstructive hydrocephalus. Suture diastasis is often striking; however, localized erosive changes in the sella turcica may be seen: demineralization of the dorsum, the sella, or the planum, identical to that seen in the adult (Fig. 12.7), or amputation of the dorsum sellae when hydrocephalus is due to a posterior fossa tumor. A pointed or demineralized dorsum may suggest a suprasellar germinoma or a parasellar metastasis from a medulloblastoma (Figs. 12.2, 12.7b). Sellar deformity, typical of aqueduct stenosis, may also be demonstrated in older children.

Excision of the tumor responsible for raised intracranial pressure, or shunt procedures, leads very rapidly to recalcification of the walls of the sella turcica especially in children (Fig. 12.2).

D. Sella Turcica in Craniostenosis

The appearance of the sella turcica in craniostenosis was discussed in Chap. 5, i.e., it may be enlarged and show very dense margins. This appearance indicates intrasellar extension of the subarachnoid spaces secondary to increased cerebrospinal fluid in the basal cisterns (Fig. 5.26). Radiologic analysis of the sella turcica may be difficult if the pituitary fossa lies very low on the lateral projection. This position results from an increase in the angle between the middle and anterior cerebral fossae (Fig. 12.14); premature closure of the coronal suture causes a deepening of the middle cerebral fossa (Grundy et al. 1970).

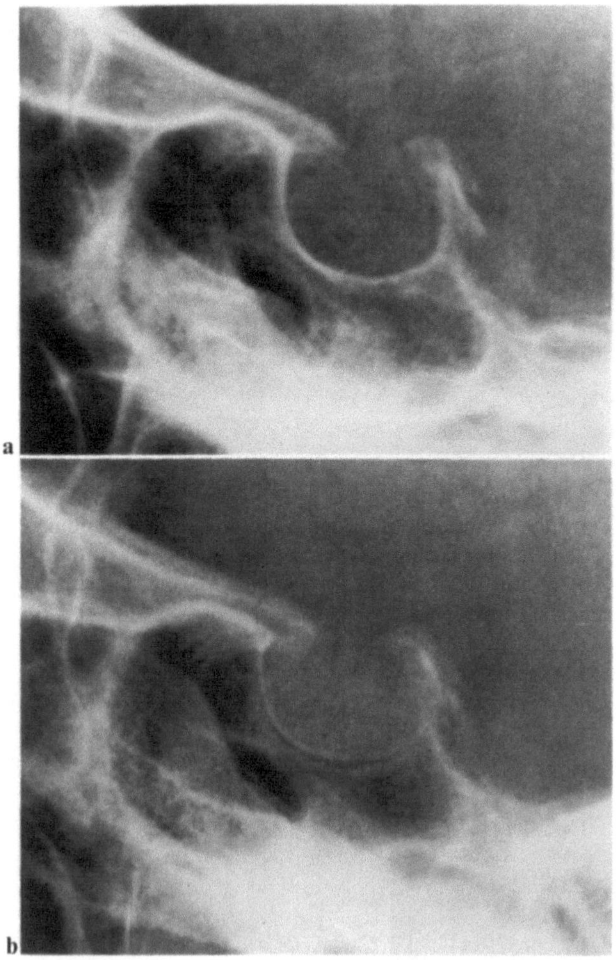

Fig. 12.1 a, b. Raised intracranial pressure; hemangioblastoma of the cerebellar hemisphere.

a Lateral projection: inconspicuous thinning of the posterior part of the floor of the sella.

b One month later: generalized demineralization of the sella is now evident

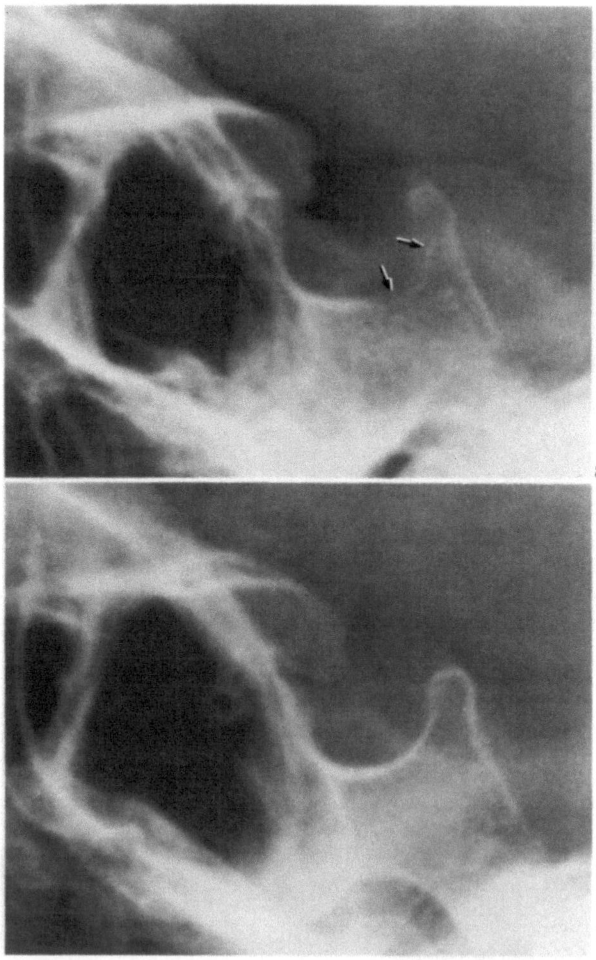

Fig. 12.2 a, b. Raised intracranial pressure; fourth ventricle and suprasellar germinomas in a 9-year-old girl.

a Lateral projection: erosion of the cortical bone of the posterior part of the floor and the dorsum sellae *(arrows)*.

b Six weeks after excision of the tumor: remineralization of the sella turcica

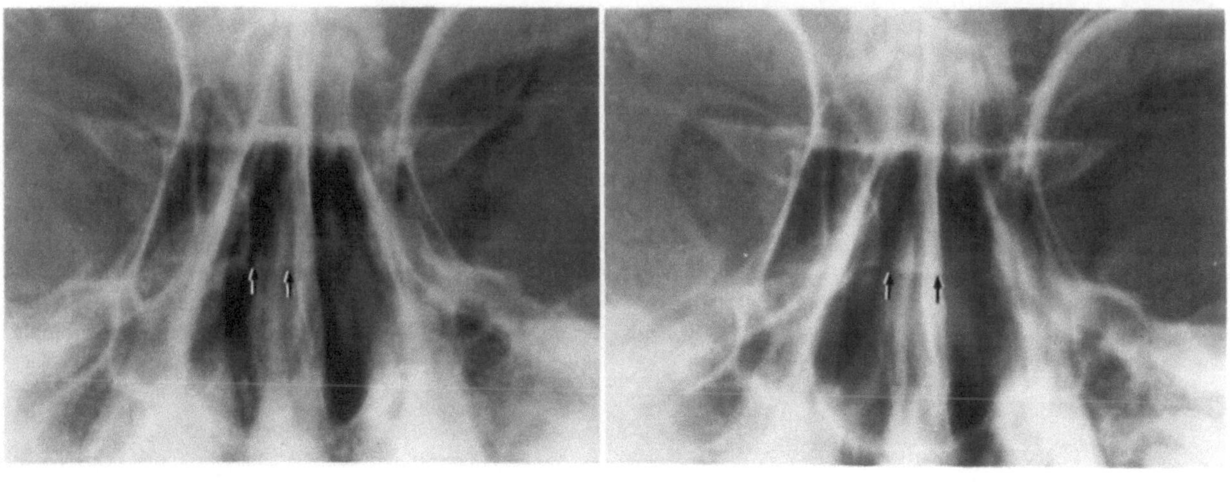

Fig. 12.3 a, b. Long-standing raised intracranial pressure in a 28-year-old patient.

a Frontal view: the floor of the sella is "rubbed out" *(arrows)*.

b Frontal view 3 weeks after relief of increased intracranial pressure: the floor is now seen clearly *(arrows)*

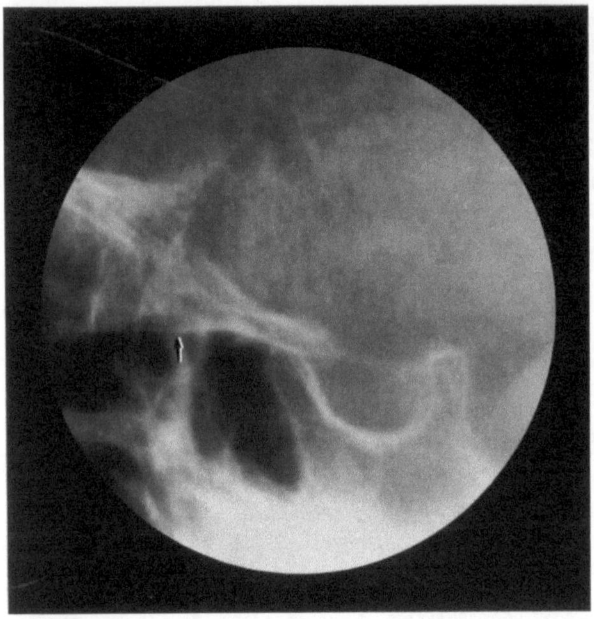

Fig. 12.4. Long-standing raised intracranial pressure. Frontal convexity meningioma. Shortening of the planum sphenoidale *(arrow)*. The sella turcica is normal

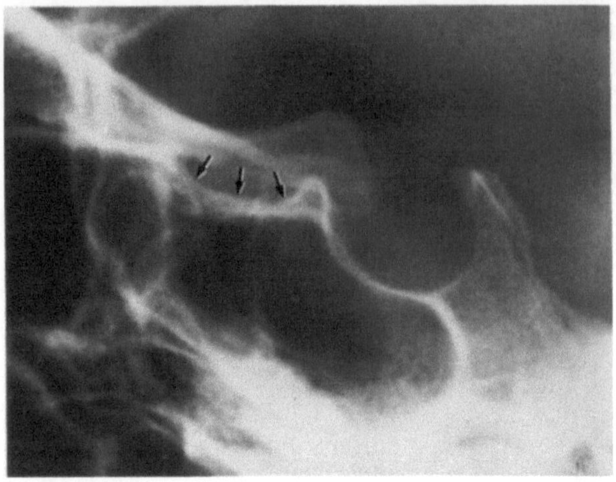

Fig. 12.5. Temporal oligodendroglioma. Slight thinning of the anterior surface of the dorsum and of the anterior wall of the sella. Marked scalloping of the chiasmatic sulcus *(arrows)*

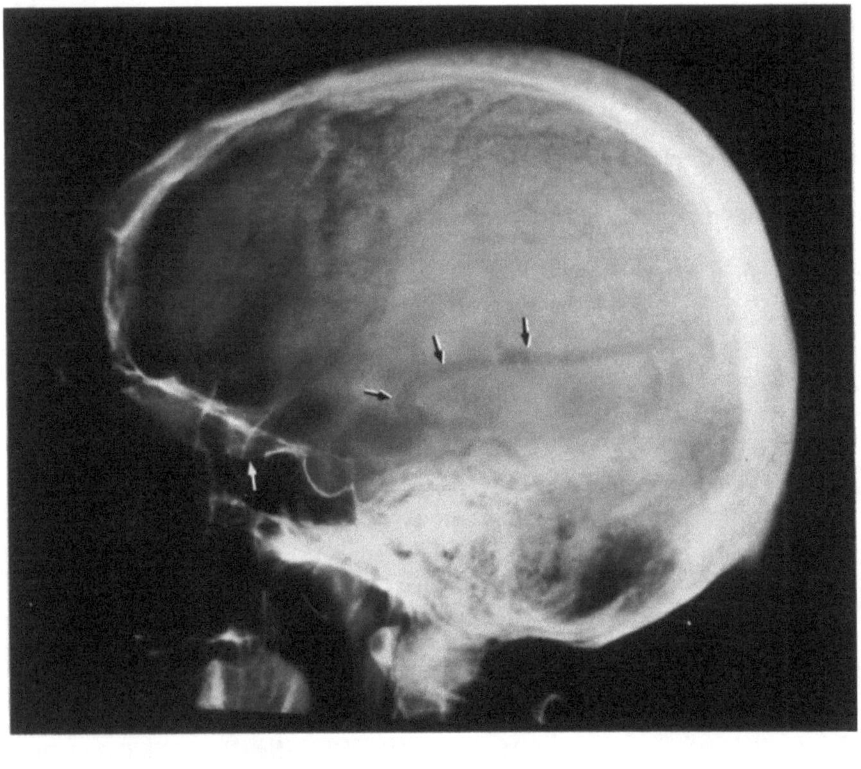

Fig. 12.6. Large, mixed, pial, and dural malformation. Deepening of the grooves of the middle meningeal artery *(black and white arrows)*. Note blurred appearance of the anterior part of the planum sphenoidale *(white arrow)*

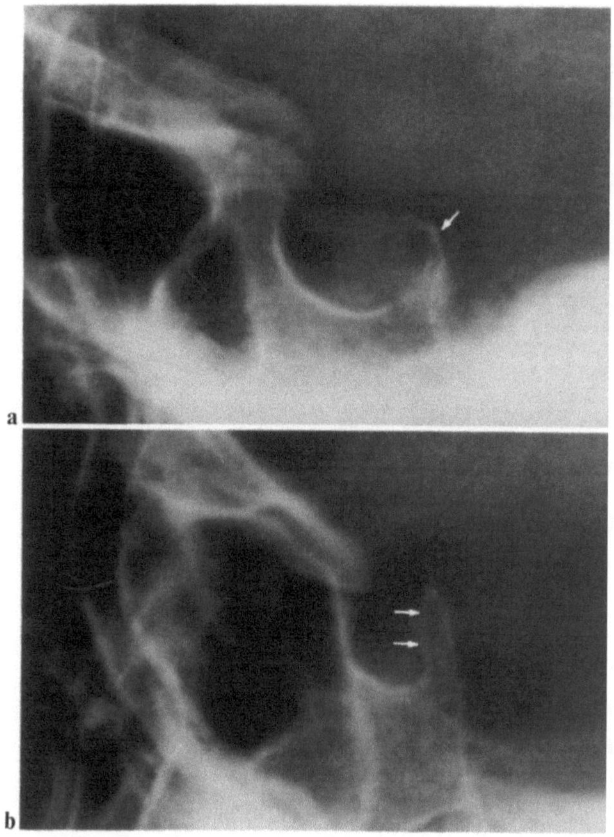

Fig. 12.7a, b. Two patients with raised intracranial pressure.
a Long-standing raised intracranial pressure. Temporal astrocytoma. The tip of the dorsum sellae is pointed and slightly demineralized *(arrow)*.
b Raised intracranial pressure. Medulloblastoma with suprasellar metastasis in a 8-year-old boy. Isolated demineralization of the dorsum *(arrows)*

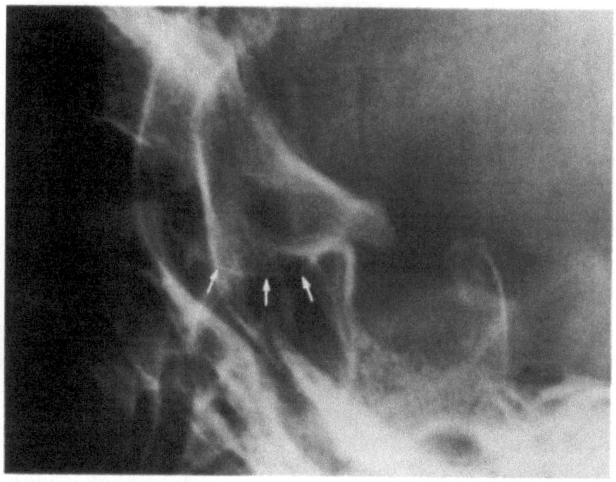

Fig. 12.8. Severely raised intracranial pressure in a 8-year-old boy. Vermis astrocytoma. Demineralization of the sella; note also blurring of the planum sphenoidale *(arrows)*

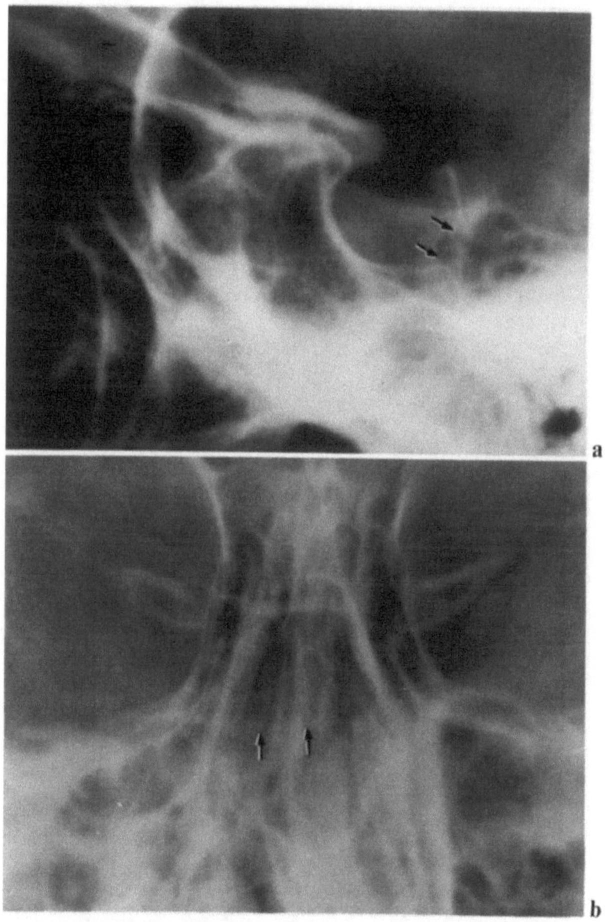

Fig. 12.9a, b. Raised intracranial pressure; colloid cyst of the third ventricle in a 18-year-old patient.
a Lateral projection: slight demineralization of the anterior surface of the base of the dorsum *(arrows)*.
b Frontal projection: diagnosis of raised intracranial pressure is evident; the sellar floor is blurred *(arrows)*

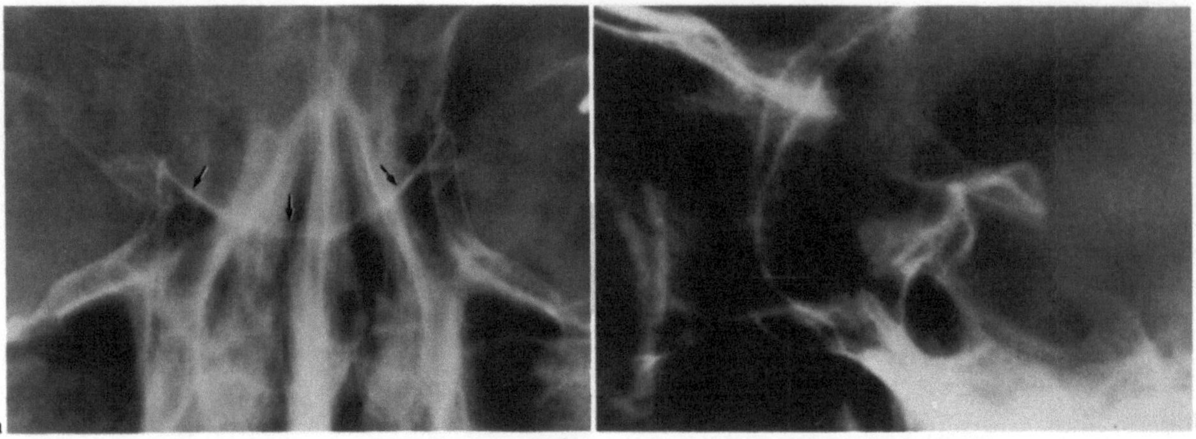

a b

Fig. 12.10a, b. Massive obstructive hydrocephalus with enormous dilatation of the frontal horns in a 35-year-old patient.

a Frontal projection: scalloping of the planum sphenoidale *(arrows)*; the floor of the sella is not shown.

b Lateral projection: marked changes in the presellar region; the planum sphenoidale and the posterior part of the orbital roofs are indistinct. The margins of the sella are generally demineralized and the dorsum is shortened

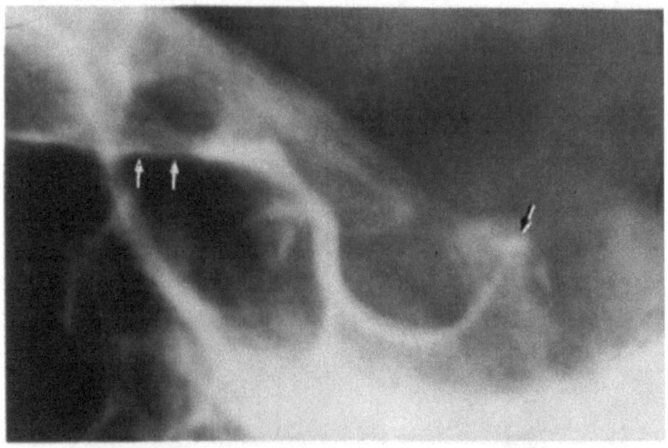

◁ **Fig. 12.11.** Chronic hydrocephalus in a 23-year-old patient; note the truncated dorsum *(black and white arrow)* and thinning of the planum sphenoidale *(white arrows)*

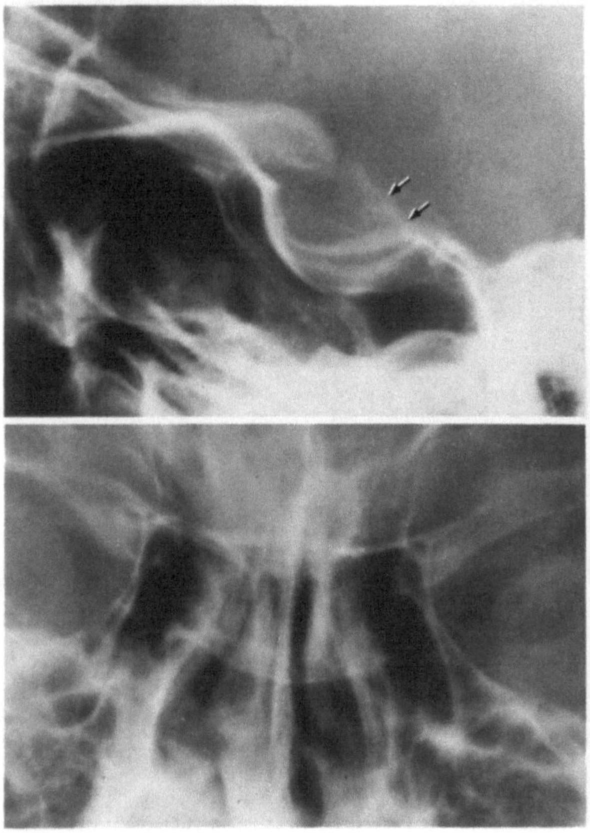

a

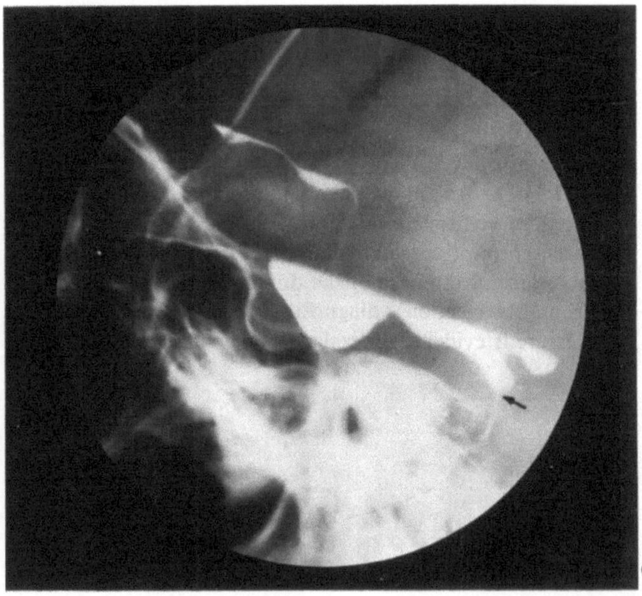

c

Fig. 12.12a–c. Aqueduct stenosis in a 28-year-old woman.

a Lateral projection: the anterior wall of the sella is prolonged into a nearly vertical chiasmatic sulcus; the dorsum sellae is thin and bent forward *(arrows)*; the anterior clinoid processes have a dumpy appearance.

b Frontal projection: the floor of the sella appears widened and is slightly concave superiorly

c Lateral ventriculogram: the dilated third ventricle has pushed the dorsum forward; note the aqueduct stenosis *(arrow)*

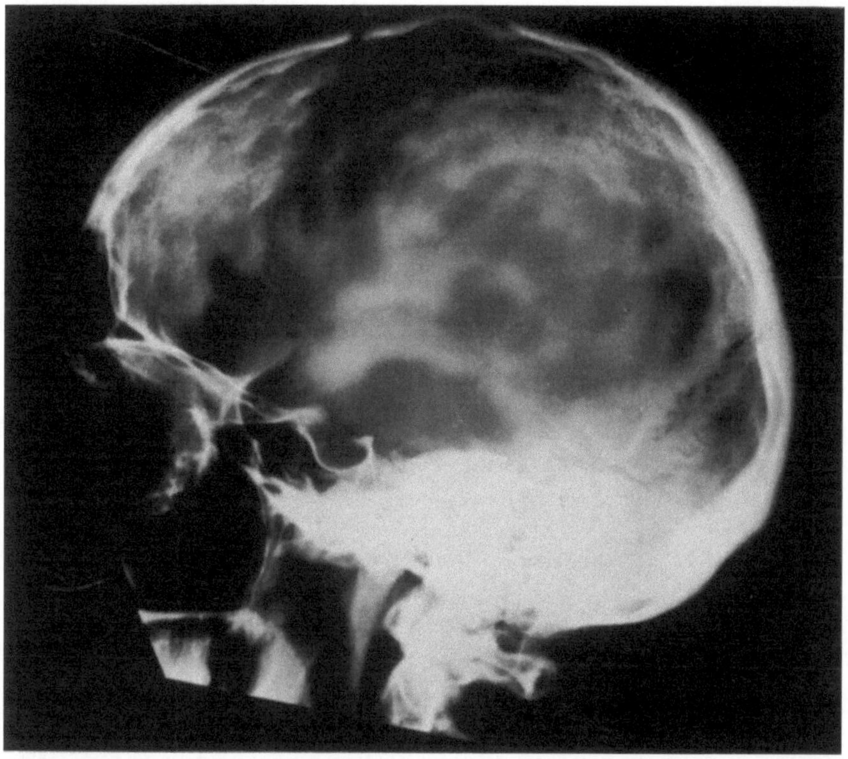

Fig. 12.13. Aqueduct stenosis in a 18-year-old patient. Prominent convolutional markings; widening of the coronal suture; elongated chiasmatic sulcus; truncated dorsum sellae; note that the lamina dura of the sella is not demineralized

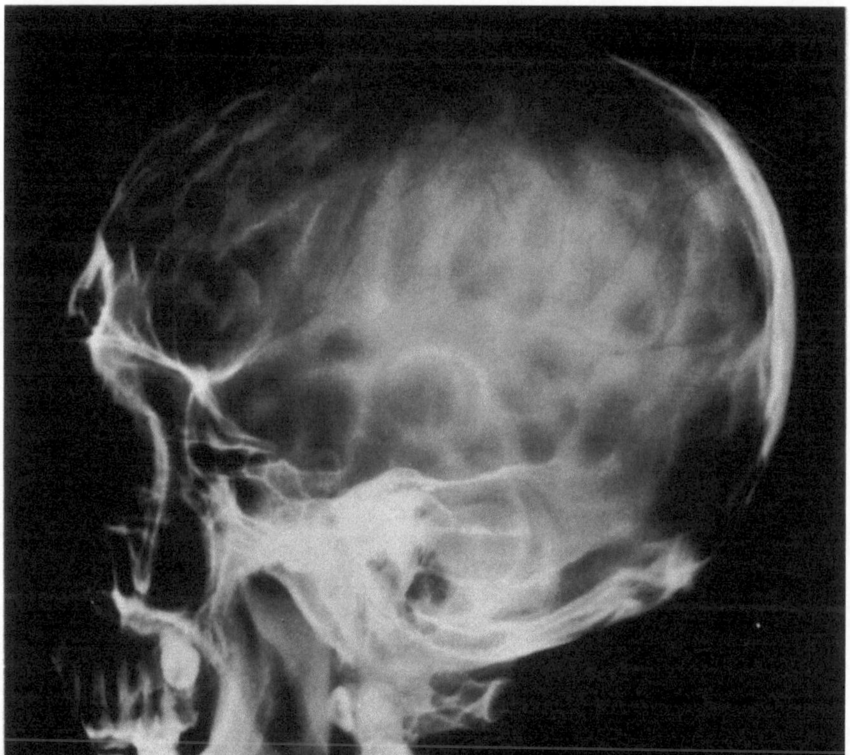

Fig. 12.14. Craniostenosis. Prominent convolutional markings; the sella turcica appears very low in position and is obscured by projected crests of the skull vault between deep digital markings

Fig. 17.14. Aqueductal stenosis in a 10-year-old patient. Prominent [illegible] down [illegible] internal [illegible] [illegible] of [illegible] [illegible] chiasmatic cistern, [illegible] [illegible] [illegible] [illegible] the [illegible] of the [illegible] [illegible] [illegible]

Chapter 13 Generalized Diseases and Changes in the Sella Turcica

A. Congenital Anomalies of the Sella Turcica

I. Congenital Skull Dysplasias[44]

1) Cleidocranial Dysostosis

Anomalies of the sellar region are minimal; the sphenoid and the sphenoid sinus are poorly developed and consequently the sella turcica is small (Job et al. 1965; Keats 1967; Buurman et al. 1978).

2) Osteopetrosis

In cases of osteopetrosis (Marble bone disease, Albers-Schönberg disease) bones are thick and have increased density. They are also fragile. The sphenoid appears dense and thick; the sella turcica is small.

3) Achondroplasia

In the skull, the bones of enchondral origin, i.e., the bones of the base, particularly the basisphenoid and the occipital bones, show defective development. The basisphenoid is poorly developed and the basal angle is reduced (Langer et al. 1967) (Fig. 13.1).

4) Apert's Syndrome

This congenital malformation has an autosomal dominant inheritance pattern. The anterior cranial fossa appears shortened (Campbell 1971). The lesser wings of the sphenoid and the planum sphenoidale show an upward displacement; as a consequence the chiasmatic sulcus becomes vertical and perpendicular to the planum (Fig. 13.2).

5) Osteogenesis Imperfecta (Lobstein's Disease)

The bones of the vault and base are thin and fragile, allowing flattening of the skull base (Fig. 13.3).

II. Neurofibromatosis[45]

Bony abnormalities around the sella turcica in neurofibromatosis may be related to congenital dysplasic bone lesions or to tumors arising in the vicinity.

1) Dysplasic Bone Changes

Essentially, these lesions affect the sphenoid body (Binet et al. 1969) with unilateral hypoplasia of the greater sphenoid wing, elevation of the lesser wing, widening of the superior orbital fissure, and enlargement of the orbit. The defect in the posterior orbital wall brings the temporal lobe in direct contact with the eye, which in some cases explains pulsating exophthalmos. The sella turcica is sometimes enlarged, with ipsilateral downward tilting of the sellar floor (Pribram and du Boulay 1971) (Fig. 13.4). This unilateral depression of the floor may sometimes be related to medial intrasellar position of the carotid siphon (Binet et al. 1969).

---

44 Refs. [128, 166, 218, 467, 484, 538, 811, 1027]

45 Refs. [76, 206, 651]

2) Bone Changes Related to Neighboring Tumors

Here, one is dealing essentially with spongioblastomas of the optic pathways which affect the optic canal and the chiasmatic sulcus (Davidson 1966) (see Chap. 8). Meningiomas in the sellar region (Fig. 8.10) are often found in neurofibromatosis, as are hypothalamic hamartomas.

III. Chromosomal Syndromes[46]

Trisomy 21 (mongolism, Down's syndrome) may provoke changes in the sella turcica: the basisphenoid is short (du Boulay 1965). Various authors such as Timme (1921), Stevenson and Stulz (1923) and du Boulay (1965) have described excavation of the presellar region (Fig. 13.5). According to Kier (1968), these appearances could correspond to an infantile sella turcica. In the case of primary target gland insufficiency widening of the sella turcica may be seen for the same reasons as in primary hypothyroidism (Hubble 1963).

For Turner's and Klinefelter's syndromes see this chapter Sect. C.III.2.

IV. Mental Retardation

We have observed blistering of the planum sphenoidale in patients with severe mental retardation (Bonneville and Dietemann to be published) (Fig. 13.6). In other such cases, a small sella turcica was seen (Fig. 13.7).

V. Meningoencephaloceles[47]

Meningoencephaloceles are herniations of the meninges, with or without nervous tissue, through an osseous defect. Basal meningoencephaloceles are rare (about 10% of encephaloceles). They are located in the midline on the cribriform plate and sphenoid body. They have been categorized by Pollock et al. (1968) into transsphenoid, spheno-orbital, sphenoethmoid, transethmoid, and sphenomaxillary. The sella turcica and the presellar region are only affected in cases of transsphenoid and sphenoethmoid meningoencephaloceles.

Meningoencephaloceles are congenital, occurring very early in embryonic development at the end of the 1st month: the absence of separation between the neuroectoderm (ventral surface of the neural tube) and the superficial ectoderm prevents the interposition of the mesoderm between these two ectodermal layers. As the bones of the skull base derive from the mesoderm, there will be a bone defect in the area of nonseparation of the two ectodermal layers.

1) Transsphenoid

Only about ten cases have been reported (Danoff et al. 1966; Pollock et al. 1968; Harwood-Nash and Fitz 1976; Silverman 1976; Manelfe et al. 1978). The bone defect is located in the floor of the sella. When the opening is wide, it can easily be identified on both lateral and frontal plain films (Fig. 13.8). Sometimes the defect is small, and a routine lateral projection may appear normal. The base view and tomograms will show the shape, size, and position of the bone defect. The meningoencephalocele is recognized by the presence of a soft tissue mass either within the sphenoid sinus or in the nasopharynx.

2) Sphenoethmoid

Few cases have been reported (Pollock et al. 1968; Kane and Lore 1975; Sakoda et al. 1979); the bone defect is localized in the planum sphenoidale and in the posterior part of the cribriform plate. The herniation occurs between the sphenoid sinus and ethmoid sinuses. The meningoencephalocele is recognized by the presence of a soft tissue mass in the posterior part of the nasal cavities.

Meningoencephaloceles may be associated with other malformations such as agenesis of the corpus callosum, median cleft lip and palate (Manelfe et al. 1978; Sakoda et al. 1979).

46 Refs. [124, 485, 518, 690, 818, 910, 950]
47 Refs. [14, 29, 84, 139, 194, 475, 541, 600, 732, 743, 819, 922, 979]

VI. Persistent Infundibular Recess

The infundibular recess, from which the neurohypophysis derives, normally disappears in the embryo of 45 mm. Two recent publications have reported possible persistence of the infundibular recess in man. The patient described by Kühne and Schwartz (1975) had a large sella turcica; however, this was apparently secondary to noncommunicating hydrocephalus. In the case reported by Cabanes (1978), the sella turcica appeared normal.

VII. Vascular Anomalies

1) Transsphenoidal Communicating Artery [48]

To date, nine cases of transsphenoidal intercarotid anastomosis have been reported, all of which were associated with unilateral agenesis of the cervical and petrosal portions of the internal carotid artery (Lie 1968; Harispe et al. 1970; Smith et al. 1972; Faivre et al. 1978; Janicki et al. 1979; Kishmore et al. 1979; Schmidt 1979; Staples 1979). These intercarotid anastomoses are wide, ensuring adequate blood supply to the intracranial carotid system on the side of the absent extracranial portion of the internal carotid artery.

Three types may be identified, by their topographical features:

1. Retrosellar anastomoses (Lie 1968; Harispe et al. 1970) pass behind the base of the dorsum sellae. On plain films or tomograms of the skull, the dorsum sellae appears normal.

2. Infrasellar anastomoses (Faivre et al. 1978) cross through the cancellous bone below the posteroinferior part of the floor of the sella.

3. Transsellar anastomoses are the most numerous, passing directly in contact with the anteroinferior part of the floor of the sella, where they may excavate a transverse groove. The lateral projection shows a sella turcica of normal size, with considerable thinning of the anteroinferior part of the floor, which may simulate a pituitary microadenoma. Lateral tomography indicates that this thinning involves the entire width of the floor (Fig. 13.9). The intercavernous segment of the internal carotid artery is dilated, and the carotid sulcus is more pronounced (Janicki et al. 1979).

These abnormal vessels may correlate with the persistence of anastomoses which are present in the embryo of 4–5 mm (Padget 1948). The persistence and development of these anastomoses could be induced by agenesis of an internal carotid artery.

48 Refs. [282, 391, 459, 487, 502, 567, 704, 710, 845, 885, 896, 1019]

2) Persistent Trigeminal Artery [49]

This carotid-basilar anastomosis is found in 0.2% of individuals. It arises from the intracavernous internal carotid artery, runs backward, passing the lateral aspect of the dorsum sellae, and ends in the basilar artery between the origin of the superior cerebellar and posterior cerebral arteries. The trigeminal artery rarely produces changes in the sella turcica. Radiologic abnormalities observed include erosion of the lateral aspect of the dorsum sellae; Pribram and du Boulay (1971) reported a case in which the dorsum was truncated on one side. We have seen a case in which there was a horizontal groove at the base of the dorsum sellae hollowed out by the trigeminal artery (Fig. 13.10).

49 Refs. [397, 533, 822]

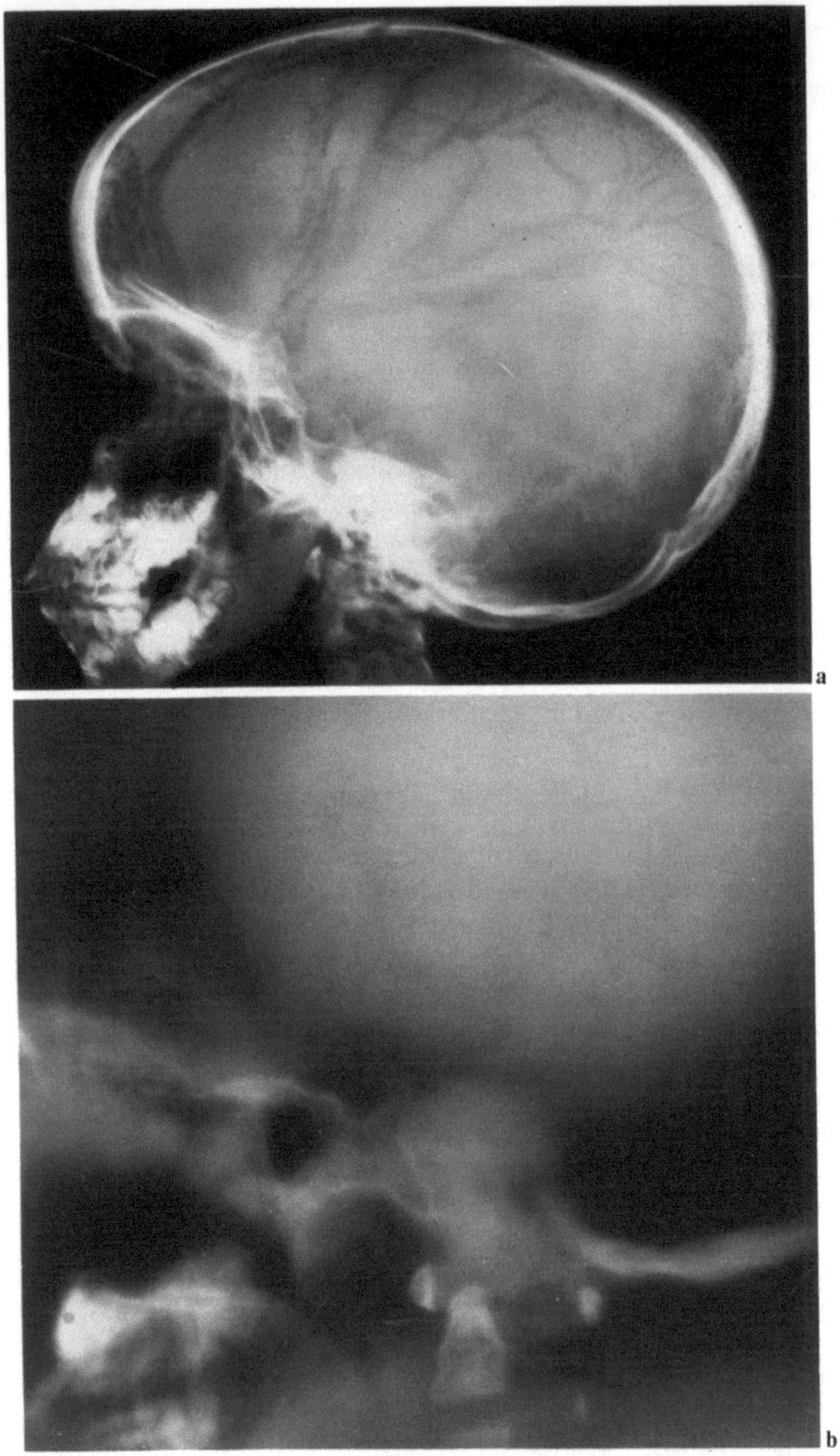

Fig. 13.1 a, b. Achondroplasia.
a Lateral view
b Lateral midsagittal polytomogram: sella turcica is not enlarged; basal angle is reduced. Note short clivus and dens. Slight atlantoaxial dislocation is also noted (Courtesy, J. Metzger)

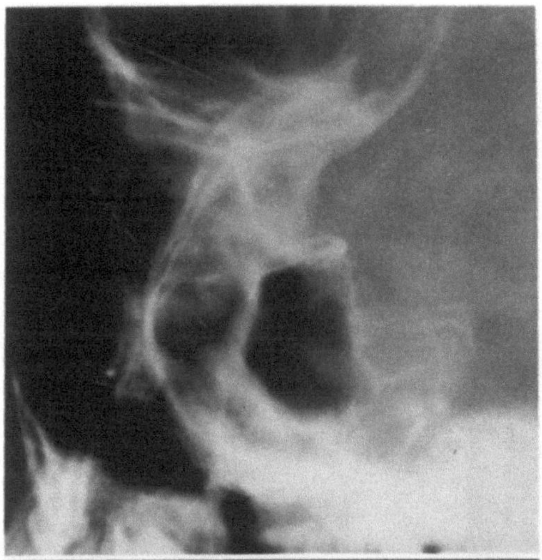

a

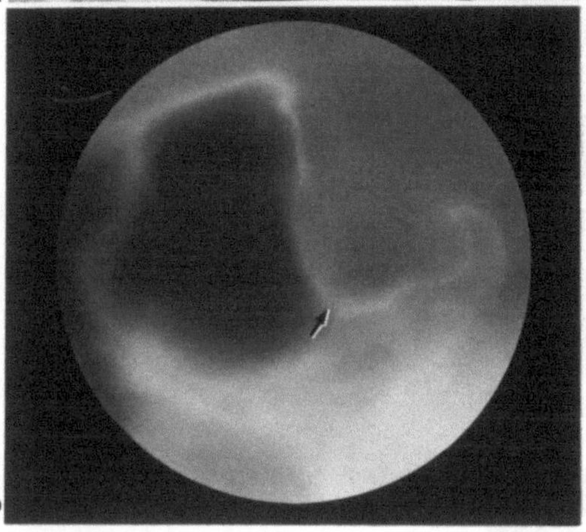

b

Fig. 13.2a, b. Apert's syndrome in a 13-year-old child.
a Lateral view: anterior cranial fossa appears shortened. Note the high position of orbital roofs and lesser sphenoid wings. Chiasmatic sulcus is perpendicular to the planum sphenoidale.
b Lateral polytomogram of sella turcica: angulated sella turcica with vertical chiasmatic sulcus and short dorsum sellae. Small bony defect of the floor of the sella *(arrow)*

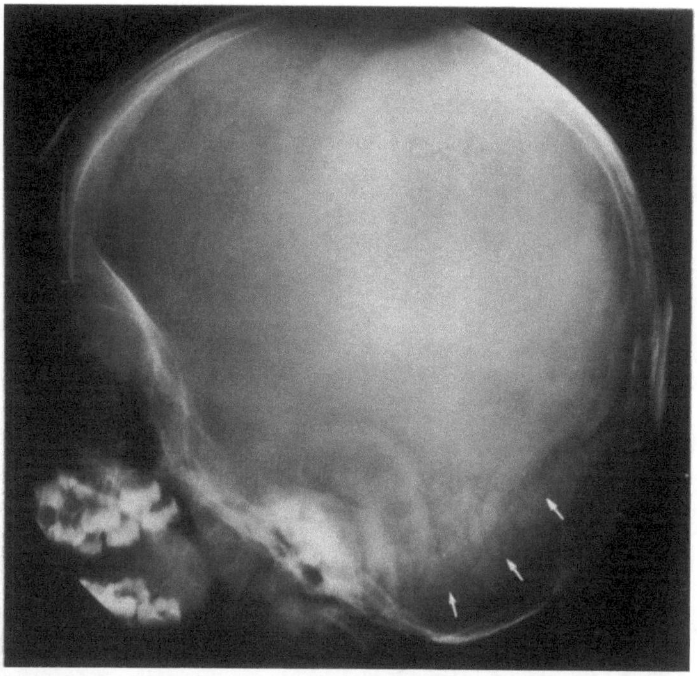

Fig. 13.3. Osteogenesis imperfecta in a 2-year-old child. Poorly mineralized skull base; sellar region appears flattened; cranial vault is thin. Wormian bones *(arrows)* and wide anterior fontanelle

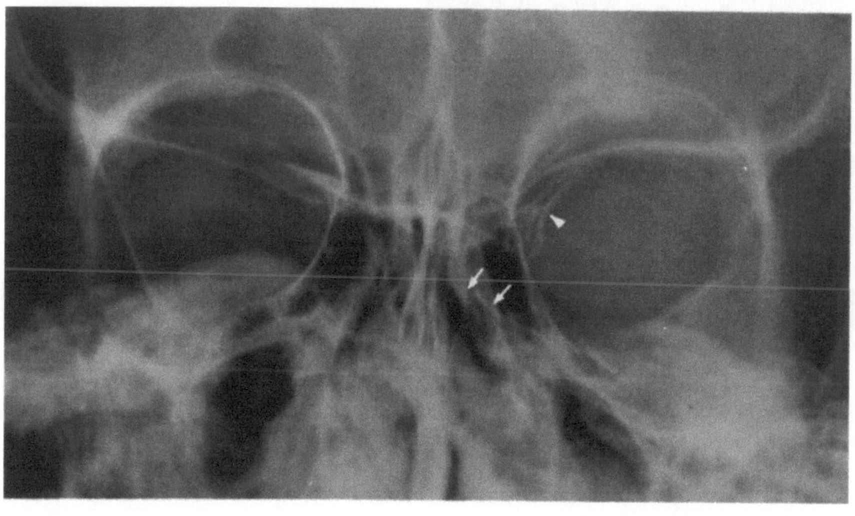

Fig. 13.4. Neurofibromatosis. Frontal projection shows unilateral downward tilting of floor of the sella *(arrows)*. Hypoplasia of greater sphenoid wing and elevation of lesser wing are noted on left *(arrowhead)*

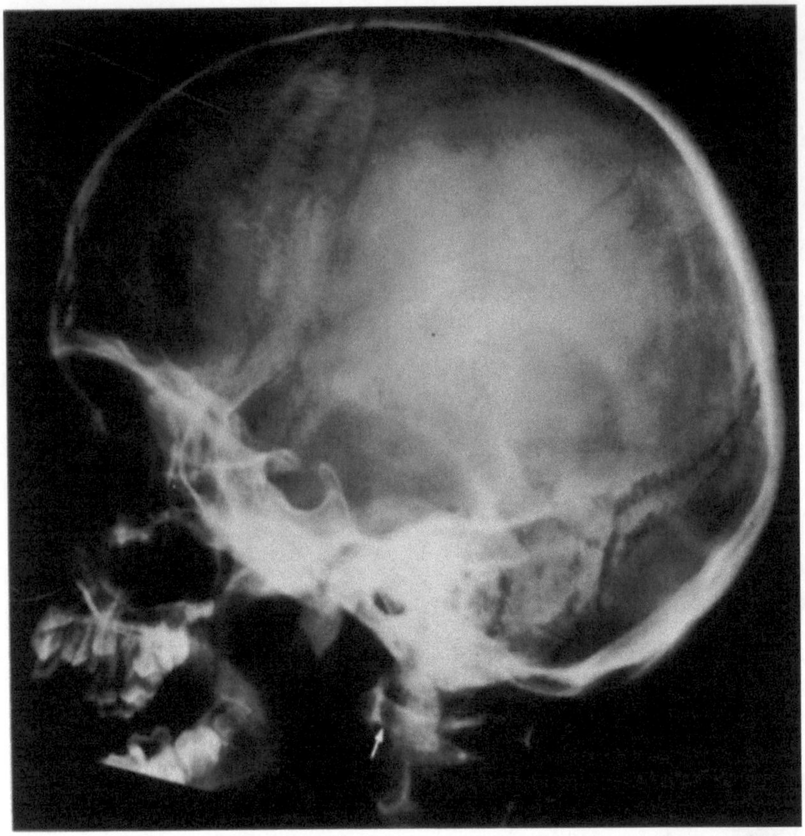

Fig. 13.5. Down's syndrome in a 12-year-old child. Lateral projection shows deepening of chiasmatic sulcus. Note also microcrania, delayed closure of sutures, and slight atlantoaxial dislocation *(arrow)*

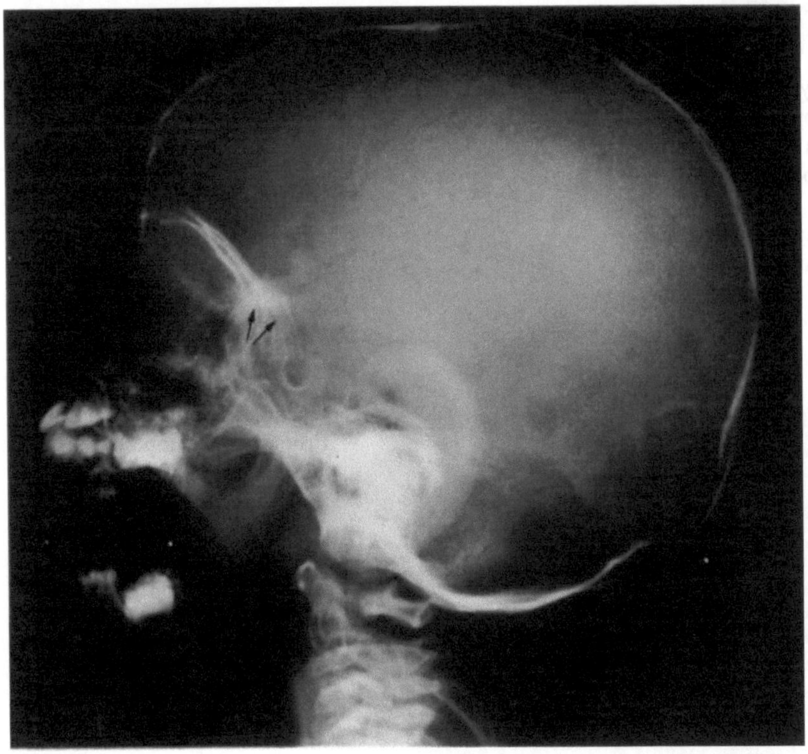

Fig. 13.6. Mental retardation in a 3-year-old child. Note upward convexity (blistering) of the planum sphenoidale *(arrows)*

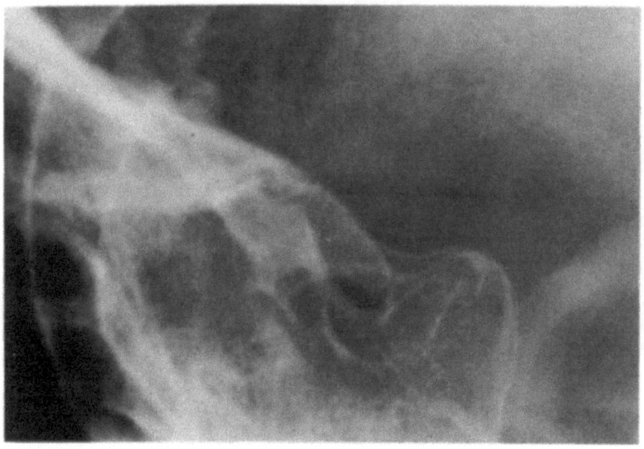

Fig. 13.7. Mental retardation in a 6-year-old child. Very small sella turcica with thick dorsum sellae

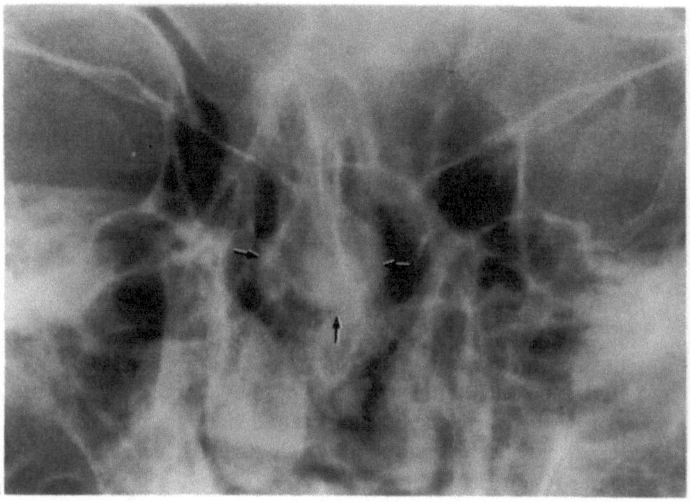

a

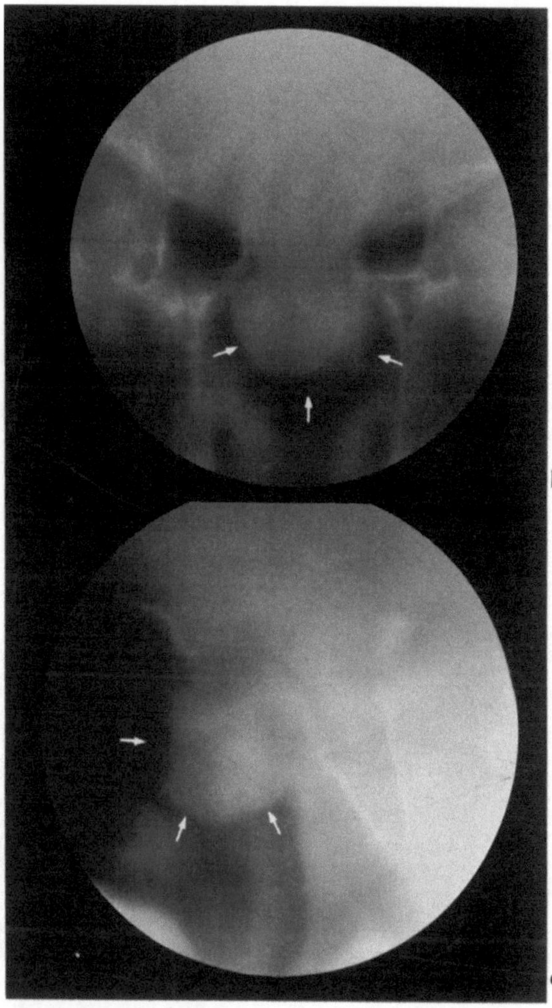

b

c

Fig. 13.8a–c. Transsphenoid meningoencephalocele in a 57-year-old male.

a Frontal projection: hypertelorism; floor of the sella is not seen. Note the presence of a soft tissue mass *(arrows)*.

b Frontal, and **c** lateral tomograms reveal a midline bony defect in floor and anterior wall of sella turcica. Dorsum sellae is thinned and displaced backward. Soft tissue mass, corresponding to downward herniation of anterior part of the third ventricle, is noted in the sphenoid sinus and nasopharynx *(arrows)* (Courtesy, C. Manelfe)

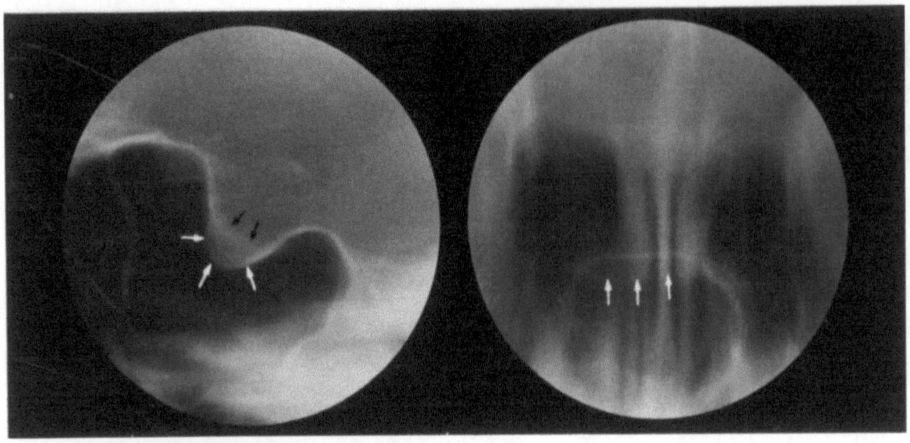

a b

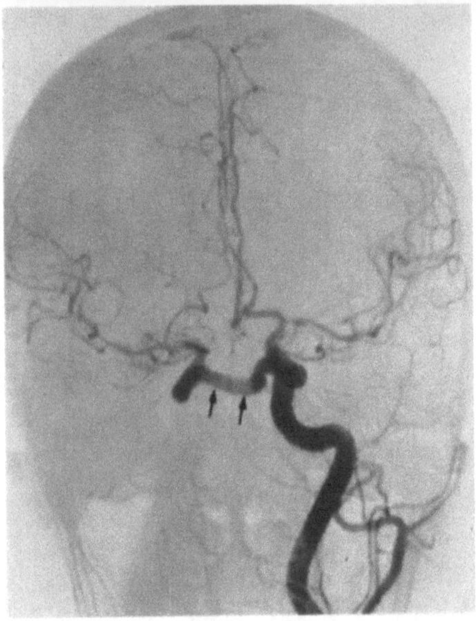

c

Fig. 13.9 a–c. Transsellar communicating artery.
a Lateral tomogram: localized bulging and thinning of the anteroinferior part of the floor of the sella *(arrows)*.
b Frontal linear tomogram: horizontal soft tissue mass in the sphenoid sinus below the floor of the sella indicates the transsellar communicating artery *(arrows)*.
c Left carotid angiogram: frontal projection; large communicating artery between left and right carotid siphons *(arrows)*. Left internal carotid artery supplies right and left cerebral hemispheres (Courtesy R.C. Schmidt)

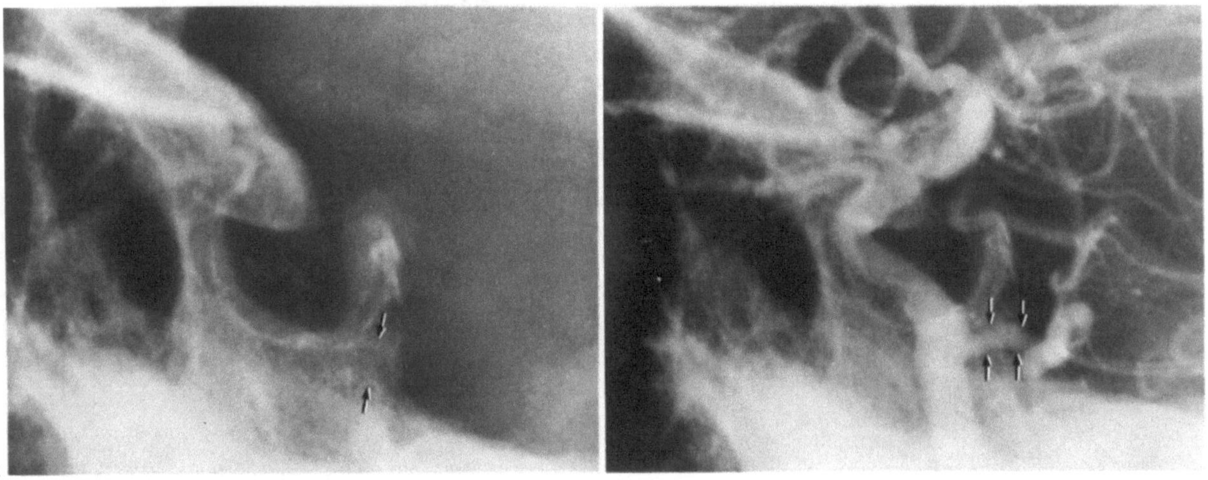

a b

Fig. 13.10 a, b. Persistent trigeminal artery.
a Magnified lateral projection shows horizontal groove with upper and lower dense margins at the base of the dorsum sellae *(arrows)*.
b Carotid angiogram shown by a lateral projection. Persistent trigeminal artery crosses the lateral aspect of the dorsum sellae *(arrows)*

B. Metabolic Diseases

I. Mucopolysaccharidoses [50]

Metabolic disorders characterized by an accumulation of mucopolysaccharides in various body tissues, the mucopolysaccharidoses have been divided into six different groups. Changes in the skull and particularly in the sella turcica are usually found in mucopolysaccharidosis type I (Hurler's syndrome) and type II (Hunter's syndrome), but also in type VI (Maroteaux-Lamy syndrome).

1) Hurler-Hunter Syndrome (Gargoylism)

The sella turcica is elongated. There is deepening of the planum sphenoidale and the chiasmatic sulcus, with spreading and thinning of the anterior clinoid processes, together with poor pneumatization of the sphenoid sinus (Neuhauser et al. 1968) (Fig. 13.11). This appearance is not specific but is common (Smith et al. 1952). Changes in skull appear only after the age of 1 (Horrigan and Baker 1961).

Three theories have been advanced to explain changes in the sphenoid: (1) Neuhauser et al. (1968) demonstrated that sellar changes could be related to suprasellar arachnoid cysts. (2) Grossman and Winchester (1971) described a case in which sellar changes seemed to be associated with dilatation of the anterior recesses of the third ventricle as part of communicating hydrocephalus. Hydrocephalus is quite frequent in this condition, being related to meningeal thickening due to an accumulation of mucopolysaccharides (Magle 1950). (3) Other authors (Harwood-Nash and Fitz 1976)

50 Refs. [350, 433, 594, 602, 660, 673, 883]

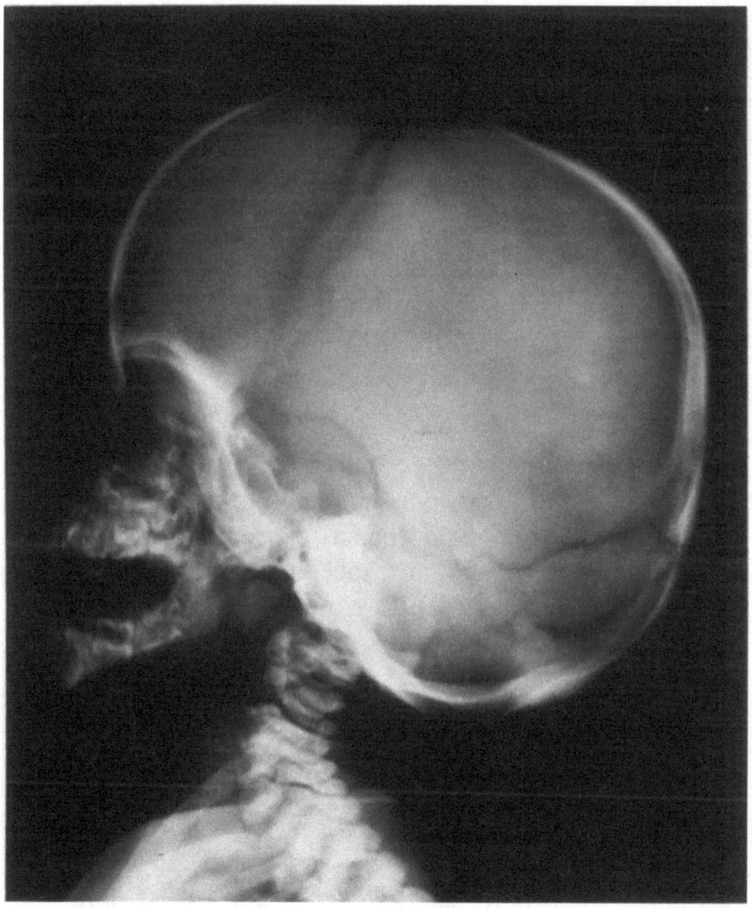

Fig. 13.11. Hurler-Hunter syndrome. Macrocrania. Sellar region is elongated with deepening of the chiasmatic sulcus. Anterior clinoid processes are tapered (Courtesy, C. Manelfe)

believe that these changes in the sphenoid bone are directly related to a bone growth disorder.

2) Maroteaux-Lamy Syndrome

The sella turcica appears elongated without other bone changes (Maroteaux et al. 1963; Grossman and Winchester 1971).

II. Hypophosphatasia and Hyperphosphatasia

Hypophosphatasia is transmitted in a recessive inheritance pattern. There is decreased activity of alkaline phosphatase in serum and in numerous tissues. Changes in the region of the sella are characterized by the presence of a wide sphenofrontal suture (McPherson et al. 1972; Harwood-Nash and Fitz 1976). In cases of chronic idiopathic hyperphosphatasia (familial osteoectasia), thickening of the cranial vault and skull base is seen (Caffey 1972).

III. Idiopathic Infantile Hypercalcemia and Hypervitaminosis D

Increased bone density is noted in the skull base and in the cranial vault (Shiers et al. 1957).

IV. Fluorine Intoxication

Increased density of the skull base is often present (Bishop 1936; Calenoff 1962).

V. Alcoholism

Chronic alcoholics may present a demineralized dorsum sellae, occasionally with erosion of the lamina dura (Albert and Le May 1968).

C. Endocrine Diseases

I. Hypopituitarism

1) Hypopituitary Dwarfism [51]

Patients with idiopathic hypopituitarism and growth hormone deficiency alone, or with deficiency of other adenohypophyseal hormones, often have a small sella turcica (Fisher and Di Chiro 1964; Chouraki et al. 1978). In some cases of pituitary dwarfism the sella turcica is only a slight depression in the upper surface of the body of the sphenoid: such a sella turcica is referred to as a "vestigial sella" (Pribram and du Boulay 1971) (Fig. 13.12). The discovery of a small sella turcica in dwarfed children raises the possibility of hypopituitary dwarfism (Underwood et al. 1973). Parks et al. (1978) have recently reported three cases of familial hypopituitarism with an enlarged sella turcica. In certain forms of moderate hypopituitarism an infantile type sellar region with a poorly developed planum sphenoidale and long chiasmatic sulcus, a poorly developed sphenoid sinus and lack of fusion of the spheno-occipital synchondrosis are seen (Fig. 13.13).

2) Congenital Absence of Pituitary Gland [52]

Absence of the adenohypophysis is well recognized in anencephalic subjects. However, the pituitary gland has also been found to be absent in live normocephalic newborn infants (Brewer 1957). In some cases the sella turcica is normal in size (Steiner and Boggs 1965), but in others it is very small (Blizzard and Alberts 1956; Dunn 1966; Sipponen et al. 1978).

---

51 Refs. [103, 108, 156, 290, 295, 966]
52 Refs. [17, 80, 110, 177, 256, 430, 771, 880, 907]

3) Postpartum Pituitary Necrosis (Sheehan's Syndrome)[53]

Fisher and Di Chiro (1964) believe that decrease in volume of the pituitary gland when the sella turcica has already reached full development (after the age of 6) does not cause a subsequent decrease in size of the hypophyseal fossa. However, certain workers (Meador and Worrell 1966; Boukhris et al. 1975), have emphasized the increased frequency of small sellae in patients with postpartum pituitary necrosis; these authors take only the lateral area of the sella turcica into account, however (Pribram and du Boulay 1971). It appears very unlikely that the size of the sella turcica might decrease after postpartum necrosis. Indeed, in 80% of cases, adenohypophyseal necrosis only involves a small part of the anterior lobe of the pituitary gland (Sheehan 1940); the decrease in volume of the adenohypophysis is certainly a promoting factor in the development of certain cases of empty sella (see Chap. 6).

II. Small Sella Turcica and Infertility

A small sella turcica is often observed in patients with primary anovulation caused by hypothalamic suppression (Fig. 13.14). We share the views of Vezina (1973) as regards this issue.

III. Primary Target Gland Insufficiency

Prolonged primary end organ failure (hypothyroidism, hypogonadism, and hypoadrenalism) may induce secondary hyperplasia or adenoma of the adenohypophysis (Leiba et al. 1969; Danziger et al. 1979).

53 Refs. [102, 393, 622, 865, 866, 867]

1) Hypothyroidism[54]

Hypothyroidism is the primary peripheral endocrine deficiency which most often causes changes in the sella turcica. Enlargement of the pituitary fossa is well known in primary congenital hypothyroidism in children, but may also be seen in adult acquired primary hypothyroidism (Leiba et al. 1969; Jackson and Hall 1970; Guinet et al. 1972; Lawrence et al. 1973). In 80% of cases of childhood hypothyroidism, the sella turcica undergoes changes (Bellini and Neves 1956; Fossati et al. 1977); these are quite characteristic (Fig. 13.15). The sella turcica is often only slightly enlarged, but the pituitary fossa is rounded in a very regular manner as if drawn with a compass; the outlines are well defined and the cortical bone of the floor of the sella is conserved, sometimes appearing dense; the dorsum sellae is vertical (Bellini and Neves 1956). Certain authors speak of a "cherry" sella or in small children of a "bowl-like" sella (Swischuk and Sarwar 1977). This sellar configuration is associated with other cranial changes: diastasis of sutures, delayed closure of the spheno-occipital synchondrosis, dense diploe, absence of cranial sinuses, and small lower jaw with open angle (Bellini and Neves 1956). Prompt, appropriate treatment results in a reduction in size of the sella turcica. Pre- and post-treatment radiographs allow assessment of the effectiveness of treatment (McCarten and Kuhns 1976; Swischuk and Sarwar 1977). Lack of replacement therapy may, in some cases, give rise to a pituitary adenoma; the sella turcica appears clearly enlarged, with destruction of its walls (Fossati et al. 1977; Samaan et al. 1977). In exceptional cases, congenital hypothyroidism may cause enlargement of the sella turcica in children, due not only to hyperplasia of TSH-secreting cells but also by hormonal overlap of FSH-LH-secreting cells which, therefore, cause precocious puberty (Van Wyk and Grumbach 1960; Hubble 1963).

54 Refs. [54, 99, 103, 106, 123, 199, 243, 303, 357, 370, 443, 444, 453, 542, 559, 590, 611, 638, 677, 714, 716, 779, 823, 913, 931, 974, 1040]

2) Hypogonadism [55]

Enlargement of the sella turcica after castration was reported as early as 1910 by Tandler and Grass. Turner's and Klinefelter's syndromes are two recognized causes of primary hypogonadism. The sella turcica may be enlarged, but in some cases this merely shows lateralized thinning in the anteroinferior portion of its floor (Danziger et al. 1979). In many cases of Turner's and Klinefelter's syndromes, the sella turcica appears normal (Fisher and Di Chiro 1964); a large sphenoid sinus is often seen (Kowowicz and Rzymski 1975) (Fig. 13.16).

3) Hypoadrenalism [56]

Dluhy et al. (1978) reported two cases of primary adrenal insufficiency (Addison's disease) with sellar enlargement. ACTH-secreting adenomas developing after bilateral adrenalectomy (Nelson's syndrome) in patients with Cushing's disease, i.e., bilateral adrenal hyperplasia secondary to excess secretion of ACTH by the pituitary, are probably adenomas existing prior to the adrenalectomy, growth of which was accelerated by removal of the adrenal glands.

55 Refs. [87, 104, 150, 199, 488, 559]
56 Refs. [16, 199, 214, 234, 559, 643, 653, 669]

IV. Hyperparathyroidism

Resorption and reparative fibrosis give the appearance of mottled demineralization of the vault and skull base (Fig. 13.17).

V. Cushing's Syndrome

Cushing's syndrome is accompanied by demineralization of the vault and the skull base, particularly the dorsum sellae (Fig. 13.18) (see also Chap. 6).

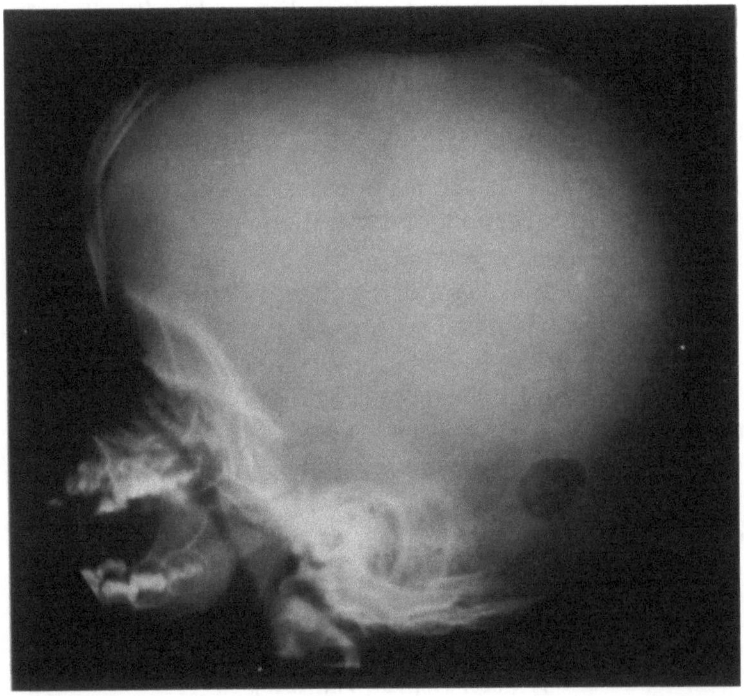

Fig. 13.12. Hypopituitarism in a 14-month-old infant. Lateral projection shows a very shallow "vestigial" sella. Shunted hydrocephalus with wide anterior fontanelle

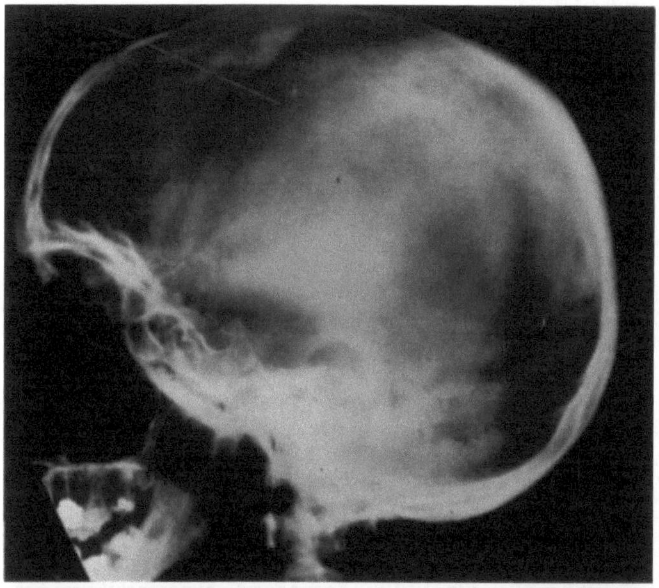

Fig. 13.13. Hypopituitarism in a 31-year-old male. Sella turcica is not enlarged, but has an infantile configuration: poor posterior development of the planum sphenoidale with long chiasmatic sulcus; spheno-occipital synchondrosis is open; poorly aerated sphenoid bone. Note also thin cranial vault in area of bregma and slight atlantoaxial dislocation

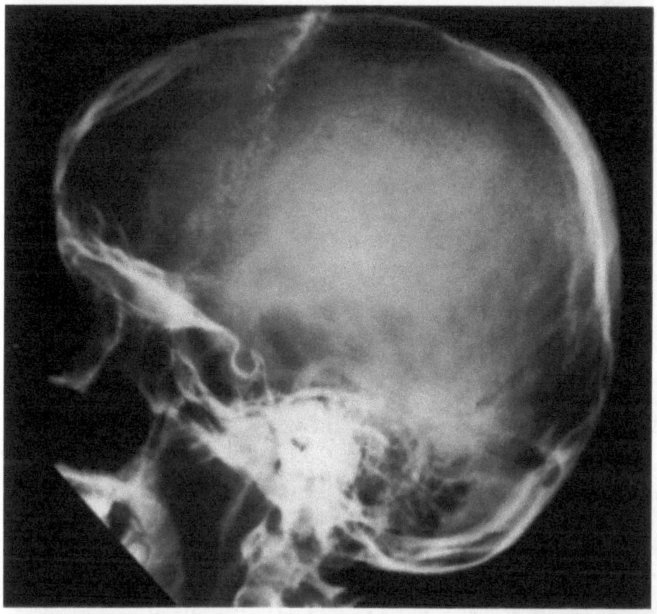

Fig. 13.14. Primary amenorrhea in a 17-year-old female. Lateral projection shows a very small sella turcica

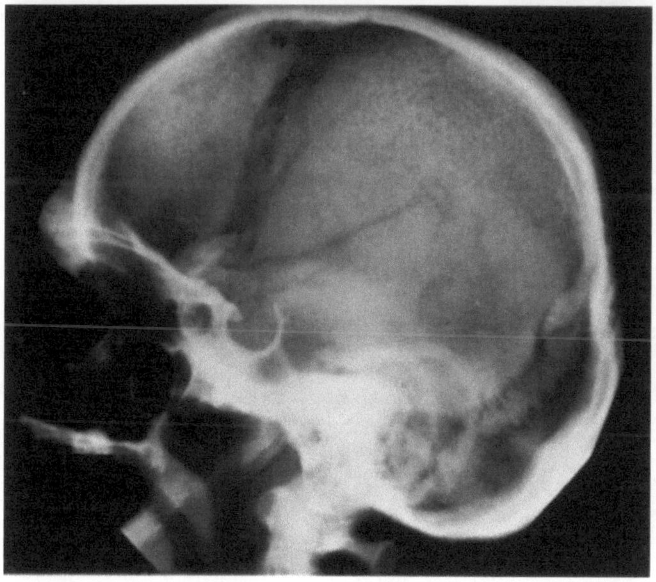

Fig. 13.15. Primary hypothyroidism in a 45-year-old male. Rounded enlargement of the sella turcica; cherry-shaped sella. The dorsum sellae is thin, but the cortical bone is preserved and even dense. Poor pneumatization of frontal and sphenoid sinus and mastoid cells. Delayed closure of the lambdoid suture

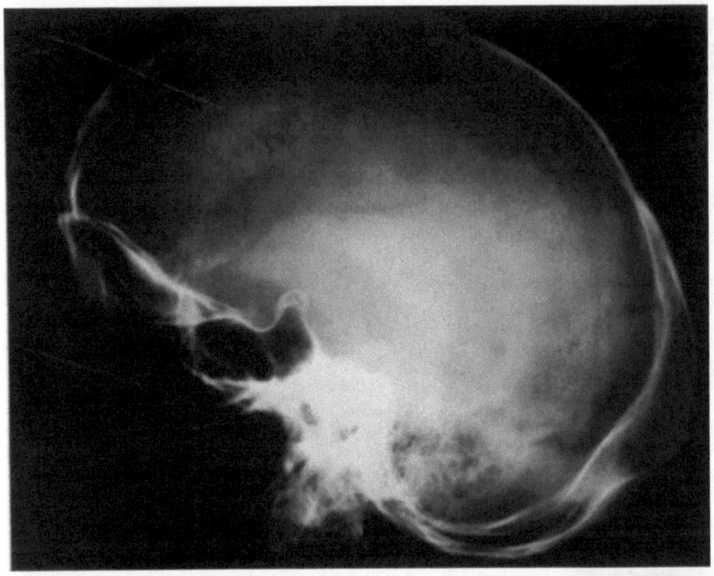

Fig. 13.16. Hypogonadism and Turner's syndrome in a 25-year-old female. Extensive sphenoid sinus with marked pneumatization of the dorsum sellae giving appearance of a small sella. Note also reduced basal angle and basilar invagination

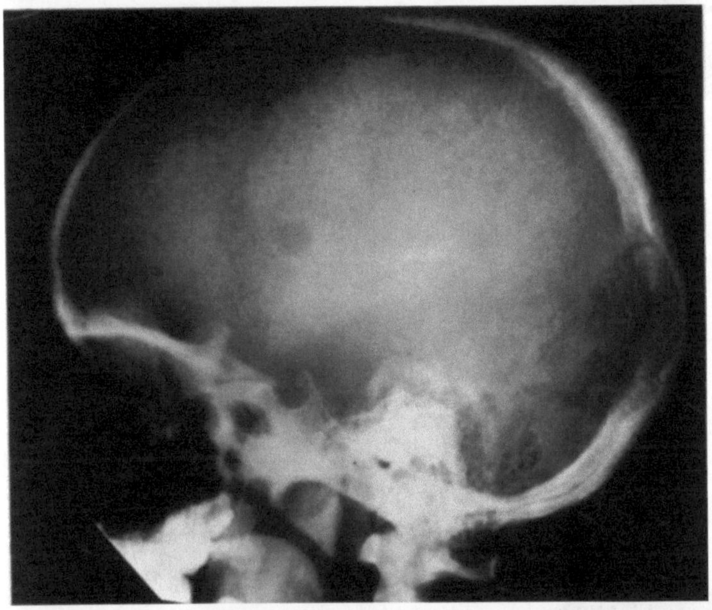

Fig. 13.17. Secondary hyperparathyroidism: mottled demineralization of cranial vault and skull base; slight sclerosis of dorsum sellae

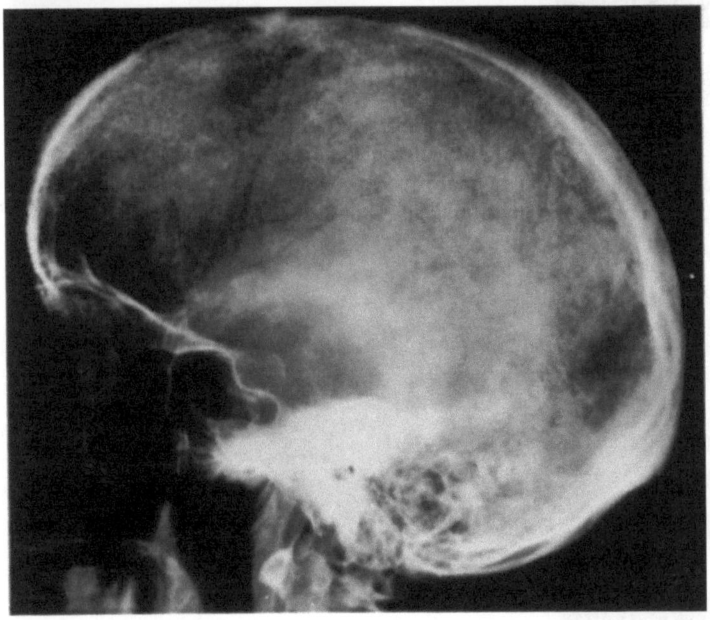

Fig. 13.18. Cushing's syndrome in a 51-year-old female. Marked demineralization of cranial vault and severe demineralization of dorsum sellae

D. Hematologic Diseases

I. Histiocytosis X

Eosinophilic granuloma usually affects the cranial vault of the frontal region. It may, however, develop in the body of the sphenoid. Harwood-Nash and Fitz (1976) described eosinophilic granuloma destroying the dorsum sellae and causing hyperostosis of the clivus.

II. Multiple Myeloma

Multiple myeloma usually produces lytic lesions of the cranial vault. Primary sellar lesions may rarely be encountered, causing destruction of the walls of the sella (Pribram and du Boulay 1971) (Fig. 13.19).

E. Neoplasms

I. Secondary Neoplasms[57]

Secondary neoplasms are the most frequent bone tumors in the sellar region. These metastases, generally of extracranial origin, may cause either lytic or sclerotic lesions of the body of the sphenoid and the neighboring regions of the skull base. Osteolytic metastases rapidly destroy the floor of the sella, and often the clivus. There may be expansion of the tumor toward the sphenoid sinus (Kistler and Pribram 1975) (Fig. 13.20). Osteoblastic metastases, which form about half, usually arising from carcinoma of the prostate or breast, may create a number of diagnostic problems.

57 Refs. [833, 863]

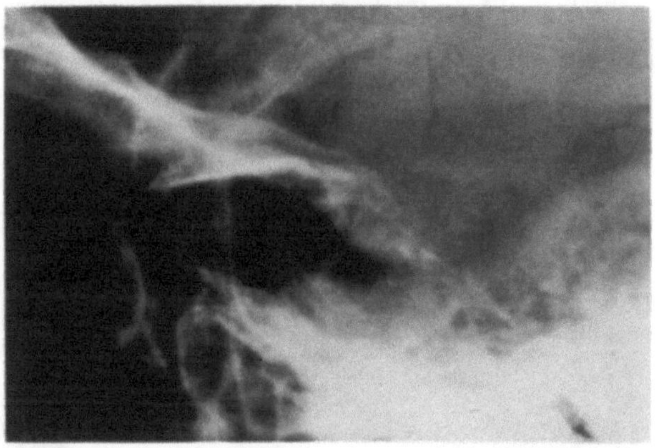

Fig. 13.19. Multiple myeloma. Destruction of margins of sella and soft tissue mass in posteroinferior part of sphenoid sinus

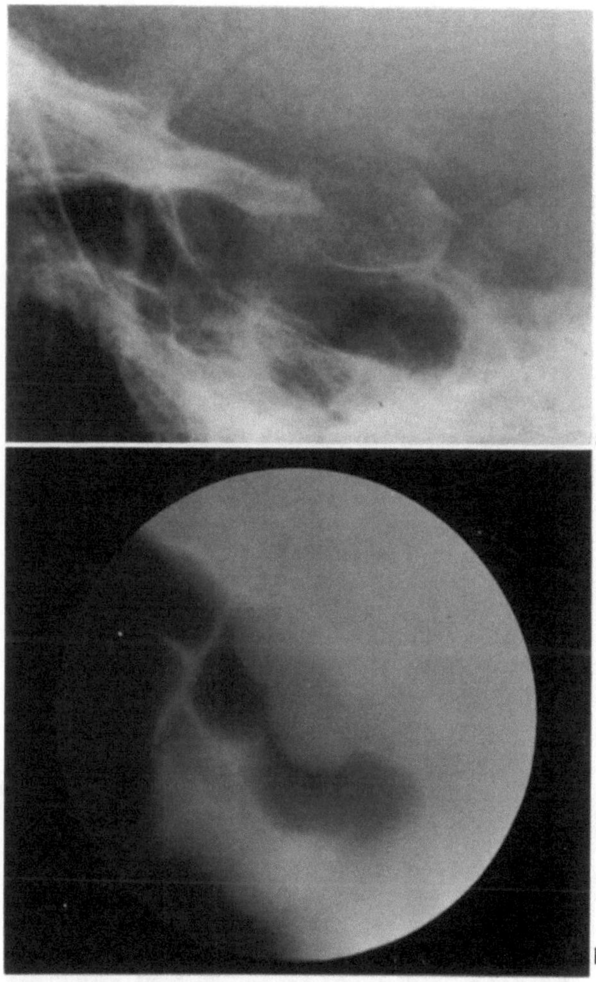

a

b

Fig. 13.20a, b. Metastatic carcinoma of breast. ▷
a Lateral projection
b Lateral tomogram: destruction of chiasmatic sulcus and anterior wall of the sella turcica. Lobulated soft tissue mass within sphenoid sinus

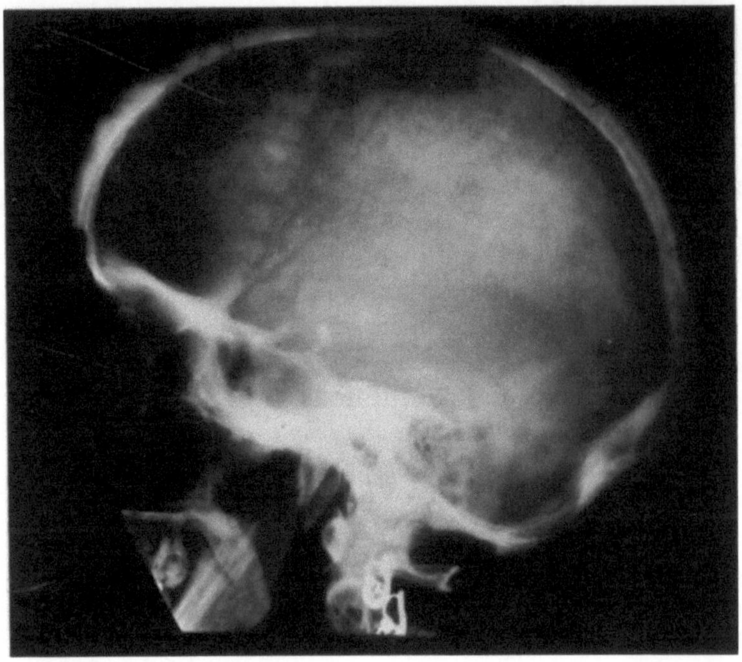

Fig. 13.21. Metastatic carcinoma of breast. Sclerosis of dorsum sellae. Clouding of the posterior part of sphenoid sinus. Note also the presence of lytic and sclerotic lesions involving the cranial vault

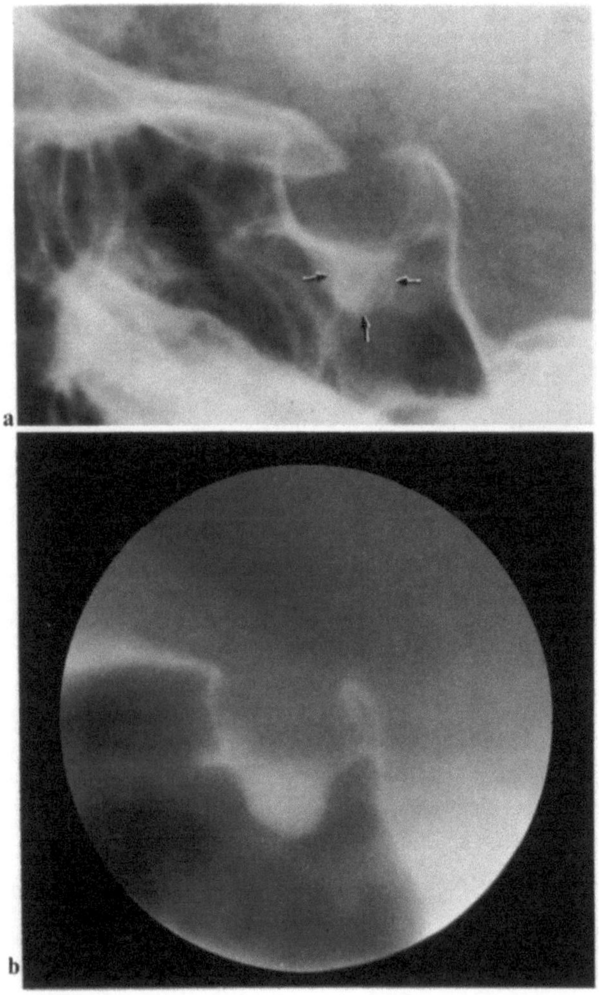

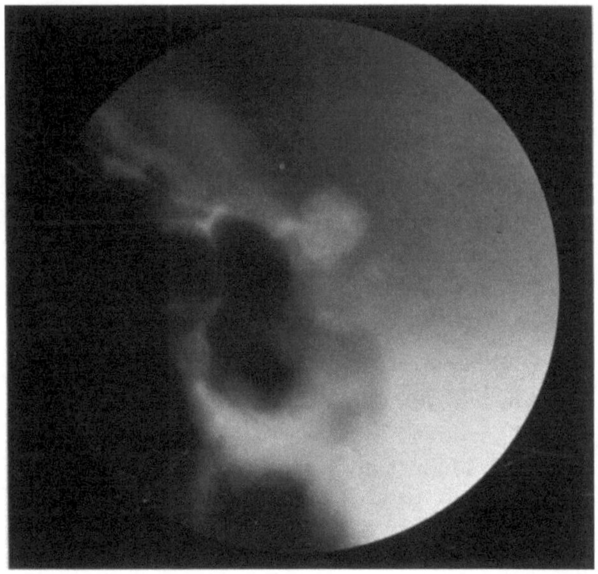

Fig. 13.23. Osteoma of anterior clinoid process (lateral tomogram)

◁ **Fig. 13.22 a, b.** Osteoma of sellar floor.
a Lateral projection
b Lateral tomogram

These metastases at first cause localized bone sclerosis (Fig. 13.21), followed by diffuse sclerosis of the sphenoid bone and clivus. This hyperostosis may suggest a presellar meningioma or meningioma *en plaque* but fibrous dysplasia, Hodgkin's disease, invasion by carcinoma of the nasopharynx, or chronic infectious disease may also be considered.

II. Primary Neoplasms

1) Malignant[58]

Osteogenic sarcomas are rare. Amine and Sugar (1976) reported a case after radiotherapy for a pituitary adenoma.

Rhabdomyosarcomas may develop in the sphenoid (Harwood-Nash and Fitz 1976).

Giant cell tumors develop in the cartilaginous ossification centers. Pitkethly and Kemp (1969) described two cases arising in the postsphenoid with destruction of the dorsum sellae and the floor of the sella. The sphenoid sinus, and the floor of the middle cranial fossa may be invaded and destroyed (Geissinger et al. 1970).

2) Benign[59]

Osteomas may occur in the bony structures of the region of the sella and sphenoid sinus (Figs. 13.22, 13.23, 8.12, 11.5).

F. Infectious Diseases

Osteomyelitis of the sella turcica and the presellar region are usually secondary to adjacent sinusitis (see Chap. 11). Hematogenous dissemination is seen only rarely. Witcombe and Cremin (1978) reported four cases of primary tuberculous osteitis of the sphenoid bone. From a radiologic standpoint, the lesions were lytic in two patients, and mixed, i.e., showing osteolytic changes and osteosclerosis, in two others. Pendergrass et al. (1956) reported a case in which osteosclerosis was predominant, while Nemir et al. (1979) reported a case of nasal tuberculosis spreading of the skull base, with considerable sclerosis in the sphenoid and ethmoid bones.

G. Fractures of the Sellar Region

Fractures involving the region of the sella are rare. Recognition of these fractures is nonetheless important because of their numerous and serious complications.

I. Mechanism of Fractures of the Sellar Region

Certain fractures are extensions toward the middle cranial fossa of temporal and frontal vault fractures; others are caused by direct trauma to the middle cranial fossa, through a fractured mandible (Carlson et al. 1973); still others are direct trauma caused by foreign bodies which have entered by the orbit or nasal fossa. Sometimes these fractures are isolated, then they could be due to shearing forces applied at the center of the skull base (Dublin and Poirier 1976).

II. Clinical Picture

It has been known for some time that sphenoid fractures are observed after severe cranial trauma. However, systematic tomographic exploration of the sellar region in craniofacial trauma often allows detection of fractures of the walls of the sphenoid sinus, the floor of the sella, and planum sphenoidale without clinical evidence. Sometimes attention is directed to the skull base by palsies of the cranial nerves contained in the cavernous sinus, by rhinorrhea, or by epistaxis.

---

58 Refs. [12, 143, 263, 265, 327, 735]
59 Refs. [899, 1028]

III. Complications [60]

Immediate or delayed complications are characterized by their severity.

1) Rhinorrhea

Rhinorrhea may be immediate or delayed. Detected and treated late, rhinorrhea may be complicated by meningitis (Lewin and Cairns 1951).

2) Vascular Lesions

Fractures of the sella turcica may cause stenosis or even occlusion of the intracranial internal carotid artery (Dublin and Poirier 1976). These fractures may also create false carotid aneurysms which can rupture toward the sphenoid sinus (epistaxis), the subarachnoid space (Gouaze et al. 1969) or, most commonly towards the cavernous sinus, causing a carotid-cavernous fistula (Archer and Sundaram 1977). Loop et al. (1964) reported a traumatic occlusion of the basilar artery associated with a fracture of the dorsum sellae.

3) Hypothalamic and Pituitary Lesions

Traumatic lesions of the hypothalamus and neurohypophysis may cause diabetes insipidus, which persists only in patients with severe hypothalamic lesions (Crompton 1971; Jazra and Berndt 1978). Adenohypophyseal lesions are rare: traumatic rupture of the long hypophyseal portal veins in the small circular opening in the diaphragma sellae causes ischemic necrosis in the anterior lobe of the pituitary gland (Daniel et al. 1959; Altman and Pruzanski 1961; Kornblum and Fisher 1969). Post-traumatic hypopituitarism occurs only with extensive necrosis.

4) Pneumocephalus

A fracture in the region of the sella may render the sphenoid and ethmoid sinuses in continuity with the subarachnoid spaces.

60 Refs. [10, 179, 193, 342, 461, 516, 565, 580, 720, 777, 794]

IV. Radiology [61]

An air-fluid level in the sphenoid sinus, or pneumocephalus are suggestive of a fracture of the skull base. The sella turcica must always be examined very carefully with lateral and frontal polytomograms. It is possible to distinguish the following four types of fracture.

1) Fractures of the Planum Sphenoidale and Chiasmatic Sulcus

These are often vault fractures extending toward the base. They may extend toward the optic canal and injure the optic nerve (Fig. 8.21). These fractures may pose differential diagnostic problems with the unfused planum sphenoidale (Smith and Kier 1971) (Figs. 8.22, 5.47). Frontal tomograms show clearly the fracture, associated with opacity of the sinus, representing blood within it (Fig. 8.20).

Recurrent meningitis or persistent rhinorrhea often lead to the discovery of old fractures of the sellar region. It is therefore possible to show either images of deepening (Fig. 8.23) of the sella, with occasionally a bone defect and the presence of an adjacent soft tissue mass (Fig. 13.24), or blistering and post-traumatic hyperostosis of the planum, simulating a meningioma (Bonneville et al. 1978) (Fig. 8.18).

2) Fractures of the Anterior Wall and Floor of the Sella Turcica

These are often temporal fractures extending toward the floor of the middle cranial fossa. However, these fractures are sometimes isolated. Extending fractures are usually transverse (Fig. 13.25). Detection of a fracture of the intersinusal septum necessitates a search for an associated fracture of the floor of the sella or the planum sphenoidale (Fig. 8.19).

61 Refs. [19, 60, 125, 144, 244, 245, 268, 699, 886, 975]

3) Fractures of the Dorsum Sellae

These are less common and usually consist of transverse fractures at the base of the dorsum sellae.

4) Comminuted Fractures

Carlson et al. (1973) described a comminuted fracture of the sphenoid with upward and backward displacement of the sella turcica.

a 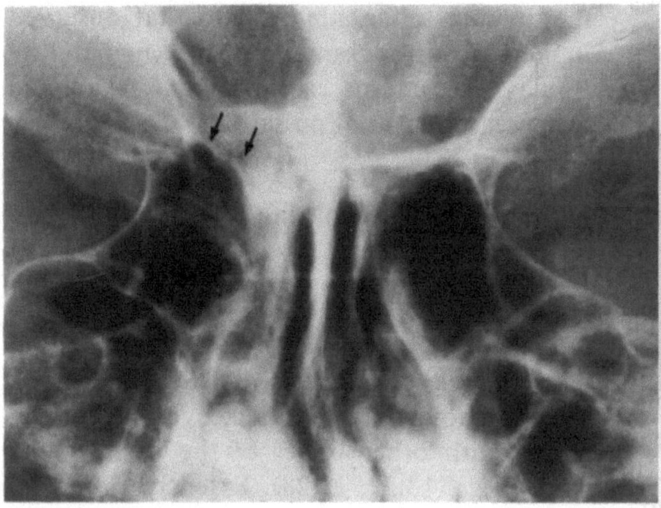

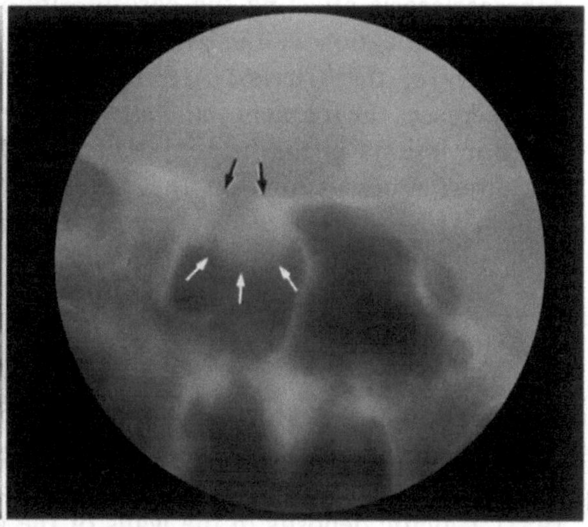 b

Fig. 13.24a, b. Old fracture of planum sphenoidale in a 35-year-old male with meningitis.

a Frontal projection: right side of planum sphenoidale is not well defined *(arrows)*.

b Frontal tomogram: Bony defect on right of planum sphenoidale *(black and white arrows)* and adjacent soft tissue mass *(white arrows)*

a 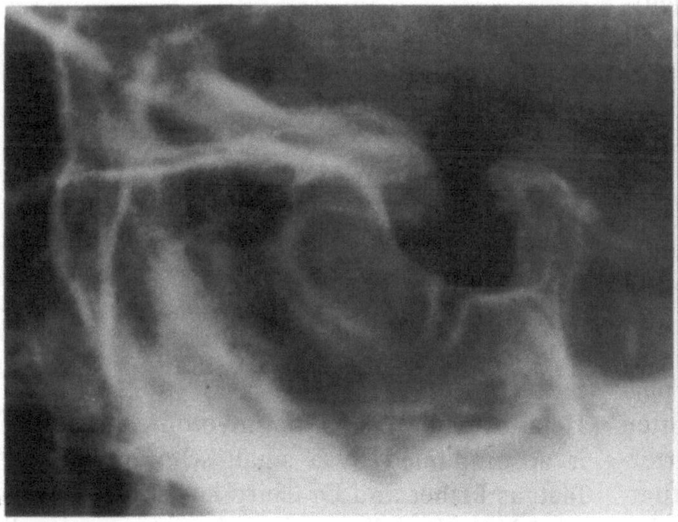

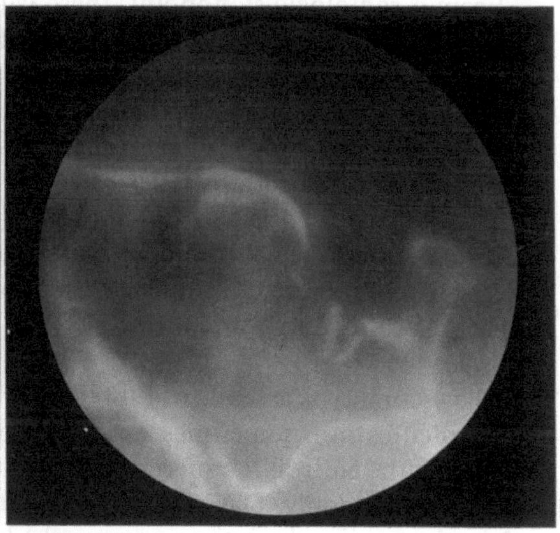 b

Fig. 13.25a, b. Fracture of the floor of the sella. Transient post-traumatic diabetes insipidus.

a Lateral projection

b Lateral tomogram: clouding of sphenoid sinus; comminuted fracture of floor of sella

H. Miscellaneous

I. Arterial Hypertension and Atherosclerosis [62]

Atherosclerosis affecting the carotid siphons causes calcification of the arterial walls and tortuosity of the arteries. Arterial hypertension increases the frequency and size of atherosclerotic lesions in the arterial walls; it also has direct repercussions on the walls of the pituitary fossa.

1) Arterial Calcifications

Calcification of the carotid siphon is frequent, the frequency increasing with age: it is present in 80% of patients over 70 years of age (Nerli 1955). Calcific deposits in the walls of the arteries develop principally in the intracavernous carotid siphon (Cole and Davies 1963). Calcification of the supraclinoid portion of the internal carotid artery is less common. Calcification is frequently bilateral.

From a radiologic standpoint, these calcifications overlie the sella turcica on the lateral projection. On the frontal view, they are projected lateral to the floor of the sella.

Di Chiro and Libow (1971) distinguished two types of calcifications (Fig. 13.26):

The first type consists of small areas of calcification, composed of irregular, scattered deposits. These represent intimal calcification of atherosclerotic lesions and directly reflect atherosclerosis.

The second type consists of thin, dense, parallel calcification, sometimes extensive, often involving the entire circumference of the arteries, often bilaterally. The calcification is situated in the media, the intima commonly being intact. Cerebral circulation is, therefore, less abnormal in the presence of this second type of calcification.

62 Refs. [101, 168, 226, 293, 315, 416, 462, 672, 783, 970]

2) Vascular Tortuosity

Tortuous carotid siphons, which are occasionally dilated, often erode the carotid sulcus (see Figs. 5.21, 5.22, 5.23, 5.24).

3) Arterial Hypertension

Hypertensive patients may show erosion of the cortical bone of the anterior surface of the dorsum sellae, identical to that observed in increased intracranial pressure (Fry and du Boulay 1965). This erosion could be related to increased arterial pulsations.

II. Fibrous Dysplasia

The frontal, ethmoid, and sphenoid bones are most often involved. Lesions of the vault are often lytic, whereas those affecting the skull base are sclerotic (Leeds and Seaman 1962). The sphenoid may be affected alone, or together with other bones of the base or the vault (Figs. 13.27, 13.28). Extensive osteoclerosis affects the entire sphenoid body, the sphenoid sinus is opaque and the size of the pituitary fossa is reduced. The thickened sellar region appears very dense and there is extensive blistering of the presellar area. Mixed forms simulating Paget's disease may be seen in both vault and base.

III. Paget's Disease

Paget's disease always involves the skull base. The sphenoid appears thick and dense. The sphenoid sinus is encroached upon by the abnormal bone (Fig. 13.29).

IV. Dystrophia Myotonica [63]

It has often been said that myotonic dystrophy is accompanied by a small sella turcica. In fact, as Fisher and Di Chiro have demonstrated (1964), the sella turcica often appears small on the lateral view but has a normal volume. Moreover, the hypophysis appears normal at autopsy.

63 Refs. [78, 127, 149, 225, 295, 450, 471, 550, 803, 891, 996, 1032]

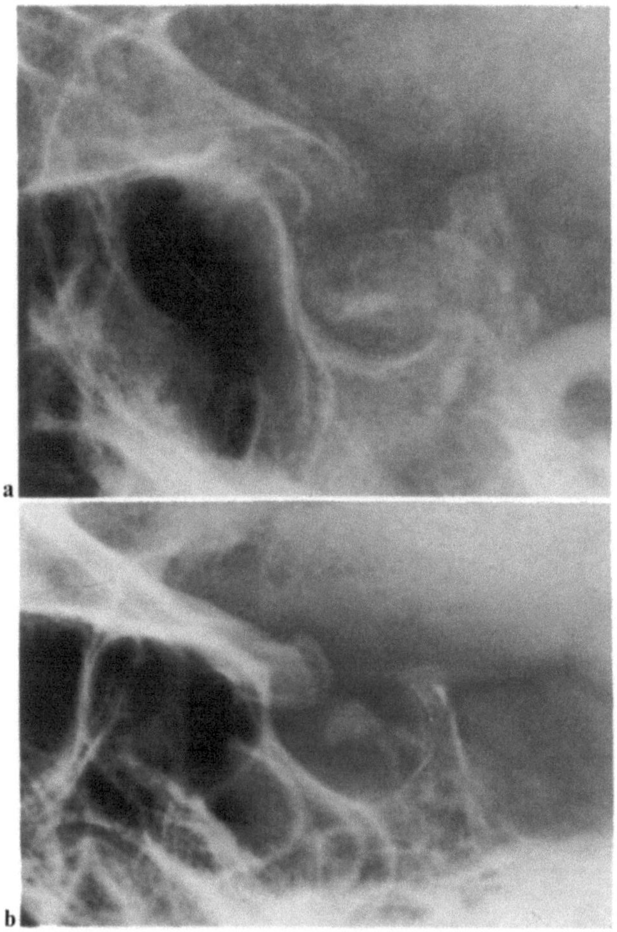

Fig. 13.26a, b. Carotid siphon calcification.
a Tubular carotid siphon calcification of medial type.
b Scattered irregular siphon calcification of intimal type

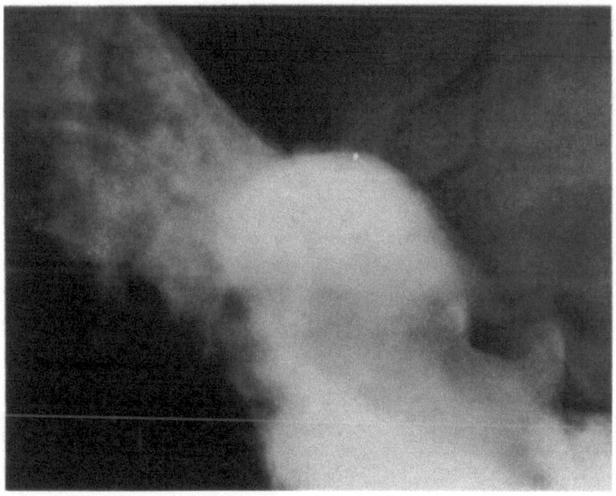

Fig. 13.27. Fibrous dysplasia. Thickening and sclerosis of the skull base and dense blistered planum sphenoidale. The sphenoid sinus is invaded; the sella turcica is small (Courtesy, D. Binnert)

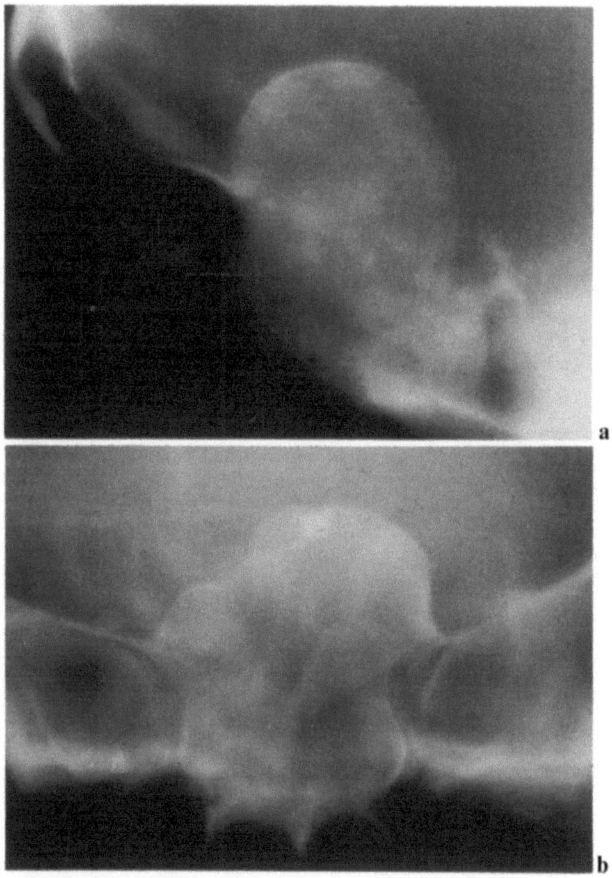

Fig. 13.28a, b. Fibrous dysplasia of sphenoid bone.
a Lateral tomogram
b Frontal tomogram: marked regular, asymmetric blistering of planum sphenoidale. Increased bony density. Anterior part of sphenoid sinus is involved

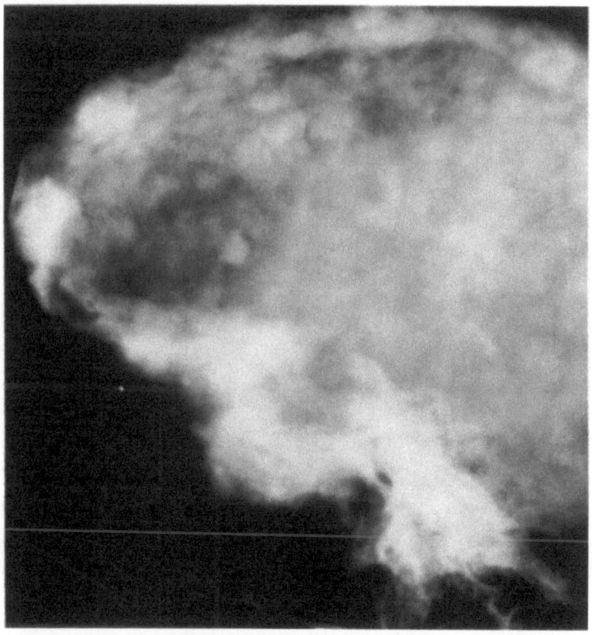

Fig. 13.29. Advanced Paget's disease. Dense sclerosis is visible in the skull base with marked thickening of margins of sella turcica. Basal angle is increased

Chapter 14 Exercises and Pitfalls

This chapter has a question and answer format.
Cover the right-hand side of each page before checking your answer.

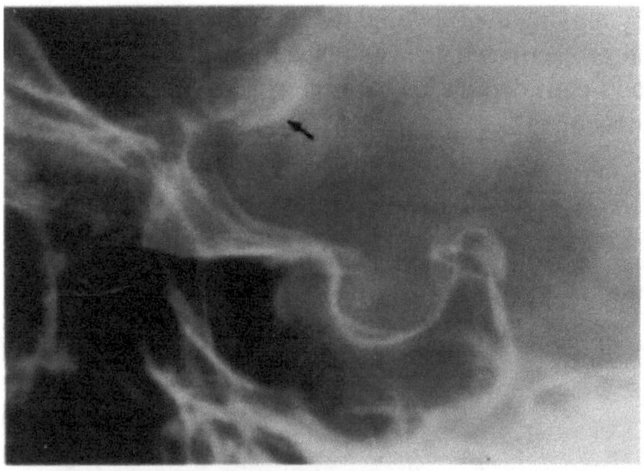

Fig. 14.1a. What does the shadow projected above and anterior to the presellar region *(arrow)* represent?

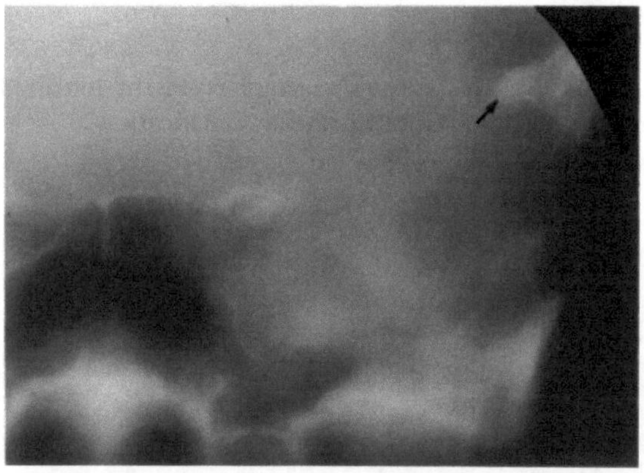

Fig. 14.1b. It is the normal pterion. The junction of the frontal, sphenoid, and temporal bones very frequently creates a more or less pronounced bony thickening which appears on the frontal tomogram as a triangular opacity whose apex points medially *(arrow)*

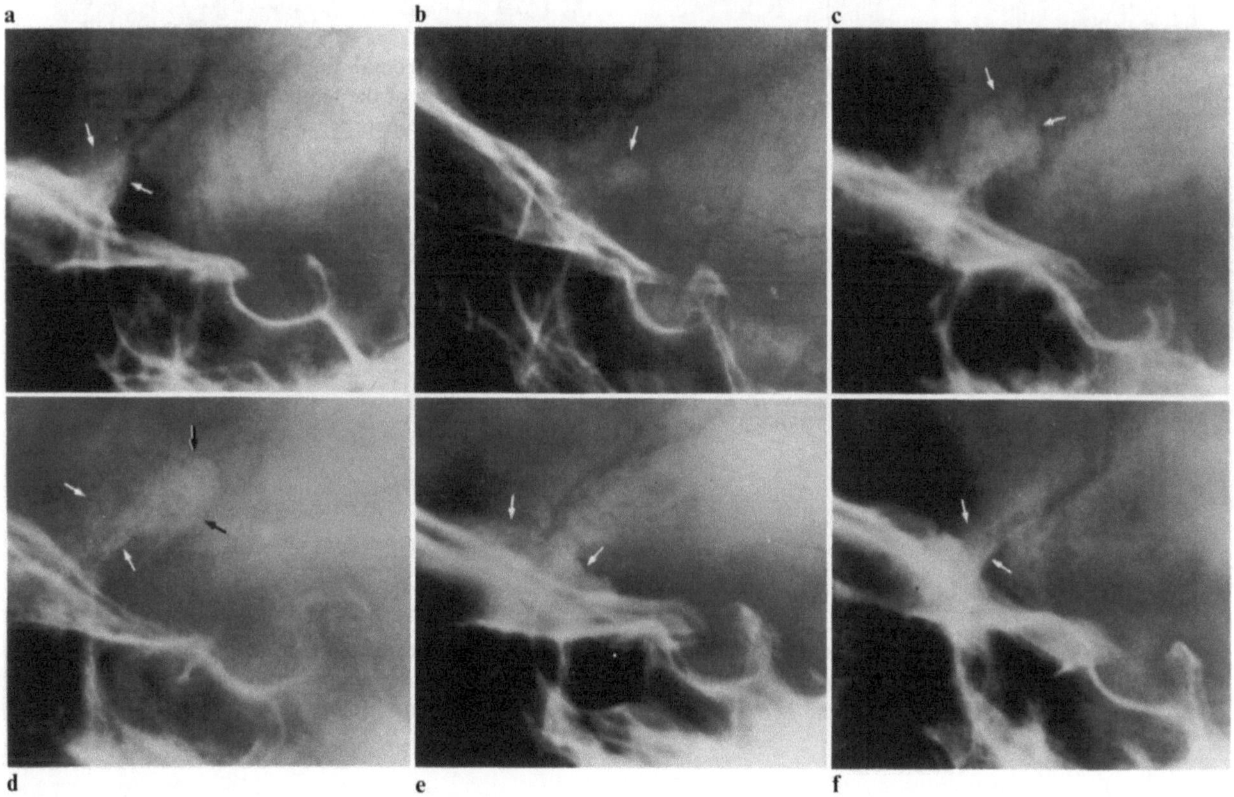

Fig. 14.2a–f. Different appearances of the pterion on the lateral projection. Note that the pterion is projected as a more or less wide and dense shadow which could, in certain cases, suggest the hyperostosis of a meningioma *(arrows)*

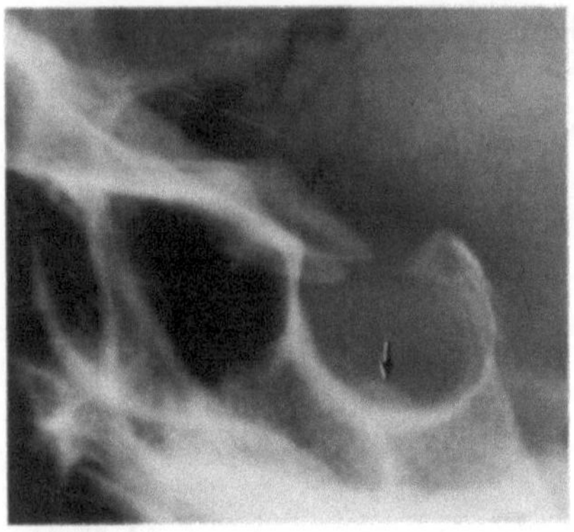

Fig. 14.3. Does this shadow *(arrow)* represent localized thickening of the floor or intrasellar calcification?

Neither; it is a localized thickening of the inner table of the temporal bone. It is relatively easy to confirm this by moving the tube slightly or by obtaining the other lateral projection. If the shadow is part of the skull vault it shows considerable movement with respect to the midline structures.

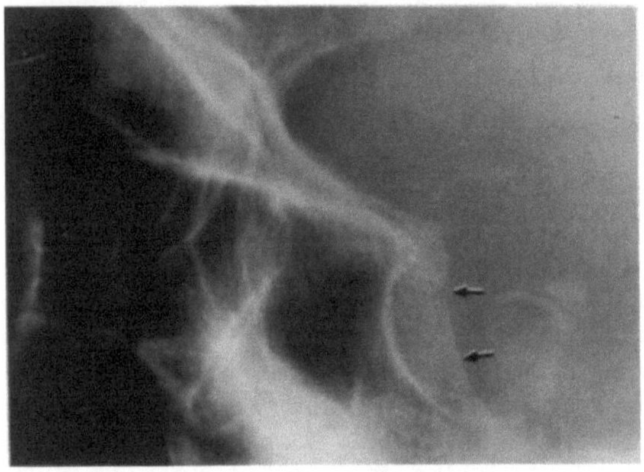

Fig. 14.4a. What does the dense shadow with a well-defined posterior margin overlying the posterior half of the sella turcica *(arrows)* represent?

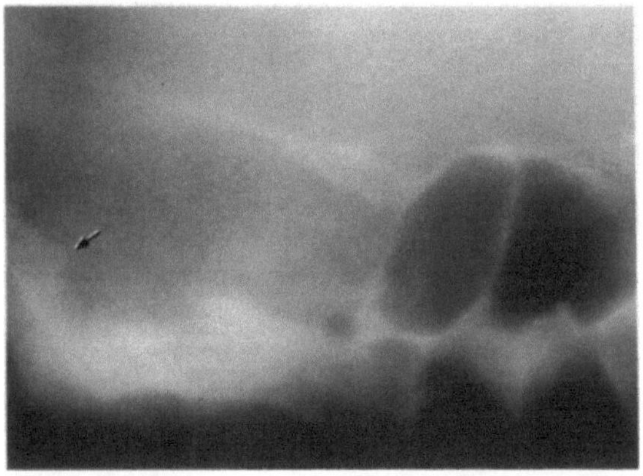

Fig. 14.4b. The frontal tomogram shows a bony crest on the inner table of the temporal bone *(arrow)*, without pathologic significance

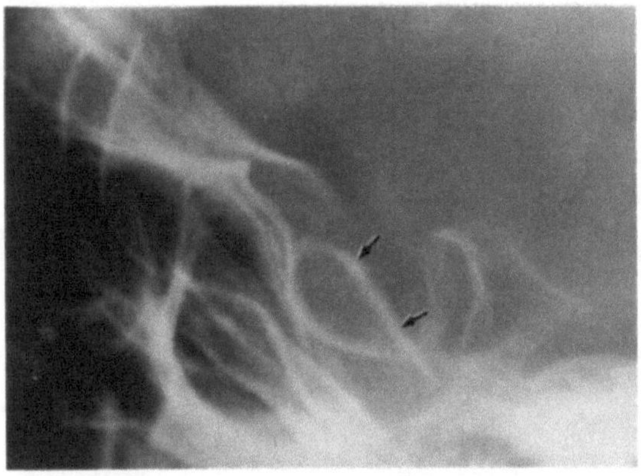

Fig. 14.5a. A shadow with a very dense posterior edge overlies the sella turcica *(arrows)*: what might it represent?

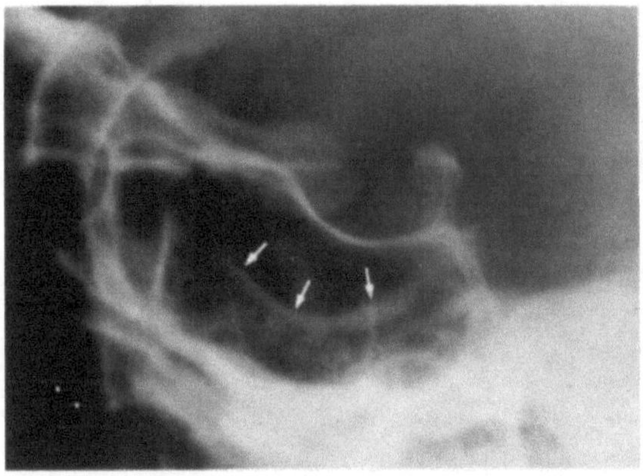

Fig. 14.5b. Once again, it is a simple bony crest on the inner table of the squamous temporal bone; this appearance is very common (other case) *(arrows)*

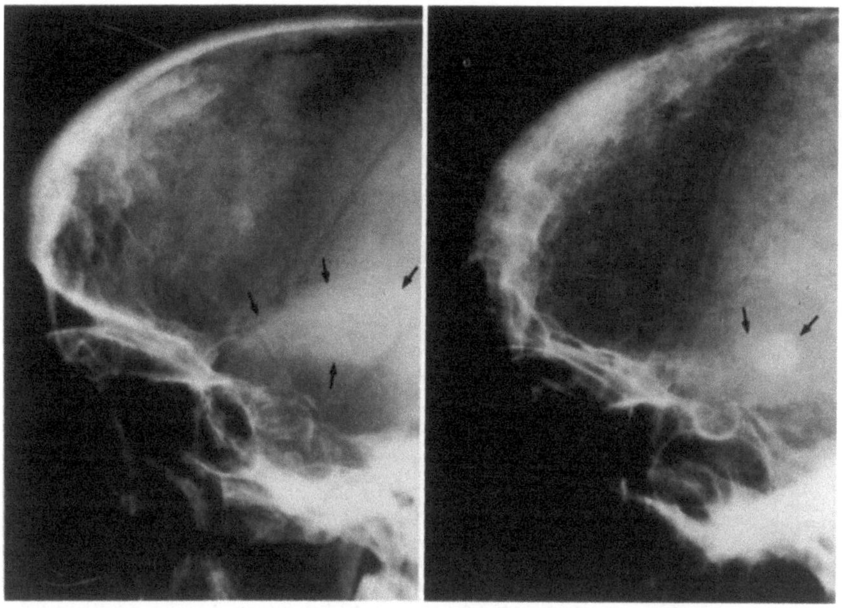

Fig. 14.6. On the *left,* note extensive calcification of the carotid siphons over the sella turcica, but also less dense overlying shadows *(arrows).* On the *right,* above the dorsum sellae, note a dense, homogeneous opacity with regular edges *(arrows).* What is the significance of these shadows?

In both cases, you have undoubtedly noticed considerable hyperostosis frontalis interna. This hyperostosis extends backward into the temporal region, giving suprasellar opacities

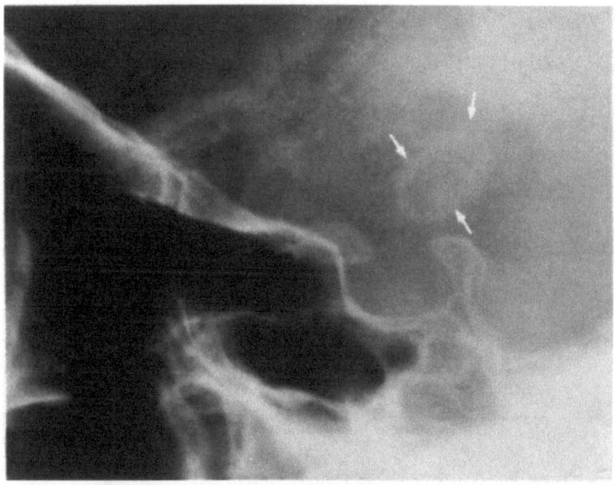

Fig. 14.7a. Is this oval suprasellar shadow another localized form of hyperostosis temporalis internal *(arrows)?*

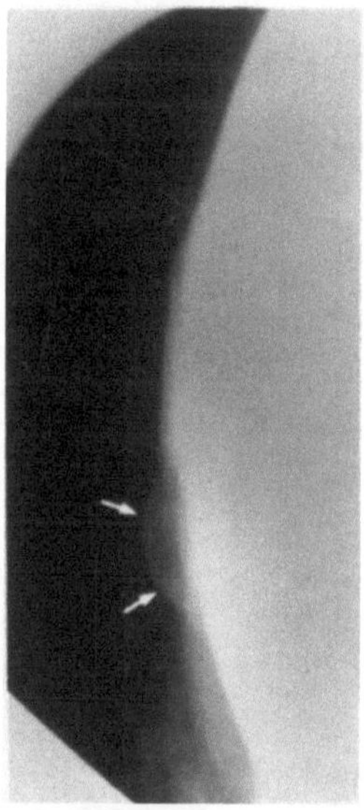

Fig. 14.7b. No; in this case there is a small exostosis of the outer table *(arrows)*

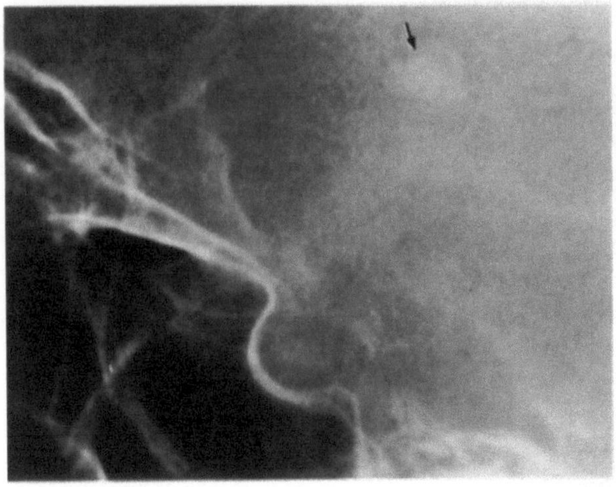

Fig. 14.8a. And this? *(arrow)*

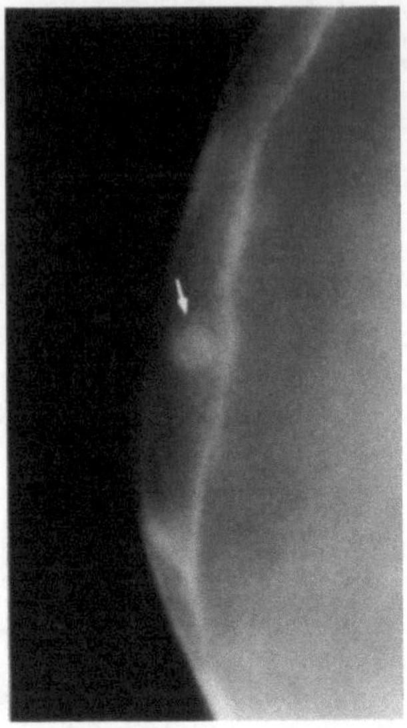

Fig. 14.8b. Obviously, a suprasellar shadow may also be the result of a diploic lesion *(arrow)*

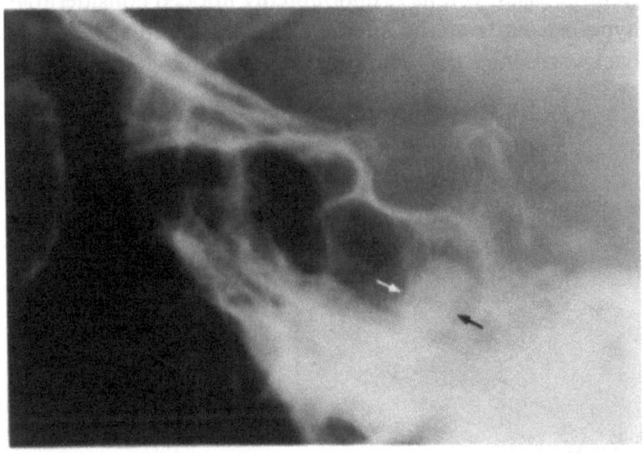

Fig. 14.9. How can the shadow overlying the sphenoid sinus in this 67-year-old male patient be explained? *(arrows)*

Have you noticed the abnormal thickening of the floor of the temporal fossa (just below the *arrows)?* You must, therefore, have thought of the possibility of osteoblastic metastases of the skull vault and base from prostatic carcinoma

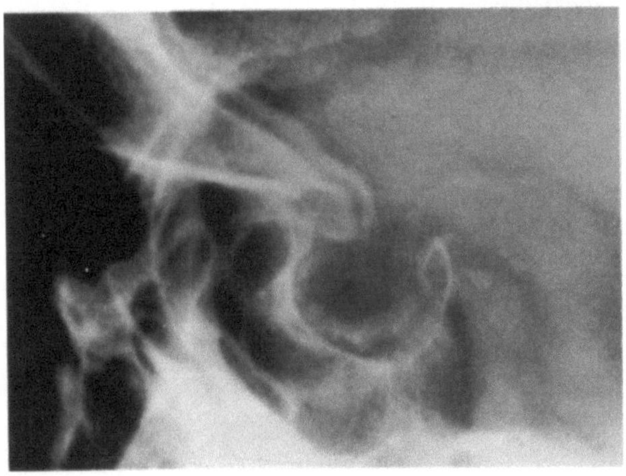

Fig. 14.10a. Thick, curvilinear calcification is projected over the sella parallel to its floor: what might it be?

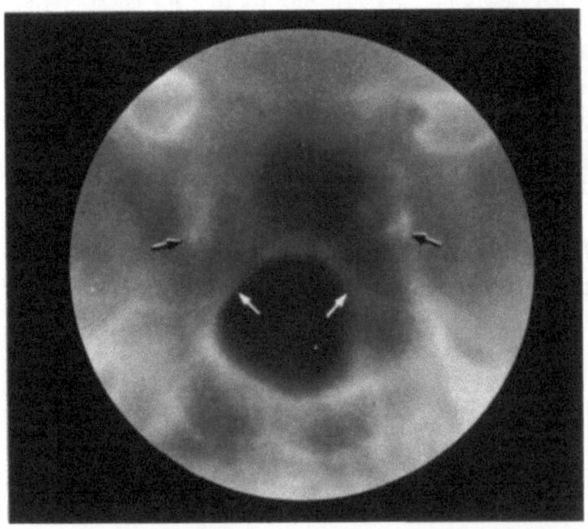

Fig. 14.10b. Scattered calcifications of the carotid siphon, shown by tomography *(black and white arrows)*. Narrowness of the floor of the sella related to enlargement of both carotid sulci *(white arrows)*

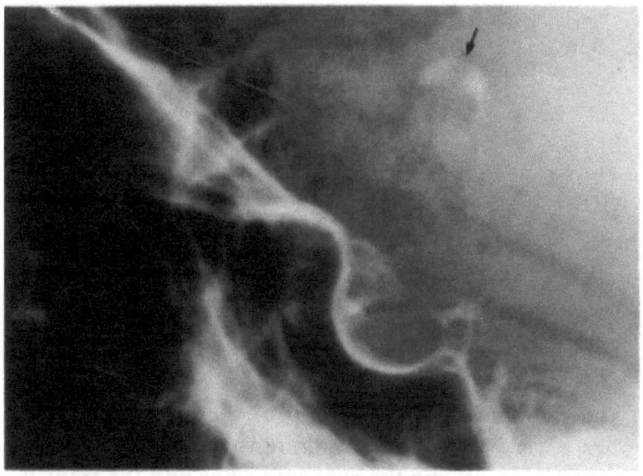

Fig. 14.11. This dense, semilunar shadow *(arrow)* is not part of the inner, outer, or middle table. So what might it be?

The morphology and thickness of this shadow must raise the possibility of a calcified vascular lesion: an aneurysm of the middle cerebral artery

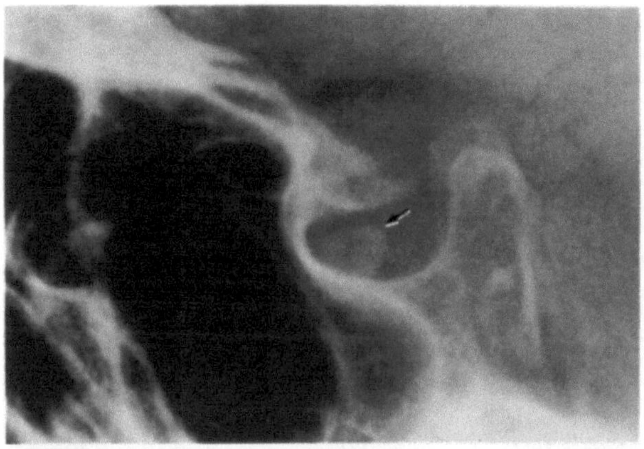

Fig. 14.12a. What might be the origin of this shadow overlying the sella turcica *(arrow)?*

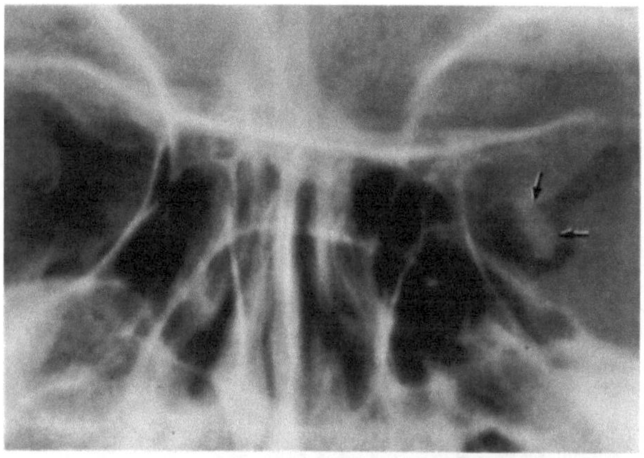

Fig. 14.12b. It is dural calcification near the superior orbital fissure (*arrows;* see also Figs. 5.70, 5.71)

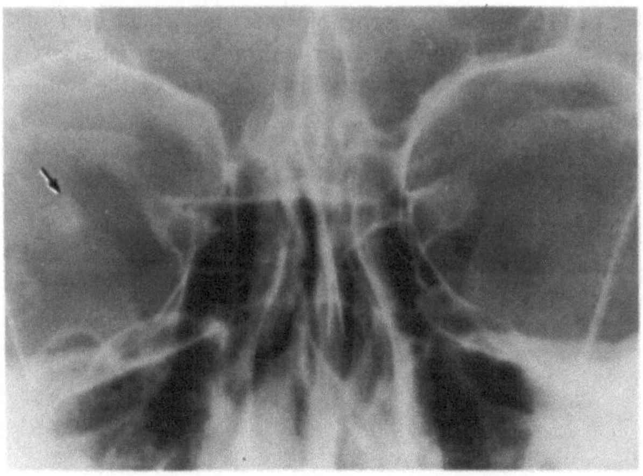

Fig. 14.13a. What is the origin of this shadow *(arrow)?*

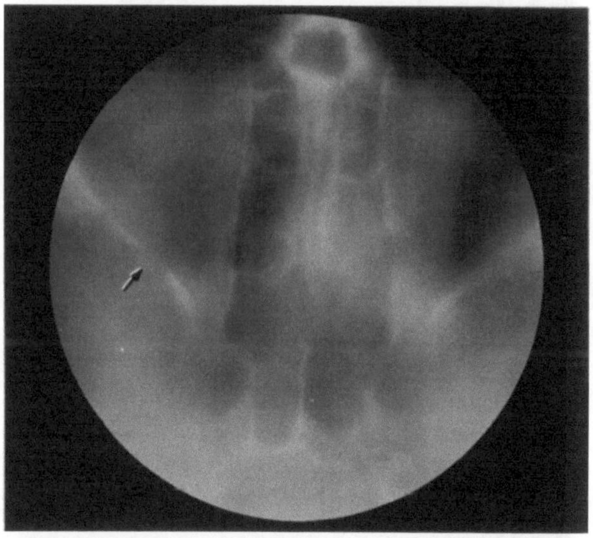

Fig. 14.13b. It is in fact bilateral, an observation which should bring to mind the possibility of a normal calcification of the dura mater of the greater wing of the sphenoid, in contact with the superior orbital fissure. This is best confirmed by axial tomograms *(arrow)*

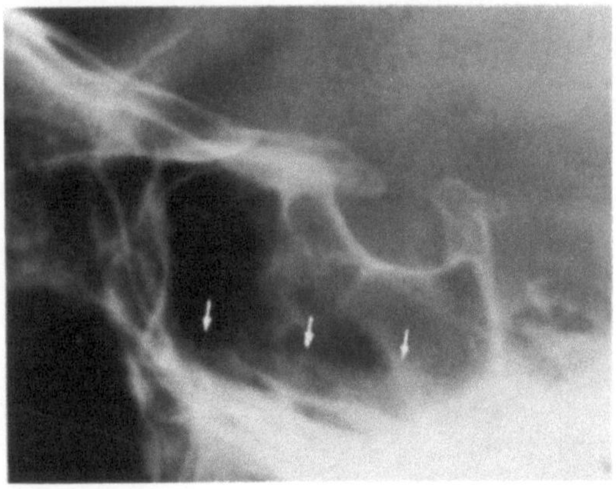

Fig. 14.14. Head injury: a horizontal level in the inferior part of the sphenoid sinus *(arrows)* must lead to an investigation of a possible fracture of the skull base. What do you think?

If you examine carefully the tip of the *arrows*, you will notice a very thin, clear, radiotransparent line between the shadow overlying the sphenoid sinus and the translucency of the sinus itself. The presence of this line indicates that the opacity is not located in the sphenoid sinus; it is the zygomatic arch

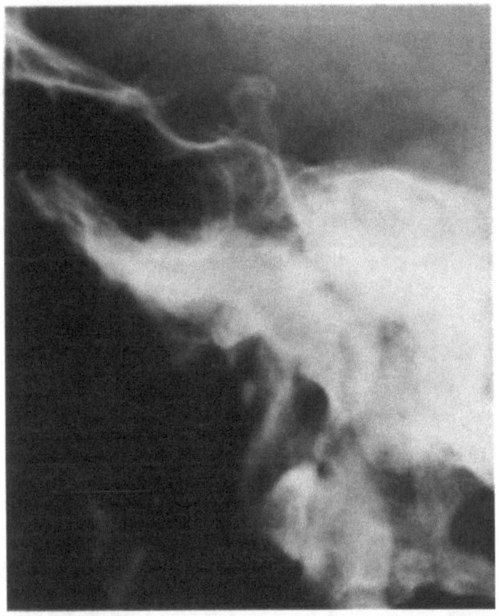

Fig. 14.15a. What is unusual about this plain film?

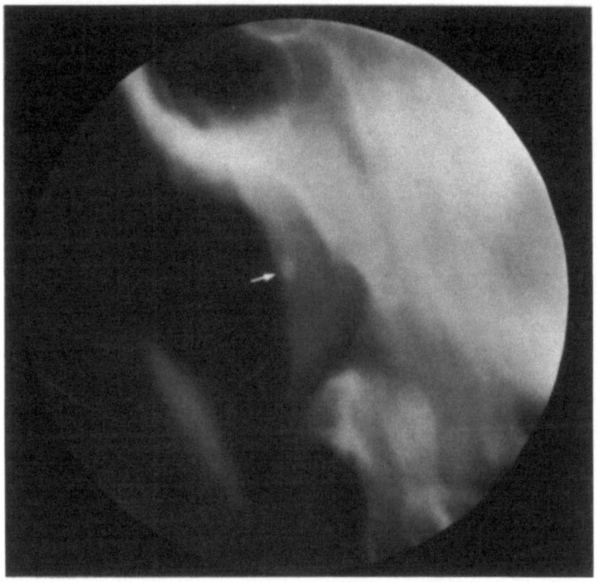

Fig. 14.15b. Below the sphenoid there is a small area of calcification near the posterior wall of the nasopharynx. This picture represents a calcified remnant of the notochord *(arrow)*

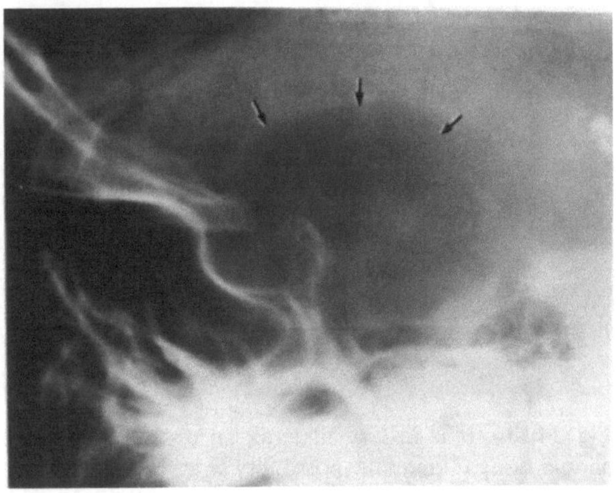

Fig. 14.16. To what does this suprasellar translucency *(arrows)* correspond?

Simply a pronounced thinning of the temporal squama

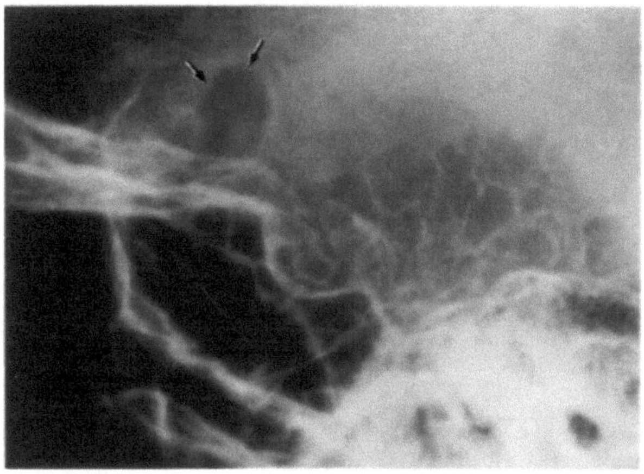

Fig. 14.17a. Above the presellar region there is a translucent image with even, dense edges *(arrows)*; is this cholesteatoma of the skull vault?

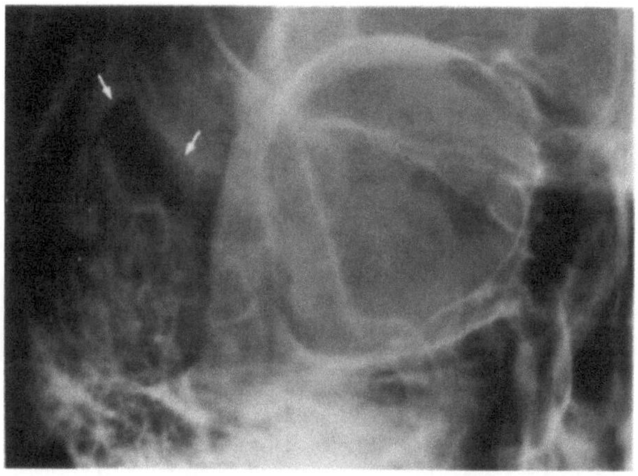

Fig. 14.17b. Extensive pneumatization of mastoid air cells which spread forward in the temporal squama; in addition a vast paramastoid air cell even further forward *(arrows)*

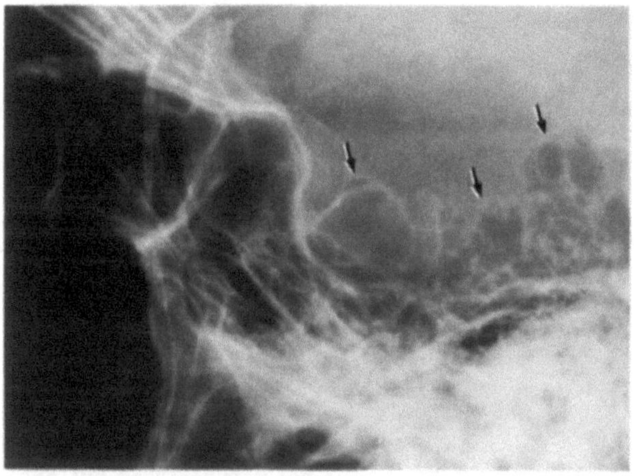

Fig. 14.18a. What do you think of this hyperpneumatization of the mastoid cells? *(arrows)*

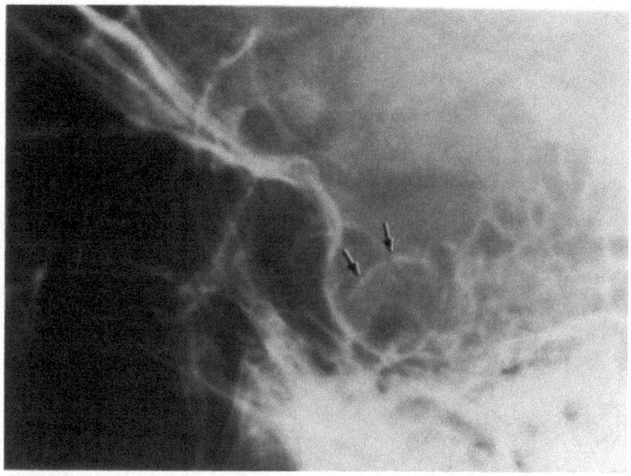

Fig. 14.18b. The margin of a mastoid air cell partially masks calcifications of the carotid siphon which appear after modification of the angulation of the X-ray beam *(arrows)*; note in addition a vertical chiasmatic sulcus

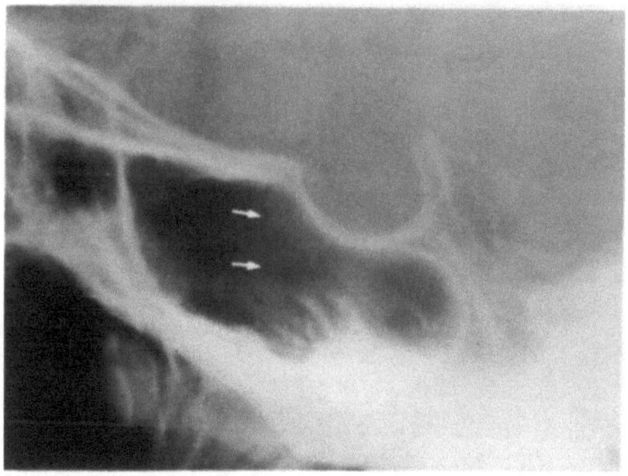

Fig. 14.19. Is this sinuous shadow *(arrows)* overlying the sphenoid sinus a fracture of the skull base?

No! The sphenoid sinus is normally translucent. It is the squamosal suture

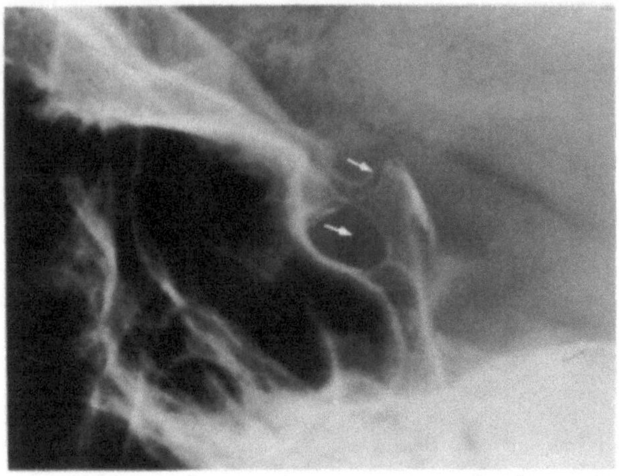

Fig. 14.20. And this line crossing the sella *(arrows)?*

Is the groove on the skull vault of the middle temporal artery

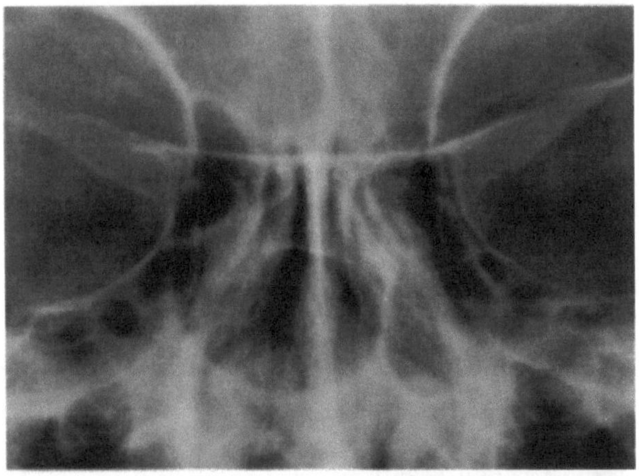

Fig. 14.21. The left half of the floor of the sella is poorly defined; what do you think about this?

The left half of the floor is not demineralized but only blurred by superimposition of an enlarged turbinate

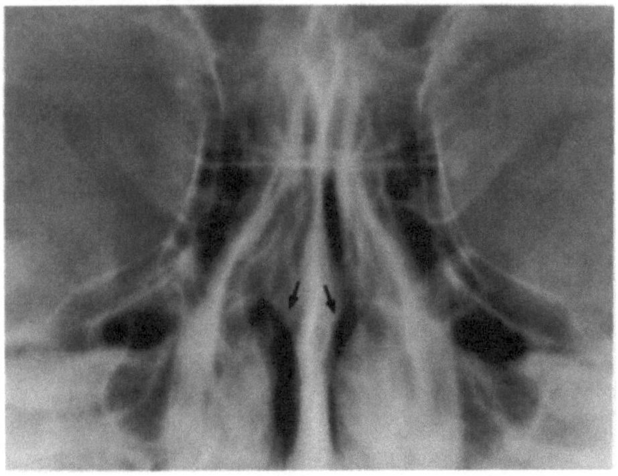

Fig. 14.22a. Could this regular midline depression of the floor of the sella correspond to an anatomic variation *(arrows)?*

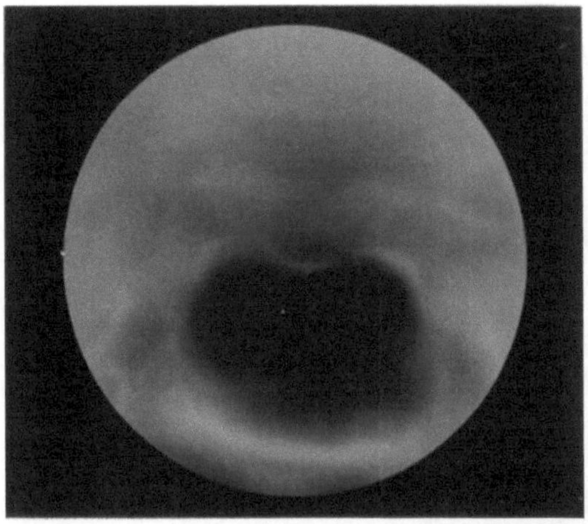

Fig. 14.22b. Yes, it could: nonpathologic accentuation of the fossula hypophyseos in the posterior part of the floor

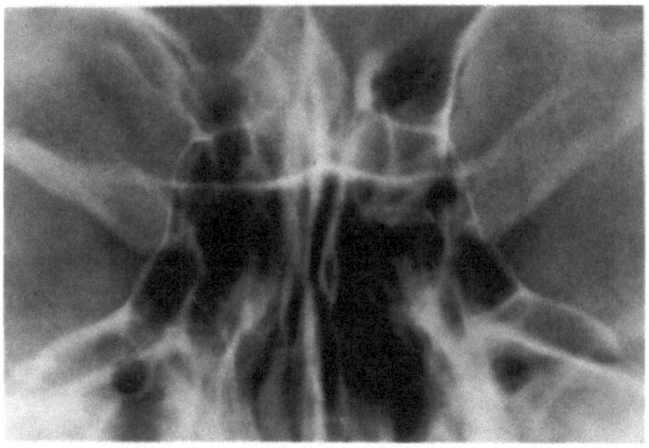

Fig. 14.23a. In a 25-year-old female patient, the floor of the sella is poorly defined on the frontal view. Why?

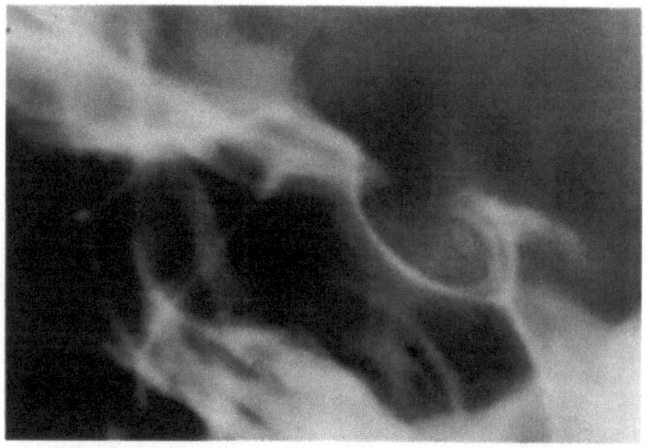

Fig. 14.23b. Lack of visibility of the sellar floor on the frontal view always merits particular attention; it is necessary, however, to compare both the frontal and lateral projections. Here we are dealing with a special configuration of the sella turcica (see also Fig. 5.5)

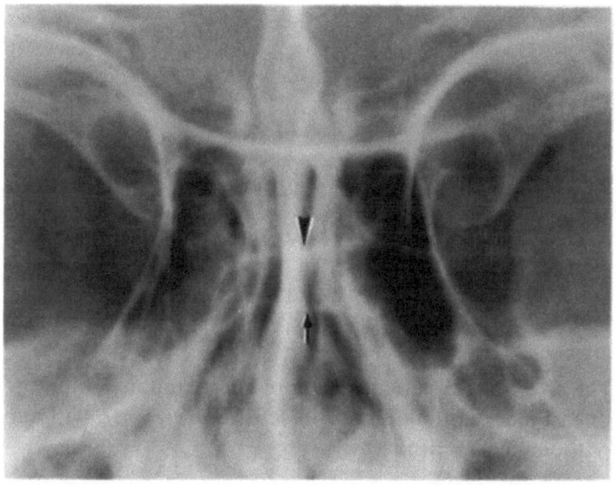

Fig. 14.24a. Where is the floor of the sella?

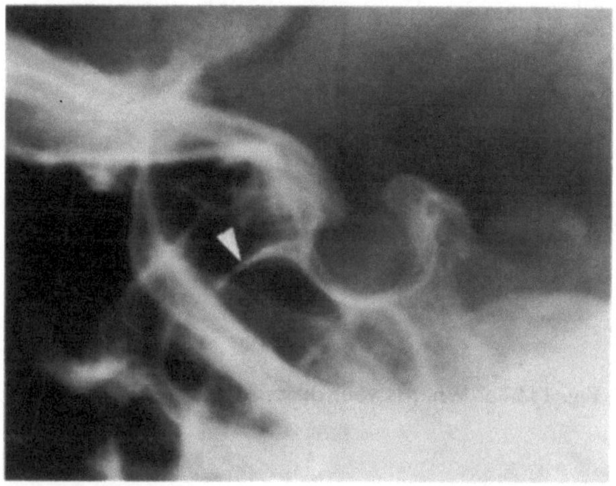

Fig. 14.24b. The floor is indicated by the *arrow*. The *arrowheads* indicate a septum of the sphenoid sinus roughly parallel to the floor of the sella turcica

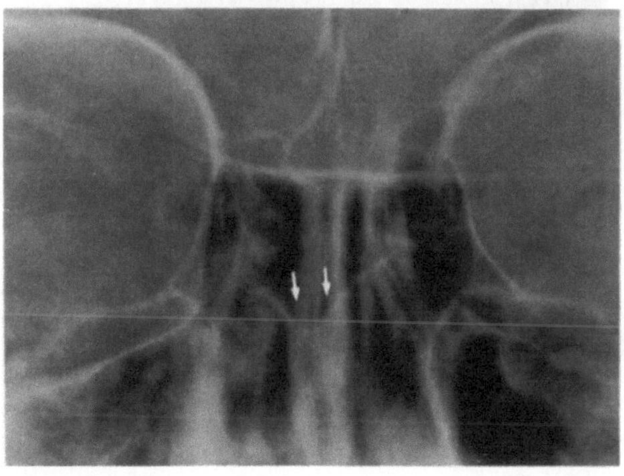

Fig. 14.25. What does the asymmetric V-shaped shadow represent?

It is not a fracture of the floor of the sella, but rather unusual septation of the anterior part of the sphenoid sinus. The floor seems poorly mineralized between the two septa *(arrows)*

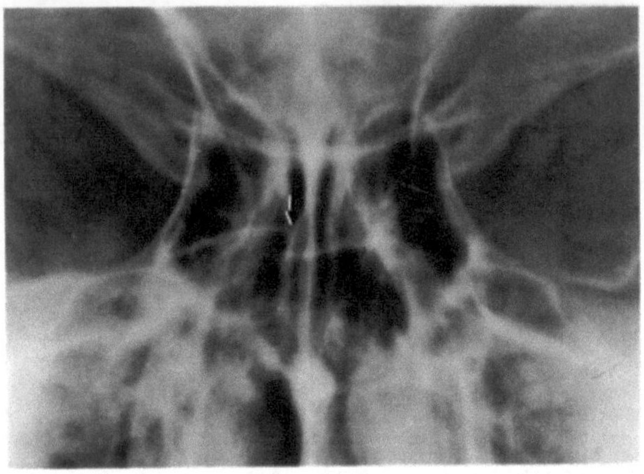

Fig. 14.26. Is this a traumatic lesion of the floor of the sella *(arrow)*?

The upper line extends too far laterally to be part of the floor. Once again this is a horizontal septum of the sphenoid sinus; the subjacent floor of the sella is normal

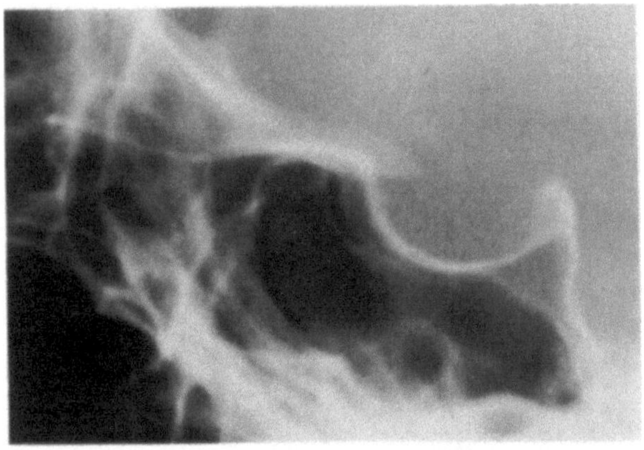

Fig. 14.27a. What is your diagnosis?

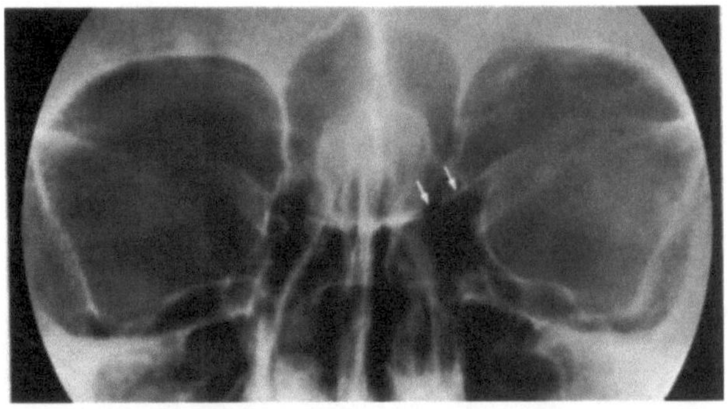

Fig. 14.27b. The anterior part of the planum is blurred, due to marked obliquity of its left side *(arrows)*, without pathologic significance

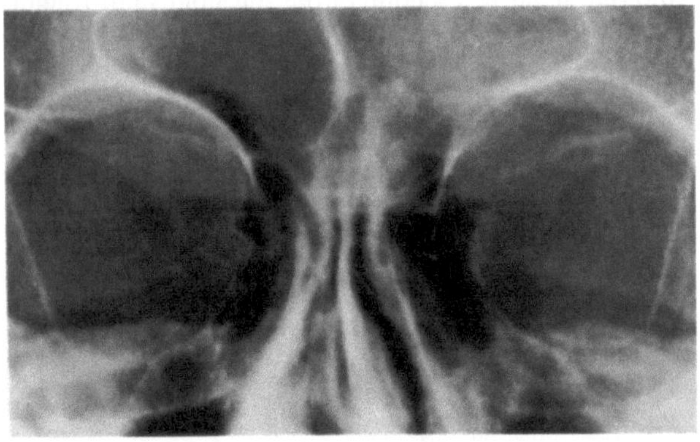

Fig. 14.28a. Why is the planum sphenoidale poorly defined?

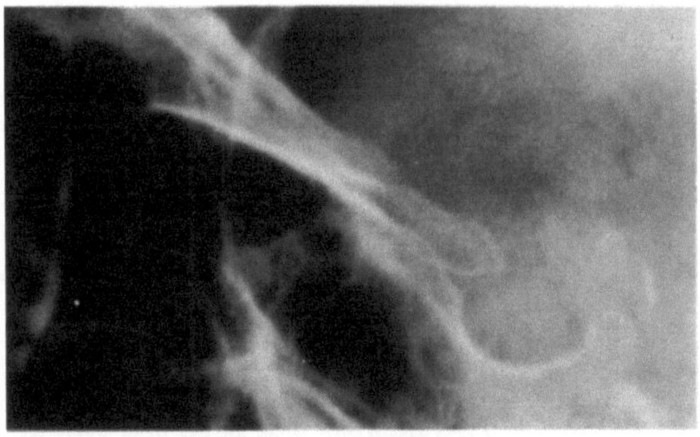

Fig. 14.28b. Because it lies very obliquely

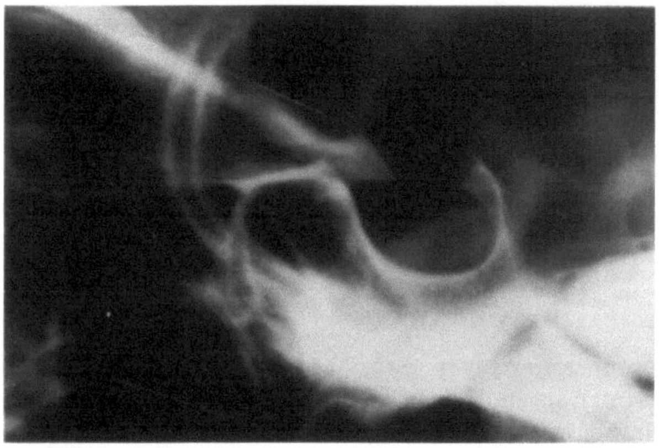

Fig. 14.29. Is this short planum necessarily pathologic?

Apart from increased intracranial pressure, anterior craniostenosis can also cause shortness of the planum sphenoidale

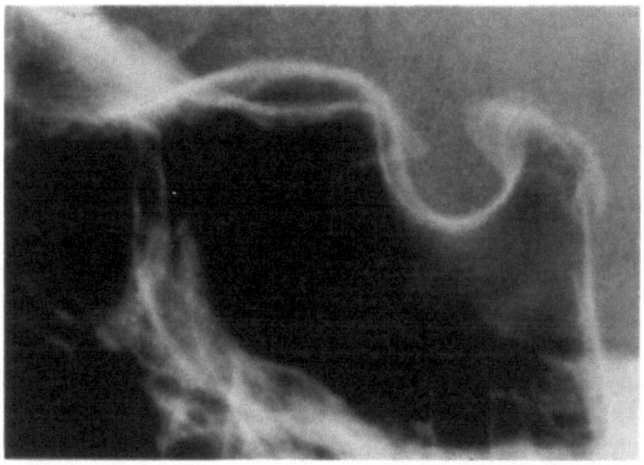

Fig. 14.30. What do you think of the presellar region?

The planum sphenoidale is convex. This blistering may be considered normal when the sphenoid is hyperpneumatized. Note that pneumatization is extensive, reaching the anterior clinoid processes and dorsum sellae

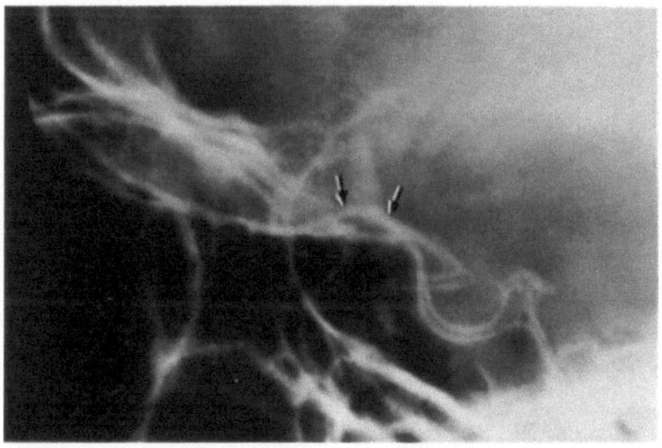

Fig. 14.31. Is the pneumatization of the sphenoid sufficient to explain the blistering of the planum *(arrows)?*

If you answered yes or no you have fallen into the trap. The *arrows* indicates the pneumatized lesser wings of the sphenoid. The planum sphenoidale below is absolutely flat

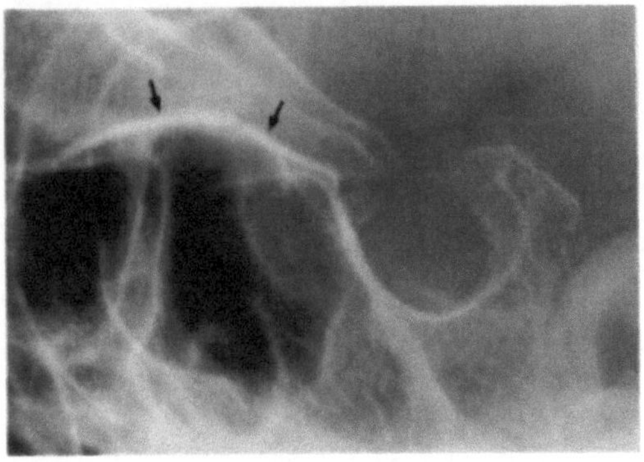

Fig. 14.32. Pneumatization of the sphenoid is presellar here; must this blistering of the planum *(arrows)* be considered pathologic?

Another trap; it is always necessary to distinguish the planum sphenoidale from the chiasmatic sulcus. Here the illusion of blistering is given by the harmonious, curved alignment of the flat planum sphenoidale *(anterior arrow)* and the flat chiasmatic sulcus *(posterior arrow)*

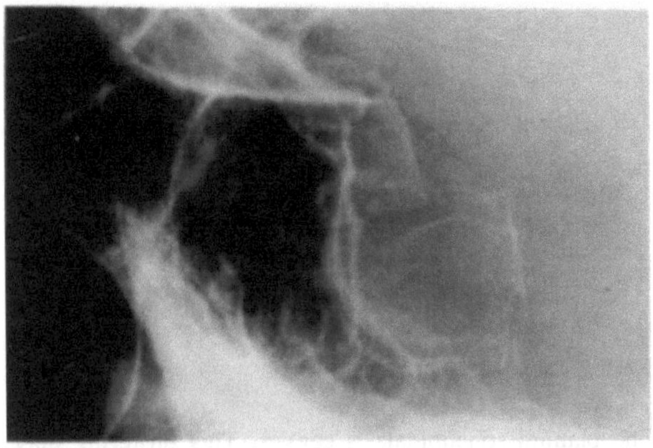

Fig. 14.33. What is your diagnosis in this 70-year-old female patient?

The sella turcica is demineralized and there is calcification of the carotid siphons. Verticalization of the chiasmatic sulcus gives the illusion of enlargement of the sella turcica (see also Figs. 5.6, 5.7)

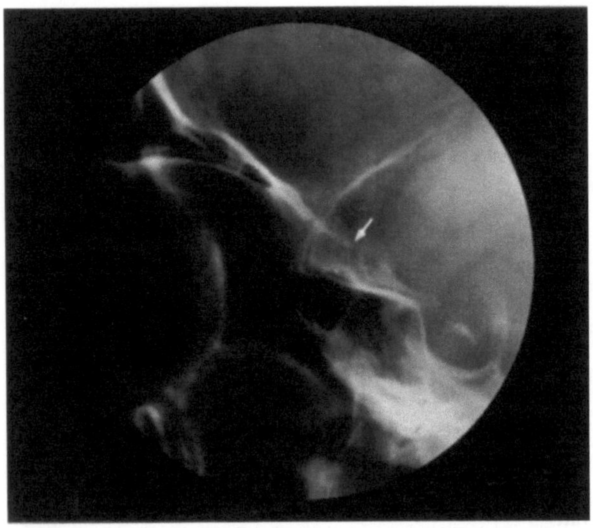

Fig. 14.34. There appears to be a distinct break in the orbital roofs *(arrow)*: is this pathologic erosion?

All the anterior part of the skull base is constitutionally thin. The appearance of a break in the orbital roofs results from an abrupt change in direction of these structures, which are here very thin. Calcification of carotid siphons

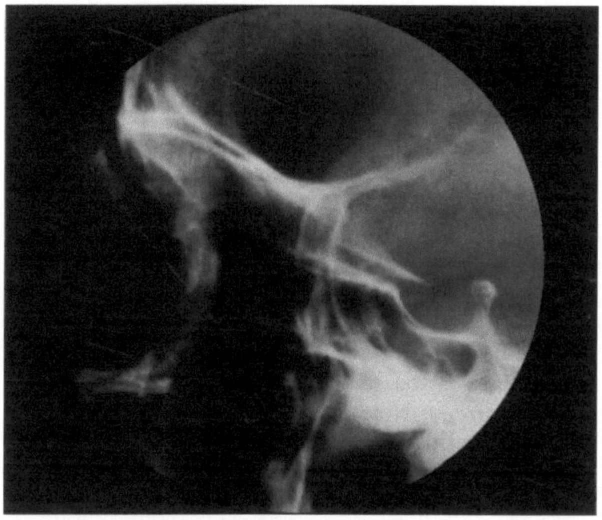

Fig. 14.35. Are these very pointed, flattened anterior clinoid processes pathologic?

The sella turcica itself is normal. As an isolated observation, this appearance of the anterior clinoid processes constitutes a simple variation and does not indicate increased intracranial pressure (see also Fig. 5.53)

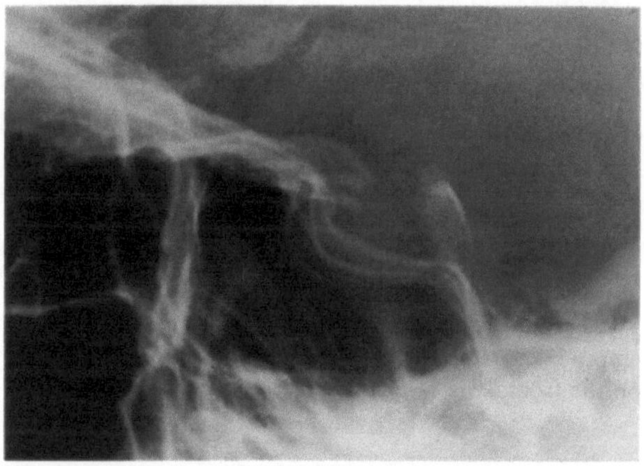

Fig. 14.36. What do you think of this double floor?

Small sella on one side (see also Figs. 5.9, 5.10). Asymmetric anterior clinoid processes

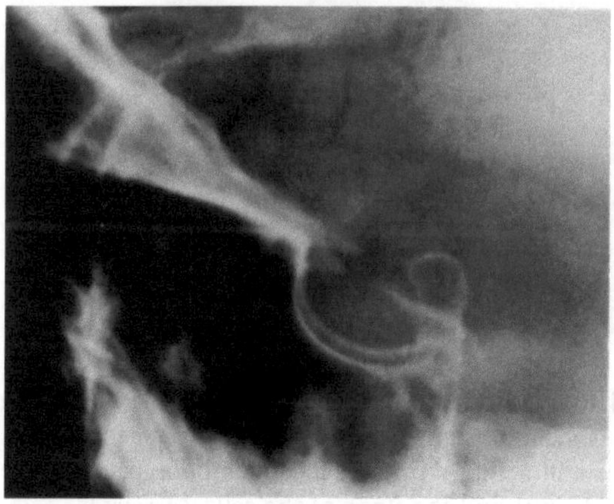

Fig. 14.37. What is your diagnosis?

Normal; the double floor is probably not pathologic since the line representing the lower part of the floor of the sella is thicker than the upper line. On the other hand there is typical asymmetry of the dorsum sellae and posterior clinoid processes which must not be considered as pathologic truncation (see also Fig. 5.61)

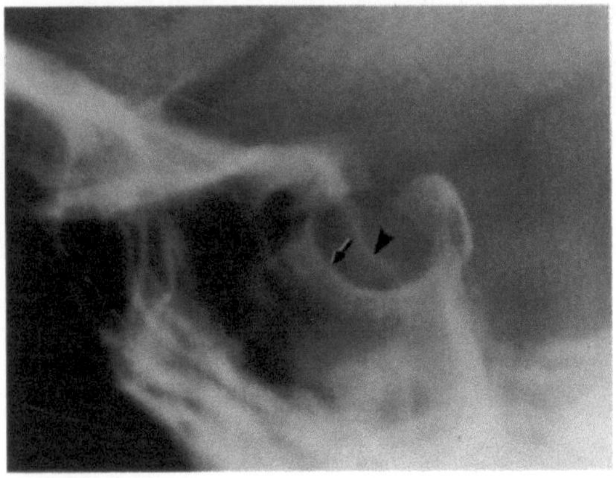

Fig. 14.38. Explain this appearance of a double floor.

We are dealing with a small sella on one side: one half of the fossa is too small *(arrowhead)* and the other is normal *(arrow)*. Note opacity of the sinus, which is the result of sinusitis

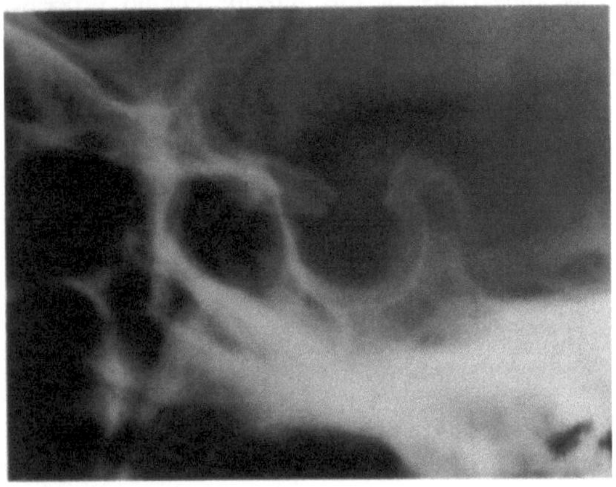

Fig. 14.39a. What do you think of this double floor?

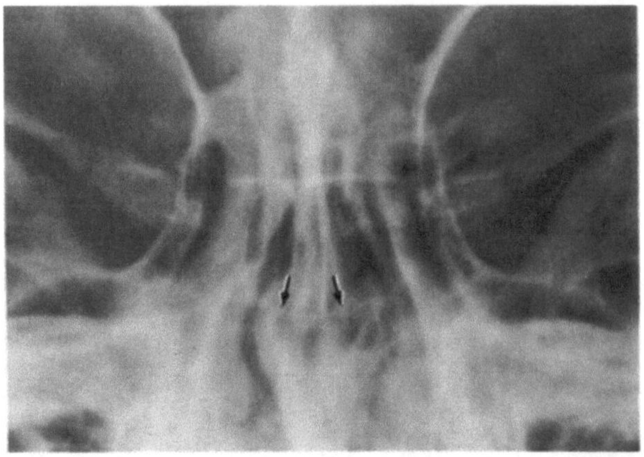

Fig. 14.39b. This is the typical appearance of an early empty sella. Note the very regular concavity of the floor on the frontal view *(arrows)* (see also Figs. 5.12, 5.13)

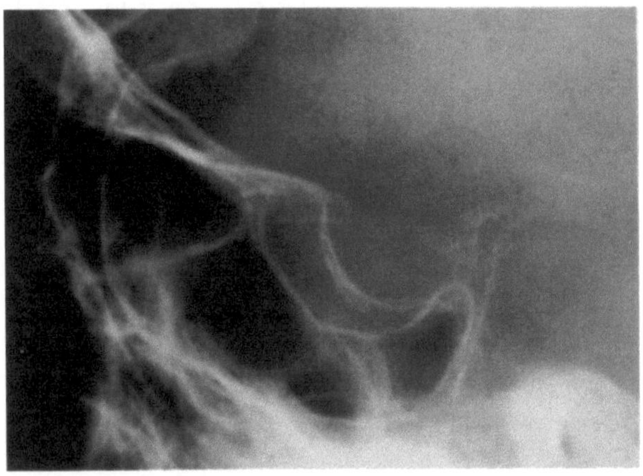

Fig. 14.40a. And this?

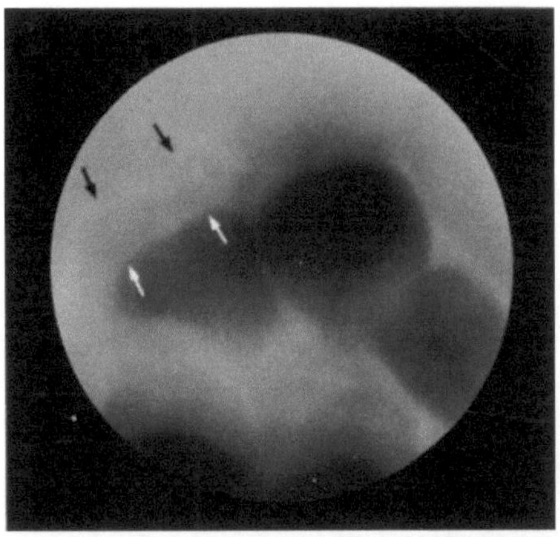

Fig. 14.40b. This is a false double floor: the right side of the sphenoid sinus is not completely pneumatized. There is still a thin strip of cancellous bone between the sphenoid sinus and the floor of the sella *(arrows)*

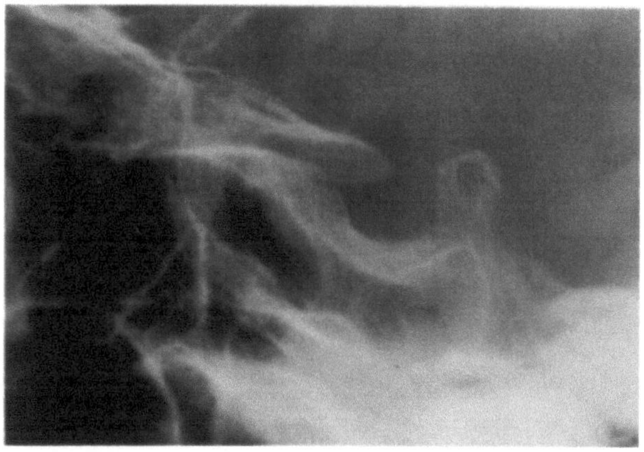

Fig. 14.41a. Is this pseudodouble floor due to a sclerotic reaction to chronic sphenoid sinusitis?

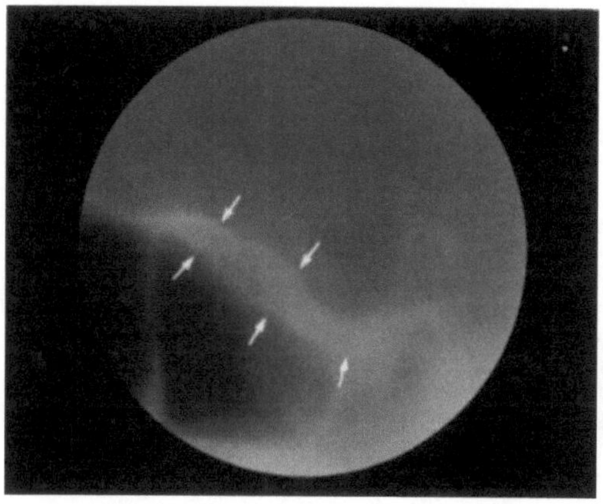

Fig. 14.41b. No. Here the floor is doubled from front to back by a layer of dense cancellous bone limited below by the cortex of the sphenoid sinus *(arrows)*

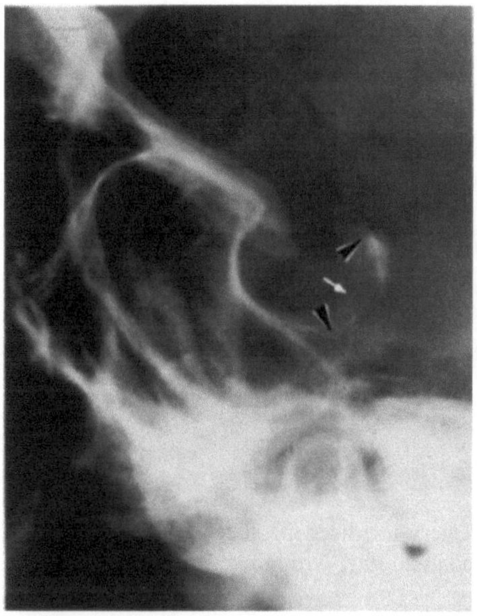

Fig. 14.42a. Backward displacement and distinct thinning of one side of the dorsum sellae *(arrowheads)*, the midline portion of which is in normal position *(white arrow)*. What does it evoke?

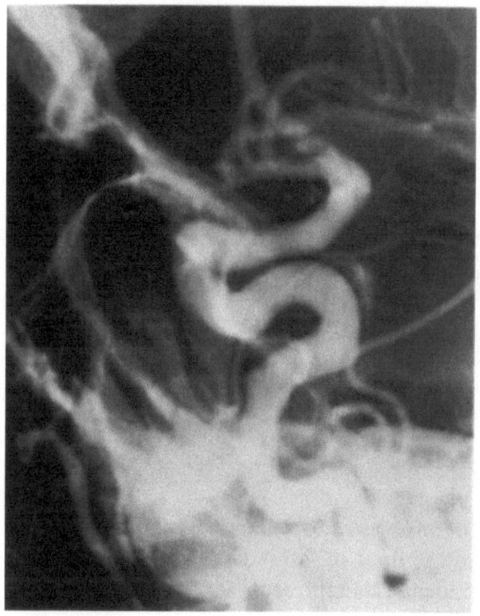

Fig. 14.42b. This is the imprint of an ectatic carotid siphon

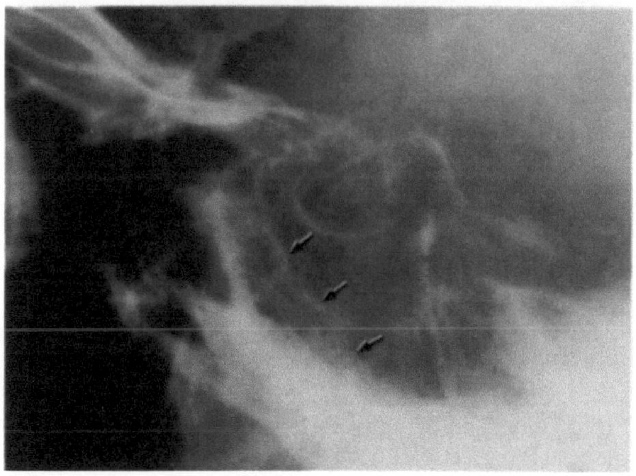

Fig. 14.43. A true or false double floor?

A false double floor caused by deepening of the carotid sulcus *(arrows)* in a hypertensive patient

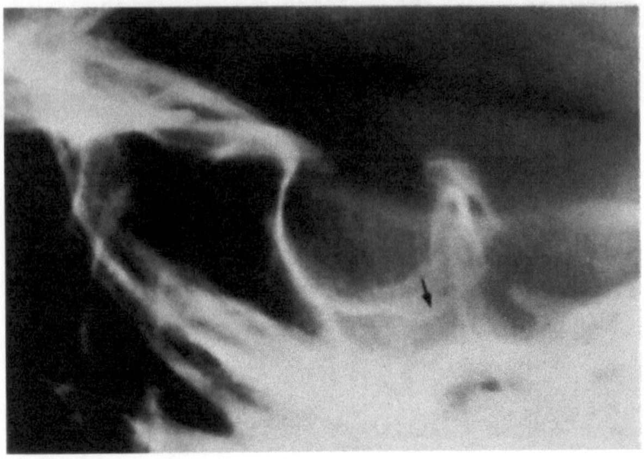

Fig. 14.44. Is this depression of the posterior part of the floor of the sella pathologic *(arrow)?*

It is the normal fossula hypophyseos (see also Fig. 5.80)

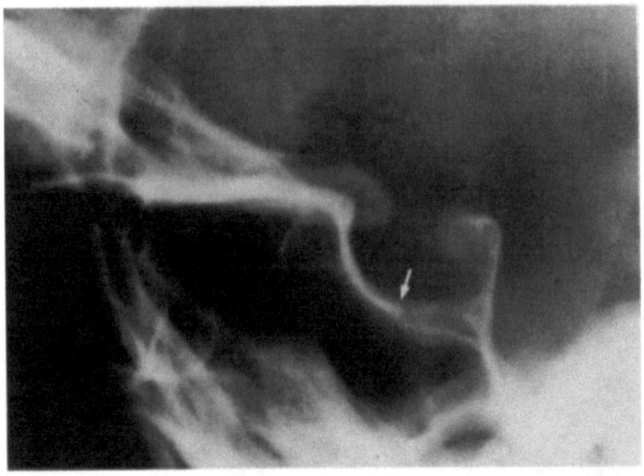

Fig. 14.45a. There appears to be an abrupt break in the cortical bone of the floor of the sella *(arrow)*. What does the shadow opposite this represent?

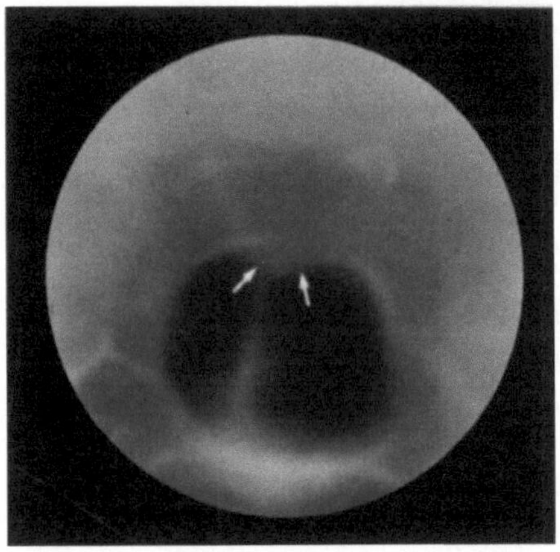

Fig. 14.45b. A very localized persistence of an island of cancellous bone in contact with posterior part of the floor of the sella *(arrows)*

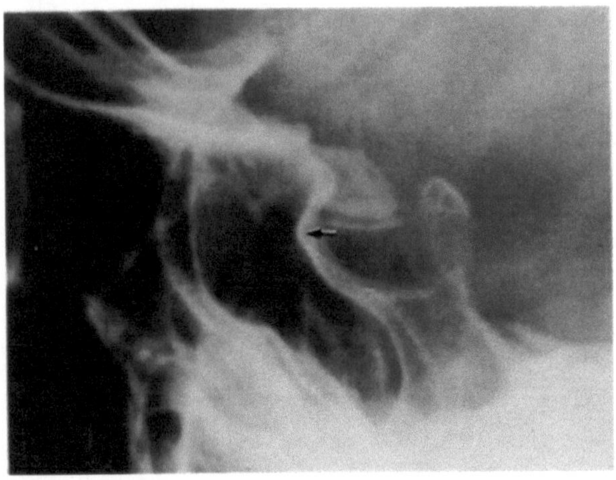

Fig. 14.46a. What is this very localized notch in the anterior wall of the sella *(arrow)?*

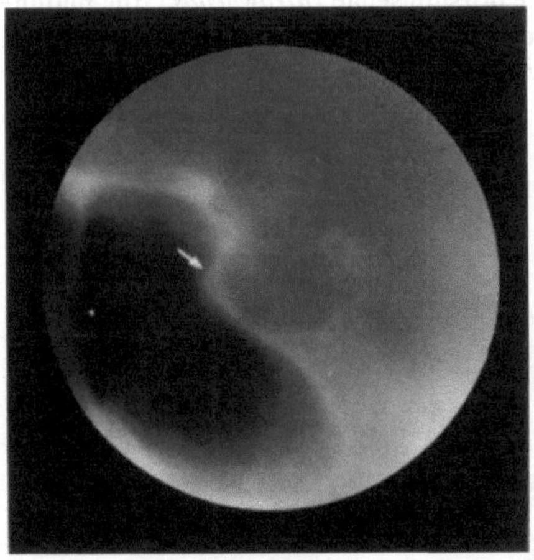

Fig. 14.46b. Tomography shows that the anterior wall of the sella is perfectly smooth. The radiographic appearance once again represents the persistence of a very small island of cancellous bone *(arrow)*

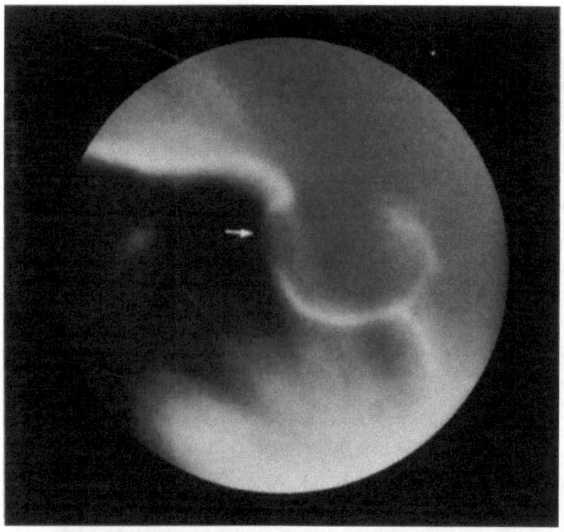

Fig. 14.47a. Very localized semilunar depression with cortical thinning in the upper part of the anterior wall of the sella turcica *(arrow)*: does this picture, which was discovered during investigation for galactorrhea, indicate the presence of a microadenoma?

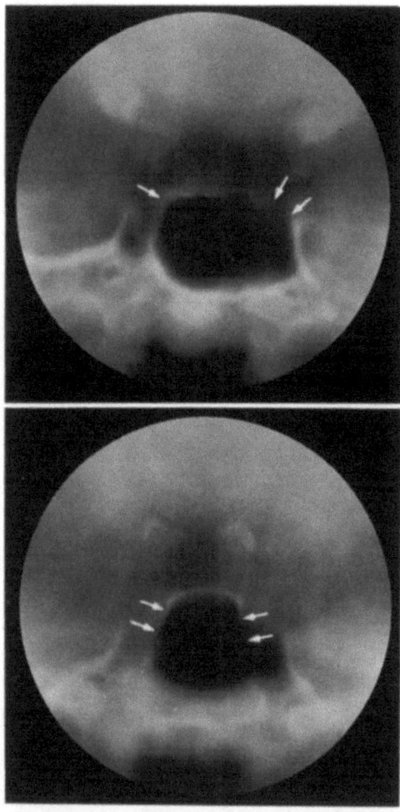

b

c

Fig. 14.47b, c. No; frontal tomograms show this appearance to be related to a very medial position of the carotid siphons *(arrows)*

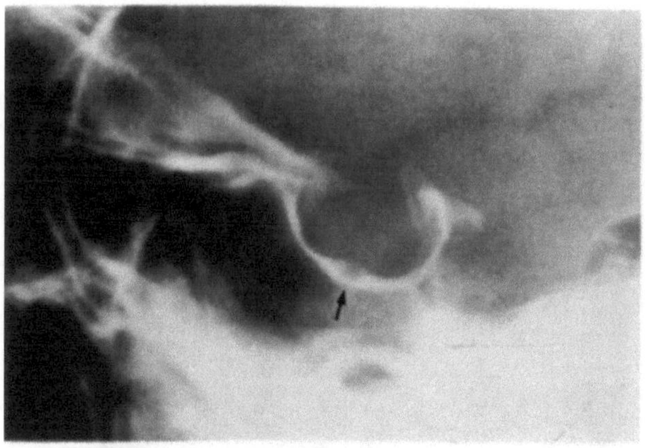

Fig. 14.48a. Is this a localized blistering of the floor of the sella *(arrow)*?

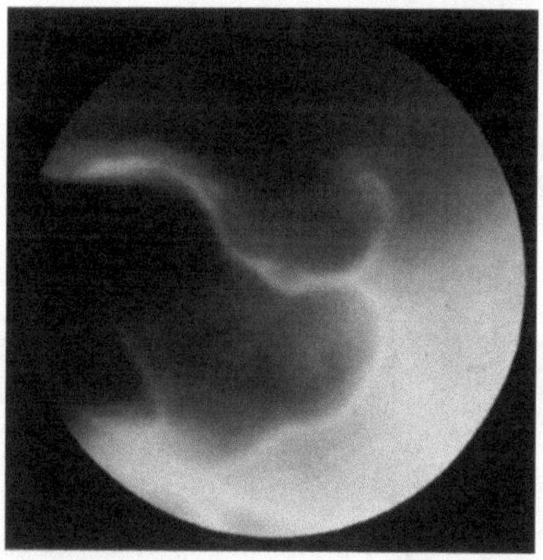

Fig. 14.48b. Actually this is a nonpathologic, localized doubling of the cortical bone of the floor

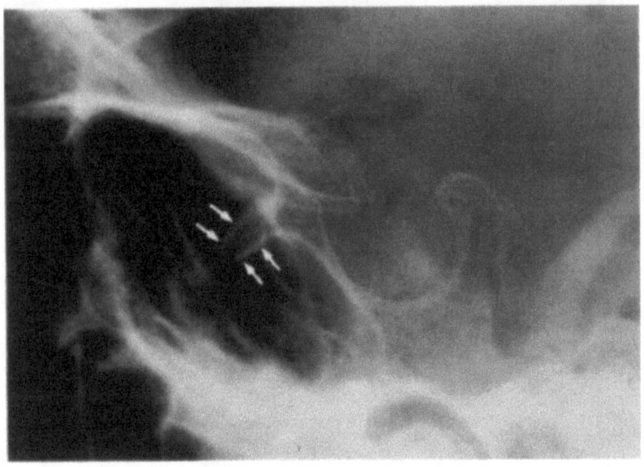

Fig. 14.49a. What is this channel *(arrows)?*

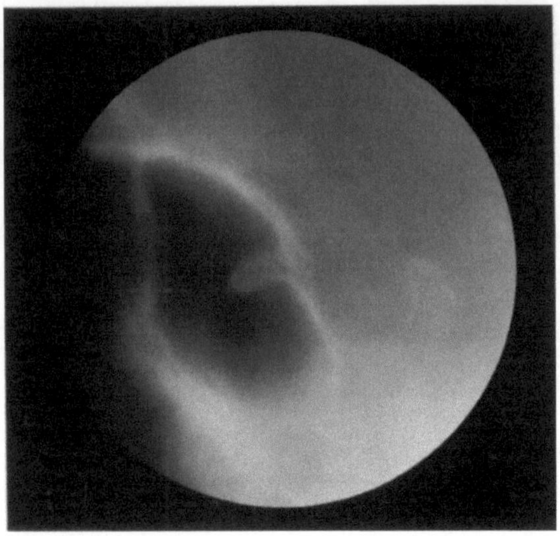

Fig. 14.49b. This is a trap. The appearance of a channel is given by a transverse crest corresponding to remnants of the intersphenoid synchondrosis (see also Fig. 2.3)

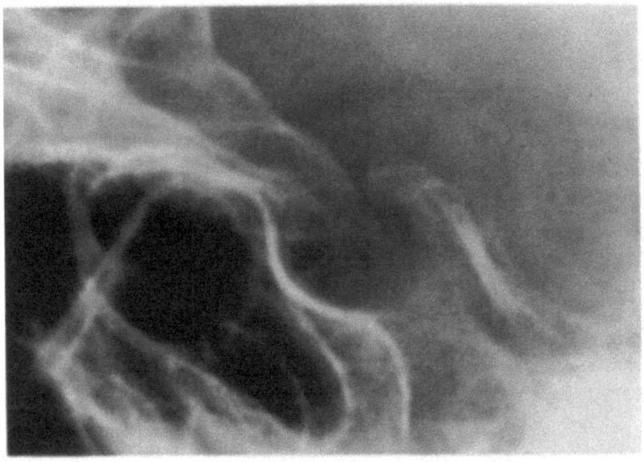

Fig. 14.50. Is the anteroposterior diameter of the sella turcica increased?

No; when the dorsum sellae is demineralized, the calcified petroclinoid ligaments must not be mistaken for the dorsum sellae

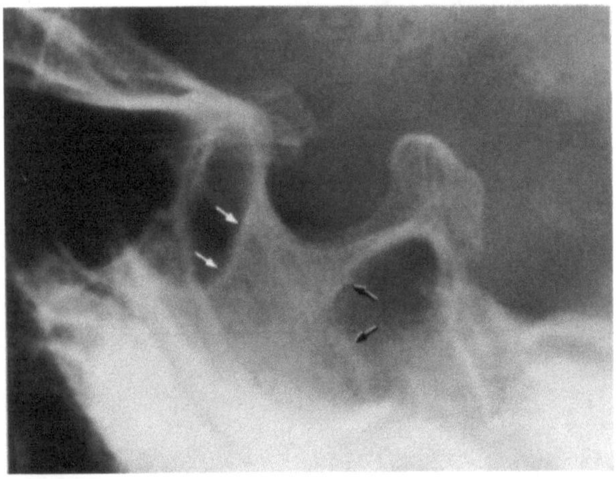

Fig. 14.51a. What does this shadow in the sphenoid sinus *(arrows)* represent?

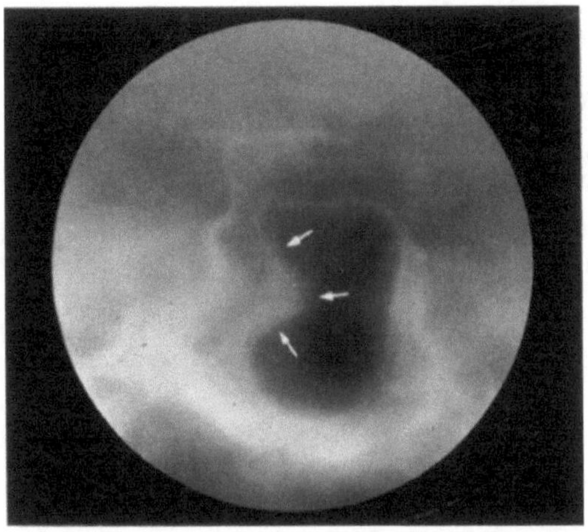

Fig. 14.51b. Frontal tomogram shows a partial absence of pneumatization of the sphenoid *(arrows)*

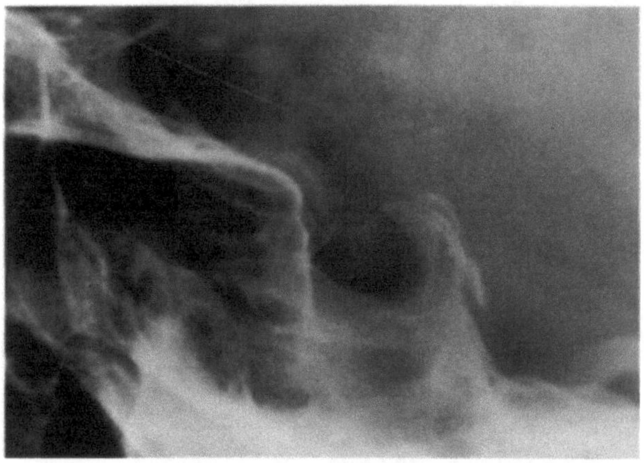

Fig. 14.52a. Inhomogeneous appearances of the posterior part of the sphenoid sinus. The cortical bone of the floor of the sella seems to be blurred. Is this a malignant infrasellar lesion?

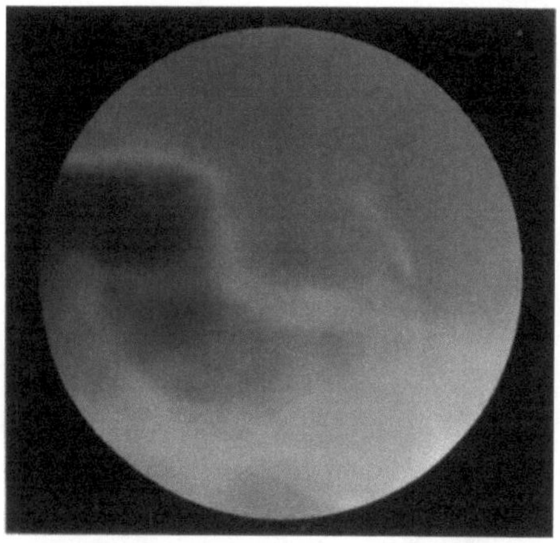

Fig. 14.52b. Tomography is reassuring: this is sphenoid sinusitis, and a thin plate of cancellous bone is seen between the sinus and the floor

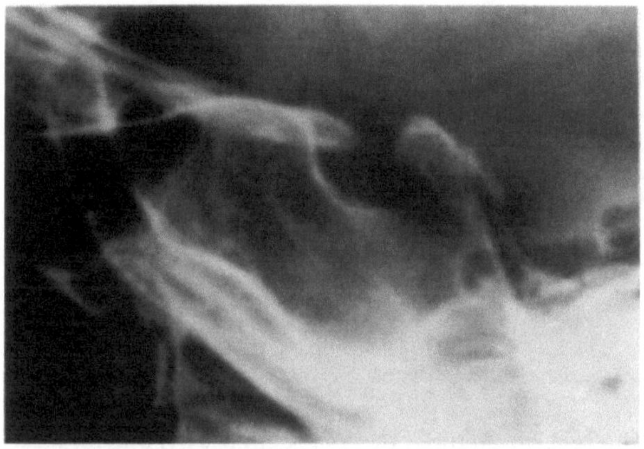

Fig. 14.53a. Homogeneous clouding of the sphenoid sinus and total disappearance of the cortical bone of the posterior part of the floor of the sella. Are we dealing with a malignant lesion of the sphenoid sinus?

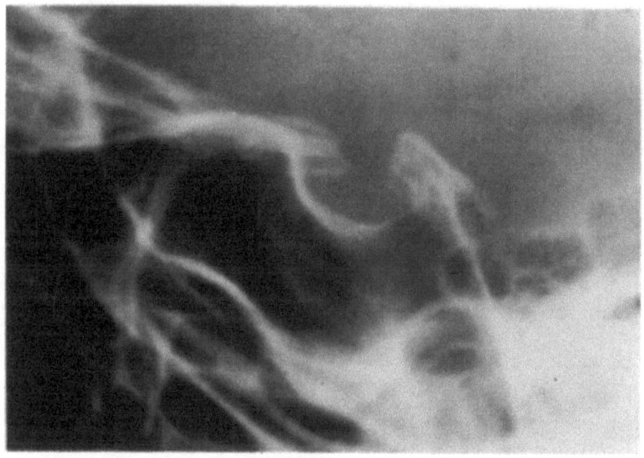

Fig. 14.53b. After antibiotic therapy, there is clearing of the sphenoid sinus and remineralization of the floor of the sella: subacute sinusitis (see also Fig. 11.1)

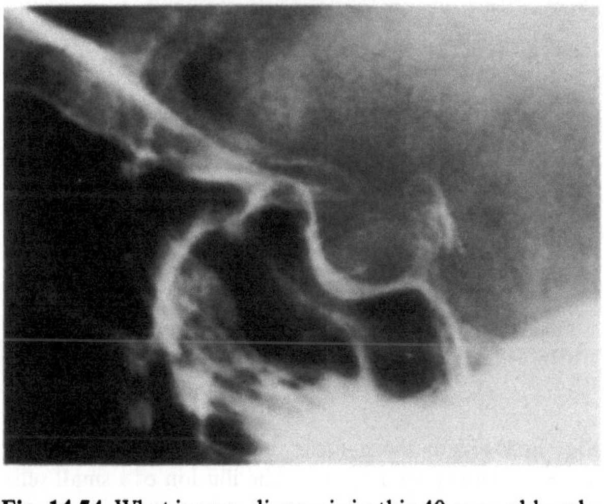

Fig. 14.54. What is your diagnosis in this 40-year-old male patient complaining of chronic headaches?

Longstanding increased intracranial pressure: the planum sphenoidale is shortened, no longer reaches anteriorly to the greater wings of the sphenoid, which are also considerably demineralized. The sella turcica itself is almost normal, however; large frontal astrocytoma

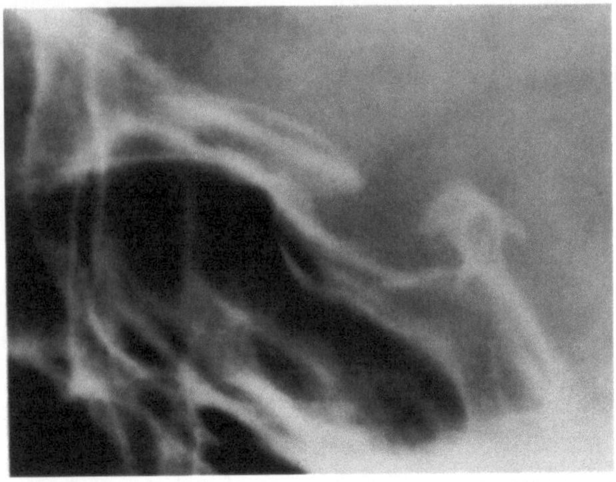

Fig. 14.55a. The lamina dura of the floor of the sella seems irregular and demineralized: what do you think about this?

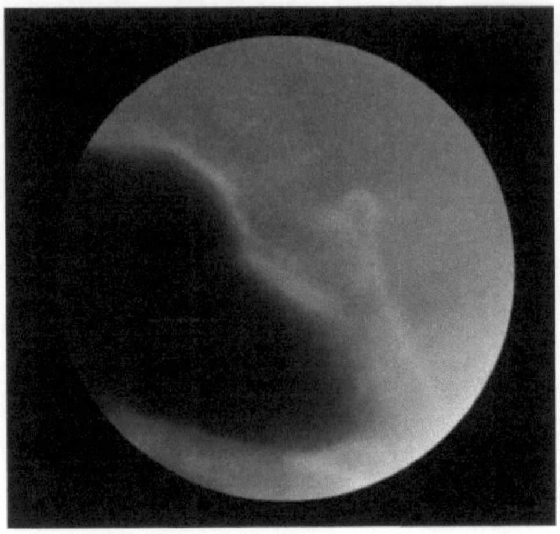

Fig. 14.55b. The plain film also shows blistering of the planum sphenoidale, which may suggest a post-traumatic change; the lateral tomograms demonstrate an old fracture of the floor

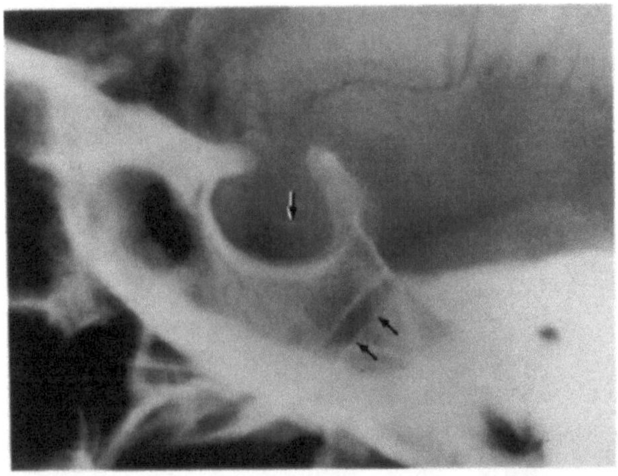

Fig. 14.56. 18-year-old girl; a thin, horizontal linear shadow overlies the sellar area *(black and white arrow)*. Could this be carotid siphon calcification?

As a rule, no. Actually this young girl has hypothyroidism. Therefore, there is definitely pathologic vascular calcification. Observe also the appearance of the spheno-occipital synchondrosis which is still wide *(black arrows)*

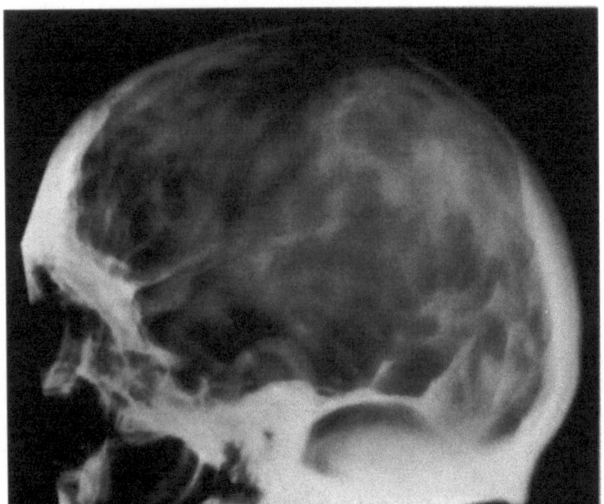

Fig. 14.57a. Craniostenosis. Does one ordinarily find a small sella turcica with this disease?

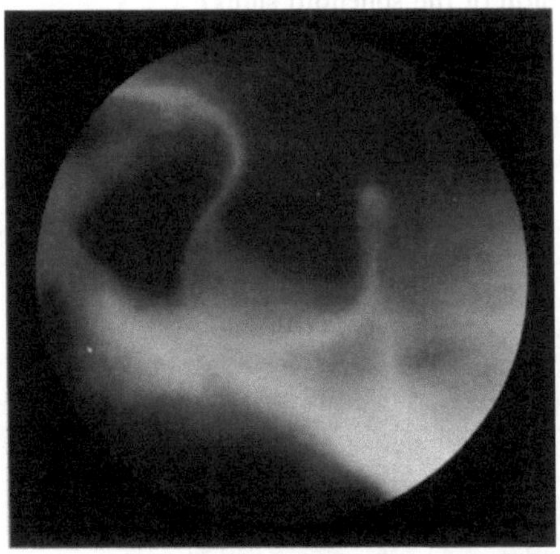

Fig. 14.57b. On the contrary, one very often finds an enlarged, empty sella turcica. The illusion of a small sella turcica is given by a bony crest between prominent convolutional impressions overlying the sella

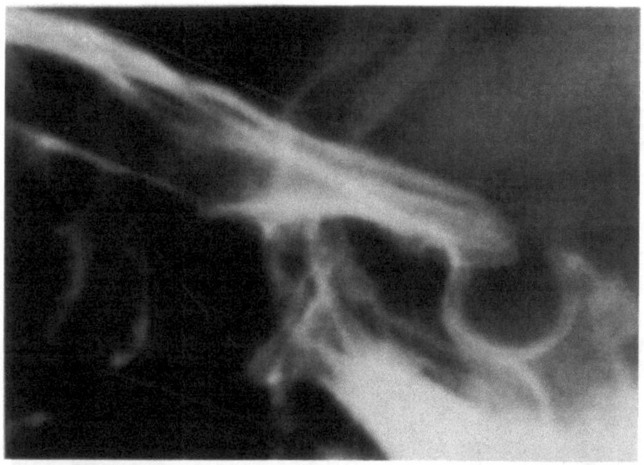

Fig. 14.58a. Loss of visual acuity in a 45-year-old female. What is your diagnosis?

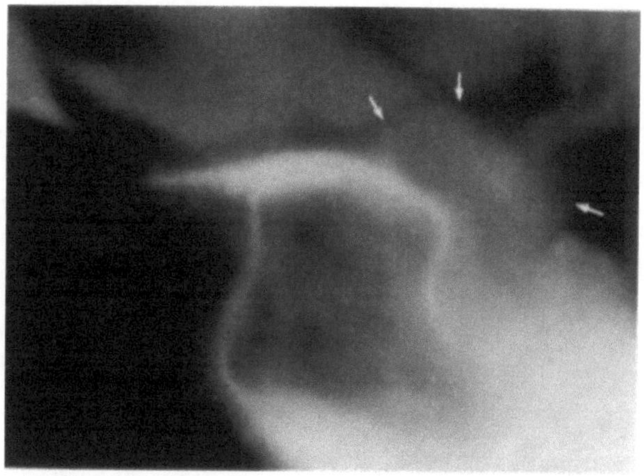

Fig. 14.58b. The thickness of the planum sphenoidale is uneven from front to back; it indicates the attachment of a meningioma *(arrows)*

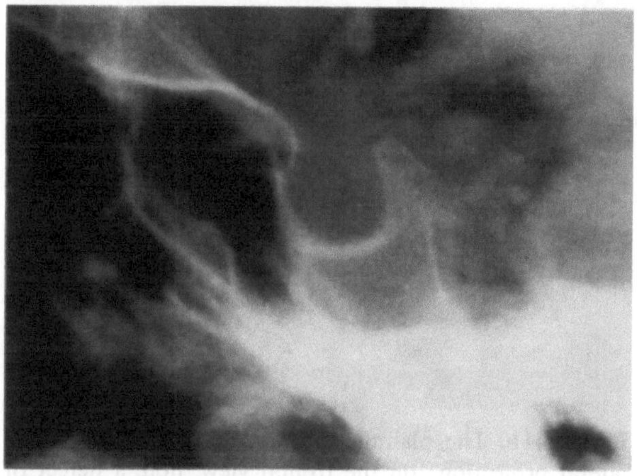

Fig. 14.59a. This 8-year-old child shows delayed growth and diabetes insipidus. What do you think of the sella turcica?

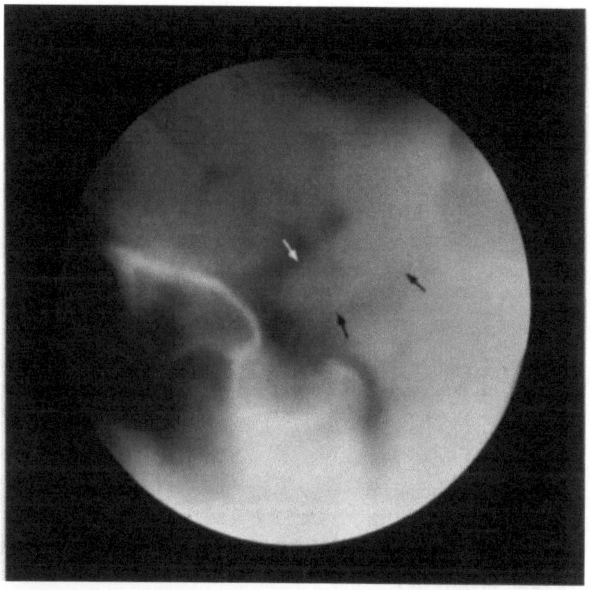

Fig. 14.59b. It is normal. However, pneumoencephalography has shown a hypothalamic glioma *(arrows)*. If the clinical findings raise the suspicion of a sellar region tumor, normal plain radiography never precludes further investigations, especially in children

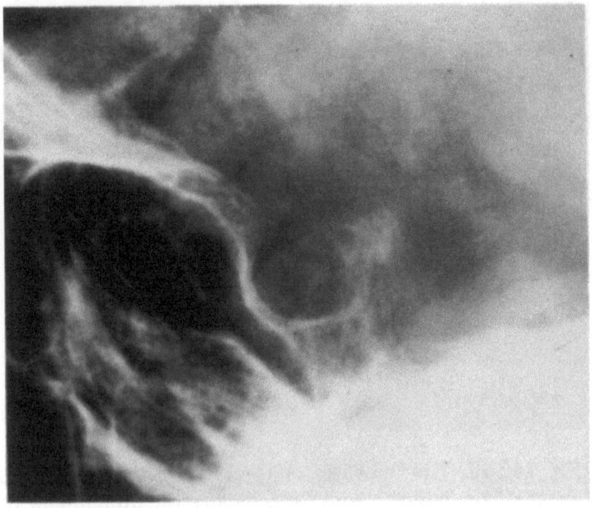

Fig. 14.60a. Decline of visual acuity and bitemporal hemianopia in a 64-year-old female. What do you see?

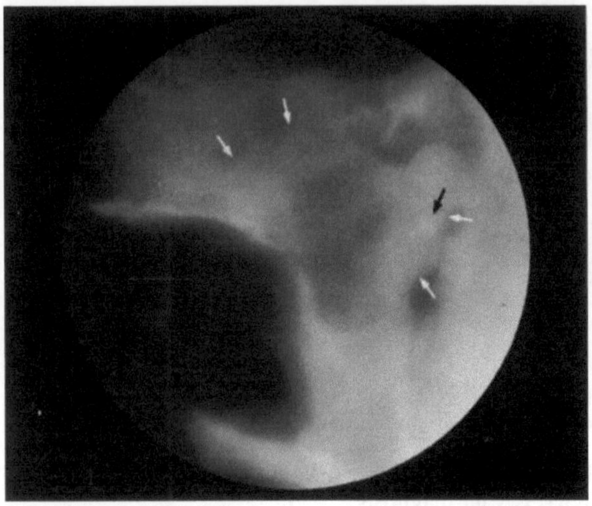

Fig. 14.60b. Demineralization of the sella turcica and the chiasmatic sulcus. In this context, considerable importance is attached to suprasellar calcification *(black arrow)*, indicating a craniopharyngioma *(white arrows)*

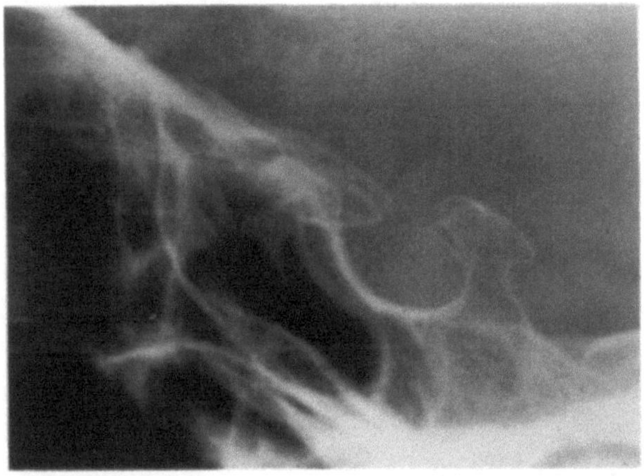

Fig. 14.61a. Epilepsy 5 years after head trauma. What is your diagnosis?

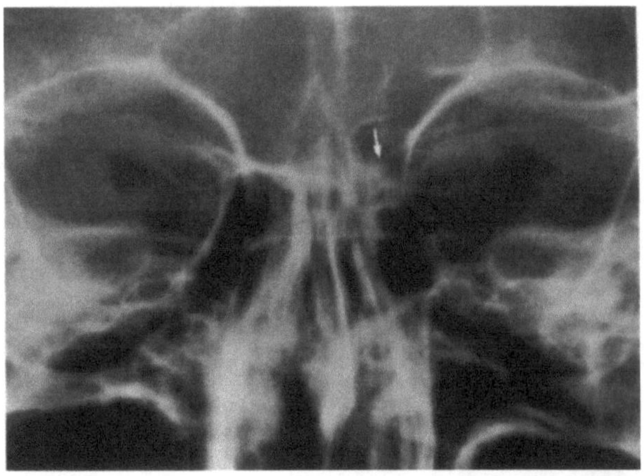

Fig. 14.61b. The planum sphenoidale is blurred on the lateral projection; on the frontal view, note a defect of the left half of the planum representing an old fracture *(arrow)*

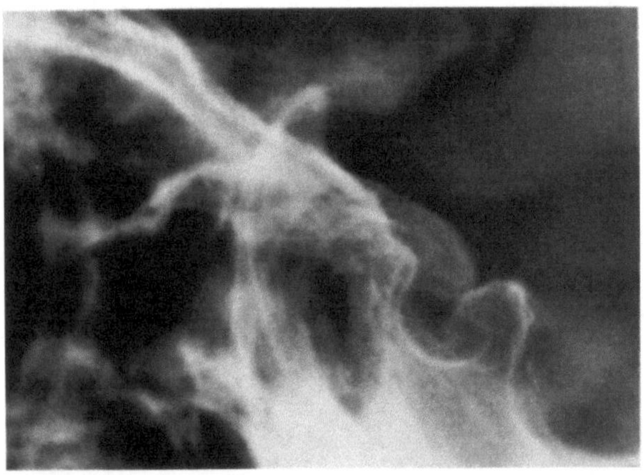

Fig. 14.62. Headache and loss of vision in a 27-year-old male; what is your diagnosis here?

A meningioma of the planum sphenoidale must be the first suggestion; note increased density and blistering of the planum; in fact, this was also post-traumatic change (see also Fig. 8.18)

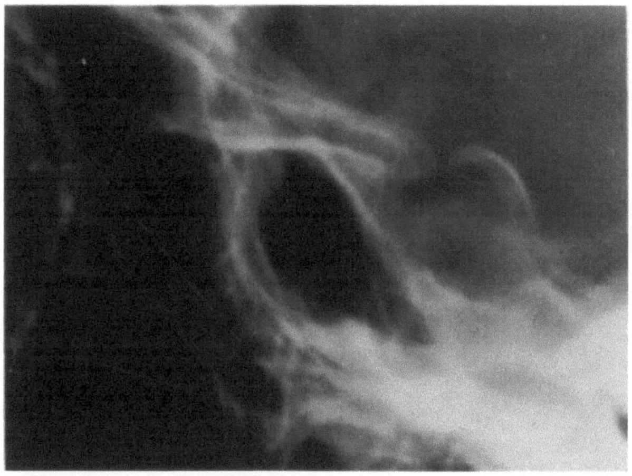

Fig. 14.63a. Chance discovery of an abnormal sella turcica with trauma. No clinical findings to suggest endocrine or visual disease. What do you think?

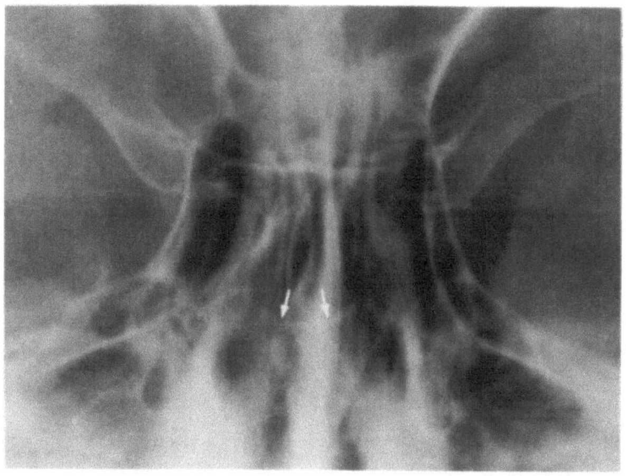

Fig. 14.63b. This lateral plain film is difficult to interpret. However, the frontal view gives us the diagnosis immediately. This very regular downward curvature of the floor *(arrow)* undoubtedly indicates an empty sella turcica

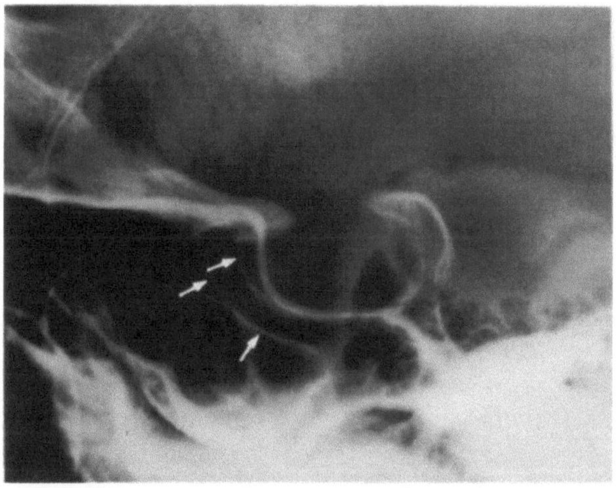

Fig. 14.64a. Plain films of the sella turcica taken for primary infertility in a 27-year-old female with hyperprolactinemia. What do you do?

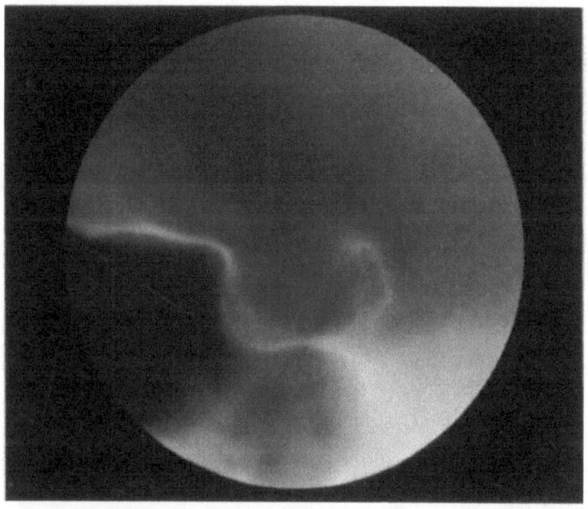

Fig. 14.64b. The two lines located in front of the anterior wall of the sella correspond to the carotid sulcus *(arrows)*. On the plain film the sella turcica is, therefore, normal. Clinical findings lead to tomography; depression and thinning of the sellar floor indicate without doubt the presence of a microadenoma

Chapter 15 Advances in CT of the Pituitary Gland

Conventional radiologic examination of the sella turcica still remains in 1980 the simplest and fastest approach to the pituitary and its surroundings; however, it must be complemented by a CT scan if one has reason to suspect some anomaly of the sellar region, be this based on radiologic, clinical, or biologic observations. In these cases radiologic examination of the sella turcica is replaced by that of the pituitary gland, possible nowadays with high resolution scanners. The recent introduction of this technique and the continuing improvements being made, make it impossible to write a monograph on CT of the sellar area at present: this chapter, then, constitutes simply a half-open window looking out onto the fascinating world revealed by this evolving and exciting technique.

A. Method of CT Examination

While the study of suprasellar expansion of pituitary tumors has been possible since the earliest scanners, a satisfactory exploration of the normal pituitary and of intrasellar lesions requires a high resolution scanner.

For adenomas with suprasellar expansion, axial and coronal slices of 5 mm thickness with sagittal reconstructions are generally sufficient. For intrasellar adenomas it is necessary to realize 1.5 mm thick axial and coronal slices to avoid partial volume phenomena and to permit high quality sagittal reconstructions. Direct sagittal slices are more difficult to realize. Obtaining a localizer scan (scout-view) allows optimal gantry angulation and a reduction in the number of slices.

If the clinical and biologic picture strongly suggests a pituitary adenoma, one does not need to realize slices before injecting the contrast medium. Generally an initial rapid injection of 60 ml contrast medium is followed by a 100 ml perfusion in order to maintain a satisfactory constant plasma iodine concentration throughout the procedure.

B. Results

I. Normal Results
(Figs. 15.1, 15.2, 15.3)

Without contrast medium injection the normal pituitary gland has a density similar to that of cerebral parenchyma. After injection it enhances strongly (mean, 45 HU), but a little less than the cavernous sinus (mean, 55 HU). With axial slices it is difficult to distinguish the external limit of the pituitary gland from the cavernous sinus. We have never noticed any density difference between the anterior and posterior lobes. On slices of the chiasmatic cistern the pituitary stalk is always well visualized, as is the supraclinoid part of the internal carotid arteries.

On coronal slices it is often easier to distinguish between the pituitary gland and the cavernous sinus. The superior face of the pituitary gland is generally flat. The height of the pituitary gland is easily measurable. The internal carotid arteries can frequently be seen in their intracavernous portion as hyperdense images. The oculomotor nerves show up as small rounded defects in the cavernous sinus. An intracavernous extension of the subarachnoid spaces is frequently found, more often bilaterally, at the inferior part of the cavernous sinus. Using the large window permits an excellent study of the bone walls of the sella turcica.

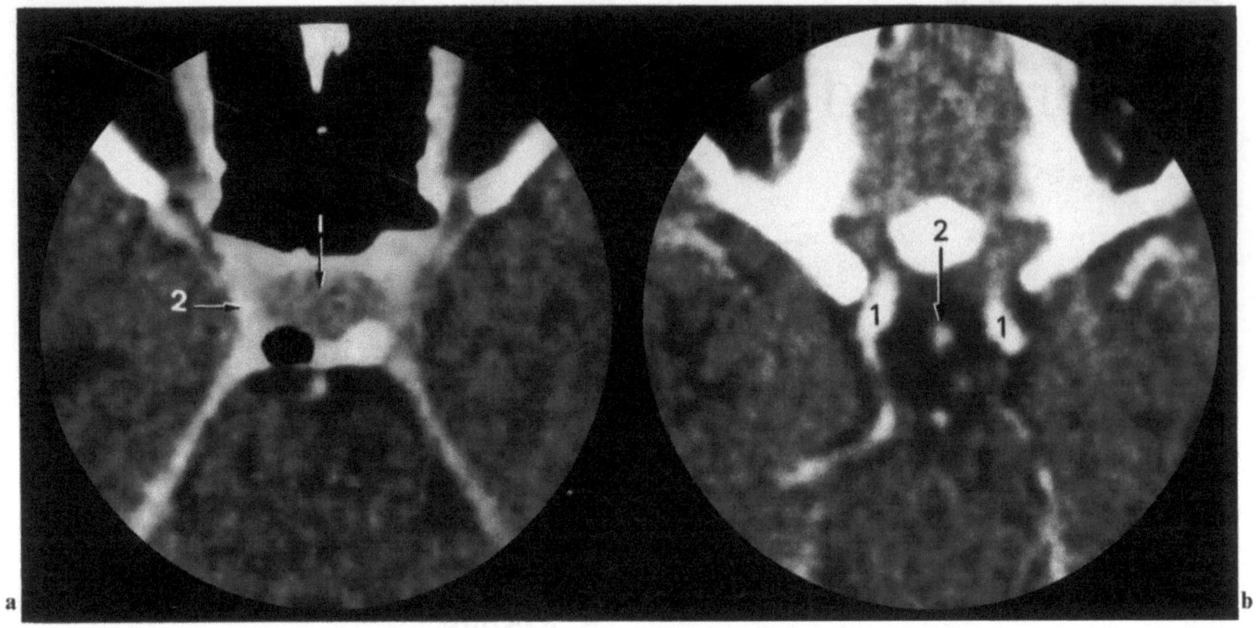

II. CT of Pituitary Adenomas

1) Adenomas with Suprasellar Expansion (Figs. 15.4, 15.5)

Most adenomas with suprasellar expansion enhance less than the normal pituitary gland. To the eye, however, they appear very dense, silhouetted as they are against the chiasmatic cistern and cerebral parenchyma. This aspect is variable depending on their structure, which might be solid, cystic, hemorrhagic, or necrotic. They are often surrounded by a hyperdense ring; capsular calcifications are more often apparent than with plain skull roentgenography. The extent of the suprasellar expansion can be appreciated perfectly with coronal slices and sagittal reconstructions.

2) Intrasellar Adenomas (Figs. 15.7, 15.8, 15.9, 15.10)

Adenomas as small as 2 mm in diameter are detectable nowadays with the scanner. Their density is generally the same as that of adenomas with suprasellar expansion, but their hypodensity relative to the normal pituitary gland makes them appear as a rounded defect in enhancement. Even when very small they modify the aspect of the sellar diaphragm, which becomes convex. Adenomas with diameters of 5–10 mm, which are almost always situated inferiorly, push the normal pituitary gland towards the chiasmatic cistern. The normal pituitary gland thus becomes apparent on axial slices as a rounded dense image in the inferior part of the chiasmatic cistern and should not be mistaken for the adenoma itself. Coronal slices are the most useful for detecting the smallest adenomas, but sagittal reconstructions are always of great help in confirming the diagnosis.

In certain cases, the contours of the adenoma are less well delimited, but it is always possible to show up their hypodensity relative to the healthy pituitary gland: these aspects correspond, perhaps, to adenomatous hyperplasia (Fig. 15.6).

3) Treated Adenomas

Adenomas treated medically with bromocriptine are modified in their volume and in their density, which is much inferior to that of untreated adenomas (Fig. 15.11).

After surgery the scanner shows up the secondary empty sella turcica well, and possibly also the remaining normal pituitary gland (Figs. 15.12, 15.13, 15.14). Residual tumor growths can be found when postoperative hormonal levels remain high.

The scanner also makes it possible to diagnose a primary empty sella turcica and to visualize the remaining pituitary gland flattened against the walls of the sella turcica (Figs. 15.15, 15.16)

Fig. 15.1a, b. Normal CT of the sellar area with intravenous contrast medium; 1.5 mm section thickness.
a Axial cut through the sella. Enhancement of the hypophysis (*1*) is nearly uniform (mean 45 HU). Enhancement of the cavernous sinus (*2*) is more pronounced (mean 55 HU).
b Axial cut through the chiasmatic cistern. Carotid arteries (*1*) are well outlined. The pituitary stalk (*2*) appears as a small round density in the low density area of the chiasmatic cistern

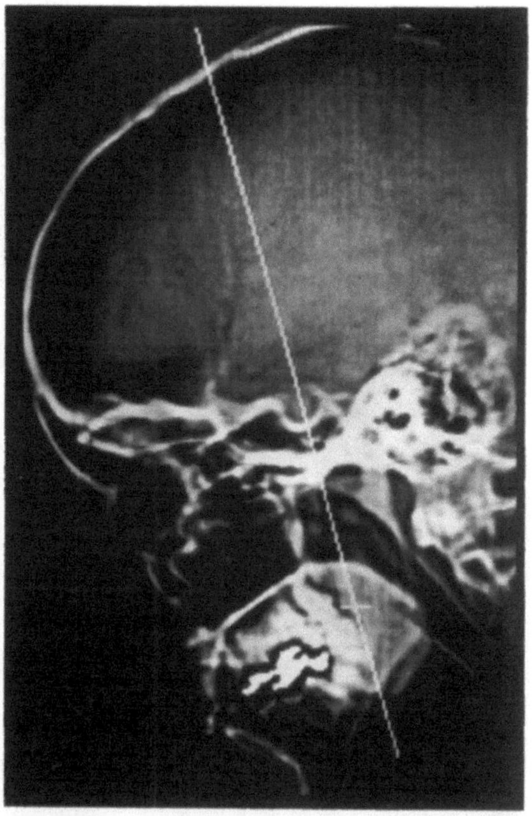

Fig. 15.2. Lateral localizer scan allowing optimal gantry angulation for coronal cuts.

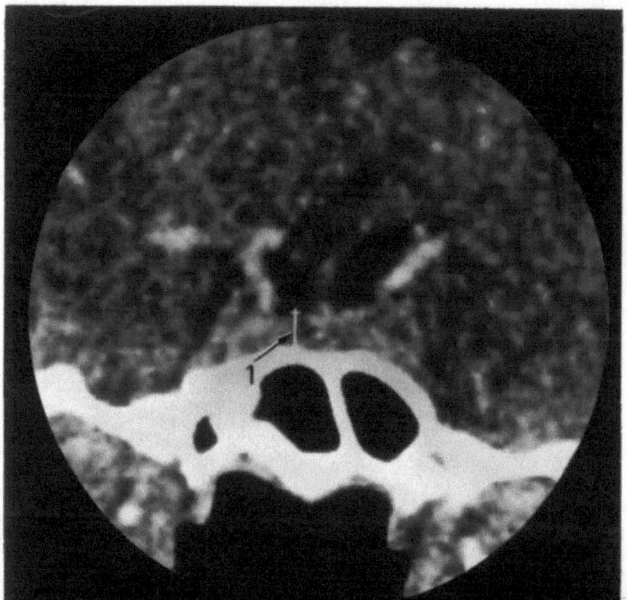

a

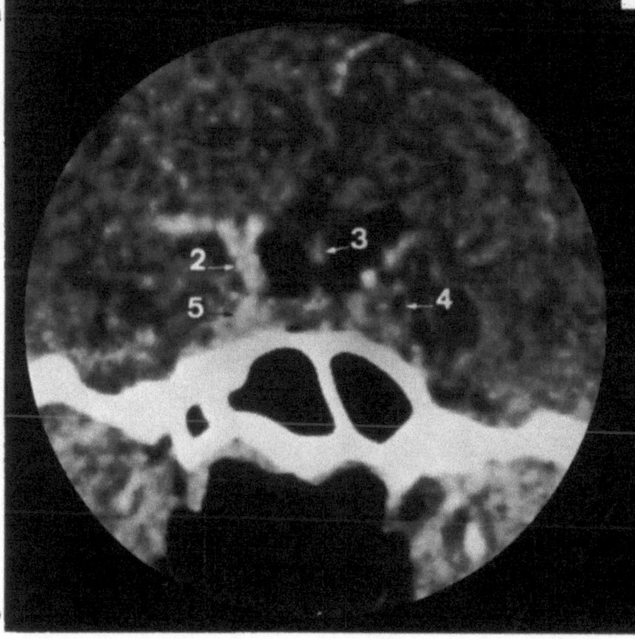

b

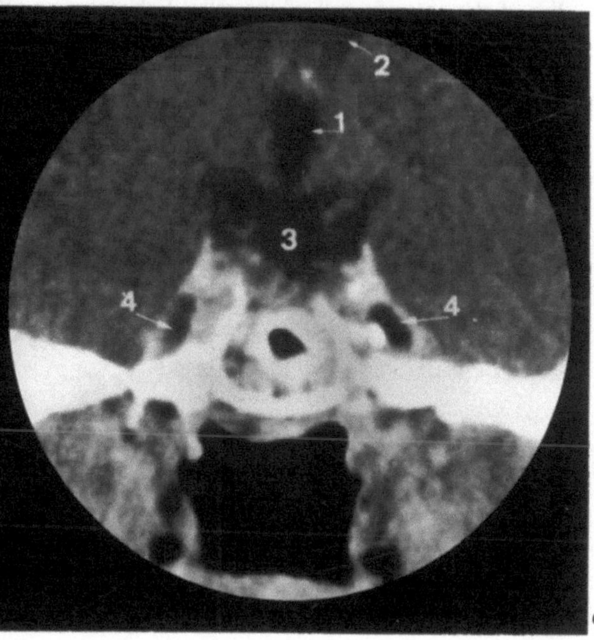

c

Fig. 15.3
a, b Normal coronal cuts through the sellar floor following infusion of contrast medium. The height of the pituitary gland (*1*) may be measured. *2*, carotid siphon; *3*, pituitary stalk; *4*, oculomotor nerve in cavernous sinus; *5*, intracavernous carotid artery.
c Cut through the posterior part of the sphenoid body. *1*, third ventricle; *2*, lateral ventricle; *3*, basal cisterns; *4*, extension of the subarachnoid spaces in the cavernous sinus

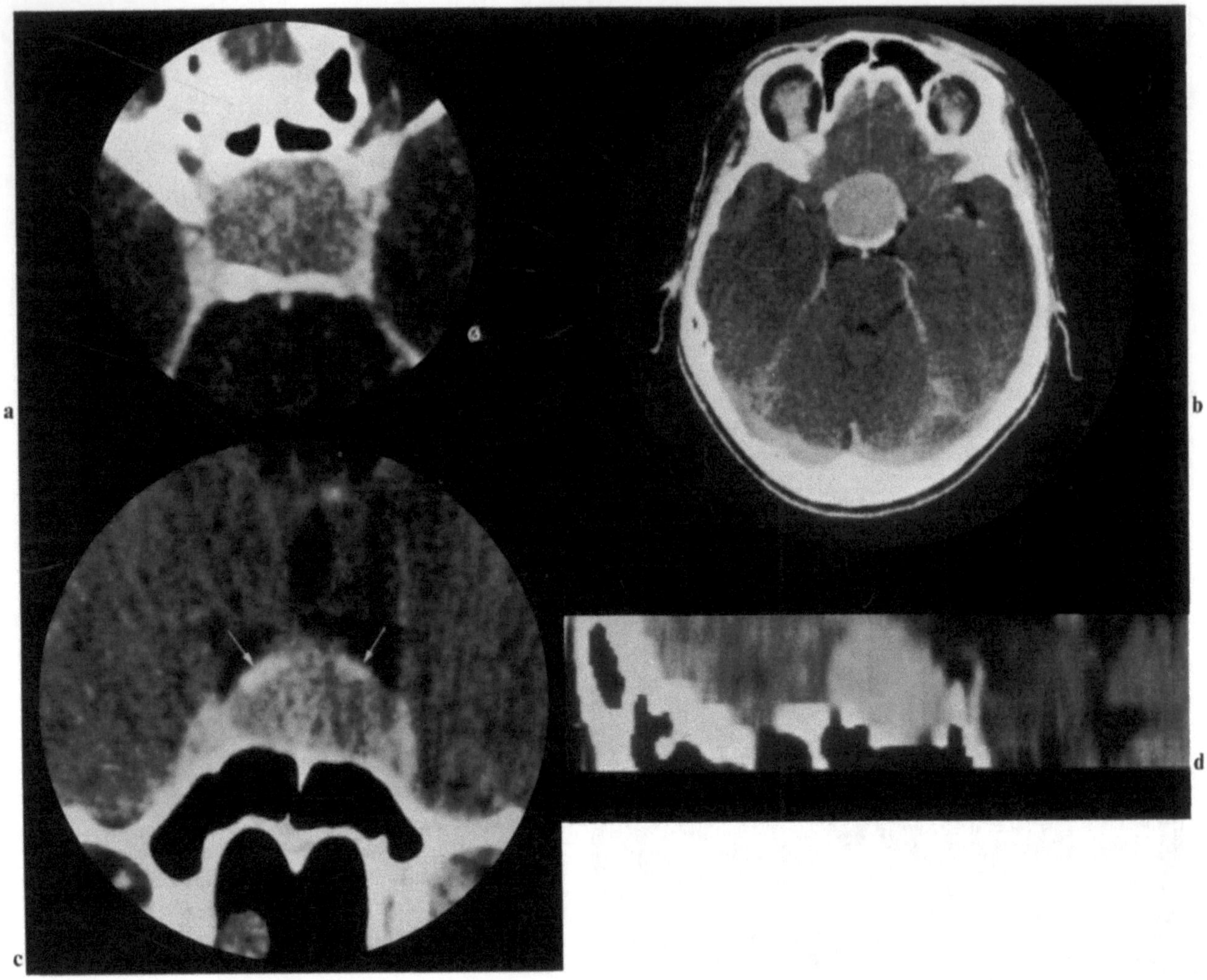

Fig. 15.4a–d. Huge prolactinoma. 45-year-old female with amenorrhea for 10 years and bitemporal hemianopia.

- **a** Axial cut through the sella turcica, which is enlarged. Thinning of the left part of the dorsum sellae and bulging of the anterior wall of the sella. Enhancement of the content of the pituitary fossa is less pronounced than usual for the normal pituitary gland; this is confirmed by the important difference between the attenuation values of the cavernous sinuses and of the adenoma.
- **b** Axial cut through the suprasellar area. Suprasellar extension of the tumor.
- **c** Coronal cut. Capsular calcifications *(arrows)* outline the peripheral limits of the adenoma.
- **d** Midsagittal reconstruction

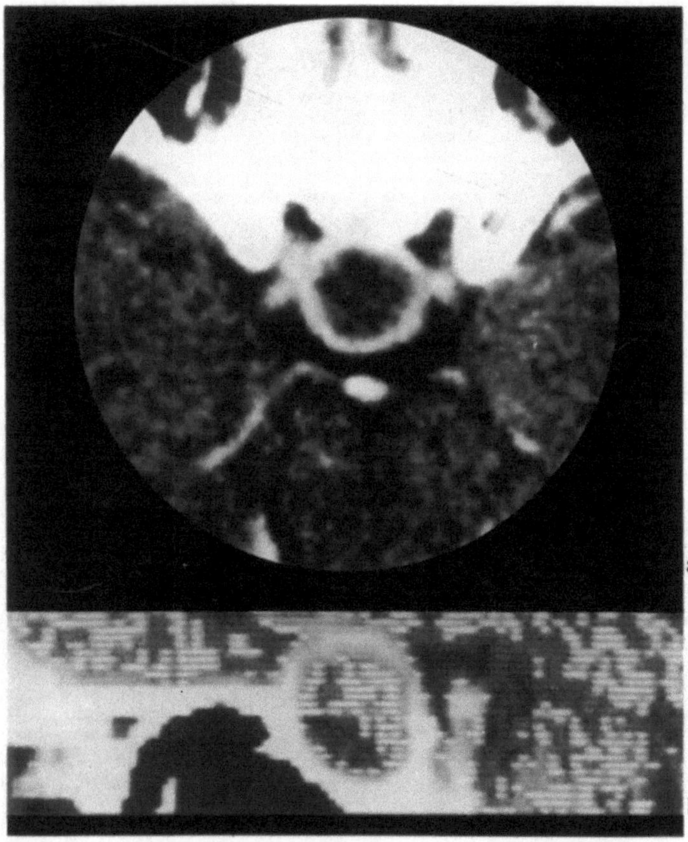

Fig. 15.5a, b. Nonsecreting adenoma. 65-year-old female. Hypopituitarism.

a Axial cut. Capsular enhancement. Low density of the central part of the adenoma.

b Midsagittal reconstruction. Suprasellar extension of the adenoma is well outlined. The anterior and inferior part of the adenoma has a very low density corresponding to partial necrosis of the tumor

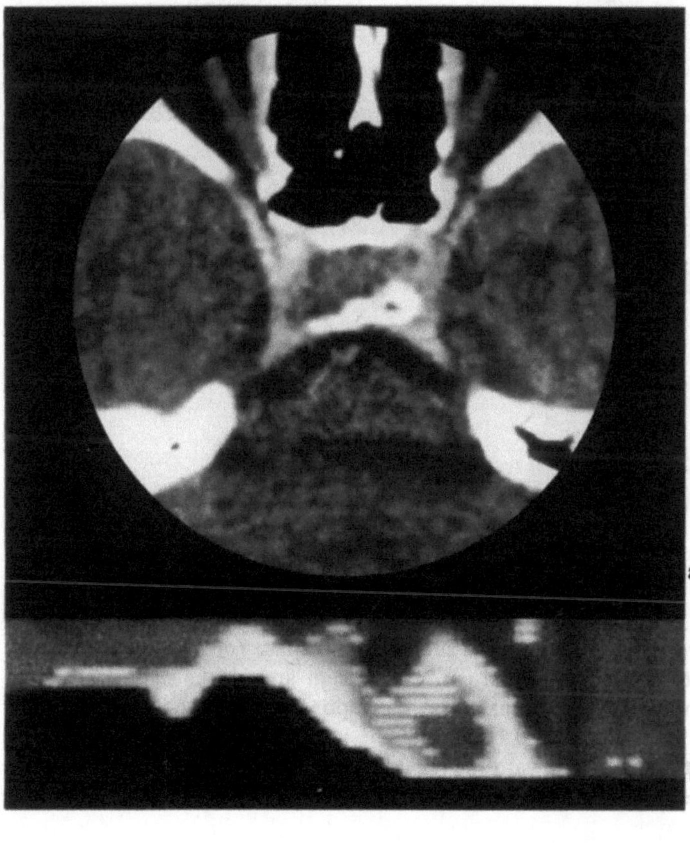

Fig. 15.6a, b. Amenorrhea-galactorrhea. 25-year-old female. Hyperprolactinemia (90 ng/ml).

a Axial cut. The content of the pituitary fossa is less dense than usually, except its left part, but there is no rounded defect in enhancement. Thinning of the right part of the dorsum.

b Sagittal reconstruction. Relative hypodensity of the postero-inferior part of the sella. Adenomatous hyperplasia? (not verified)

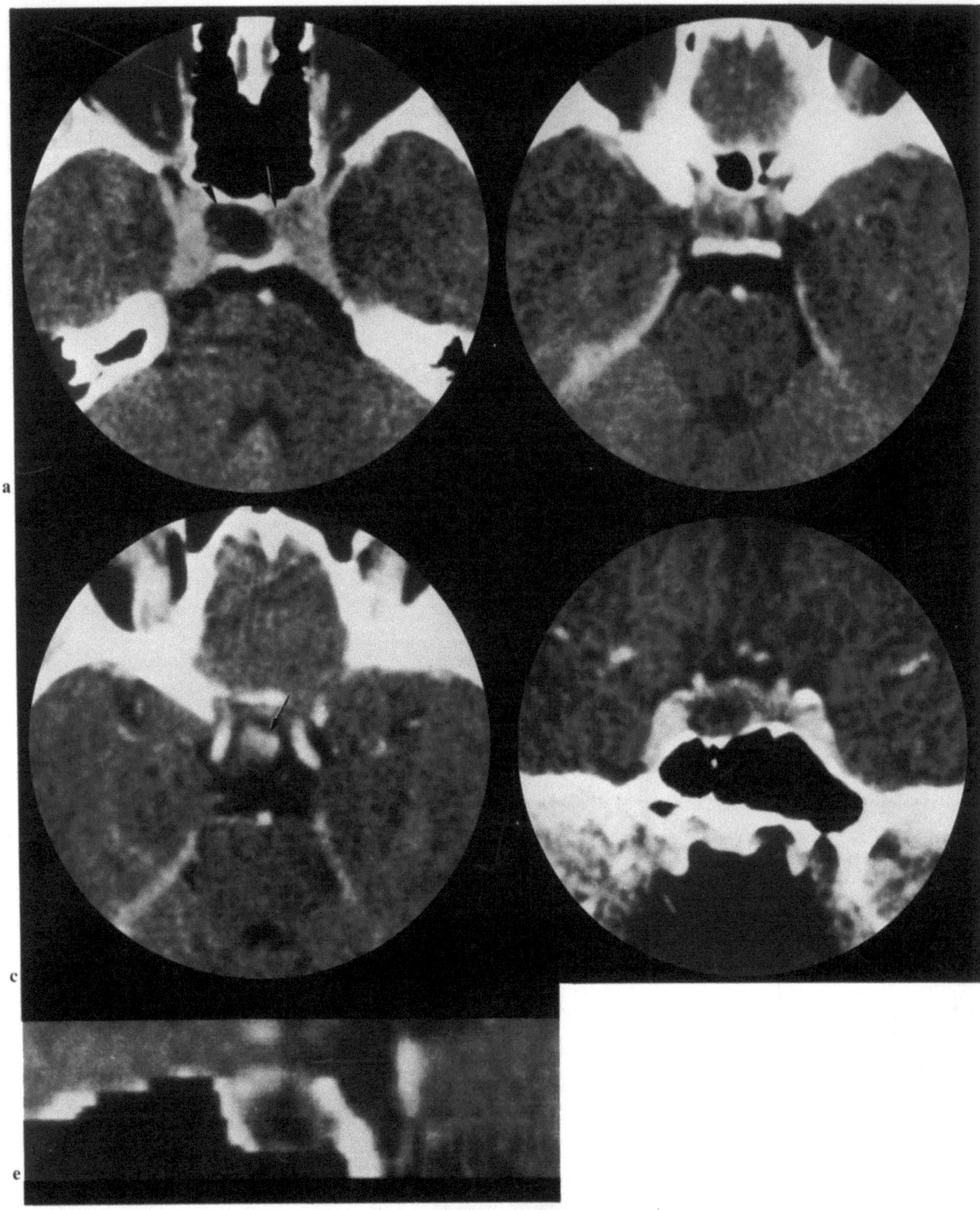

Fig. 15.7a–e. Prolactinoma (necrotic at surgery). 28-year-old female with amenorrhea-galactorrhea and elevated prolactin level (220 ng/ml).

a Axial cut through the lower part of the pituitary fossa. Defect in enhancement in the right part of the pituitary gland *(arrowhead)* (20 HU). Compare with the normal pituitary gland on the left side *(arrow)* (45 HU).

b, c Axial cuts through the upper part of the pituitary fossa and the suprasellar cistern. Swelling of the non-adenomatous portion of the pituitary gland toward the chiasmatic cistern results in an area of increased density within the subarachnoid space *(arrow)*.

d Coronal cut. Translucent area in the right part of the pituitary fossa corresponding to the adenoma and upward bulging of the non-adenomatous pituitary gland toward the chiasmatic cistern.

e Sagittal reconstruction

a 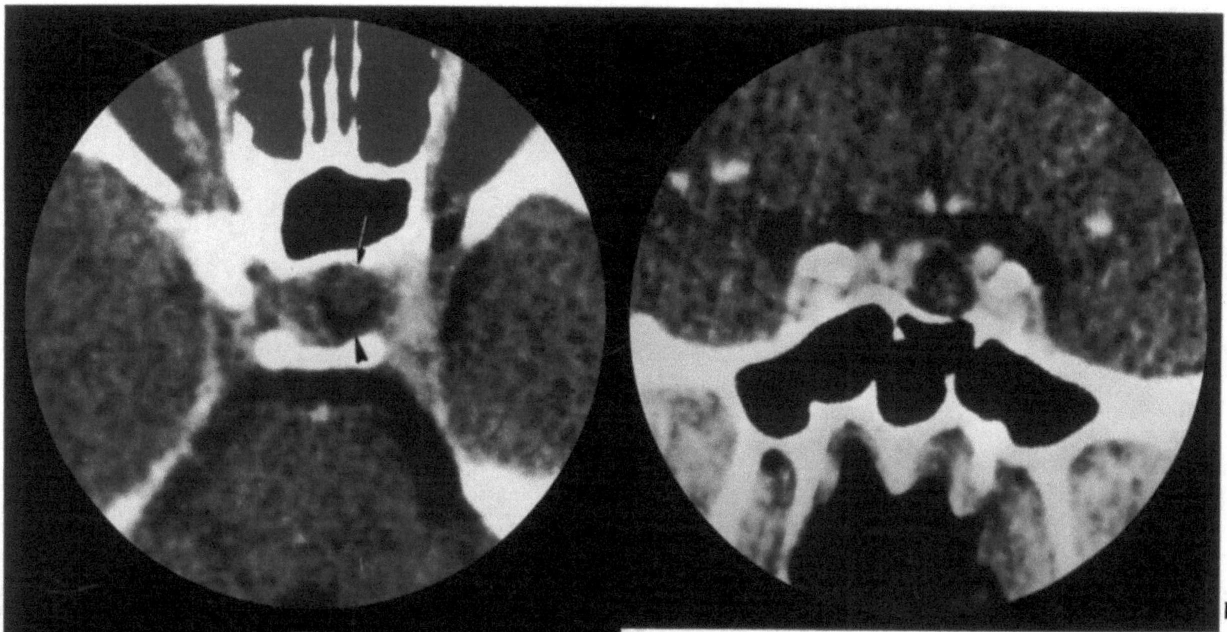 b

c 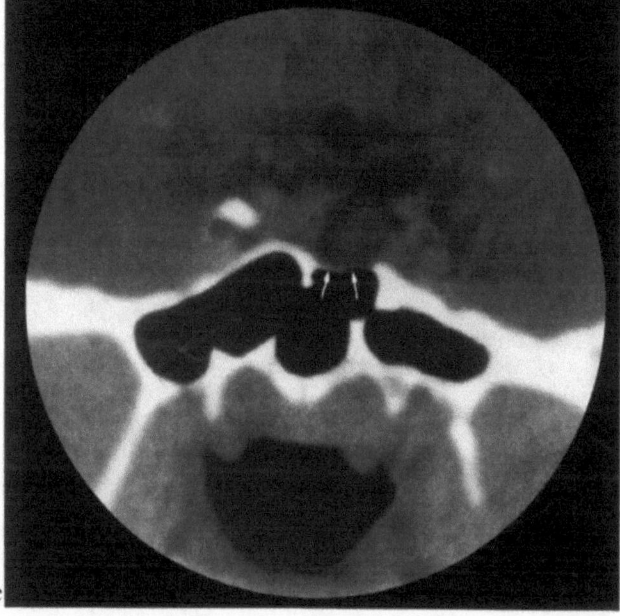

Fig. 15.8a–c. Prolactinoma. 20-year-old female. Irregular menses, galactorrhea, and anovulation. Hyperprolatinemia (60 mg/ml).

a, b Axial and coronal cuts. The mild enhanced area (20 HU) corresponding to the microadenoma *(arrow)* is surrounded by a low density ring (4 HU) *(arrowhead)*; spontaneous decrease in size of the adenoma.

c Coronal section. Large window. Depression and thinning of the left part of the sellar floor *(arrows)*

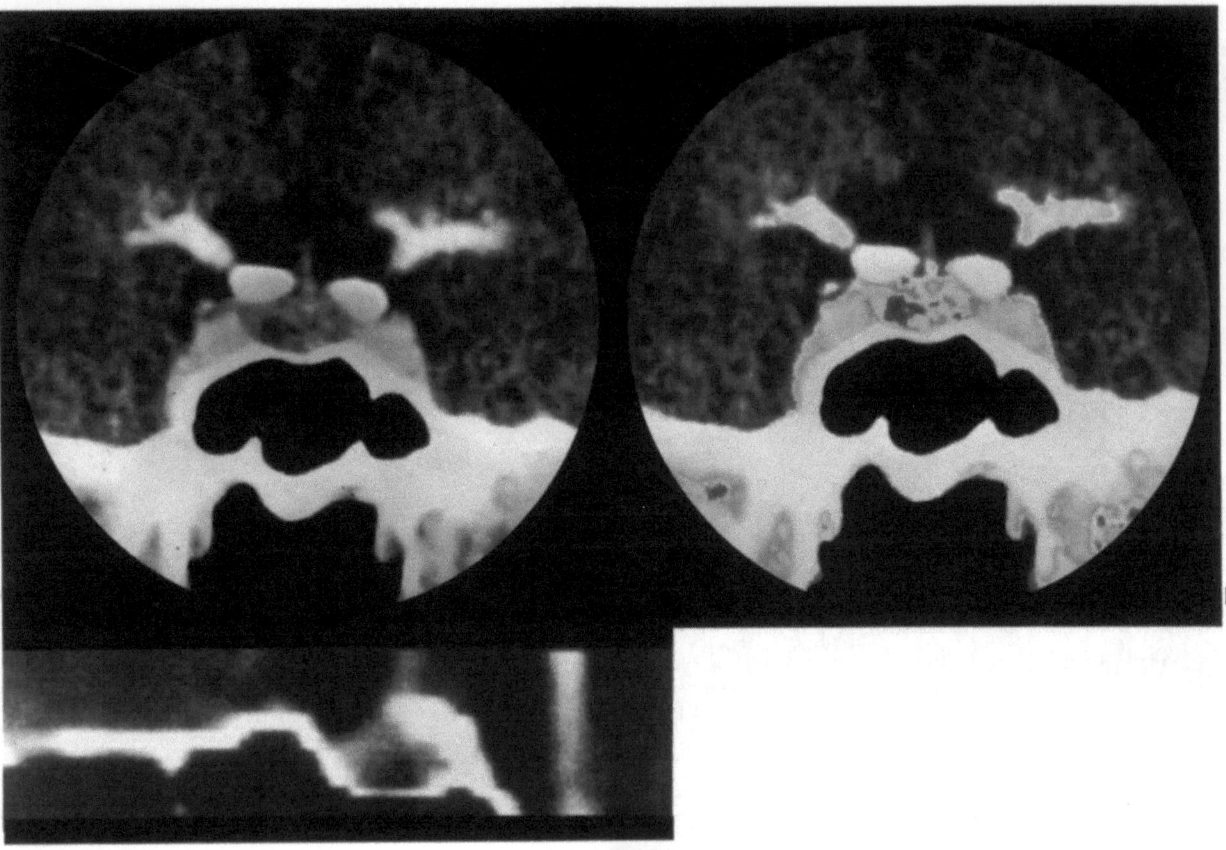

a b c

Fig. 15.9a–c. Prolactinoma. 25-year-old female. Amenorrhea-galactorrhea. Hyperprolactinemia (70 ng/ml).

a, b Coronal cuts: limited defect in enhancement in the right part of the pituitary gland.

c Sagittal reconstruction

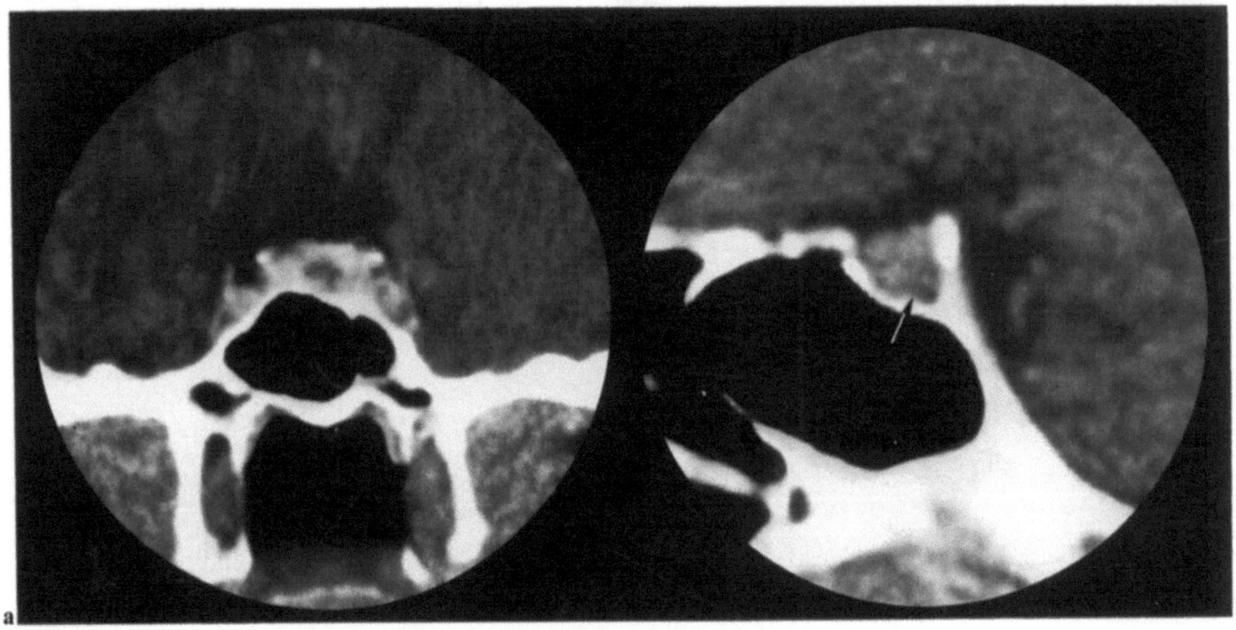

a b

Fig. 15.10a, b. Prolactinoma. Small rounded defect in enhancement in the posterior part of the pituitary gland in the midline *(arrow)*

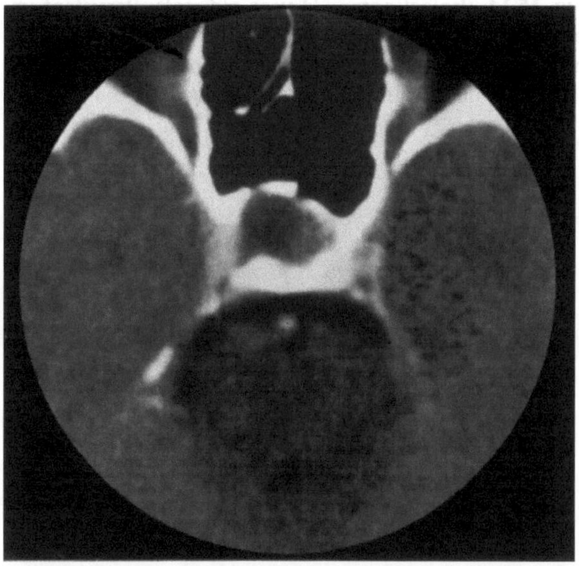

Fig. 15.11. Prolactinoma. 28-year-old female. After 4 years of bromocriptine therapy the patient is totally asymptomatic; the prolactin level is normal and a frank remineralization of the sella turcica is observed. Bulging of the anterior wall of the sella is seen in front of an area of very low density (5 HU) corresponding to the adenoma treated with bromocriptine

a

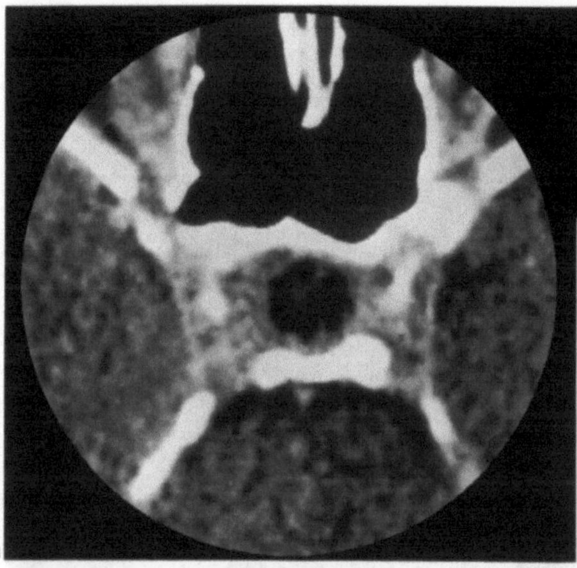

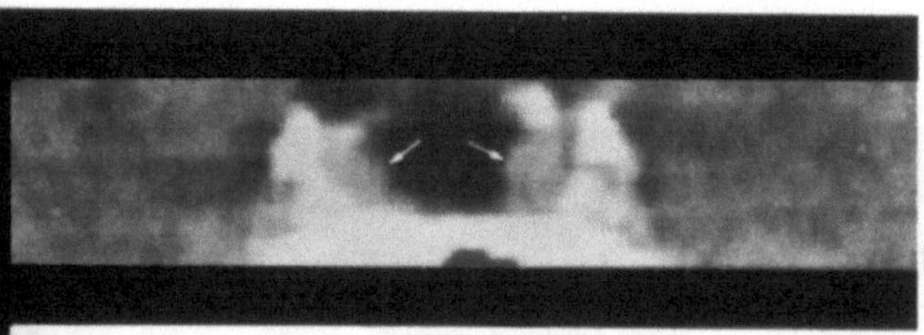

b

Fig. 15.12a, b. STH-secreting adenoma, after trans-sphenoidal surgery. Postoperative empty sella. Persistent ring of normal pituitary tissue

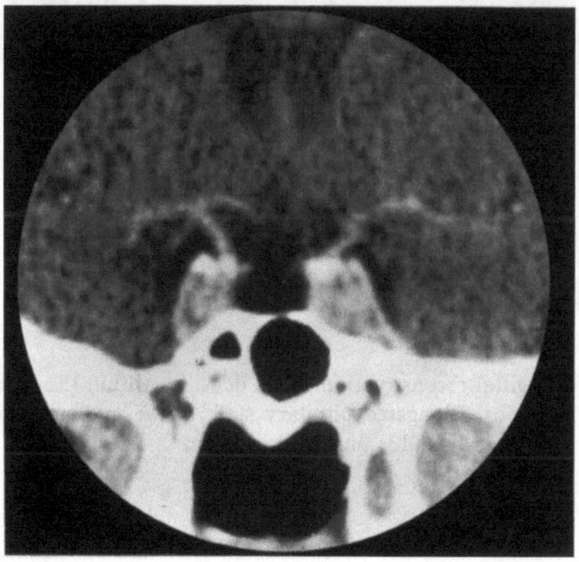

Fig. 15.13. Postoperative empty sella

Fig. 15.14. Prolactinoma. 24-year-old male. Persistent impotence and hyper-prolactinemia after subfrontal removal of a large adenoma. Residual adenomatous tissue in the right part of the pituitary fossa *(arrows)*

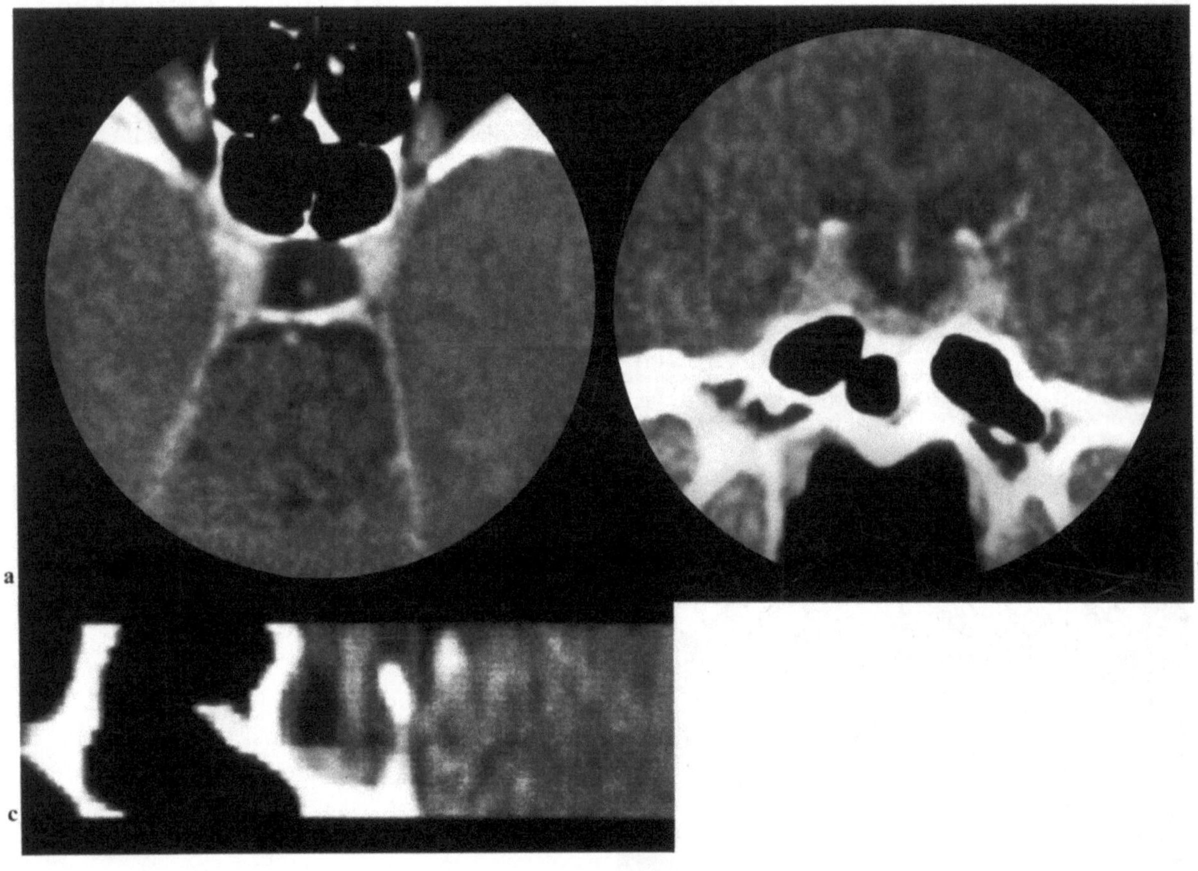

Fig. 15.15 a–c. Primary empty sella turcica. 55-year-old female. Headache and obesity.
a Axial cut.
b Coronal cut.
c Sagittal reconstruction. CSF densities within the sella turcica. Elongated pituitary stalk in the midline. The pituitary gland is pushed downward

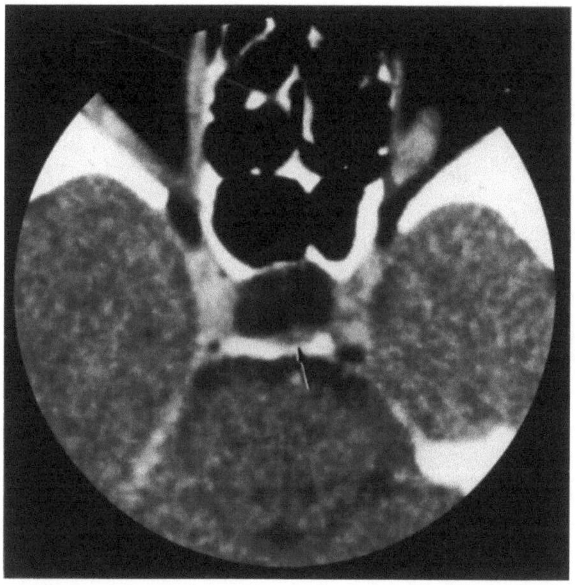

Fig. 15.16. Primary empty sella turcica. Low density area within the sella turcica corresponding to the extension of the subarachnoid space. The pituitary gland is pushed against the dorsum sellae *(arrow)*

C. Bibliography

Aubin ML, Bentson J, Vignaud J (1978) Tomodensitométrie de la tige pituitaire. Neuroradiology 5:153–160

Bajraktari X, Bergström M, Brismar K, Goulatia R, Greitz T, Grepe A (1977) Diagnosis of intrasellar cisternal herniation (empty sella) by computer assisted tomography. Comput Assist Tomography 1(1):105–116

Belloni G, Baciocco A, Borrelli P, Sagui G, Di Rocco C, Maira G (1978) The value of CT for the diagnosis of pituitary microadenomas in children. Neuroradiology 15:179–181

Citrin CM, Davis DO (1977) Computerized tomography in the evaluation of pituitary adenomas. Invest Radiol 12:27–35

Drayer B, Kattah J, Rosenbaum A, Kennerdell J, Maroon J (1979) Diagnostic approaches to pituitary adenomas. Neurology 29:161–169

Drayer BP, Rosenbaum AE, Reigel DB, Bank WO, Deeb ZL (1977) Metrizamide computed tomography cisternography: pediatric applications. Radiology 124:349–357

Drayer BP, Rosenbaum AE, Kennerdel JS, Robinson AG, Bank WO, Deeb ZL (1977) Computed tomographic diagnosis of suprasellar masses by intrathecal enhancement. Radiology 123:339–344

Gardeur D, Nachanakian A, Kulesza E, Metzger J (1979) La tomodensitométrie dans les adénomes hypophysaires. Ann Radiol (Paris) 22:489–499

Gyldensted C, Karle A (1977) Computed tomography of intra- and juxtasellar lesions. Neuroradiology 14:5–13

Hatam A, Bergström M, Greitz T (1979) Diagnosis of sellar and parasellar lesions by computed tomography. Neuroradiology 18:249–258

Hayman LA, Evans RA, Hinck VC (1979) Rapid high dose (RHD) contrast conputed tomography of perisellar vessels. Radiology 131:121–123

Jacobs L, Weisberg LA, Kinkel WR (1980) Computerized tomography of the orbit and sella turcica. Raven Press, New York

Miller JH, Pena AM, Segall HD (1980) Radiological investigation of sellar region masses in children. Radiology 134:81–87

Naidich TP, Pinto RS, Kushner MJ, Lin JP, Kricheff II, Leeds NE, Chase NE (1976) Evaluation of sellar and parasellar masses by computed tomography. Radiology 120:91–99

Osborn AG, Anderson RE (1978) Direct sagittal computed tomographic scans of the face and paranasal sinuses. Radiology 129:81–87

Pinto RS, Handel SF, Sadhu VK (1979) CT metrizamide cisternography in the recognition of intrasellar cistern. AJR 133:320–321

Post MJD, David NJ, Glaser JS, Safran A (1980) Pituitary apoplexy: diagnosis by computed tomography. Radiology 134:665–670

Reich NE, Zelch JV, Alfidi RJ, Meaney TF, Duchesneau PM, Weinstein MA (1976) Computed tomography in the detection of juxtasellar lesions. Radiology 118:333–335

Rozario R, Hammerschlag SB, Post KD, Wolpert SM, Jackson I (1977) Diagnosis of empty sella with CT scan. Neuroradiology 13:85–88

Sheldon P, Molyneux A (1979) Metrizamide cisternography and computed tomography for the investigation of pituitary lesions. Neuroradiology 17:83–87

Syverisen A, Haughton VM, Williams AL, Cusick JF (1979) The computed tomographic appearance of the normal pituitary gland and pituitary microadenomas. Radiology 133:385–391

Wiggli U, Benz UF (1978) Normal computed tomography anatomy of the suprasellar subarachnoid space. Radiology 128:65–70

Wolpert SM, Post KD, Biller BJ, Molitch ME (1979) The value of computed tomography in evaluating patients with prolactinomas. Radiology 131:117–119

References

1. Abrikossoff A (1926) Über Myome, ausgehend von der quergestreiften willkürlichen Muskulatur. Virchows Arch [Pathol Anat] 260:215–233
2. Acheson RM, Archer M (1959) Radiological studies of the growth of the pituitary fossa in man. J Anat 93:52–67
3. Acquaviva R, Tamic PM, Thevenot C, Fournie R, Grau P, Bru P, Le Bascle J, Esplan JFR (1965) Los condromas intracraneales. Revision de la literatura a proposito de dos casos. Rev Esp Otoneurooftalmol Neurocir 24:15–34
4. Adler H, Kaplan G (1956) Improvement of osseous changes in sella turcica following irradiation for pituitary tumor. Radiology 66:856–858
5. Ahmad K, Fayos JV (1978) Pituitary fibrosarcoma secondary to radiation therapy. Cancer 42:107–110
6. Alajouanine T, De Martel T, Oberling C, Guillaume J (1930) Chordome de la région sellaire. Considérations anatomiques et cliniques. Rev Neurol (Paris) 1:1221–1227
7. Alajouanine T, Thurel R, Nehlil J, Baldacci Y (1955) L'anévrysme du siphon carotidien et son retentissement osseux. Rev Neurol (Paris) 92:249–253
8. Albert M, Lemay M (1968) Demineralisation of the dorsum sellae associated with alcoholism. Br J Radiol 41:331–332
9. Algyogyi H (1909) Ein Fall von radiologisch nachgewiesenen Tumor der Hirnbasis. ROEFO 14:257–262
10. Altman R, Pruzanski W (1961) Post-traumatic hypopituitarism. Ann Intern Med 55:149–154
11. Ambrose J (1973) Radiological diagnosis of pituitary tumors. In: Jenkins JS (ed) Pituitary tumors. Butterworths, London
12. Amine ARC, Sugar O (1976) Suprasellar osteogenic sarcoma following radiation for pituitary adenoma. J Neurosurg 44:88–91
13. Amirchanjan SE, Lebedev AN (1979) Röntgendiagnostik der Trigeminus Neurinome des Ganglion Gasseri. Radiol Diagn (Berl) 6:78–85
14. Anderson FM (1947) Intranasal (sphenopharyngeal) encephalocele; a report of a case with intracranial repair and a review of the subject. Arch Otolaryngol 46:644–654
15. Anderson RD (1976) Tortuosity of the cavernous carotid arteries causing sellar expansion simulating pituitary adenoma. AJR 126:1203–1210
16. Andresen JH, Hedemand N, Hermansen K (1978) Das Nelson Syndrom. Frühdiagnose kleinster Veränderungen der Sella Turcica. Radiologe 18:138–139
17. Angevine DM (1938) Pathologic anatomy of hypophysis and adrenals in anencephaly. Arch Pathol 26:507–518
18. Arai H (1907) Der Inhalt des Canalis Cranio-Pharyngeus. Anat Hefte 33:411–451
19. Archer CR, Sundaram M (1977) Uncommon sphenoidal fractures and their sequelae. Radiology 122:157–161
20. Arey LB (1950) The craniopharyngeal canal reviewed and reinterpreted. Anat Rec 106:1–16
21. Arey LB (1965) Developmental anatomy. Saunders, Philadelphia London
22. Arnould G, Lepoire J, Leclere J, Lamaze R, Aubrege A (1973) La selle turcique vide. A propos d'une observation. Rev Otoneuroophtalmol 45:247–258
23. Arseni C, Maretsis M (1972) Craniopharyngioma. Neurochirurgia (Stuttg) 15:25–32
24. Arseni C, Gritescu M, Gritescu A, Mihaila G (1970) Intrasellar aneurysms simulating hypophyseal tumors. Eur Neurol 3:321–329
25. Asenjo AR, Chiorino R (1956) Mixocondroma endocraneamo (Estudio anatomo-clinico de dos casos). Rev Esp Otoneurooftalmol Neurocir 15:15–21
26. Askenasy HM, Israeli J, Karny H, Dujovny M (1971) Intrasellar abscess simulating pituitary adenoma. Neurochirurgia (Stuttg) 14:34–37
27. Atwell WJ (1926) The development of the hypophysis cerebri in man, with special reference to the pars tuberalis. Am J Anat 37:159–194
28. Atwell WJ, Woodworth EA (1926) The relative volumes of the three epithelial parts of the hypophysis cerebri. Anat Rec 33:377–386

28a. Augier M (1931) Squelette cephalique. In: Poirier P, Charpy A (eds) Traité d'anatomie humaine, vol 1, 4th ed. Masson, Paris

29. Avanzini G, Crivelli G (1970) A case of sphenopharyngeal encephalocoele. Acta Neurochir (Wien) 22:205–212
30. Avman N, Erdogan G, Kanpolat Y (1978) Pituitary pseudotumor. Surg Neurol 9:107–110
31. Azar Kia B, Krishnan UR, Schechter MM (1975) Neonatal craniopharyngioma. J Neurosurg 42:91–93
32. Bailey P (1933) Crâniopharyngiomas in intracranial tumors. Tindall & Cox, Baillière London
33. Bajraktari X (1977) Skull changes with intrasellar cisternal herniation (empty sella). Neuroradiology 13:89–96
34. Bakay L (1950) Results of 300 pituitary adenoma operations. J Neurosurg 7:240–255

35. Banna M (1973) Craniopharyngioma in adults. Surg Neurol 1:202–204
36. Banna M (1974) Arachnoid cysts in the hypophyseal area. Clin Radiol 25:323–326
37. Banna M (1976) Craniopharyngioma: based on 160 cases. Review article. Br J Radiol 49:206–223
38. Banna M (1977) Erosion of the pituitary fossa by benign nasal polyps. J Can Assoc Radiol 28:300–302
39. Banna M, Hoare RD, Stanley P, Till K (1973) Craniopharyngioma in children. J Pediatr 83:781–785
40. Banna M, Schatz SW, Molot MJ, Groves J (1976) Primary intrasellar germinoma. Br J Radiol 49:971–973
41. Banna M, Nicholas W, McLachlan M (1978) The borderline pituitary fossa in patients with amenorrhea and/or galactorrhea. Neuroradiology 16:440–442
42. Barnett DJ (1959) Radiologic aspects of craniopharyngiomas. Radiology 72:14–18
43. Barr HWK, Blackwood W, Meadows SP (1971) Intracavernous carotid aneurysms. A clinical pathological report. Brain 94:607–622
44. Barred PA, Robinson GC, Hardwick DF, Sovereign AE (1968) Congenital osteopetrosis: an unusual cause of hydrocephalus. Can Med Assoc J 98:362–365
45. Bartlett JR (1971) Craniopharyngiomas. A summary of 85 cases. J Neurol Neurosurg Psychiatry 34:37–41
46. Basauri L, Castro M, Garcia A, Palma A, Palma L (1977) The empty sella syndrome. Analysis of 10 cases. Acta Neurochirur (Wien) 38:111–120
47. Basso A, Guitelman A, Amezua L, Molocznik Z, Mancini A, Montez M (1975) Adénomes de l'hypophyse à prolactine. Possibilités actuelles de diagnostic et de traitement. Neurochirurgie 21:111–120
48. Bastide P (1968) Les épistaxis traumatiques par rupture de la carotide interne intracaverneuse dans le sinus sphénoïdal. Thèse, Nancy
49. Batznitzky S, Sarwar M, Leeds NE, Schechter MM, Azar-Kia B (1974) Colloid cysts of the third ventricle. Radiology 112:327–341
50. Becker LE, Yates AJ, Hoffman HJ, Norman MG (1975) Intracranial chordomas in infancy. J Neurosurg 42:349–352
51. Becker MH, Ngo N, Beranbaum JL (1968) Mycotic infection of the paranasal sinuses. Radiographic manifestations. Radiology 90:49–51
52. Bedfort MD (1966) The cavernous sinus. Br J Ophtalmol 50:41–46
53. Belanger WG, Dyke CG (1943) Roentgen diagnosis of malignant nasopharyngeal tumors. AJR 50:9–18
54. Bellini MA, Neves I (1956) Skull in childhood myxedema; its roentgen appearance. AJR 76:495–498
55. Belza J (1966) Double midline intracranial tumors of vestigial origin; contiguous intrasellar chordoma and suprasellar craniopharyngioma. J Neurosurg 25:199–204
56. Benedetti A, Carbonin C, Colombo F (1977) Possible aetiopathogenic correlation between primary empty sella and arachnoid cyst. Acta Neurochirur (Wien) 38:269–278
57. Berger PE, Harwood-Nash DC, Fitz CR (1976) The dorsum sellae in infancy and childhood. Pediatr Radiol 4:214–220
58. Bergerhoff W (1956) Über röntgenologische Sellamessungen. ROEFO 85:695–708
59. Bergerhoff W (1964) Atlas anatomischer Varianten des Schädels im Röntgenbild. Springer, Berlin Heidelberg New York
60. Bergeron RT, Rumbaugh CL (1971) Skull trauma. In: Newton TH, Potts DG (eds) Radiology of the skull and brain, vol 1, book 1. 763–818 Mosby, Saint-Louis
61. Bergland RM, Ray BS (1969) The arterial supply of the human optic chiasm. J Neurosurg 31:327–334
62. Bergland RM, Torack RM (1969) An ultrastructural study of follicular cells in the human anterior pituitary. Am J Pathol 57:273–297
63. Bergland RM, Ray BS, Torack RM (1968) Anatomical variations in the pituitary gland and adjacent structures in 225 human autopsy cases. J Neurosurg 28:93–99
64. Bergström K, Lundberg PO, Sjövall A (1973) The volume of the sella turcica in patients with anorexia nervosa. Eur Neurol 9:183–189
65. Berke JP, Buxton LF, Kokmen E (1975) The empty sella. Neurology (Minneap) 25:1137–1143
66. Berkmen YM, Blatt ES (1968) Cranial and intracranial cartilaginous tumours. Clin Radiol 19:327–333
67. Bernard JD, Aguilar MJ (1969) Localized hypothalamic histiocytosis X. Report of a case. Arch Neurol 20:368–372
68. Bernasconi V, Pagni CA (1969) Medial displacement of the cavernous portion of the internal carotid artery. Its diagnosis significance. Acta Neurochirur (Wien) 20:15–29
69. Bernasconi V, Giovanelli MA, Papo I (1972) Primary empty sella. J Neurosurg 36:157–161
70. Berry RG, Schlesinger NS (1959) Rathke cleft cysts. Arch Neurol 1:48–58
71. Besser GM (1949) The pituitary fossa: Normal or abnormal. Br J Radiol 49:652–653
72. Bhatia ML, Mishra SC (1967) Intracranial extensions of juvenile angiofibroma of the nasopharynx. J Laryngol Otol 81:1395–1403
73. Bigos T, Robert F, Pelletier G, Hardy J (1977) Cure of Cushing's disease by transphenoidal removal of a microadenoma from pituitary gland despite a radiographically normal sella turcica. J Clin Endocrinol Metab 45:1251–1260
74. Binet EF (1974) The base of the adult skull. Semin Roentgenol 9:137–150
75. Binet EF, Moro JJ (1972) A tomographic study of the base of the skull. Med Radiogr Photogr 48:30–60
76. Binet EF, Kieffer SA, Martin SH, Peterson HO (1969) Orbital dysplasia in neurofibromatosis. Radiology 93:829–833
77. Bishop PA (1936) Bone changes in chronic fluorine intoxication: Roentgenographic study. AJR 35:577–585
78. Black WC, Ravin A (1947) Studies in dystrophia myotonica. Autopsy observations in five cases. Arch Pathol 44:176–191
79. Bleisch VR, Robbins SL (1952) Sarcoid-like granulomata of the pituitary gland. Arch Intern Med 89:877–892

80. Blizzard RM, Alberts M (1956) Hypopituitarism, hypoadrenalism and hypogonadism in newborn infant. J Pediatr 48:782–792
81. Bloch H, Joplin GF (1959) Some aspects of radiological anatomy of pituitary gland and its relationship to surrounding structures. Br J Radiol 32:527–532
82. Block MA, Goree JA, Jimenez JP (1973) Craniopharyngioma with optic canal enlargement simulating glioma of the optic chiasm. J Neurosurg 39:523–527
83. Bloom DL (1965) Mucoceles of the maxillary and sphenoid sinuses. Radiology 85:1103–1109
84. Blumenfeld R, Skolnick EM (1965) Intranasal encephaloceles. Arch Otolaryngol 82:527–531
85. Bochu M, Crastes G (1974) La selle turcique normale. Etude radiographique. Lyon Med 231:797–805
86. Bochu M, Goutelle A, Sassolas G, Joyeux O (1974) Les selles turciques vides. Etude radiographique. Rev Otoneuroophthalmol 46:407–415
87. Bokelmann O (1934) Die spezielle Anatomie der Sella Turcica und ihre klinische Bedeutung für die Erkennung der Hypophysengröße, zugleich ein Beitrag zur Frage der Beziehungen der Hypophysengröße sowie Größe und Form der Sella zum anatomischen und funktionellen Hypogenitalismus. ROEFO 49:364–396
88. Boldrey EB, McNally WJ (1941) Chordoma of the basi-occiput and basi-sphenoid: report of 4 cases. Arch Otolaryngol 33:391–400
89. Bollatti A, Giunta F, Lenzi A, Marini G (1973) Third ventricle intrinsic craniopharyngioma. J Neurosurg Sci 17:316–317
90. Bonnet P (1955) La loge caverneuse et les syndromes de la loge caverneuse. Clin Ophtalmol 1:2–17
91. Bonneville JF, Bugault R, van Effenterre R, Pertuiset B, Metzger J (1976) Delineation of pituitary adenomas by angiotomography. Neuroradiology 11:49–51
92. Bonneville JF, Dietemann JL (to be published) Blistering of the planum sphenoidale in children with mental retardation
93. Bonneville JF, Cabanis E, Bugault R, Millet D, Netter A, Metzger J (1974) Diagnostic radiologique des petits adénomes hypophysaires. Rev Otoneuroophtalmol 46:331–335
94. Bonneville JF, Tuetey JB, Vernier F, Jacquet G, Steimle R (1978) Post-traumatic blistering of the planum sphenoidale. J Neuroradiol 5:237–246
95. Bonneville JF, Sauget Y, Bihr E, Tuetey JB (1979) La région pré-sellaire: anatomie et sémiologie radiologiques. Feuill Radiol 111:205–218
96. Bonneville JF, Mawazini H, Belloir A, Runge M, Manzoni JM, Dietemann JL, Monnier G (1980) Calcified remnants of the notochord in the roof of the pharynx. Radiology 137
97. Bonneville JF, Runge M, Belloir A, Manzoni JM, Dietemann JL (to be published) Physiological dural calcifications in the area of the superior orbital fissure
98. Boone MLM, Harle TS (1968) Malignant tumors of the paranasal sinuses. Semin Roentgenol 3:202–213
99. Borg SA, Fitzer PM, Young LW (1975) Roentgenologic aspects of adult cretinism. Two cases reports and review of the literature. AJR 123:820–828
100. Bosma JF (1976) Development of the basicranium. DHEW Publication No. (NIH) 76–989, Bethesda, Maryland
101. Boström K, Hassler O (1965) Radiological study of arterial calcification – 2. Intracranial arteries. Neurology (Minneap) 15:1168–1172
102. Boukhris R, Ben Khalifa F, Slim R, Ben Ayed H (1975) Sella turcica in Sheehan's syndrome. Ann Endocrinol (Paris) 36:343–344
103. Boukhris R, Sellami M, Guedri H, Ben Cheihk M, Ben Khalifa F, Ben Ayed H (1978) Variations in the volume of the sella turcica in primary hypothyroidism and hypopituitarism. Nouv Presse Med 7:210–211
104. Bower BF (1968) Pituitary enlargement secondary to untreated primary hypogonadism. Ann Intern Med 69:107–109
105. Boyce DW, Huckman MS (1976) Contiguous internal carotid arteries in empty sella syndrome. Radiology 120:120–121
106. Boyce R, Beadles CF (1893) Enlargement of hypophysis cerebri in myxedema with remarks up on hypertrophy of hypophysis associated with changes in thyroid body. J Pathol Bacteriol 1:223–239
107. Bradfield JS, Perez CA (1972) Pineal tumors and ectopic pinealomas. Radiology 103:399–406
108. Brasel A, Wright JC, Wilkins L, Blizzard RM (1965) An evaluation of seventy-five patients with hypopituitarism beginning in childhood. Am J Med 38:484–498
109. Braun JP (1966) Examen radiologique de la selle turcique normale et pathologique. Fr Med 5:191–199
110. Brewer DB (1957) Congenital absence of the pituitary gland and its consequences. J Pathol Bacteriol 73:59–67
111. Bricaire H, Luton JP, Turpin G (1974) Tumeurs de l'hypophyse au cours des syndromes de Cushing. Rev Otoneuroophtalmol 47:287–294
112. Brisman R, Hughes JEO, Mount LA (1969) Cerebrospinal fluid rhinorrhea and empty sella. J Neurosurg 31:538–543
113. Brisman R, Hughes JEO, Holub DA (1972) Endocrine fonction in nineteen patients with empty sella syndrome. J Clin Endocrinol Metabol 34:570–573
114. Brismar K, Bajraktari X, Goulatia R, Efendic S (1978) The empty sella syndrome. Intrasellar cisternal herniation in "normal" patients and in patients with communicating hydrocephalus and intracranial tumors. Neuroradiology 17:35–43
115. Bruneton JN, Drouillard JP, Boedec R, Tavernier J (1978) Etude radiologique de la selle turcique normale. Intérêt de l'examen tomographique. Rev Fr Endocrinol Clin 19:539–548
116. Bruneton JN, Drouillard JP, Sabatier JC, Elie GP, Tavernier JE (1979) Normal variants of the sella turcica. Comparison of plain radiographs and tomograms in 200 cases. Radiology 131:99–104
117. Buchsbaum HW, Colton RP (1967) Anterior third ventricular cysts in infancy. Case report. J Neurosurg 26:264–266
118. Buckman MT, Husain M, Carlow TJ, Peake GT

(1976) Primary empty sella syndrome with visual field defects. Am J Med 61:124–128
119. Bugault R (1974) Les adénomes à prolactine. Bilan neuroradiologique. A propos de 75 cas sur 662 aménorrhées – galactorrhées isolées ou associées. Thèse Médecine, Paris
120. Bull J (1956) The normal variations in the position of the optic recess of the third ventricle. Acta Radiol 46:72–80
121. Burrows EH (1964) So-called J-sella. Br J Radiol 37:661–669
122. Burston J, John R, Spencer H (1962) "Myoblastoma" of the neurohypophysis. J Pathol Bacteriol 83:455–461
123. Burt AS, Cohen RB (1953) Pituitary changes in thyroid aplasia: possible significance. Report of case with autopsy. Lab Invest 2:357–367
124. Burt AS, Reiner L, Cohen RB, Sniffen RC (1954) Klinefelter's syndrome: report of an autopsy with particular reference to the histology and histochemistry of the endocrine glands. J Clin Endocrinol Metab 14:719–728
125. Busch G (1963) Fraktur der Sella Turcica. ROEFO 99:420–421
126. Busch W (1951) Die Morphologie der Sella Turcica und ihre Beziehungen zur Hypophyse. Virchows Arch [Pathol Anat] 320:437–458
127. Butt EM, Hall EM, Courville CB (1939) Progressive muscular dystrophy (dystrophia myotonica); report of case with autopsy. Bull Los Angeles Neurol Soc 4:58–68
128. Buurman R, Gundbach K, Schmidt-Hoberg W, Vogel H (1978) Die Dysplasia cleidocranialis. Zwei Einzelbeobachtungen. ROEFO 128:60–66
129. Cabanes J (1978) Asymptomatic persistence of infundibularis recessus. Case report. J Neurosurg 49:769–722
130. Caffey J (1972) Pediatric X-ray diagnosis, 6th edn. Year Book Medical Publishers, Chicago
131. Calenoff L (1962) Osteosclerosis from intentional ingestion of hydrofluoric acid. AJR 87:1112–1114
132. Calkins RA, Pribram HFW, Joynt RJ (1968) Intrasellar arachnoid diverticulum. A case report. Neurology (Minneap) 18:1037–1040
133. Camins MB, Mount LA (1974) Primary suprasellar atypical teratoma. Brain 97:447–456
134. Camp JD (1923) The normal and pathologic anatomy of the sella turcica as revealed at necropsy. Radiology 1:65–73
135. Camp JD (1924) Normal and pathological anatomy of sella turcica as revealed by roentgenograms. AJR 12:143–156
136. Camp JD (1931) Roentgenologic observations in meningiomas of olfactory groove and meningiomas arising from tuberculum sellae. Proc Staff Meet Mayo Clin 6:221–224
137. Camp JD (1949) Roentgenologic observations concerning erosion of the sella turcica. Radiology 53:666–674
138. Camp JD (1950) Significance of intracranial calcification in roentgenologic diagnosis of intracranial neoplasms. Radiology 55:659–668
139. Campbell JA (1971) Craniofacial anomalies. In: Newton TH, Potts DG (eds) Radiology of the skull and brain, vol 1, book 2. 571–633. Mosby, Saint-Louis
140. Campbell JA, Bull JWD (1973) Craniopharyngioma in a predominantly adult population. A radiological review. J Can Assoc Radiol 23:182–191
141. Canuyt G, Terracol J (1925) Le sinus sphenoidal – Anatomie et exploration, chirurgie, 1st vol. Masson, Paris
142. Caplan RH, Dobben GD (1969) Endocrine studies in patients with the "empty sella syndrome" Arch Intern Med 123:611–619
143. Cares HL, Bakay L (1971) Giant cell lesions of the skull. Acta Neurochir (Wien) 25:1–8
144. Carlson GO, Haverling M, Molin C (1973) Isolated fracture of the base of the skull within the sella region. Acta Radiol [Diagn] (Stockh) 14:662–666
145. Cartens M (1979) Die Selladiagnostik. ROEFO 71:257–272
146. Cashion EL, Young JM (1971) Intraventricular craniopharyngioma. Report of two cases. J Neurosurg 34:84–87
147. Castellano F, Ruggiero G (1953) Meningiomas of the posterior fossa. Acta Radiol [Suppl] (Stockh) 104:136–153
148. Castro L, de la Pava S, Webster JH (1969) Esthesioneuroblastomas. A report of 7 cases. AJR 105:7–13
149. Caughey JE (1952) Bone changes in the skull in dystrophia myotonica. J Bone Jt Surg [B] 34:343–351
150. Caughey JE (1961) Hypogonadism and pituitary tumors. NZ Med J 60:482–486
151. Cave AJE (1931) The craniopharyngeal canal in man and mammal. J Anat 65:363–367
152. Chambers AA, Lukin R, Tsunekawa N (1976) Calcification in a chromophobe adenoma. J Neurosurg 44:623–625
153. Check JH, Rakoff AE, Goldfarb AF, King LC (1977) Amenorrhea – galactorrhea associated with the empty sella syndrome. Am J Obstet Gynecol 128:688–689
154. Chereau O (1978) Les principaux aspects de la séméiologie radiologique de la selle turcique. Thèsis, Paris
155. Cherington M, Schneck SA (1966) Clivus meningiomas. Neurology (Minneap) 16:86–91
156. Chouraki JP, Calzada LD, Job JC (1978) Measurements of the sella turcica in idiopathic hypophyseal nanism. Arch Fr Pediatr 35:529–534
157. Christiaens JL, Cousin R, Dhellemmes P, Cecille JP, Asfour M (1976) Kyste colloïde intrasellaire. Neurochirurgie 22:649–651
158. Christiaens JL, Dupont A, Lefebvre J, Dobbelaere P, Laine E (1976) Adénome hypophysaire géant et calcifié. Neurochirurgie 22:645–647
159. Chutorian AM, Schwartz JF, Evans RA, Carter S (1964) Optic gliomas in children. Neurology (Minneap) 14:83–95
160. Chynn KY (1966) Neuroradiologic exploration in intra- and parasellar conditions. Radiol Clin North Am 4:93–116

161. Ciric I (1977) On the origin and nature of the pituitary gland capsule. J Neurosurg 46:596–600
162. Cloutier MD, Hayles AB, Sprague RG (1966) Pituitary tumor and Cushing's syndrome. Am J Dis Child 112:596–599
163. Cogan DG, Poppen JL, Hicks SP (1961) Ganglioneuroma of chiasm and optic nerves. Arch Ophthalmol 65:481–482
164. Cohen DN, Steinberg M, Buchwald R (1974) Suprasellar germinomas: diagnostic confusion with optic gliomas. Case report. J Neurosurg 41:490–493
165. Cohen GH, Dorfman S, Norwood C (1978) Tomographic diagnosis of pituitary microadenomas in Forbes-Albright syndrome (amenorrhea – galactorrhea). Am J Obstet Gynecol 130:822–824
166. Cohen ME, Rosenthal AD, Matson DD (1967) Neurological abnormalities in achondroplastic children. J Pediatr 7:367–376
167. Coin CG, Malkasian DR (1971) Clivus. In: Newton TH, Potts DG (eds) Radiology of the brain and skull, vol 1, book 1. Mosby, Saint-Louis
168. Cole M, Davies H (1963) Carotid siphon calcification. Br J Radiol 36:289–293
169. Concha S, Hamilton BPM, Millan JC, McQueen JD (1975) Symptomatic Rathke's cleft cyst with amyloid stroma. J Neurol Neurosurg Psychiatry 38:782–786
170. Congdon CC (1952) Benign and malignant chordomas. A clinical anatomical study of twenty cases. Am J Pathol 28:793–821
171. Congdon ED (1920) The distribution and mode of origin of septa and walls of the sphenoid sinus. Anat Rec 18:97–115
172. Contreras C, Copty M, Langelier R, Gagne F (1977) Traumatic suprasellar arachnoid cyst. Surg Neurol 8:196–198
173. Cooper PR, Ransohoff J (1972) Craniopharygioma originating in the sphenoid bone. J Neurosurg 36:102–106
174. Cope ZV (1917) The internal structure of the sphenoid sinus. J Anat Physiol 51:127–137
175. Corenblum B (1978) Bromocriptine in pituitary tumours. Lancet II:786
176. Corning HK (1925) Lehrbuch der Entwicklungsgeschichte des Menschen. Bergmann, München
177. Covell WP (1927) Quantitative study of hypophysis of human anencephalic fetus. Am J Pathol 3:17–28
178. Crastes G (1973) Contribution à l'étude anatomoradiologique de la selle turcique normale. Thèsis, Lyon
179. Crompton MR (1971) Hypothalamic lesions following closed head injury. Brain 94:165–175
180. Cronquist S, Furst E (1959) Pre- and post-therapeutical radiological changes in chromophobe adenoma of pituitary. A survey in connection with a clinical examination of 121 cases. Radiologe 4:182–189
181. Currarino G, Weinberg A (1974) Os supra petrosum of Meckel. AJR 121:139–142
182. Curtis J, Jimenez JA (1968) Intracranial aneurysm with panhypopituitarism with associated aneurysms of thoracic and abdominal aortas. JAMA 203:180–182
183. Cushing H (1912) The pituitary body and its disorders. Clinical states produced by disorders of the hypophysis cerebri. Lippincott, Philadelphia London, p 97
184. Cushing H (1929) The chiasmal syndrome of primary optic atrophy and bitemporal field defects in adult patients with a normal sella turcica. Trans XIII Concilium Ophtalmol Hollandia 3:97–184
185. Cushing H (1932) Basophilic adenomas of pituitary body, their clinical manifestations (pituitary basophilism). Bull Johns Hopkins Hosp 50:137–195
186. Cushing H (1932) Intracranial tumors. Thomas, Springfield, Ill
187. Cushing H, Eisenhardt L (1929) Meningiomas arising from the tuberculum sellae with the syndrome of primary optic atrophy and bitemporal field defects combined with a normal sella turcica in a middle aged person. Arch Ophthalmol 1:168–206
188. Cushing H, Eisenhardt L (1938) Meningiomas. Thomas, Springfield, III
189. Dahlin DC, McCarty CS (1952) Chordoma: a study of fifty-nine cases. Cancer 5:1170–1178
190. Dahlstrom R, Acers TE (1975) Chiasmatic arachnoiditis and empty sella. Ann Ophthalmol 4:73–76
191. Dan NG, Smith DE (1975) Pituitary hemangioblastoma in a patient with von Hippel-Lindau disease. J Neurosurg 42:232–235
192. Dandy WE (1933) Benign tumors of the third ventricle of the brain: diagnosis and treatment. Thomas, Springfield, Ill
193. Daniel PM, Prichard M, Treip CS (1959) Traumatic infarction of the anterior lobe of the pituitary gland. Lancet II:927–929
194. Danoff D, Serbu J, French LA (1966) Encephalocoele extending into the sphenoid sinus. Report of a case. J Neurosurg 24:684–686
195. Dany A, Delcour J, Laine E (1963) Les méningiomes du clivus. Neurochirurgie 9:249–277
196. Dany A, Tapie P, Pefferkorn JP, Bebear JP, Pradet JM, Claude M, Vallat JN (1973) Selle turcique vide ou selle turcique ouverte? – A propos de trois observations personnelles. Neurochirurgie 19:225–231
197. Danzinger J, Bloch S (1974) Suprasellar arachnoid pouches. Br J Radiol 47:448–451
198. Danzinger J, Allen JL, Bloch S (1974) Intracranial chordomas. Clin Radiol 25:309–316
199. Danzinger J, Wallace S, Handel S, Samaan NB (1979) The sella turcica in primary end organ failure. Radiology 131:111–115
200. Daughaday WH (1974) The adenohypophysis. In: Williams RH (ed) Textbook of endocrinology. Saunders, Philadelphia, pp 31–79
201. Davenport CB, Renfroe O (1940) Adolescent development of sella turcica and frontal sinus based on consecutive roentgenograms. AJR 44:665–679
202. David M, Constans JP (1957) Les chondromes intracrâniens. A propos de deux nouveaux cas de chondromes primitifs du crâne. Sem Hop Paris 33:1195–1214
203. David M, Dilenge D (1957) La selle turcique (radiologie normale et pathologique). Les monographies médicales et scientifiques. Heures de France. Paris

204. David M, Messimy R, Berd H, Pertuiset B (1960) Tumeur dite de la placode olfactive (esthesioneuroepithelioma) à development simultanément cerebral et nasal. Presse Med 68:1981-1984
205. David M, Bernard-Weil E, Pradat P (1962) Adénomes hypophysaires chez l'enfant. Neurochirurgie 8:49–59
206. Davidson KC (1966) Cranial and intracranial lesions in neurofibromatosis. AJR 98:550–556
207. Davis DO (1969) Neuroradiological diagnosis of sellar and parasellar lesions. Clin Neurosurg 17:160–165
208. Dawson BH (1958) The blood vessels of the human optic chiasma and their relation to those of the hypophysis and hypothalamus. Brain 81:207–217
209. Dawson BH, Kothandaram P (1972) Acute massive infarction of pituitary adenomas. A study of five patients. J Neurosurg 37:275–279
210. Dechaume J, Wertheimer P (1936) Les méningiomes rétro-sellaires. Sud Med Chir 68:1072–1091
211. Deck MDF (1973) Radiographic and radioisotopic techniques in diagnosis of pituitary tumors. In: Kohler PO, Ross GT (eds) Diagnosis and treatment of pituitary tumors. Excerpta Medica, Amsterdam, American Elsevier, New York
212. Decker RE, Mardayat M, Marc J, Rasool A (1979) Neurosarcoidosis with computerized tomographic visualization and transsphenoidal excision of a supra- and intra-sellar granuloma. Case report. J Neurosurg 50:814–816
213. Deery EM (1929) Note on calcification in pituitary adenomas. Endocrinology 13:455–458
214. De Gennes JL, Turpin G (1978) Adénomes corticotropes avec mélanodermie (syndrome de Nelson). Etude clinique et biologique de 2 cas avant et après nécrose spontanée intercurrente. Nouv Presse Med 7:1185–1189
215. Delaney P (1977) Neurologic manifestations in sarcoidosis. Review of the literature with a report of 23 cases. Ann Intern Med 87:336–345
216. Delarue J, Chomette G, Pinaudeau Y, Brocheriou C, Auriol M (1964) Les metastases hypophysaires. Frequence. Etude histopathologique. Arch Anat Pathol 8:179–182
217. Delorme G, Tessier JP (1978) Manuel d'électroradiologie, tome 1; Bases physiques et biologiques. Masson, Paris
218. Dennis JP, Rosenberg NS, Alvord ED (1961) Megalencephaly, internal hydrocephalus and other neurological aspects of achondroplasia. Brain 84:427–434
219. Destee A, Mizon JP, Froissart M, Gentit F, Deramon H, Remond A (1977) Sellar lithiasis. Nouv Presse Med 6:3118
220. Devic M, Goutelle A, Plauchu G, Bocquet B, Sindou M, Haguenauer J (1973) Les adénomes hypophysaires invasifs. Rev Otoneuroophtalmol 45:41–51
221. Devilliers HH (1956) Abscess formation in the pituitary fossa associated with a pituitary adenoma. J Neurosurg 13:208–210
222. Di Chiro G (1953) Cordomi intracranii. Studio radiografico. Radiologia 2:29–39
223. Di Chiro G (1960) The width (third dimension) of the sella turcica. AJR 84:26–37
224. Di Chiro G, Anderson WB (1965) The clivus. Clin Radiol 16:211–223
225. Di Chiro G, Caughey JE (1960) Skull changes in eighteen cases of dystrophia myotonica. Acta Radiol 54:22–28
226. Di Chiro G, Libow LS (1971) Carotid siphon calcification and cerebral blood flow in the healthy aged male. Radiology 99:103–107
227. Di Chiro G, Lindgren E (1952) Bone changes in cases of supra-sellar meningioma. Acta Radiol 38:133–138
228. Di Chiro G, Nelson KB (1962) The volume of the sella turcica. AJR 87:989–1008
228a. Dietemann JL, Bonneville JF, Lang J, Francke JP, Clarisse J The sellar spine (to be published)
229. Divitiis E, Cerillo A (1973) Adénome hypophysaire à développement intraorbitaire. Neurochirurgie 19:561–566
230. Dixon FW (1937) Comparative study of sphenoidal sinus: study of 1600 skulls. Ann Otol Rhinol Laryngol 46:687–698
231. Dixon FW (1959) The clinical significance of anatomical arrangement of paranasal sinuses. Ann Otol Rhinol Laryngol 67:736–741
232. Djindjian R, Merland JJ (1978) Super-selective arteriography of the external carotid artery. Springer, Berlin Heidelberg New York
233. Djindjian R, Theron J, Lamour Y, Vignaud J (1972) Les metastases intrasellaires. A propos d'un cas. Nouv Presse Med 1:2389–2391
234. Dluhy RG, Moore TJ, Williams GH (1978) Sella turcica enlargement and primary adrenal insufficiency. Ann Intern Med 89:513–514
235. Dodd GD, Jing BS (1977) Radiology of nose, paranasal sinuses and nasopharynx. Williams & Wilkins, Baltimore
236. Domingue JN, Wilson CB (1977) Pituitary abscesses. Report of seven cases and review of the litterature. J Neurosurg 46:601–608
237. Domingue JN, Wing DS, Wilson CB (1978) Coexisting pituitary adenomas and partially empty sella. J Neurosurg 48:23–28
238. Doron Y, Behar A, Beller AJ (1965) Granular cell "myoblastoma" of the neurohypophysis. J Neurosurg 2:95–99
239. Dowgjallo N (1929) Über die Nerven der harten Hirnhaut des Menschen und einiger Säuger. Z Gesamte Anat 89:453–466
240. Drager GA (1944) Nerve fibers within the craniopharyngeal canal. Anat Rec 88:235–241
241. Drake CG, Vanderlinden RG, Amacher AL (1968) Carotid-ophtalmic aneurysms. J Neurosurg 29:24–31
242. Dreyfus G, Fischold H, Zara M, Fissore A (1950) La tomographie frontale de la loge hypophysaire en endocrinologie. Presse Med 58:691–693
243. Dreyfus G, Fischold H, Zara M, Frank LJ (1950) Absence des sinus crâniens dans le myxoedeme. Ann Endocrinol (Paris) 11:423–426

244. Dublin AB, Poirier VC (1976) Fracture of sella turcica. AJR 127:969–972
245. Dubois PJ, Maroon JC, Heinz ER, Markarian PB, Hydovitz JD (1977) Spontaneous fracture of the dorsum sellae associated with post-irradiation empty sella syndrome. Surg Neurol 8:216–218
246. Dubois PJ, Orr DP, Hoy RJ, Herbert DL, Heintz ER (1979) Normal sellar variations in frontal tomograms. Radiology 131:105–110
247. du Boulay GH (1957) The radiological evidence of raised intra-cranial pressure in children. Br J Radiol 30:375–377
248. du Boulay GH (1965) Principles of x-ray diagnosis of the skull. Butterworth, London
249. du Boulay GH, El Gammal T (1966) Classification, clinical value, and mechanisms of sella turcica changes in raised intra-cranial pressure. Br J Radiol 39:422–442
250. du Boulay GH, Trickey SE (1962) Case reports: calcification in chromophobe adenoma. Br J Radiol 35:793–795
251. du Boulay GH, Trickey SE (1967) The choice of radiological investigation in the management of tumors around the sella. Clin Radiol 18:349–365
252. du Boulay GH, Trickey SE (1970) The sella in aqueduct stenosis and communicating hydrocephalus. Br J Radiol 43:319–326
253. Duchen LW (1966) Metastatic carcinoma in the pituitary gland and hypothalamus. J Pathol Bacteriol 91:347–355
254. Duffy WC (1920) Hypophyseal duct tumors; a report of three cases and a fourth case of cyst of Rathke's pouch. Ann Surg 72:537–540
255. Dufour M (1976) Exploration radiologique mòderne de la selle turcique. Cours de perfectionnement post-universitaire. Journ Natl Radiol
256. Dunn JM (1966) Anterior pituitary and adrenal absence in a live-born normocephalic infant. Am J Obstet Gynecol 96:893–894
257. Dupuis M (1974) Le canal craniopharyngien, aspects radiologiques. Thèse Médecine, Strasbourg
258. Dussault J, Plamondon C, Volpe R (1969) Aneurysms of the internal carotid artery simulating pituitary tumors. Can Med Assoc J 101:785–790
259. Dutou L, Colon J (1964) La largeur réelle de l'hypophyse normale chez l'adulte. Ann Radiol (Paris) 7:560–581
260. Duvernoy H (1958) Contribution à l'étude de la vascularisation de l'hypophyse. Thèse Médecine, Paris
261. Duvernoy H, Koritke JG (1960) Sur la vascularisation de l'hypophyse. J Med Lyon 54:478–488
262. Dyke CG (1941) Roentgen ray diagnosis of skull and intra-cranial contents. In: Golden R (ed) Diagnostic Roentgenology. T Nelson, New York
263. Echols DH (1945) Giant - cell tumors of the sphenoid bone; report of a case. J Neurosurg 2:16–20
264. Ehlers N, Malmros R (1973) The suprasellar meningioma. A review of the literature and presentation of a series of 31 cases. Acta Ophtalmol [Suppl] (Copenh) 121:1–74
265. Eller JL, Decker JT, Brittis AC (1968) Roentgen therapy for a giant cell tumor of the sphenoid bone. A case report. Clin Radiol 37:36–44
266. Engels EP (1958) The roentgen appearance of the carotid sulcus of the sphenoid bone. Acta Radiol [Diagn] (Stockh) 49:113–116
267. Engels EP (1958) Roentgenographic demonstration of a hypophyseal subarachnoid space AJR 80:1001–1004
268. Engels EP (1961) Basal skull fractures involving the sella turcica. Clin Radiol 12:177–178
269. Epstein BS (1951) Shortening of the posterior wall of the sella turcica caused by dilatation of the third ventricle or certain suprasellar tumors. AJR 65:49–55
270. Erdelyi J (1935) Die Röntgendiagnostik der Hypophysengeschwülste. ROEFO 51:125–147
271. Erdheim J (1903) Zur normalen und pathologischen Histologie der Glandula thyreoidea, parathyreoidea, und Hypophysis. Beitr Pathol Anat 33:158–236
272. Erdheim J, Stumme E (1909) Über die Schwangerschaftsveränderung der Hypophyse. Beitr Pathol Anat 46:1–122
273. Espinasse PG (1933) The development of the hypophysioportal system in man. J Anat 68:11–18
274. Esplan J (1963) Les chondromes intra-crâniens. Thèsis, Marseille
275. Etter LE (1961) New technique for accurately projecting the dorsum sellae into the foramen magnum. Radiology 77:418–425
276. Etter LE (1968) The paranasal sinuses: normal roentgen anatomy. Semin Roentgenol 3:123–132
277. Evans DC, Netzky MG, Allen VE, Kasantikul V (1979) Empty sella secondary to suprasellar colloid cyst of foregut (respiratory) origin. J Neurosurg 51:114–117
278. Evans RA, Schwartz JF, Chutorian AM (1963) Radiologic diagnosis in pediatric ophtalmology. Radiol Clin North Am 1:459–495
279. Fager CA, Carter H (1966) Intrasellar epithelial cysts. J Neurosurg 24:77–81
280. Faglia G, Ambrosi B, Beck-Peccoz P, Giovanelli M (1973) Disorders of growth hormone and corticotropin regulation in patients with empty sella. J Neurosurg 38:59–64
281. Fairburn B, Larkin IM (1964) A cyst of Rathke's cleft. J Neurosurg 21:223–225
282. Faivre J, Vallee B, Carsin M, Scarabin JM, Simon J (1978) Agenesis of the cervical and petrosal portions of the left internal carotid artery, with reconstitution of the cavernous portion through a transsphenoidal intercarotid anastomosis. J Neuroradiol 5:133–138
283. Falconer MA, Bailey IC, Duchen LW (1968) Surgical treatment of chordoma and chondroma of the skull base. J Neurosurg 29:261–275
284. Farinas PL (1939) Value of X-ray examination of the sella turcica in the sagittal positions. Radiology 32:411–415
285. Fassbender CW, Häussler G, Stössel HG (1961) Schädelbasis Chondrome mit intrakranieller Ausdehnung. ROEFO 94:718–723
286. Fathie K (1965) Large aneurysm of internal carotid

artery simulating an eosinophilic adenoma. J Iowa Med Soc 55:246–249
287. Faure J, Binnert D, Michotey P, Salamon G (1972) Etude radio-anatomique des branches collatérales de l'artère carotide interne (dans son segment intracaverneux). Neurochirurgie 18:561–579
288. Fawcett E (1909) Notes on the development of the human sphenoid. J Anat 44:207–222
289. Ferner H (1960) Die Hypophysenzisterne des Menschen und ihre Beziehung zum Entstehungsmechanismus der sekundären Sellaerweiterung. Z Anat Entwicklungsgesch 121:407–416
290. Ferrier PE, Stone EF (1969) Familial pituitary dwarfisms associated with an abnormal sella turcica. Pediatrics 43:858–865
291. Finn JE, Mount LA (1974) Meningiomas of the tuberculum sellae and planum sphenoidale. Arch Ophthalmol 92:23–27
292. Firooznia H, Pinto RS, Lin JP, Baruch HH, Zausner J (1976) Chordoma: radiological evaluation of 20 cases. AJR 127:797–805
293. Fisher CM, Gore I, Okabe N, White PD (1965) Calcification of the carotid siphon. Circulation 32:538–548
294. Fisher MS (1978) An unusual bone change in acromegaly. Skel Radiol 3:177–178
295. Fisher RL, Di Chiro GH (1964) The small sella turcica. AJR 91:996–1008
296. Fleischer AS, Reagan TH, Ransohoff J (1972) Primary carcinoma of the pituitary with metastasis to the brainstem. J Neurosurg 36:781–784
297. Floyd HL, Pribram HFW, Velo AG (1970) Primary cerebrospinal fluid fistula. AJR 110:88–91
298. Foley KM, Posner JB (1975) Does pseudotumor cerebri cause the empty sella syndrome? Neurology (Minneap) 25:565–569
299. Forbes AF, Henneman PH, Griswald GC, Albright F (1954) Syndrome characterized by galactorrhea, amenorrhea, and low urinary F.S.H.; Comparison with acromegaly and normal lactation. J Clin Endocrinol Metab 14:265–271
300. Forster HW, Bouzarth WF (1967) Meningioma of the tuberculum sellae as a cause of optic atrophy. Am J Ophtalmol 64:908–910
301. Fortier-Beaulieu M, Vallee G, Michel JR (1960) Radiologie de la selle turcique. Fr Med 2:91–101
302. Fortuna A, Gambacorta D (1971) Cylindroma in the region of the Gasserian ganglion. J Neurosurg 34:427–431
303. Fossati P, Lefebvre J, Asfour M, Granier F, Fourlinnie JC, L'Hermite M, Golstein J (1977) The sella turcica and pituitary function in congenital hypothyroidism due to thyroid ectopia or hypogenesis. Nouv Presse Med 6:4177–4182
304. Fournier AM, Denizet D (1965) La selle turcique en omega. Marseille Med 102:503–509
305. Fowler FD, Matson DD (1957) Gliomas of optic pathways in childhood. J Neurosurg 14:515–528
306. Francis CC (1948) Growth of the human pituitary fossa. Hum Biol 20:1–20
307. Francis RS (1975) Lipoid proteinosis. A case report. Radiology 117:301–302
308. Frazer JE (1940) A manual of embryology. The development of the human body. Baillière, Tindall & Cox, London
309. Frazier CH (1930) Series of pituitary pictures. Commentaries on pathologic, clinical and therapeutic aspects. Arch Neurol Psychiatry 23:656–695
310. Frazier CH, Alpers BJ (1931) Adamantinoma of the craniopharyngeal duct. Arch Neurol Psychiatry 26:905–965
311. Frazier CH, Alpers BJ (1934) Tumors of Rathke's cleft. Arch Neurol Psychiatry 32:973–984
312. Fredy D, Missir O, Levesque M, Bories J (1973) La selle turcique normale en radiologie osseuse standard et tomographique. Feuill Electroradiol 75:1–17
313. Freundlich IM, Holes PJ (1963) A nasopharyngeal fibroma presenting with pituitary and optic nerve symptomatology. AJR 89:41–44
314. Friedmann G, Marjuth F (1961) Intrasellaere Liquorzystem. Zentralbl Neurochir 21:33–41
315. Fry KI, du Boulay GH (1965) Some observations on the sella in old age and arterial hypertension. Br J Radiol 38:16–22
316. Fujioka M, Young LW (1978) The sphenoidal sinuses: Radiographic patterns of normal development and abnormal findings in infants and children. Radiology 129:133–136
317. Gaal A (1935) Die diagnostische Bedeutung der an der Sellaseitenwand vorkommenden Karotis Impressionen. ROEFO 52:156–162
318. Gabriele OF (1960) Mucormycosis. AJR 83:227–235
319. Gabriele OF (1968) The empty sella syndrome. AJR 104:168–170
320. Gabrielsen TO, Kingman AF Jr (1964) Osteocartilaginous tumors of the base of the skull. Report of a unique case and review of the literature. AJR 91:1016–1063
321. Gammal TE, Allen MB (1972) The intrasellar subarachnoid recess. Some clinical and radiologic observations. Acta Radiol [Diagn] (Stockh) 13:401–412
322. Gammal TE, Allen MB (1972) Further consideration of sellar changes associated with increased intracranial pressure. Br J Radiol 45:561–569
323. Ganguly A, Stanchfield JB, Roberts TS, West CD, Tyler FH (1976) Cushing's syndrome in a patient with an empty sella turcica and a microadenoma of the adenohypophysis. Am J Med 60:306–309
324. Gardeur D, Nachanakian A, Millard JC, Metzger J, Poisson M, Mashaly R (1978) Neuroradiological aspects and therapeutic incidence of a case of posterior pituitary tumor (choristoma). J Neuroradiol 5:321–326
325. Geehr RB, Allen WE, Rothman SLG, Spencer DD (1978) Pluridirectionnal tomography in the evaluation of pituitary tumors. AJR 130:105–109
326. Gefferth K (1937) Über das Sellaröntgenbild der Frühgeburten. Arch Kinderheilkd 111:87–93
327. Geissinger JD, Siqueira EB, Ross ER (1970) Giant cell tumors of the sphenoid bone. J Neurosurg 32:665–670
328. Gemzell C (1975) Induction of ovulation in infertile

women with pituitary tumors. Am J Obstet Gynecol 121:311–315
329. Geraud J, Lazorthes G, Roulleau J (1955) Les crâniopharyngiomes. Diagnostic radiologique. Rev Otoneuroophtalmol 27:106–120
330. Ghatak NR, Hirano A, Zimmermann HM (1969) Intrasellar germinomas: a form of ectopic pinealoma. J Neurosurg 31:670–675
331. Gibbered FB, Miller TN, Morgan AD (1973) Glioblastoma of the optic chiasm. Br J Ophtalmol 57:788–791
332. Gifford RD, Goree JA, Jimenez JP (1971) Tumor bulge into the sphenoid sinus. A roentgen sign of parasellar meningioma. AJR 112:324–328
333. Glancy JJ (1967) Some radiological aspects of acromegaly. Aust Radiol 11:226–233
334. Glasauer FE, Tandon PN (1969) Trigeminal neurinoma in adolescents. J Neurol Neurosurg Psychiatry 32:562–568
335. Glass SJ, Davis S (1944) Granuloma of the pituitary associated with panhypopituitarism. J Clin Endocrinol Metab 4:489–491
336. Glasser SP, Earll JM (1968) Pituitary "stone" an unusual calcification. JAMA 203:129
337. Glazer N, Hauser H, Slade H (1956) Granular-cell tumor of the neurohypophysis. AJR 76:324–326
338. Goldberg GM, Eshbaugh DE (1960) Squamous cell nests of the pituitary gland as related to the origin of craniopharyngiomas. A study of their presence in the newborn and infants up to age four. Arch Pathol 70:293–298
339. Goldberg MB, Sheline GE, Malamud N (1963) Malignant intracranial neoplams following radiation therapy for acromegaly. Radiology 80:465–470
340. Gordy PD (1965) Neurinoma of the gasserian ganglion. Report of a case and review of the literature. J Neurosurg 22:90–94
341. Gouaze A, Salles M, Rouzaud M, Kiffer A, Jezegabel C (1967) Anévrysmes géants des siphons carotidiens et acromégalie. Rev. Otoneuroophtalmol 29:200–202
342. Gouaze A, Salles M, Santini JJ (1969) Les lésions anévrysmales traumatiques du siphon carotidien par fracture de la base. Neurochirurgie 15:234–240
343. Green ML, Childrey JH (1939) Intracranial chondroma. Case report. J Nerv Ment Dis 22:350–356
344. Greenhouse AH (1974) Pituitary sarcoma: a possible consequence of radiation. JAMA 190:269–273
345. Griscom NT, Oh KS (1970) The contracting skull: inward growth of the inner table as a physiologic response to the diminution of intracranial content in children. AJR 110:106–110
346. Grisoli F, Leclercq TA, Robert F, Hardy J (1974) Les métastases hypophysaires dans le cancer du sein. Neurochirurgie 20:583–592
347. Grisoli F, Guibout M, Jaquet P, Michotey P, Mouly A, Lucas C (1978) Les adénomes à prolactine péripubertaires. 8 observations. Nouv Presse Med 7:1819–1825
348. Grisoli F, Leclercq TA, Guibout M, Jaquet P, Heim M, Hassoun J, Lissitzky JC, Farnarier P (1979) Maladie de Cushing et tumeurs de l'hypophyse. A propos de 8 observations. Neurochirurgie 25:3–10
349. Grossman CB (1975) Dynamic roentgenographic changes in the empty sella syndrome. Radiology 116:341–344
350. Grossman H, Winchester PH (1971) Mucopolysaccharidoses. In: Newton TH, Potts DG (eds) Radiology of the skull and brain, 1st vol. Mosby, Saint-Louis, 685–690
351. Grover WD, Rorke LB (1968) Invasive craniopharyngioma. J Neurol Neurosurg Psychiatry 31:580–582
352. Grundy L, Goree JA, Jimenez JP (1970) Oxycephaly in the adult simulating pituitary tumor. Clinical and roentgenologic manifestations. AJR 108:762–766
353. Grzybowski J (1932) L'innervation de la dure-mère crânienne chez l'homme. Arch Anat Histol Embryol 14:387–428
354. Guibout M, Jaquet P, Grisoli F, Louvet JP, Combes R, Angeletti S, Mouly A (1979) Microadénomes hypophysaires lors de la maladie de Cushing. Evolution après exérèse transphenoidale. 4 observations. Nouv Presse Med 8:2959–2963
355. Guillaume J, Garrigues P, Lazorthes Y, Rouleau J, Manelfe C (1977) Selle turcique vide primitive. A propos de 8 observations et revue de la littérature. J Radiol Electrol 58:5–12
356. Guinet P, Tourniaire J (1974) L'adénome hypophysaire à cellules thyréotropes. Rev. Otoneuroophtalmol 47:253–264
357. Guinet P, Tourniaire J, Orgiazzi J, Robert M (1972) Myxoedème périphérique: aménorrhée – galactorrhée, image radiologique d'adénome hypophysaire; hyperthyreostimulie freinable. Ann Endocrinol (Paris) 33:376–389
358. Guiot G (1958) Adénomes hypophysaires. Masson Paris
359. Guiot G, Oproiu A, Hertzog E, Fredy D (1969) Adénomes hypophysaires. Encycl Med Chir 17340 A10, Paris
360. Guiot G, Olson D, Hertzog E (1971) Kystes arachnoidiens intrasellaires. Neurochirurgie 17:539–547
361. Haas L (1925) Erfahrungen auf dem Gebiet der radiologischen Sella Diagnostik. ROEFO 33:469–495
362. Haas L (1927) Verfahren zur sagittalen Aufnahme der Sellagegend. ROEFO 36:1198–1204
363. Haas L (1934) Einzelheiten aus der Röntgendiagnostik der Sella Turcica. ROEFO 50:468–469
364. Haas L (1935) Einzelheiten aus der Röntgendiagnostik der Sella Turcica. ROEFO 51:147–152
365. Haas L (1935) Einzelheiten aus der Röntgendiagnostik der Sella Turcica. Mitteilung. Mittlere Klinoidfortsätze und Foramina carotico-clinoidea. ROEFO 52:186–187
366. Haas L (1935) Einzelheiten aus der Röntgendiagnostik der Sella Turcica: Über eine Spalte der Tuberkulumgegend. ROEFO 52:188–189
367. Haas L (1937) Über die Entstehung der Sellavergrößerungen extrasellaren Ursprungs. ROEFO 55:458–464
368. Haas LL (1954) The size of the sella turcica by age and sex. AJR 72:754–761

369. Haberfeld W (1909) Die Rachendachhypophyse, andere Hypophysengangreste und deren Bedeutung für die Pathologie. Ziegler Beitr Pathol Anat 46:133–240
370. Hadden DR, Alexander F (1968) Adult cretinism with lingual thyroid and pituitary fossa enlargement. Proc R Soc Med 61:655–656
371. Hägerstrand J, Schönebeck J (1969) Metastases to the pituitary body. Acta Pathol Microbiol Scand 75:64–70
372. Hakuba A, Katsuyama J, Matsuoka Y, Shim JH, Nishimura S (1975) Sphenoid sinus mucoceles. Report of two cases. J Neurosurg 43:368–373
373. Hamberger CA, Hammer G, Norlen G (1961) Transantrosphenoidal hypophysectomy. Arch Otolaryngol 74:2–8
374. Hamby WB (1934) Spongioblastoma bipolare in region of hypothalamus associated with infantilism and without dwarfism. Arch Neurol Psychiatry 31:1258–1265
375. Hamilton AE, Rubinstein LJ, Polle GJ (1973) Primary intracranial esthesioneuroblastoma (olfactory neuroblastoma). J Neurosurg 38:548–556
376. Hammer G, Rådberg C (1958) Anatomisch röntgenologische Studien über die Verhältnisse zwischen Sella Turcica und Sinus Sphenoidalis. Arch Ohren Nasen Kehlkopfheilkd 173:278–281
377. Hammer G, Rådberg C (1961) The sphenoidal sinus. An anatomical and roentgenologic study with reference to transphenoidal hypophysectomy. Acta Radiol [Diagn] (Stockh) 56:401–422
378. Hammer G, Rådberg C, Röckert H (1964) The sella turcica after transsphenoidal hypophysectomy. A radiological and X-ray microscopical study. Acta Otolaryngol [Suppl] (Stockh) 188:119–124
379. Hanafee W, Bilodeau L, Alberti J, Wilson G (1967) Pseudo suprasellar tumors. AJR 100:631–638
380. Hankinson J, Banna M (1976) Pituitary and parapituitary tumors. Saunders, London
381. Hans F, Ethyl S (1967) Chordoma of the clivus. AJR 100:639–649
382. Hardy J (1971) Transsphenoidal hypophysectomy. J Neurosurg 34:582–594
383. Hardy J (1973) Transsphenoidal surgery of hypersecreting pituitary tumors. In: Kohler PO, Ross GT (eds) Diagnosis and treatment of pituitary tumors. Excerpta Medica Amsterdam American Elsevier New York
384. Hardy J (1974) Les microadénomes hypophysaires sécrétants. Rev Otoneuroophtalmol 46:347–354
385. Hardy J (1979) Microsurgical exploration of a normal sella turcica for a microadenoma. 2nd European Workshop on pituitary adenomas, September 20–22, Paris
386. Hardy J, Robert F (1969) Un méningiome de la selle turcique, variété sous-diaphragmatique. Exérèse par voie transsphénoidale. Neurochirurgie 15:535–544
387. Hardy J, Bertrand C, Maltais R, Robert F, Thierry A (1966) Volumineux chondro-sarcome calcifié de la région sellaire. Exérèse transsphénoidale sous contrôle radioscopique télévisé. Neurochirurgie 12:491–502
388. Hardy J, Robert F, Somma M, Vezina JL (1973) Acromégalie-gigantisme. Traitement chirurgical par exérèse transsphénoidale de l'adénome hypophysaire. Neurochirurgie [Suppl 2] 19:35–56
389. Hardy J, Grisoli F, Leclercq TA, Somma M (1975) Transsphenoidal surgical hypophysectomy in advanced cancer of the breast (experience of 160 cases). Nouv Presse Med 4:2387–2390
390. Hare HF, Silveus E, Smedal MI (1941) Roentgenologic diagnosis of pituitary tumors. Radiology 52:193–198
391. Harispe L, Creissard P, Martinez G, Metzger J (1970) Une artère communicante anormale unissant les deux artères carotides internes intracrâniennes. J Radiol Electrol Med Nucl 51:759–762
392. Harland WA (1953) Granular-cell myoblastoma of hypophyseal stalk. Cancer 6:1134–1138
393. Harlin RS, Givens JR (1968) Sheehan's syndrome associated with eclampsia and a small sella turcica. South Med J 61:909–911
394. Harris FS, Rhoton AL Jr (1976) Anatomy of the cavernous sinus. A microsurgical study. J Neurosurg 45:169–180
395. Harris GW, Donovan BT (1966) The pituitary gland. Butterworths, London
396. Harris RD, Seljeskog EL, Murray KJ, Chou SN, Cunningham WP, Douglas SD (1978) Surface topography of normal and neoplastic human anterior pituitary cells maintained in vitro. J Neurosurg 49:169–178
397. Harrison CR, Luttrell C (1953) Persistent carotid basilar anastomosis – Three arteriographically demonstrated cases with one anatomical specimen. J Neurosurg 10:205–215
398. Harrison DFN (1961) A case of primary chordoma of the sphenoidal sinus. J Laryngol Otol 75:429–432
399. Hartmann E, Gilles E (1959) Roentgenologic diagnosis in ophthalmology. Lippincott, Philadelphia
400. Harwood-Nash DC (1970) Axial tomography of the optic canals in children. Radiology 96:367–374
401. Harwood-Nash DC, Fitz CR (1976) Neuroradiology in infants and children. Mosby, St Louis
402. Hass GM (1934) Chordomas of the cranium and cervical portion of the spine. Arch Neurol Psychiatry 32:300–327
403. Hassler O (1967) Calcifications in the intracranial arachnoid. A microradiological and histological study of the occurrence and appearance of calcifications, with special reference to the trigeminal nerve. J Neurosurg 27:336–345
404. Hasso NA, Bentson JR, Wilson GH, Vignaud J (1975) Neuroradiology of the sphenoidal region. Radiology 114:619–627
405. Hastrup J (1966) Chromophobe adenoma of the pituitary with extensive calcifications. Acta Neuropathol (Berl) 6:98–100
406. Hatam A, Greitz T (1972) Ectasia of cerebral arteries in acromegaly. Acta Radiol [Diagn] (Stockh) 12:410–419
407. Hempel KJ (1970) Zur Pathologie der Sellarregion. Radiologe 10:425–429

408. Herishanu Y (1970) Craniopharygioma in aged patients. J Am Geriatr Soc 18:261–263
409. Herlant M, Linquette M, Laine E, Fossati P, May JP, Lefebvre J (1966) Adenome hypophysaire à cellules thyréotropes. Ann Endocrinol (Paris) 27:181–189
410. Herschberg AD, Creff A (1956) Modifications radiologiques de la selle turcique au cours du syndrome hyperhormonal de la ménopause et de la castration. Sem Hop Paris 32:499–503
411. Hertz H, Rosendal T (1956) Roentgen changes in the cranium in 153 intracranial tumors in children aged 0–15 years. Acta Radiol [Diagn] [Suppl] (Stockh) 141:1–54
412. Hertzog E, Bamberger C, Guiot G (1970) Les aspects neuroradiologiques des adénomes hypophysaires. Ann Radiol (Paris) 13:765–783
413. Hervet E, Barrat J, Pigne A, Darbois Y, Faguer C (1975) Adénome à prolactine; hypophysectomie pendant la grossesse. Nouv Presse Med 4:2393–2395
414. Heuer GJ, Dandy WE (1916) Roentgenography in the localization of brain tumor. Bull Johns Hopkins Hosp 27:311–322
415. Higinbotham NL, Philipps RF, Farr HW, Hustu HO (1967) Chordoma: thirty-five year study at memorial hospital. Cancer 20:1841–1850
416. Hilton GF, Hoyt WH (1966) An arteriosclerotic chiasmal syndrome. JAMA 196:1018–1020
417. Hinck VC, Hopkins CE (1965) Concerning growth of sphenoid sinus. Arch Otolaryngol 82:62–66
418. Hingorani RK (1970) Nasopharyngeal chordoma. J Laryngol Otol 84:1281–1286
419. Hirsh O (1910) Endonasal method of removal of hypophyseal tumors with a report of two cases. JAMA 55:772–774
420. Hirsch O (1958) Pathology of the sphenoidal sinus. Arch Otolaryngol 67:85–92
421. Hirschfeld A (1951) Une tumeur cartilagineuse de la base du crâne (enchondrome). CR Soc Biol (Paris) 3:94
421a. Hirst LW, Miller NR, Allen ES (1979) Sphenoidal pneumosinus dilatans with bilateral optic nerve meningiomas. Case report. J Neurosurg 51:402–407
422. Hitchon PW, Graf CJ, McCabe BF, Namba K (1979) Trigeminal neurilemmona presenting as a sphenoid sinus mass. Surg Neurol 11:305–307
423. Hodgson SF, Randall RV, Holman CB, McCarty CS (1972) Empty sella syndrome. Med Clin North Am 56:897–898
424. Hoff JT, Patterson RH (1972) Crâniopharyngiomas in children and adults. J Neurosurg 36:299–302
425. Holman CB (1959) Roentgenologic manifestations of glioma of optic nerve and chiasm. AJR 82:462–471
426. Holman CB, Miller WE (1965) Juvenile nasopharyngeal fibroma – roentgenologic characteristics. AJR 94:292–298
427. Holman CB, Olive I, Svien HJ (1961) Roentgenologic features of neurofibromas involving the gasserian ganglion. AJR 86:148–153
428. Holmes GW, Robbins LL (1955) Roentgen interpretation, 6th edn. Lea & Febiger, Philadelphia
429. Holstein J, Herlbauer R (1973) Normvarianten der Sella Turcica: Große, freie, mittlere Klinoidfortsätze. ROEFO 119:760–761
430. Holt E (1921) Absence of the pars buccalis of the hypophysis in a 40 mm pig. Anat Rec 22:207–215
431. Hooper AC (1971) Sphenoidal defects. A possible cause of cerebrospinal fluid rhinorrhea. J Neurol Neurosurg Psychiatry 34:739–742
432. Hora JF (1965) Primary aspergillosis of the paranasal sinuses and associated areas. Laryngoscope 75:768–773
433. Horrigan WD, Baker DH (1961) Gargoylism: a review of the roentgen skull changes with description of a new finding. AJR 86:473–477
434. Horwitz T (1941) Chordal ectopia and its possible relationship to chordoma. Arch Pathol 31:354–362
435. Hotton F, Kleiner S, Benchekroun S, L'Hermitte M (1976) Etude radiologique de la selle turcique dans les aménorrhées. J Belge Radiol 59:335–340
436. Houdart R, Thurel C, Rey A, Chai N (1975) Artériographic dans les tumeurs de l'hypophyse. Anévrysme pseudotumoral. Neurochirurgie 21:163–168
437. Hovelacque A (1927) Anatomie des nerfs crâniens et rachidiens et du système grand sympathique chez l'homme. Doin, Paris
438. Hoyt WF, Baghdassarian SA (1969) Optic glioma of childhood. Natural history and rational conservative management. Br J Ophtalmol 53:793–798
439. Hoyt WF, Meshel LG, Lessell S, Schatz N, Suckling R (1973) Malignant optic glioma of adulthood. Brain 96:121–130
440. Hrdlicka A (1898) Dimensions of normal pituitary fossa or sella turcica in the white and the negro races. An anatomical study of fifty-seven normal skulls of white and sixteen normal skulls of colored individuals. Arch Neurol Psychopathol 1:679–698
441. Hsu T, Shapiro JR, Tyson JE, Leddy AL, Paz-Guevera AT (1976) Hyperprolactinemia associated with empty sella syndrome. JAMA 235:2002–2004
442. Hubble D (1963) Precocious menstruation in a mongoloid child with hypothyroidism. Hormonal overlap. J Clin Endocrinol Metab 23:1302–1305
443. Hubble D (1967) Diagnosis of hypopituitarism in childhood. Arch Dis Child 42:228–231
444. Huber P, Rivoir R (1973) The influence of intraventricular pressure on the size and shape of the anterior part of the third ventricle. Neuroradiology 5:33–36
445. Hullay J (1965) Planum (jugum) sphenoidale meningioma. Acta Neurochir (Wien) 12:717–747
446. Iancu TC, Almagor G, Friedman E, Hardoff R, Front D (1978) Chronic familial hyperphosphatasemia. Radiology 129:669–676
447. Ingraham FD, Scott HW (1946) Craniopharyngiomas in children. J Pediatr 29:95–116
448. Inoue K, Sugiyama Y, Ishii R (1978) Intrasellar fibromyxochondroma. Surg Neurol 10:167–170
449. Irwin GL (1960) Roentgen determination of the time of closure of the sphenooccipital synchondrosis. Radiology 75:450–453
450. Isherwood I, Mawdsley C (1963) Familial aspects of radiologic changes in dystrophia myotonica. Acta Radiol [Diagn] (Stockh) 1:110–117

451. Israel H (1970) Continuing growth in sella turcica with age. AJR 108:516–527
452. Jackson A, Hood TR (1958) Sarcoidosis with involvement of pituitary gland. Ann Intern Med 49:467–471
453. Jackson IMD, Hall R (1970) Pituitary enlargement resulting from primary thyroid disease. Proc R Soc Med 63:578
454. Jacobson HG, Lubetsky HW, Shapiro JH, Carton CA (1959) Intracranial meningiomas. A roentgen study of 126 cases. Radiology 72:356–367
455. Jakubowski J, Kendall B (1978) Coincidental aneurysms with tumors of pituitary origin. J Neurol Neurosurg Psychiatry 41:972–979
456. Jakumeit HD (1971) Neuroblastoma of the olfactory nerve. Acta Neurochir (Wien) 25:99–108
457. Jakumeit HD, Zimmermann V, Guiot G (1974) Intrasellar gangliocytomas. J Neurosurg 40:626–630
458. James AE, McMillan AS Sen, McMillan AS Jr, Momose KJ (1968) Thornwaldt's cyst. Br J Radiol 41:902–904
459. Janicki PC, Limbacher JP, Guinto FC Jr (1979) Agenesis of the internal carotid artery with a primitive transsellar communicating artery. AJR 132:130–132
460. Janny P, Heldt N, Flori B (1974) Les tumeurs à cellules granuleuses de la neuro-hypophyse. A propos d'une observation. Rev Otoneuroophtalmol 46:321–329
461. Jazra S, Berndt V (1978) Das posttraumatische Diabetes-insipidus-syndrom. Dtsch Med Wochenschr 103:114–120
462. Jefferson A (1955) Hypertensive cerebral vascular disease and intracranial tumor; a study in diagnosis. Q J Med 24:245–268
463. Jefferson G (1937) Compression of the chiasm, optic nerves, and optic tracts by intracranial aneurysms. Brain 60:444–497
464. Jefferson G (1940) Extrasellar extensions of pituitary adenomas; president's address. Proc R Soc Med 33:433–458
465. Jelsma F, Ross PJ (1966) Traumatic intracranial arachnoidal cyst involving the Gasserian ganglion. Case report. J Neurosurg 25:439–441
466. Jezegabel C (1960) Manifestations oculaires du pneumosinus dilatans. Etude clinique et étiopathogénique. Arch Ophtalmol 20:28–47
467. Job JC, Nahum M, Faure C, Rosier A (1965) Cleidocranial dysostosis: its polymorphism. Arch Fr Pediatr 22:669–672
468. Jones HM (1973) Cartilaginous tumors of the head and neck. J Laryngol Otol 87:135–151
469. Jones JR, Facog C, Kemmann E (1976) Sella turcica abnormalities in an anovulatory population. Obstet Gynecol 48:76–78
470. Jordan RM, Kendall JW, Kerber CW (1977) The primary empty sella syndrome. Analysis of the clinical characteristics, radiographic features, pituitary function and cerebrospinal fluid adenohypophysial hormone concentrations. Am J Med. 62:569–580
471. Jungmayer L (1951) Myotonische Dystrophie und Sellaveränderung. Klin Wochenschr 29:205–208
472. Kageyama N (1971) Ectopic pinealoma in the region of the optic chiasma. Report of five cases. J Neurosurg 35:755–759
473. Kajtar T, Tomkin GH (1971) Emergency hypophysectomy in pregnancy after induction of ovulation. Br Med J 4:88–90
474. Kamrin RP, Potanos JN, Pool JL (1964) An evaluation of the diagnosis and treatment of chordoma. J Neurol Neurosurg Psychiatry 27:157–165
475. Kane AM, Lore J Jr (1975) Meningoencephalocele of the paranasal sinuses. Laryngoscope 85:2087–2091
476. Kaplan HA, Brocuder J, Krieger AJ (1976) Intercavernous connections of the cavernous sinuses. The superior and inferior circular sinuses. J Neurosurg 45:166–168
477. Kaufman B (1968) The "empty" sella turcica – manifestation of the intrasellar subarachnoid space. Radiology 90:931–941
478. Kaufman B, Chamberlain WB Jr (1972) The ubiquitous "empty" sella turcica. Acta Radiol [Diagn] (Stockh) 13:413–425
479. Kaufman B, Sandstrom PH, Young HF (1970) Alteration in size and configuration of the sella turcica as the result of prolonged cerebrospinal fluid shunting. Radiology 97:537–542
480. Kaufman B, Pearson OH, Chamberlain WB (1973) Radiographic features of intrasellar masses and progressive asymmetrical non-tumorous enlargements of the sella turcica, the "empty sella". In: Kohler PO, Ross GT (eds) Diagnosis and treatment of pituitary tumors. Excerpta Medica, Amsterdam, American Elsevier, New York
481. Kaufman B, Nulsen FE, Weiss MH, Brodkey JS, White RJ, Sykora GF (1977) Acquired spontaneous, non traumatic normal pressure cerebrospinal fluid fistulas originating from the middle fossa. Radiology 122:379–387
482. Kaufman HH (1969) Nontraumatic cerebrospinal fluid rhinorrhea. Arch Neurol 21:59–65
483. Kay S, Lees JK, Stout AP (1950) Pituitary chromophobe tumors of nasal cavity. Cancer 3:695–704
484. Keats TE (1967) Cleidocranial dysostosis; some atypical roentgen manifestations. AJR 100:71–74
485. Keats TE, Burns TW (1964) The radiographic manifestations of gonadal dysgenesis. Radiol Clin North Am 2:297–313
486. Keibel F, Mall FP (1910) Handbuch der Entwicklungsgeschichte des Menschen, part 1. p 431
487. Keller N, Galatius-Jensen F (1975) Enlarged sella turcica associated with a rare vascular anomaly. Br J Radiol 48:936–937
488. Kelly LW Jr (1963) Ovarian dwarfism with pituitary tumor. J Clin Endocrinol Metab 23:50–53
489. Kendall BE, Lee BCP (1977) Cranial chordomas. Br J Radiol 50:687–698
490. Kenter EJW (1970) Posterior cavernous syndrome caused by pachymeningitis due to Besnier – Boeck – Schaumann's disease (sarcoidosis). Psychiatr Neurol Neurochir 73:135–144

491. Kepes J (1978) Transitional cell tumor of the pituitary gland developing from a Rathke's cleft cyst. Cancer 41:337–343
492. Keravel Y (1973) Contribution au diagnostic et au traitement chirurgical de la selle turcique "vide". A propos de 25 observations. Thèsis Médicine, Paris Créteil
493. Kerisit J (1967) L'atteinte néoplasique de l'hypophyse au cours des processus cancéreux. Mémoire, CES d'Anatomie Pathologique, Strasbourg
494. Key A, Retzius G (1875) Studien in der Anatomie des Nervensystems und des Bindegewebes. Erste Hälfte. Norstedt & Söner, Stockholm
495. Kier EL (1968) The infantile sella turcica. New roentgenologic and anatomic concepts based on a developmental study of the sphenoid bone. AJR 102:747–767
496. Kier EL (1969) "J" and "omega" shape of sella turcica; anatomic clarification of radiologic misconceptions. Acta Radiol [Diagn] (Stockh) 9:91–94
497. Kier EL (1971) The fetal skull. In: Newton TH, Potts DG (eds) Radiology of the skull and brain, vol 1, book 1. 99–117. Mosby, Saint-Louis
498. King LS, Butcher J (1944) Osteochondroma of base of skull. Arch Pathol 37:282–285
499. Kinnman J (1977) Surgical aspects of the anatomy of the sphenoidal sinuses and the sella turcica. J Anat 124:541–553
500. Kino M (1977) Significance of frontal imaging of the sella turcica. Jpn J Clin Radiol 22:361–364
501. Kirkwood JR, Margolis MT, Newton TH (1971) Prostatic metastasis to the base of the skull simulating meningioma en plaque. AJR 112:774–778
502. Kishmore PRS, Kaufman AB, Melichar FA (1979) Intrasellar carotid anastomosis simulating pituitary microadenoma. Radiology 132:381–383
503. Kistler M, Pribram HW (1975) Metastatic disease of the sella turcica. AJR 123:13–21
504. Klebs E (1864) Ein Fall von Ecchondrosis sphenooccipitalis amylacea. Virchows Arch [Pathol Anat] 31:396–400
505. Kleinberg DL, Noel GL, Frantz AG (1977) Galactorrhea: a study of 235 cases, including 48 with pituitary tumors. N Engl J Med 296:489–600
506. Kleinsasser O, Friedmann G (1958) Die Knorpelgeschwülste der Schädelbasis. Dtsch Z Nervenheilkd 177:378–404
507. Klinger M (1951) Cartilage tumors of the base of the skull with intracranial extension. Acta Neurochir (Wien) 1:337–380
508. Klöppner K (1934) Die Sella turcica des Neugeborenen im Röntgenbild (Größe, Kontur und Formvarianten der Sella Turcica). ROEFO 50:370–379
509. Kodama G (1965) Developmental studies on the presphenoid in the human sphenoid bone. Okajimas Folia Anat Jpn 41:159–177
510. Kodama G (1971) Developmental studies on the body of the human sphenoid bone. Hokkaido J Med Sci 46:313–321
511. Kodama G (1971) Developmental studies on the orbitopshenoid of the human sphenoid bone. Hokkaido J Med Sci 46:324–335
512. Kodama G, Ito S (1971) Studies on the separate bones and their developments in the extreme part of the lesser wings of the human sphenoid bone. Hokkaido J Med Sci 46:281–300
513. Kollmann J (1904) Der Canalis craniopharyngeus. Anat Anz 25:83–90
514. Korach G, Vignaud J (1977) Manuel de techniques radiographiques du crâne, Masson, Paris
515. Kornblum K (1932) Alterations in structure of sella turcica as revealed by roentgen ray. Arch Neurol Psychiatry 27:305–321
516. Kornblum K, Osmund LH (1935) Effect of intracranial tumors on the sella turcica. An analysis of 446 cases of verified intracranial tumor. Arch Neurol Psychiatry 34:111–123
517. Kornblum RN, Fisher RS (1969) Pituitary lesions in cranio-cerebral injuries. Arch. Pathol 88:242–248
518. Kosowicz J, Rzymski K (1975) Radiological features of the skull in Klinefelter's syndrome and male hypogonadism. Clin Radiol 26:371–378
519. Kovacs A (1934) Untersuchungen über die Sellagröße nach Haas bei Kindern und bei Erwachsenen. ROEFO 50:469–482
520. Kovacs K (1973) Metastatic cancer of the pituitary gland. Oncology 27:533–542
521. Kozlowski P (1961) Relation of the internal carotid siphon to the carotid sulcus. Acta Radiol [Diagn] (Stockh) 1:328–331
522. Kramer S, Lee KF (1974) Complications of radiation therapy: the central nervous system. Semin Roentgenol 9:75–83
523. Kricheff II (1979) The radiologic diagnosis of pituitary adenoma. An overview. Radiology 131:263–265
524. Kroe DJ, Hudgins WR, Simmons JCH, Blackwell CF (1968) Primary intrasellar leiomyoma. J Neurosurg 29:189–191
525. Krueger EG, Unger SM (1966) Extrasellar extension of pituitary adenomas. AJR 98:616–630
526. Krueger EG, Unger SM, Roswit B (1960) Hemorrhage into pituitary adenoma with spontaneous recovery and reossification of sella turcica. Neurology (Minneap) 10:691–696
527. Krueger T, McFarland J, Ommaya AK (1965) Pyocele of the sphenoid sinus. J Neurosurg 22:616–621
528. Kruyff E, Jeffs R (1966) Skull abnormalities associated with the Arnold Chiari malformation. Acta Radiol [Diagn] (Stockh) 5:9–24
529. Kruyff E, Munn JD (1963) Posterior fossa tumors in infants and children. AJR 89:951–965
530. Kuhne D, Schwartz RB (1975) Persisting intrapituitary recessus infundibuli. Neuroradiology 10:177–178
531. Kurlander GJ, Chua GT (1967) Roentgenology of ventriculo-atrial shunts for treatment of hydrocephalus. AJR 101:157–167
532. Laine E, Delandtsheer JM, Delandtsheer M, Galibert P (1955) A propos d'une série de trois observations de méningiomes du clivus. Neurochirurgie 1:221–226
533. Lamb J, Morris L (1961) Carotid-basilar artery: report and discussion of five cases. Clin Radiol 12:179–186

534. Landolt AM (1974) L'oncocytome, un nouveau type histologique des adénomes hypophysaires. Rev Otoneuroophtalmol 47:265–268
535. Lang EK, Bessler WT (1961) The roentgenologic features of acromegaly. AJR 86:321–328
536. Lang J (1977) Structure and postnatal organization of here to fore uninvestigated and infrequent ossifications of the sella turcica region. Acta Anat (Basel) 99:121–139
537. Lang J, Tisch-Rottensteiner KF (1977) Form and form variations of the sella turcica. Verh Anat Ges 71:1279–1282
538. Langer LO Jr, Baumann PA, Gorlin RJ (1967) Achondroplasia. AJR 100:12–26
539. Laschi G (1931) Un caso di foramen dorsi sellae. Radiol Med (Torino) 18:707–712
540. Lasjaunias P (1975) Artère méningée moyenne chez l'homme. Thesis, Paris
541. Lau BP, Newton TH (1965) Sphenopharyngeal encephalomeningocele. Radiol Clin (Basel) 34:386–398
542. Lawrence AM, Wilber JF, Hagen TL (1973) The pituitary and primary hypothyroidism. Enlargement and unusual growth hormone, secretory responses. Arch Intern Med 132:327–333
543. Lawrence WP, El Gammal T, Pool WH Jr (1974) Radiological manifestations of neurosarcoidosis: report of three cases and review of literature. Clin Radiol 25:343–348
544. Lazorthes Y, Zadeh J (1974) Fistules sphénoïdales et adénomes de l'hypophyse. Rev Otoneuroophtalmol 46:355–365
545. Le Clech G, Vallee B, Scarabin JM, Bourdiniere J, Ardouin M, Urvoy M (1978) Aspergillose pseudotumorale du sinus sphénoidal. Rev. Otoneuroophtalmol 50:421–428
546. Leclercq TA, Hardy J, Vezina JL, Mercky F (1974) Intrasellar arachnoidocele and the so-called empty sella syndrome. Surg Neurol 2:295–299
547. Le Double AF (1903) Traité des variations des os du crâne de l'homme et de leur signification au point de vue de l'anthropologie zoologique. Vigot, Paris
548. Lee KF (1976) The diagnosis value of hyperostosis in midline subfrontal meningioma. An analysis of 66 cases. Radiology 119:121–130
549. Lee KF, Lin SR (1979) Neuroradiology of sellar and juxtasellar lesions. Thomas, Springfield, Ill
550. Lee KF, Lin SR, Hodes PJ (1972) New roentgenographic findings in myotonic dystrophy. An analysis of 18 patients. AJR 115:179–185
551. Lee KF, Lin SR, Schatz NJ (1972) Neuroradiology of juxtasellar mass lesions. CRC Crit Rev Radiol Sci 3:105–169
552. Lee KF, Richter HA, Tsai FY (1975) Secondary empty sella syndrome. Acta Radiol [Suppl] (Stockh) 347:131–147
553. Lee KF, Whiteley WH, Schatz NJ, Edeiken J, Lin SR, Tsai FY (1976) Juxtasellar hyperostosis of non meningiomatous origin. J Neurosurg 44:571–579
554. Lee KF, Parke W, Lin SR, Choi HY, Schatz NJ (1978) The vasculature of the diaphragma sellae: a post-mortem injection study. Neuroradiology 16:281–283
555. Lee KF, Suh JH, Lee YE, Berry RG (1979) Meningioma of paranasal sinuses. Neuroradiology 17:165–171
556. Lee WM, Adams JE (1968) The empty sella syndrome. J Neurosurg 28:351–356
557. Leeds N, Seaman WB (1962) Fibrous dysplasia of the skull and its differential diagnosis. Radiology 78:570–582
558. Legre J, Dufour M, Debaene A, Lesto JL (1972) La selle turcique. Concours Med 94:5975–5987
559. Leiba S, Landau B, Ber A (1969) Target gland insufficiency and pituitary tumors. Acta Endocrinol (Copenh) 60:112–120
560. Lemaitre G, Dehaene JL, Genevois A, Linquette M (1974) La tomographie de la selle turcique dans les adénomes hypophysaires. Bilan de notre expérience actuelle. J Radiol Electrol 55:632–633
561. Lemay M (1967) The radiologic diagnosis of pituitary disease. Radiol Clin North Am 5:303–315
562. Leonardi M, Fabris G (1976) Le pneumosinus dilatans, signe radiologique direct des méningiomes de l'angle antérieur du chiasma. Ann Radiol (Paris) 19:803–806
563. Lepage M, Froissart M, Legars D, Dubreuil A (1975) Oedème papillaire et selle turcique vide. Rev Otoneuroophtalmol 47:285–289
564. Le Pogamp C, Grall JY, Massart C, Ramee A, Toulouse R (1979) Poussée évolutive d'adénome à prolactine au cours d'une grossesse après stérilité traitée. 2 observations. Nouv Presse Med 8:2009–2011
565. Lewin W, Cairns H (1951) Fractures of sphenoid sinus with cerebrospinal rhinorrhea. Br Med J I:1–6
566. Lewis NDC (1921) Contribution to the study of tumors from primitive notochord. Arch Intern Med 28:434–452
567. Lie TA (1968) Congenital anomalies of the carotid arteries: an angiographic study and a review of the literature, vol 1. Williams & Wilkins, Baltimore, pp 44–49
568. Lim GHK (1975) Clivus chordoma with unusual bone sclerosis and brain stem invasion. Australas Radiol 19:242–250
569. Lin SR, Bryson MM, Gobien R, Fitz CR, Lee YY (1978 a) Neuroradiologic study of hamartomas of the tuber cinereum and hypothalamus. Neuroradiology 16:17–19
570. Lin SR, Brysson MM, Gobien CR, Lee YY (1978 b) Radiologic findings of hamartomas of the tuber cinereum and hypothalamus. Radiology 127:697–704
571. Lindgren E, Di Chiro G (1951) Suprasellar tumours with calcifications. Acta Radiol [Diagn] (Stockh) 36:173–195
572. Lindholm J, Rasmussen P, Korsgaard O (1973) Intrasellar or pituitary abscess. J Neurosurg 38:616–619
573. Linquette M, Buvat J, Gauthier A, Gasnault JP, Pagniez I, Decoulx M, Laine E (1977) Apoplexie hypophysaire révélatrice d'un adénome à cellules prolactiniques au cours d'une grossesse permise par la bromocriptine. Nouv Presse Med 6:3525–3531

574. List CF (1943) Osteochondromas arising from the base of the skull. Surg Gynecol Obstet 76:480–492
575. List CF, Dowman CE, Bagchi BK, Bebin J (1958) Posterior hypothalamic hamartomas and gangliomas causing precocious puberty. Neurology (Minneap) 8:164–174
576. Lodge SD, Walker GF, Stewart MJ (1927) Aneurysm of the left internal carotid artery simulating pituitary tumour. Br Med J II:1179–1180
577. Login I, Santen RJ (1975) Empty sella syndrome-sequela of the spontaneous remission of acromegaly. Arch Intern Med 135:1519–1521
578. Lombardi G, Passerini A, Miglivacca F (1963) Intracavernous aneurysms of internal carotid artery. AJR 89:361–371
579. Lombardi G, Passerini A, Cecchini A (1968) Pneumosinus dilatans. Acta Radiol [Diagn] (Stockh) 8:535–542
580. Loop JW, White LE, Shaw CM (1964) Traumatic occlusion of the basilar artery within a clivus fracture. Radiology 83:36–40
581. Lorenz RPD (1949) Zwei neue Meßmethoden der Sella turcica im Röntgenbild durch Auswertung ihrer Beziehung zur Schädelbasis und Schädelhöhe. ROEFO 71:273–286
582. Lowman RM, Robinson F, McAllister WB (1966) Craniopharyngeal canal. Acta Radiol [Diagn] (Stockh) 5:41–54
583. Lucas C (1977) Adénomes à prolactine de l'enfant. Thèsis Médicine, Marseille
584. Lucien M (1911) Le poids, les dimensions et la forme générale de l'hypophyse humaine. CR Assoc Anat 147–158
585. Luger A (1914) Zur Kenntnis der im Röntgenbild sichtbaren Hirntumoren mit besonderer Berücksichtigung der Hypophysengangsgeschwülste. ROEFO 21:605–614
586. Lundberg PO, Gemzell C (1966) Dysplasia of the sella turcica: clinical and laboratory investigations in three cases. Acta Endocrinol (Copenh) 52:478–488
587. Luschka H (1850) Die Nerven in der harten Hirnhaut. Laupp, Tübingen
588. Luschka H (1856) Die Altersveränderungen der Zwischenwirbelknorpel. Virchows Arch [Pathol Anat] 9:34–327
589. Luschka H (1857) Über gallertartige Auswüchse am Clivus blumenbachii. Virchows Arch [Pathol Anat] 11:8–17
590. Luton JP, Seiffer J, Bellanger C, Strauch G, Mahoudeau J, Bricaire H (1973) Etude statique et dynamique de la fonction thyréotrope au cours d'une observation de myxoedème primaire avec ectopie linguale et suspicion d'adénome hypophysaire. Ann Endocrinol (Paris) 34:232–235
591. Mabrey RE (1935) Chordoma: study of 150 cases. Am J Cancer 25:501–517
592. Madonick MJ, Rubinstein LJ, Dasco MR, Ribner H (1963) Chromophobe adenoma of pituitary gland with subarachnoid metastases. Neurology (Minneap) 13:836–840
593. Magle KR (1950) Leptomeningeal changes associated with lipochondrodystrophy (gargoylism). Arch Neurol 63:282–297
594. Mahmoud MS (1958) The sella in health and disease: value of the radiographic study of the sella turcica in morbid anatomical and topographic diagnosis of intracranial tumours. Br J Radiol [Suppl] 8:1–160
595. Maisel RH, El Deeb M, Bone RC (1973) Sphenoid sinus mucoceles. Laryngoscope 83:930–939
596. Majd M, Farkas J, Lopresti JM, Chandra R, Hung W, Lussenhop AJ (1971) A large calcified craniopharyngioma in the newborn. Radiology 99:399–400
597. Majlessi H, Shariat AS, Katirai A (1978) Nasopharyngeal craniopharyngioma. Case report. J Neurosurg 49:119–120
598. Majoros M (1970) Meningioma of paranasal sinuses. Laryngoscope 80:640–645
599. Malarkey WB, Johnson JC (1976) Pituitary tumours and hyperprolactinemia. Arch Intern Med 136:40–44
600. Manelfe C, Starling-Jardim D, Touibi S, Bonafe A, David J (1978) Transsphenoidal encephalocele associated with agenesis of corpus callosum: value of Metrizamide computed cisternography. Case report. J Comput Assist Tomogr 2:356–361
601. Mansuy L, Allegre G, de Rougemont J, Thierry A, Pradon Y (1962) Le craniopharyngiome chez l'enfant. Considérations sur une série opératoire de 24 cas. Neurochirurgie 8:234–241
602. Maroteaux P, Leveque B, Marie J, Lamy M (1963) Une nouvelle dysostose avec élimination urinaire de chondroitine-sulfate B. Presse Med 71:1849–1852
603. Martin P, Cushing H (1923) Primary gliomas of chiasm and optic nerves in their intracranial portion. Arch Ophtalmol 52:209–241
604. Martinez-Farinas LO (1967) The sellar-cranial index. Radiology 88:264–267
605. Martins AN, Hayes GJ, Kempe LG (1965) Invasive pituitary adenomas. J Neurosurg 22:268–276
606. Marx H, Hesse W, Neumann H (1947) Sella turcica und hypophyse. Klin Wochenschr 24/25:299–304
607. Mathews W, Wilson CB (1974) Ectopic intrasellar chordoma. J Neurosurg 39:260–263
608. Matisonn R, Pimstone B (1973) Diabetes insipidus associated with an empty sella turcica. Postgrad Med J (S Afr) 49:274–275
609. Matsumura S, Mori S, Uozumi T (1979) Size reduction of a large prolactinoma by bromoergocryptine (B-154) treatment. 2nd European workshop on pituitary adenomas, Paris, September 20–22
610. Mayer EG (1934) Über Lageanomalien des Planum sphenoidale und ihre diagnostische Bedeutung. Roentgenpraxis 6:427–431
611. McCarten KM, Kuhns LR (1976) The area and volume of sella turcica in childhood primary hypothyroidism. Radiology 119:645–650
612. McCarty CS, Boyd AS Jr, Childs DS Jr (1970) Tumors of the optic nerve and optic chiasm. J Neurosurg 33:439–444
613. McConnell EM (1953) The arterial blood supply of the human hypophysis cerebri. Anat Rec 115:175–203
614. McCusker JJ, Caplan RM (1962) Lipoid proteinosis (lipoglycoproteinosis). Am J Pathol 40:599–613

615. McErlean DP, Doyle FH (1976) The pituitary fossa in Cushing's syndrome. A retrospective analysis of 93 patients. Br J Radiol 49:820–826
616. McKenzie KG, Sosman MC (1924) The roentgenological diagnosis of craniopharyngeal pouch tumors. AJR 11:171–176
617. McLachlan MSF, Banna M (1979) Observer variation in interpreting radiographs of the pituitary fossa. Invest Radiol 14:23–26
618. McLachlan MSF, Williams ED, Doyle FH (1968a) Applied anatomy of the pituitary gland and fossa. A radiological and histopathological study based on 50 necropsies. Br J Radiol 41:782–788
619. McLachlan MSF, Williams ED, Fortt RW, Doyle FH (1968b) Estimation of pituitary gland dimensions from radiographs of the sella turcica. Br J Radiol 41:323–330
620. McLachlan MSF, Wright AD, Doyle FH (1970) Plain film and tomographic assessment of the pituitary fossa in 140 acromegalic patients. Br J Radiol 43:360–369
621. McPherson RI, Kroeker M, Houston CS (1972) Hypophosphatasia. J Can Assoc Radiol 23:16–21
622. Meador CK, Worrell JL (1966) The sella turcica in post-partum pituitary necrosis (Sheehan's syndrome). Ann Intern Med 65:259–264
623. Meadows SP (1959) Intracavernous aneurysms of the internal carotid artery. Arch Ophtalmol 62:566–574
624. Mello LR, Tänzer A (1972) Some aspects of trigeminal neurinomas. Neuroradiology 4:215–221
625. Melsen B (1972) Time and mode of closure of the spheno-occipital synchondrosis determined on human autopsy material. Acta Anat (Basel) 83:112–118
626. Meredith JM, Mandeville FB, Kay S (1960) Osteogenic sarcoma of the skull following roentgenotherapy for benign pituitary tumor. J Neurosurg 17:792–799
627. Messimy R, Namin P, Martines N (1955) Tumeur kystique de l'hypophyse de type vésiculaire, à revêtement cilié. Troubles pigmentaires associés. Rev Neurol (Paris) 235–240
628. Messimy R, Metzger J, Helias A, Paul D, Schaison G, Cabanis EA (1974) Etude des troubles coexistant avec une "selle turcique vide", en particulier des troubles visuels. Rev. Otoneuroophtalmol 46:417–433
629. Mestdagh H, Bacqaert JP, Bailleul JP, Depreux R (1976) Morphological anatomy of the interjugochiasmatic fenestra. Bull Assoc Anat (Nancy) 60:747–756
630. Metzger J, Helias A, Fischgold H (1970) La selle turcique vide. Rev Neurol (Paris) 122:198–203
631. Metzger J, Helias A, Messimy R, Fischgold H (1970) Nouveaux cas de selles turciques dites vides avec signes neuroophtalmologiques ou rhinorrhée. Rev Neurol (Paris) 123:435–442
632. Metzger J, Bonneville JF, Bugault R (1974) Exploration radiologique de la région hypophysaire. Encycl Med Chir Paris, Glandes endocrines, 10017 E 10, 4
633. Metzger J, Bonneville JF, Bugault R (1975) Les adénomes à prolactine. Signes radiologiques. Acta Radiol [Suppl] (Stockh) 347:327–338
634. Metzger J, Weill F, Cabanis EA, Bonneville JF, Ben Hamida M (1975) Les calcifications intra-crâniennes. In: Traité de Radiodiagnostic, neuroradiologie, vol 14/1. Masson, Paris, pp 3–76
635. Metzger J, Bugault R, Bonneville JF (1976) Diagnostic radiologique des adénomes à prolactine. In: Traité de radiodiagnostic, vol 14/2. Masson, Paris, pp 278–296
636. Metzger J, Bugault R, Messimy R, Bonneville JF (1976) Pre- and supra sellar meningiomas. Revue Philips
637. Metzger J, Messimy R, Helias A (1976) La selle turcique vide. In: Metzger J, Simon J, Salamon G, Fischgold H (eds) Traité de radiodiagnostic, Neuradiologie, vol 14/2. Masson, Paris, pp 347–354
638. Middlemass IBD (1959) Bone changes in adult cretins. Br J Radiol 32:685–688
639. Milas RW, Sugar O, Dobben G (1977) Benign pituitary adenoma associated with hyperostosis of the sphenoid bone and monocular blindness. Case report. J Neurosurg 46:107–110
640. Miller WE, Holman CB, Dockerty MB, Devine KD (1969) Roentgenologic manifestations of malignant tumors of the nasopharynx. AJR 106:813–823
641. Minagi H (1971) Skull changes in endocrine disease. In: Newton TH, Potts DG (eds) Radiology of the skull and brain, vol 1, book 2. 665–673
642. Minagi H, Newton TH (1969) Cartilaginous tumors of the base of the skull. AJR 105:308–313
643. Minagi H, Steinbach HL (1968) Roentgen aspects of pituitary tumors manifested after bilateral adrenalectomy for Cushing's syndrome. Radiology 90:276–280
644. Minagi H, Margolis MT, Newton TH (1972) Tomography in the diagnosis of sphenoid sinus mucocele (S.S.M.). AJR 115:587–591
645. Mitchell SW (1968) Aneurysm of an anomalous artery causing antero-posterior division of the chiasm of the optic nerves and producing bitemporal hemianopsia. Trans Am Neurol Assoc 239–251
646. Mitts MG, McQueen JD (1965) Visual loss associated with fusiform enlargement of the intracranial portion of the internal carotid artery. J Neurosurg 23:33–37
647. Mogensen EF (1957) Chromophobe adenoma of the pituitary gland; a follow-up study on 60 surgical patients with special reference to endocrine disturbances. Acta Endocrinol (Copenh) 24:135–152
648. Mohanty S (1977) Haemorrhage into pituitary adenomas. J Neurol Neurosurg Psychiatry 40:987–991
649. Molitch ME (1977) Coexisting primary empty sella syndrome and acromegaly. Clin Endocrinol (Oxf) 7:261–263
650. Momose KJ (1976) Quantitative analysis of the pituitary fossa and gland in Cushing's disease. Acta Radiol [Suppl] (Stockh) 347:537–552
651. Montaut J, Picard L, Weber M, Barroche G, Galoisy CH (1974) Meningioma of sella turcica tuberculum and Recklinghausen's neurofibromatosis. A propos of an infantile form. Rev Otoneuroophtalmol 46:47–51

652. Montrieul B, Janny P, Pignide L, Chabannes J (1965) Considérations sur les abcès de l'hypophyse. Neurochirurgie 11:366–374
653. Moore TJ, Dluhy RG, Williams GH, Cain JP (1976) Nelson's syndrome: frequency, prognosis and effect of prior pituitary irradiation. Ann Intern Med 85:731–734
654. Morato F (1939) The blood supply of the hypophysis. Anat Rec 74:297–310
655. Morley TP, Wortzman G (1965) The importance of the lateral extensions of the sphenoidal sinus in posttraumatic cerebrospinal rhinorrhea, and meningitis. Clinical and radiological aspects. J Neurosurg 22:326–331
656. Mornex R, Tommasi M, Cure M, Farcot J, Orgiazzi J, Rousset B (1972) Hyperthyroidie associée à un hypopituitarisme au cours de l'évolution d'une tumeur hypophysaire sécrétant de la T.S.H. Ann Endocrinol (Paris) 33:390–396
657. Mortada A (1967) Misinterpretation of sphenoid ridge meningiomata. Br J Ophtalmol 51:829–838
658. Mortara R, Norrell H (1970) Consequences of a deficient sellar diaphragm. J Neurosurg 32:565–573
659. Mouren P, Poinso Y, Sebahoun G, Pellissier JF (1970) Les chordomes de la base du crâne. Sem Hop Paris 46:501–510
660. Mühlendahl KE, Bradac GB (1975) Empty sella syndrome in a boy with mucopohysaccharidosis type VI (Maroteaux-Lamy). Helv Paediatr Acta 30:185–190
661. Müller H (1858) Über das Vorkommen von Resten der Chorda dorsalis bei Menschen nach der Geburt und über ihr Verhältniss zu den Gallertgeschwülsten am Clivus. Z Ration Med 2:202–230
662. Müller W, Hoffmann H (1967) Über Verkalkungen in Hypophysenadenomen. Zentralbl Neurochir 28:287–290
663. Murali R, Epstein F (1979) Diagnosis and treatment of suprasellar arachnoid cyst. Report of three cases. J Neurosurg 50:515–518
664. Naiken VS, Tellem M, Meranze DR (1961) Pituitary cyst of Rathke's cleft origin with hypopituitarism. J Neurosurg 18:703–708
665. Neelon FA, Goree JA, Lebovitz HE (1973) The primary empty sella: clinical and radiological characteristics and endocrine function. Medicine (Baltimore) 52:73–92
666. Neilson JM, Moffat AD (1963) Hypopituitarism caused by a melanoma of the pituitary gland. J Clin Pathol 16:144–149
667. Nelson DH, Meakin JW, Dealy JB Jr, Matson DD, Emerson K Jr, Thorn GW (1958) ACTH-producing tumor of the pituitary gland. N Engl J Med 259:161–164
668. Nelson DH, Meakin JW, Thorn GW (1960) ACTH-producing pituitary tumors following adrenalectomy for Cushing's syndrome. Ann Intern Med 52:560–569
669. Nelson DH, Sprunt JG, Mims RB (1966) Plasma ACTH determination in fifty-eight patients before or after adrenalectomy for Cushing's syndrome. J Clin Endocrinol Metab 26:722–728
670. Nelson PB, Robinson AG, Archer DF, Maroon JC (1978) Symptomatic pituitary tumor enlargement after induced pregnancy. Case report. J Neurosurg 49:283–287
671. Nemir RL, Branom-Genieser N, Balasubramanyam P (1979) Extensive sclerosis of the base of the skull due to primary nasal tuberculosis. Pediatr Radiol 8:42–44
672. Nerli A (1955) Le calcificazioni delle carotidi interne intracavernose. Semeiotica radiologica. Radiol Med (Torino) 41:676–686
673. Neuhauser EBD, Griscom NT, Gilles FH, Crocker AC (1968) Arachnoid cysts in the Hurler-Hunter syndrome (kystes arachnoïdiens dans le syndrome de Hurler-Hunter). Ann Radiol (Paris) 11:153–469
674. New PFJ (1966) The sella turcica as a mirror of disease. Radiol Clin North Am 4:75–80
675. Newton TH, Potts DG (1971) Radiology of the skull and brain, vol 1, book 1. Mosby, Saint-Louis
676. Newton TH, Burbenne HJ, Palubinskas AJ (1962) Primary carcinoma of the pituitary. AJR 87:110–120
677. Niepce B (1851) Enlarged sella due to cretinism (hypothyroidism). Traité du goître et du crétinisme. Ballière, Paris
678. Nilus F, Le Picard (1933) Recherches sur le sinus sphénoïdal et la selle turcique chez l'homme et les vertébrés. Rev Fr Endocrinol 11:97–107
679. Noback CR (1944) The developmental anatomy of the human osseus skeleton during the embryonic, fetal and circumnatal periods. Anat Rec 88:91–96
680. Nolte K (1977) Contribution to the diagnosis of tumors situated in basal parts of brain and skull by conventional neuroradiological methods. 13th Meeting ESPR., Stockholm, May 1976. Ann Radiol (Paris) 20:159
681. Nolte K (1979) Malignant intracranial chordoma and sarcoma of the clivus in infancy. Pediatr Radiol 8:1–6
682. Nordmark B (1949) Pressure changes in the sella turcica in the presence of glioma in the cerebral hemisphere. Acta Radiol (Stockh) 32:461–467
683. Northfield DWC, Russel DS (1966) Pubertas precox due to hypothalamic hamartoma: report of two cases surviving surgical removal of the tumor. J Neurol Neurosurg Psychiatry 30:166–173
684. Notermann J, Lerinck PL, Dor P (1972) Les tumeurs malignes de l'ethmoïde et du sphénoïde (à propos d'une série de 33 cas). Neurochirurgie 18:485–490
685. Nugent GR, Sprinkle P, Bloor BM (1970) Sphenoid sinus mucoceles. J Neurosurg 32:443–451
686. Obenchain TG, Becker DP (1972) Abscess formation in a Rathke's cleft cyst. J Neurosurg 36:354–362
687. Obrador S (1972) The empty sella and some related syndromes. J Neurosurg 36:162–168
688. Obrador S, Blazquez MG (1972) Pituitary abscess in a craniopharyngioma. A case report. J Neurosurg 36:785–789
689. Ogilvy KM, Jakubowski J (1973) Intracranial dissemination of pituitary adenomas. J Neurol Neurosurg Psychiatry 36:199–205
690. Ohsawa T, Furuse M, Kiruchi Y, Suda Y, Tamiya

T, Hikita M (1971) Roentgenographic manifestations of Klinefelter's syndrome. AJR 112:178–184
691. Okudera T (1976) Primary empty sella with special reference to roentgenological anatomy. Jpn J Clin Radiol 21:487–495
692. Olivecrona H (1932) On suprasellar cholesteatomas. Brain 55:122–131
693. Olivier L, Vila-Porcile E, Racadot O, Peillon F, Racadot J (1975) Ultrastructure of pituitary tumor cells: a critical study. In: Tixier-Vidal A, Farquhar MG (eds) The anterior pituitary. Academic Press, New York
694. Olson DR, Guiot G, Derome P (1972) The symptomatic empty sella. Prevention and correction via the transsphenoidal approach. J Neurosurg 37:533–537
695. Ommaya AK, Di Chiro G, Baldwin M, Pennybacker JB (1968) Non traumatic cerebrospinal fluid rhinorrhea. J Neurol Neurosurg Psychiatry 31:214–225
696. Onur K, Lala K, Zimmer J, Juan CS, Ruskin CT (1977) The primary empty sella syndrome in a child. J Pediatr 90:425–427
697. Oon CK (1963) The size of the pituitary fossa in adults. Br J Radiol 36:294–299
698. Oon CK, Lavender JP, Joplin GF (1962) The width of the normal pituitary gland. A comparison of two radiological methods of measurement. Br J Radiol 35:418–422
699. Ortega FJV, Longridge NS (1975) Fracture of the sella turcica. Injury 6:335–337
700. Ortiz-Suarez H, Erickson DL (1975) Pituitary adenomas of adolescents. J Neurosurg 43:437–439
701. Ougier J (1965) Contribution à l'étude des anévrysmes simulant une tumeur hypophysaire. Thèsis, Nancy
702. Overbeek WJ, Front D, Penning L (1971) Primary enlarged "empty" sella. Neurochirurgia (Stuttg), 14:110–115
703. Ozonoff MB, Burrows EH (1971) Intracranial calcification. In: Newton TH, Potts DG (eds) Radiology of the brain and skull, vol 1, book 2. Mosby, Saint-Louis, 823–873
704. Padget DH (1948) The development of the cranial arteries in the human embryo. Contrib Embryol 32:205–262
705. Paillas JE, Alliez B (1974) Intrasellar chondroma. Case report. Neurochirurgia (Stuttg) 17:136–140
706. Palacios E, MacGee EE (1972) The radiographic diagnosis of trigeminal neurinomas. J Neurosurg 36:153–156
707. Palubinskas AJ, Davies H (1959) Roentgen features of nasal accessory sinus mucoceles. Radiology 72:576–584
708. Pallo D, Guinet P, Tourniaire J, Trouillas J (1976) L'adénome à prolactine (à propos de 14 observations personnelles). Nouv Presse Med 5:264–268
709. Papo I, Beltrami CA, Salvolini U, Carusellim G (1977) Sarcoidosis simulating a glioma of the optic nerve. Surg Neurol 8:353–354
710. Parkinson D (1964) Collateral circulation of the cavernous carotid artery: anatomy. Can J Surg 7:251–268
711. Parkinson D (1965) A surgical approach to the cavernous portion of the carotid artery. Anatomical studies and case report. J Neurosurg 23:474–483
712. Parkinson D (1967) Transcavernous repair of carotid cavernous fistula. Case report. J Neurosurg 26:420–424
713. Parkinson D, Hunt B, Zeal A (1978) Lateral sellar venous anatomy, "cavernous sinus". Anat Rec 190:503–508
714. Parks JS, Tenore A, Bongiovanni AM, Kirkland RT (1978) Familial hypopituitarism with large sella turcica. N Engl J Med 298:698–702
715. Pasquet G (1976) Le sinus sphénoïdal: étude radioanatomique (à propos de 125 observations). Thèsis Médicine, Paris
716. Patel YC, Kilpatrick JA (1969) Pituitary enlargement with long-standing myxoedema. NZ Med J 70:21–23
717. Patouillard P, Vanneuville G (1972) Les parois du sinus caverneux. Neurochirurgie 18:551–560
718. Patten BM (1968) Human embryology. McGraw-Hill, New York
719. Paulus DD Jr, Dodd GD (1970) The roentgen diagnosis of tumors of the nasal cavity and accessory paranasal sinus. Radiol Clin North Am 8:343–360
720. Paxson CL, Brown DR (1976) Post traumatic anterior hypopituitarism. Pediatrics 57:893–896
721. Pecina M, Novoselac M (1970) Über einen Fall, in dem die Hypophysengrube die Nasenhöhle begrenzt. Anat Anz 127:32–37
722. Peele JC (1931) Unusual anatomical variations of the sphenoid sinuses. Laryngoscope 67:208–237
723. Peeters FLM (1973) The "empty sella". A problem in differential diagnosis. ROEFO 118:624–630
724. Pendergrass EP, Schaeffer JP, Hodes PJ (1956) The head and neck in roentgen diagnosis. Thomas, Springfield, Ill
725. Penfield W, McNaughton F (1940) Dural headache and innervation of the dura mater. Arch Neurol Psychiatry 44:43–75
726. Penkrot RJ, Bures C (1979) The "apparently" eroded dorsum sella: a new anomaly. AJR 132:1005–1006
727. Perlemuter L, Waligora J (1974) Cahiers d'anatomie. Masson, Paris (Cahier 1)
728. Pertuiset B (1975) Crâniopharyngiomas. In: Vinken PJ, Bruyn GW (eds) Handbook of clinical neurology. Excerpta Medica, Amsterdam American Elsevier, New York
729. Petit-Dutaillis D, Daum S (1950) Les méningiomes de la fosse postérieure. Rev Neurol (Paris) 83:241–255
730. Petit-Dutaillis D, Messimy R, Berdet H, Benhaim C (1951) Contribution au diagnostic des chordomes sphéno-occipitaux. Sem Hop Paris 27:2663–2676
731. Petit-Dutaillis D, Schiff-Wertheimer M, Aubry M, Metzger J (1958) Syndrome de compression d'un nerf optique par une mucocèle sphénoïdale latente. Intervention par voie para-latéro-nasale. Importance des signes radiologiques pour le diagnostic. Neurochirurgie 4:81–94
732. Pilpel R (1922) Persistenz des Canalis craniopharyn-

geus mit Rachendach Hypophyse und Kephalokele. Monatsschr Ohrenheilkd 56:793–802
733. Pincus JH, Daroff RB (1964) Sphenoid sinus mucocele. A curable cause of the ophtalmoplegic migraine syndrome. JAMA 187:459–460
734. Piraud G (1933) La notochorde. Embryologie générale et expérimentale. Vestiges et tumeurs. Thèsis Médicine, Paris
735. Pitkethly DT, Kempe LG (1969) Giant cell tumors of the sphenoid. J Neurosurg 30:301–304
736. Plair CM, Perry S (1962) Hypothalamic, pituitary sarcoidis. Arch Pathol 74:527–535
737. Platzer W (1957) Zur Anatomie der "Sellabrücke" und ihrer Beziehung zur A. Carotis interna. ROEFO 87:613–616
738. Plaut A (1952) Pituitary necrosis in routine necropsies. Am J Pathol 28:883–893
739. Plaut HF, Blatt ES (1967) Chordoma of clivus: report of four cases. AJR 100:639–649
740. Plaut MR (1978) Anatomic variations of the sella turcica. Surg Neurol 10:259–261
741. Podoshin L, Rolan L, Altman MM, (1970) Pharyngeal craniopharyngioma. J Laryngol Otol 84:93–99
742. Poirier P, Charpy A (1931) Traité d'anatomie humaine, vol 1. Masson, Paris
743. Pollock JA, Newton TH, Hoyt WF (1968) Transsphenoidal and transethmoidal encephaloceles: a review of clinical and roentgen features in 8 cases. Radiology 90:412–453
744. Popa GT, Fielding U (1930) A portal circulation from the pituitary to the hypothalamic region. J Anat 65:88–91
745. Poppen JL, King AB (1952) Chordoma: experience with 13 cases. J Neurosurg 9:139–163
746. Potter GD (1970) Sclerosis of the base of the skull as a manifestation of nasopharyngeal carcinoma. Radiology 94:35–38
747. Potter GD (1971) Sectional anatomy and tomography of the head; an atlas of the normal sectional anatomy of the head. Grune & Stratton, New York
748. Potter GD, Trokel SL (1971) Optic canal. In: Newton TH, Potts DG (eds) Radiology of the skull and brain, vol 1, book 2. 487–507. Mosby, Saint-Louis
749. Poursines Y, Paillas JE (1951) Les chordomes sphéno-occipitaux et les tumeurs du foramen ovale. Rev Otoneuroophtalmol 23:65–114
750. Prasad U, Kwi NK (1975) Nasopharyngeal craniopharyngioma. J Laryngol Otol 89:445–452
751. Pressman BD, Waldron RL, Wood EH (1975) Histocytosis-X of the hypothalamus. Br J Radiol 48:176–178
752. Pribram HW, du Boulay GH (1971) Sella turcica. In: Newton TA, Potts DG (eds) Radiology of the skull and brain, vol 1, book 1. Mosby, Saint-Louis, pp 357–405
753. Pribram HW, Swann GF (1960) The radiological changes and clinical incidence of endocrine effects in sellar and parasellar tumours. Radiology 75:877–884
754. Proetz AW (1948) The sphenoid sinus. Br Med J II:243–245
755. Prosser PR (1978) Intrasellar cysticercosis presenting as a pituitary tumor: successful transsphenoidal cystectomy with preservation of pituitary function. Am J Trop Med Hyg 27:976–978
756. Puech P, Stuhl L (1929) La selle turcique. Etude radiographique dans les tumeurs de l'hypophyse et de la région hypophysaire. Rev Neurol (Paris) 1:100
757. Racadot J (1966) Histopathologie de l'hypophyse. Rev Prat 16:22–31
758. Racadot J, Vila-Porcile E, Peillon F, Olivier L (1971) Adénomes hypophysaires à cellules à prolactine: étude structurale et ultrastructurale, corrélations anatomo-cliniques. Ann Endocrinol 32:298–305
759. Racadot J, Olivier L, Vila-Porcile E, Peillon F (1974) Le problème de la classification des adénomes de l'hypophyse: apport de l'ultrastructure. Rev Otoneuroophtalmol 47:215–222
760. Rådberg C (1963) Somes aspects of the asymetric enlargement of sella turcica. Acta Radiol [Diagn] (Stockh) 1:152–163
761. Rådberg C (1963) Appearance of sella turcica following transsphenoidal hypophysectomy. Acta Radiol [Diagn] (Stockh) 1:140–151
762. Radoievitch S, Jovanovitch E (1955) La morphologie du canal vidien et les rapports avec les sinus paranasaux chez l'homme adulte et l'enfant. Rev Laryngol Otol Rhinol (Bord) 76:481–492
763. Radoievitch S, Jovanovitch E, Lotritch N (1956) La morphologie du trou ovale et les rapports du nerf maxillaire inférieur avec le sinus sphénoïdal. Rev Laryngol Otol Rhinol (Bord) 77:11–19
764. Raiti S, Albrink MJ, Maclaren NK, Chadduck WM, Gabriele OF, Chou SM (1976) Empty sella syndrome secondary to intrasellar cyst in adolescence. Am J Dis Child 130:1009–1012
765. Ramos E Silva J (1943) Lipoid proteinosis (Urbach-Wiethe). Arch Dermatol 47:301–326
766. Rasmussen AT (1924) Quantitative study of human hypophysis cerebri or pituitary body. Endocrinology 8:509–524
767. Rasmussen AT (1928) Ciliated epithelium and mucus-secreting cells in the human hypophysis. Anat Rec 41:273–284
768. Rasmussen AT (1938) Proportions of various subdivisions of normal adult human hypophysis cerebri and relative number of different types of cells in pars distalis, with biometric evaluation of age and sex differences and special consideration of basophilic invasion into infundibular process. Res Nerv Ment Dis Proc 17:118–150
769. Raymond LA, Tew J (1978) Large suprasellar aneurysms imitating pituitary tumor. J Neurol Neurosurg Psychiatry 41:83–87
770. Reddy JB, Farb S (1963) Angiofibroma of the sphenoid sinus. A case report. Laryngoscope 73:108–119
771. Reid JD (1960) Congenital absence of pituitary gland. J Pediatr 56:658–664
772. Reimann F, Berker F, Gökmen E, Kücükcakirlar T (1978) Das Verhalten der Sella Turcica bei jugendlichen Patienten mit schwerer Eisenmangelkrankheit. ROEFO 129:598–604

773. Reinecke RD, Montgomery WWM (1964) Oculomotor nerve palsy associated with mucocele of the sphenoid sinus. Arch Ophtalmol 71:50–51
774. Reinert H (1927) Beitrag zur röntgenologischen Selladiagnostik. ROEFO 35:553–573
775. Renn WH, Rhoton AL (1975) Microsurgical anatomy of the sellar region. J Neurosurg 43:288–298
776. Reschini E, Giustina G, Canta Lamessa L, Peracchi M (1976) Hyperthyroidism with elevated plasma levels and pituitary tumor: study with somatostatin. J Clin Endocrinol Metab 43:924–927
777. Reynolds DF (1961) Traumatic effusion of sphenoid sinus. Clin Radiol 12:171–176
778. Riach ICF (1966) The sellar index. A relationship between the area of the sella turcica and the lateral area of the skull. Br J Radiol 39:824–826
779. Riach ICF (1966) The pituitary fossa in childhood with particular reference to hypopituitarism. Br J Radiol 39:241–246
780. Richards AG, Marvey PW (1954) Giant cell granuloma and other pituitary granulomata. Q J Med 23:425–437
781. Richmond IL, Wilson CB (1978) Pituitary adenomas in childhood and adolescence. J Neurosurg 49:163–168
782. Ridgway EC, Kourides IA, Kliman B, Bigos T, Maloof F (1975) Thyrotropin and prolactin pituitary reserve in the "empty sella syndrome". J Clin Endocrinol Metab 41:968–973
783. Ring BA, Eddy WM (1963) Calcification of carotid arteries. JAMA 184:866–870
784. Ring BA, Waddington M (1966) Primary arachnoid cysts of the sella turcica. AJR 98:611–615
785. Ringel SP; Bailey OT (1972) Rathke's cleft cyst. J Neurol Neurosurg Psychiatry 35:693–697
786. Ripert J (1931) Les chordomes du naso-pharynx. Thèsis Médicine, Paris
787. Rischbieth RHC, Bull JWD (1958) Significance of enlargement of superior orbital (sphenoidal) fissure. Br J Radiol 31:125–135
788. Riviere J, Manciet G, Roger P, Guerin J, Barraud AM, Mauriac L, Kern AM, Raschedi M (1979) Maladie de Cushing: traitement par exérèse sélective de l'adénome hypophysaire. 3 observations. Nouv Presse Med 8:2255–2259
789. Rizzuti RJ, Whalen JP (1972) The nasopharynx: roentgen anatomy and its alteration in the base view. Radiology 104:537–540
790. Roberson GH, Patterson AK, Deeb ME, Maisel RH, Bone RC (1976) Sphenoethmoidal mucocele: radiographic diagnosis. AJR 127:595–599
791. Robertson DM, Hetherington RF (1964) A case of ganglioneuroma arising in the pituitary fossa. J Neurol Neurosurg Psychiatry 27:268–272
792. Robertson EG (1957) Pneumo-encephalography. Thomas, Springfield, Ill.
793. Robertson WD, Newton TH (1978) Radiologic assessment of pituitary microadenomas. AJR 131:489–492
794. Robinson AE, Meares BM, Goree JA (1967) Traumatic sphenoid sinus effusion. AJR 101:795–801
795. Rosenblatt S, Hershman JM, Marder S, Garai T (1978) Chronic phenothiazine therapy does not increase sellar size (letter). Lancet II:319–320
796. Ross RJ, Greitz TVB (1966) Changes of the sella turcica in chromophobe adenomas and eosinophilic adenomas. Radiology 86:892–899
797. Ross Russel RW, Pennybacker JB (1961) Craniopharyngioma in the elderly. J Neurol Neurosurg Psychiatry 24:1–13
798. Roth M, Toman I, Hanak L, Svejdova M (1971) Die "leere" Sella. Radiologe 11:27–31
799. Rothman LM, Sher J, Quencer RM, Tenner MS (1976) Intracranial ectopic pituitary adenoma. J Neurosurg 44:96–99
800. Rothman SLG, Kier EL, Kirchner JA, Allen WE (1975) Preoperative laminographic evaluation of the sphenoid sinus for transsphenoidal surgery. Laryngoscope 85:1986–1998
801. Rothman SLG, Kier EL, Allen WE (1976) The radiology of transsphenoidal hypophysectomy: a review of 100 cases. AJR 127:601–606
802. Rougerie J, Fardeau M (1962) Les craniopharyngiomes, vol 1. Masson, Paris
803. Rouques L (1931) La myotonie atrophique. Thèsis. Legrand, Paris
804. Roussel Le Guen JC, Peillon F, Messimy R, Metzger J, Bugault R (1974) Selle turcique vide. Découverte à l'occasion du bilan d'une aménorrhée secondaire. J Gynecol Obstet Biol Reprod (Paris) 3:1135–1142
805. Rovit RL, Berry R (1965) Cushing's syndrome and hypophysis. A reevaluation of pituitary tumors and hypoadrenalism. J Neurosurg 23:270–295
806. Rovit RL, Duane TD (1969) Cushing's syndrome and pituitary tumors. Am J Med 46:416–427
807. Rovit RL, Fein JM (1972) Pituitary apoplexy. A review and reappraisal. J Neurosurg 37:280–288
808. Rowbotham GF, Clarke PRR (1956) Colloid cyst of the pituitary gland causing chiasmal compression. Br J Surg 44:107–108
809. Rowe PW, Jones TK (1966) Malignant chromophobe adenoma with extensive skull destruction. Radiology 86:532–533
810. Ruberti R, Galligioni F, Carteri A (1966) Considérations radiologiques dans les neurinomes du ganglion de Gasser. Acta Radiol [Diagn] (Stockh) 5:465–469
811. Rubin P (1964) Dynamic classification of bone dysplasias. Year Book Medical Publishers, Chicago
812. Rubin P, Kramer S (1965) Ectopic pinealoma. Radiology 85:512–523
813. Rudwan MA (1977) Pituitary abscess. Neuroradiology 12:243–248
814. Ruggiero G, Castellano F (1952) Diagnostic radiologique des méningiomes de la fosse postérieure. Rev Neurol (Paris) 86:700–703
815. Rumbaugh CL, Davis DO (1974) The normal skull: techniques and indications. Semin Roentgenol 9:91–100
816. Rumbaugh CL, Potts DG (1966) Skull changes associated with intracranial arteriovenous malformations. AJR 98:525–534
817. Russell DS, Rubinstein LJ (1977) Pathology of tumors of the nervous system. Arnold, London

818. Rzymski K, Kosowicz J (1975) The skull in gonadal dysgenesis: a roentgenometric study. Clin Radiol 26:379–384
819. Sakoda K, Ishikawa S, Uozumi T, Hirakawa K, Okazaki H, Harada Y (1979) Sphenoethmoidal meningoencephalocele associated with agenesis of corpus callosum and median cleft lip and palate. Case report. J Neurosurg 51:397–401
820. Salassa RM, Kearns TP, Kernohan JW, Sprague RG, MacCarty CS (1959) Pituitary tumors in patients with Cushing's syndrome. J Clin Endocrinol Metab 19:1523–1539
821. Salassa RM, Laws ER, Carpenter PC, Northcutt RC (1978) Transsphenoidal removal of pituitary microadenoma in Cushing's disease. Mayo Clin Proc 53:24–28
822. Saltzman GF (1959) Patent primitive trigeminal artery studied by cerebral angiography. Acta Radiol [Diagn] (Stockh) 1:329–335
823. Samaan NA, Osborne BM, MacKay B, Leavan ME, Duello TM, Halmi NS (1977) Endocrine and morphologic studies of pituitary adenomas secondary to primary hypothyroidism. J Clin Endocrinol Metab 45:903–911
824. Sansregret A, Ledoux R, Duplantis F, Lamoureux C, Chapdelaine A, Leblanc P (1969) Suprasellar subarachnoid cysts: radioclinical features. AJR 105:291–297
825. Sarwar M (1979) Primary sellar-parasellar esthesioneuroblastoma. AJR 133:140–141
826. Sarwar M, Swischuk LE, Schechter MM (1976) Intracranial chondromas. AJR 127:973–977
827. Sasaki H, Kodama G (1973) Developmental studies on the postsphenoid in the human sphenoid bone. Hokkaido J Med Sci 48:167–178
828. Sassin JF, Chutorian AM (1967) Intracranial chordoma in children. Arch Neurol 17:89–93
829. Satyamurti S, Huntington HW (1972) Granular cell myoblastoma of the pituitary. Case report. J Neurosurg 37:483–486
830. Scatliff JH, Bull JWD (1965) The radiological manifestations of suprasellar metastatic tissue. Clin Radiol 16:66–70
831. Schaeffer JP (1924) Some points in regional anatomy of optic pathway with especial reference to tumors of hypophysis cerebri and resulting ocular changes. Anat Rec 28:243–279
832. Schaeffer JP (1936) The clinical anatomy and development of the paranasal sinuses. Pa Med J 39:395–404
833. Schaerer JP, Whitney RL (1953) Prostatic metastases simulating intracranial meningioma: case report. J Neurosurg 10:546–549
834. Schaison G, Metzger J (1976) The primary empty sella and endocrine study on 12 cases. Acta Endocrinol (Copenh) 83:483–492
835. Schechter MM, Zingesser LH (1967) The radiology of aqueductal stenosis. Radiology 88:905–916
836. Schechter MM, Rovit RL, Schachter JM (1969) Rhinorrhea and hydrocephalus. Observations on spontaneous cerebrospinal fluid fistulae in patients with increased intracranial pressure. Acta Radiol [Diagn] (Stockh) 9:101–116
837. Schechter MM, Liebeskind AL, Azar-Kia B (1974) Intracranial chordomas. Neuroradiology 8:67–82
838. Scheinmann P, Ducotet-Valdant MC, Meyer B, Luton JP, Paupe J, Vialatte J (1977) Acromégalie-gigantisme. Ann Pediatr (Paris) 53:287–291
839. Scheuermann H (1932) The roentgenological picture of the normal and the pathological sella turcica. Acta Radiol [Diagn] (Stockh) 13:404–432
840. Schey WL (1973) Plain film skull roentgenographic changes in hydrocephalus. AJR 118:134–146
841. Schiefer W, Marguth F (1956) Intraselläre Aneurysmen. Acta Neurochir (Wien) 4:344–354
842. Schisano G, Olivecrona H (1960) Neurinomas of the Gasserian ganglion and trigeminal root. J Neurosurg 17:306–322
843. Schisano G, Tovi D (1962) Clivus chordomas. Neurochirurgia (Stuttg) 5:99–120
844. Schloffer H (1907) Erfolgreiche Operation eines Hypophysentumors auf nasalem Wege. Wien Klin Wochenschr 20:621–624
845. Schmidt RC (1979) Partielle Aplasie der rechten Arteria carotis interna mit einer ungewöhnlichen Kollateralarterie unterhalb der Hypophyse. ROEFO 130:572–575
846. Schneider RC, Kriss FC, Falls HF (1970) Prechiasmal infarction associated with intrachiasmal and suprasellar tumors. J Neurosurg 32:197–208
847. Schönberg D (1971) Hamartomas of the tuber cinereum in precocious puberty: endocrinology and diagnosis in vivo. Radiologe 11:319–321
848. Scholtz CL, Siu K (1976) Melanoma of the pituitary. J Neurosurg 45:101–103
849. Schüller A (1926) The sella turcica. AJR 16:336–340
850. Schüller A (1941) X-ray symptoms of intracranial hypertension. Confin Neurol 3:253–256
851. Schuster G, Westberg G (1967) Gliomas of the optic nerve and chiasm. Acta Radiol [Diagn] (Stockh) 6:221–232
852. Schwartz HG (1936) The meningeal relations of the hypophysis cerebri. Anat Rec 67:35–51
853. Scott TV, Schatz NJ, Lee KF (1974) Malignant pituitary adenoma. A cause of sphenoid hyperostosis and monocular blindness. Arch Ophtalmol 91:123–125
854. Seaman WB (1951) Olfactory esthesioneuro-epitheliomas. Radiology 57:541–546
855. Sebahoun G (1968) Les chordomes de la base du crâne. Thèsis Médicine, Marseille
856. Segall HD, Hassan G, Ling SM, Carton C (1974) Suprasellar cysts associated with isosexual precocious puberty. Radiology 111:607–616
857. Seltzer J, Hurteau EF (1957) Bilateral aneurysms of internal carotid artery within cavernous sinus. Case report. J Neurosurg 14:448–451
858. Sengupta RP, Gryspeerdt GL, Hankinson J (1976) Carotid-ophtalmic aneurysms. J Neurol Neurosurg Psychiatry 39:837–853
859. Sennett EJ (1953) Chordoma: its roentgen diagnostic aspects and its response to roentgen-therapy. Am J Roentgenol 69:613–622

860. Shanklin WM (1949) On the presence of cysts in the human pituitary. Anat Rec 104:379–407
861. Shanklin WM (1951 a) The histogenesis and histology of an integumentary type of epithelium in the human hypophysis. Anat Rec 109:217–231
862. Shanklin WM (1951 b) The incidence and distribution of cilia in the human pituitary with a description of micro-follicular cysts derived from Rathke's cleft. Acta Anat (Basel) 11:361–382
863. Shapiro R, Janzen AH (1959) Osteoblastic metastasis to floor of skull simulating meningioma in plaque. AJR 81:964–966
864. Shealy CN, Kahana L, Engel FL, MacPherson HT (1961) Hypothalamic-pituitary sarcoidosis. Am J Med 30:46–55
865. Sheehan HL (1939) Simmond's disease due to postpartum necrosis of the anterior pituitary. Q J Med 8:277–278
866. Sheehan HL (1940) Postpartum necrosis of the anterior lobe of the pituitary. Lancet II:321–324
867. Sheehan HL, Summers SVK (1949) The syndrome of hypopituitarism. Q J Med 18:319–379
868. Sheline GE (1971) Untreated and recurrent chromophobe adenomas of the pituitary. AJR 112:768–773
869. Sheline GE, Boldrey EB, Phillips TL (1964) Chromophobe adenomas of pituitary gland. AJR 92:160–173
870. Shiers JA, Neuhauser EBD, Bowman JR (1957) Idiopathic hypercalcemia. AJR 78:19–29
871. Sholkoff SD, Kerber C, Cramm RE, Silverman L (1977) Parasellar choristoma. AJR 128:1051–1052
872. Shopfner CE, Wolfe TW, O'Kell RT (1968) The intersphenoid synchondrosis. AJR 104:184–193
873. Shuangshoti S, Netsky MG (1966) Neuroepithelial (colloid) cysts of the nervous system. Further observations on pathogenesis, location, incidence, and histochemistry. Neurology (Minneap) 16:887–903
874. Shuangshoti S, Netsky M, Nashold BS Jr (1970) Epithelial cyst related to the sella turcica. Proposed origin from neuroepithelium. Arch Pathol 90:444–450
875. Silverman FN (1957) Roentgen standards for size of the pituitary fossa from infancy through adolescence. AJR 78:451–460
876. Silverman FN (1976) Some features of developmental changes observed clinically in the sphenoid and basicranium. In: Bosma JF (ed) Development of the basicranium. DHEW Publication No (NIH), 76–989, Bethesda, Maryland
877. Simms NM, Brown WE, French LA (1970) Mucocele of the sphenoid sinus presenting as an intrasellar mass. Case report. J Neurosurg 32:708–710
878. Simon HM Jr, Tingwald FR (1955) Syndrome associated with mucocele of the sphenoid sinus. Report of two cases and their radiograpic findings. Radiology 64:538–545
879. Simson LR, Lampe I, Abell MR (1968) Suprasellar germinomas. Cancer 22:533–544
880. Sipponen P, Simila S, Collan Y, Autere T, Herva R (1978) Familial syndrome with panhypopituitarism, hypoplasia of the hypophysis and poorly developed sella turcica. Arch Dis Child 53:664–667
881. Sluder G (1913) Some anatomic and clinical relations of the sphenoid sinus to the cavernous sinus and the third, fourth, fifth, sixth and vidian nerves. Ann Otol Rhinol Laryngol 22:725–768
882. Slusher M, Cowing RH (1973) The empty sella turcica. Ann Ophtalmol 4:966–968
883. Smith EB, Hempelmann TC, Moore S, Barr DP (1952) Gargoylism (dysostosis multiplex): two adult cases with one autopsy. Ann Intern Med 36:652–667
884. Smith RA, Ducy PC (1953) Pituitary cyst lined with a single layer of columnar epithelium. J Neurosurg 10:540–543
885. Smith RR, Kees CJ, Hogg ID (1972) Agenesis of the internal carotid artery with and unusual primitive collateral. Case report. J Neurosurg 37:460–462
886. Smith TR, Kier EL (1971) The unfused planum sphenoidale: differentiation from fracture. Radiology 98:305–309
887. Sobrinho LG, Nunes MCP, Santos MA, Mauricio JC (1978) Radiological evidence for regression of prolactinoma after treatment with bromocriptine. Lancet II:257–258
888. Sofferman RA, Fabian RL (1977) Traumatic carotid-cavernous sinus fistula. Laryngoscope 87:1140–1150
889. Sokolow P (1904) Der Canalis craniopharyngeus. Arch Anat Physiol Anat Abt. 71:
890. Solitaire GB, Jatlow P (1967) Adenohypophysial carcinoma. J Neurosurg 26:624–632
891. Somogyi L, Horanyi-Hechst B (1936) Zur Nosologie der myotonischen Dystrophie. Dtsch Z Nervenheilkd 140:102–112
892. Sondheimer FK (1971) Basal foramina and canals. In: Newton TH, Potts DG (eds) Radiology of the skull and brain, vol 1, book 1. 287–347 Mosby, Saint-Louis
893. Sones PJ, Heinz ER (1972) The sella turcica in multiparity; with comments on the effects of pseudotumor cerebri. Br J Radiol 45:503–506
894. Sosman MC, Vogt EC (1926) Aneurysms of the internal carotid artery and the circle of Willis, from a roentgenological viewpoint. AJR 15:122–134
895. Spencer WH, Borit A (1967) Diffuse hyperplasia of the optic nerve in von Recklinghausen's disease. Am J Ophtalmol 64:638–642
896. Staples GS (1979) Transsellar intracavernous intercarotid collateral artery associated with agenesis of the internal carotid artery. J Neurosurg 50:393–394
897. Starck D (1965) Embryologie. Ein Lehrbuch auf allgemein biologischer Grundlage. Thieme, Stuttgart
898. Stargardter F, Margolis TH (1972) Sella turcica destruction with chromophobe adenomas. AJR 115:774–776
899. Stark L, Weber F (1961) Die Osteome der Schädelbasis. Zentralbl Neurochir 21:126–144
900. Steimle R, Jacquet G, Carbillet JP, Bonneville JF (1977) Intrasellar epidermoid cyst. Acta Neurochir (Wien) 38:285–292
901. Steimle R, Royer J, Jacquet G, Bonneville JF, Abdul-Razzak A (1978) Chondroma of the sella turcica and the base of the skull. Rev Otoneurolophtalmol 50:161–167
902. Steimle R, Raffi A, Bonneville JF, Carbillet JP,

Schraub S, Jacquet G, Abdul-Razzak A (1979) Germinome (pinéalome ectopique) à double localisation: suprasellaire et cérébelleuse sans tumeur pinéale. Neurochirurgie 25:129–133
903. Stein BM, Fraser RAR, Tenner MS (1972) Tumors of the third ventricle in children. J Neurol Neurosurg Psychiatry 35:776–788
904. Steinbach HL, Russell W (1964) Measurement of the heel pad as an aid to diagnosis of acromegaly. Radiology 82:418–423
905. Steinbach HL, Feldman R, Golberg MB (1959) Acromegaly. Radiology 72:535–549
906. Steinbach HL, Noetzli M, Ozonoff MB (1963) Small pituitary fossa in Cushing's syndrome due to adrenal neoplasm. N Engl J Med 269:1286–1289
907. Steiner MM, Boggs JD (1965) Absence of pituitary gland, hypothyroidism, hypoadrenalism and hypogonadism in a 17-year-old dwarf. J Clin Endocrinol Metab 25:1591–1598
908. Stern WE (1973) Meningiomas in the cranio-orbital junction. J Neurosurg 38:428–437
909. Sternberg M (1888) Ein bisher nicht beschriebener Kanal im Keilbein des Menschen. Anat Anz. 3:784–786
910. Stevenson GS, Stultz AD (1923) Sella turcica of mongolian imbeciles. Arch Neurol Psychiatry 10:299–303
911. Stewart MJ, Burrow JF (1923/24) Ecchordosis physaliphora spheno-occipitalis. J Neurol Psychopathol 4:218–220
912. Stewart MJ, Morin JE (1926) Chordoma: a review with report of a new sacro-coccygeal case. J Pathol Bacteriol 29:41–60
913. Stockigt JR, Essex WB, West RH, Murray RML, Breidahl HD (1976) Visual failure during replacement therapy in primary hypothyroidism with pituitary enlargement. J Clin Endocrinol Metab 43:1094–1100
914. Stom BV (1968) The postnatal development of the sphenoidal sinus and its spread into the dorsum sellae and posterior clinoid process. AJR 104:177–183
915. Stough DR, Hartzog JT, Fisher RG (1970) Unusual intradural spinal metastasis of a cranial chordoma. J Neurosurg 34:560–562
916. Strand RD, Green B (1974) A "calcified suprasellar mass" in association with spondylo-metaphyseal dysplasia. Ann Radiol (Paris) 17:369–374
917. Streeter GL (1915) The development of the venous sinuses of the dura mater in the human embryo. Am J Anat 18:145–176
918. Sugita K, Hirota T, Iguchi I, Kageyama N, Ito A (1977) Transient amaurosis under decreased atmospheric pressure with sphenoidal sinus dysplasia. Case report. J Neurosurg 46:111–114
919. Sultan CH, Pages A, Dumas R, Greze J, Bonnet H, Jean R (1974) Adénomes thyrétropes réactionnels. A propos de deux cas dont l'un chirurgicalement confirmé. Ann Endocrinol (Paris) 35:585–586
920. Sunderland S (1945) The meningeal relations of the human hypophysis cerebri. J Anat 79:33–37
921. Sutton TJ, Vezina JL (1974) Co-existing pituitary adenoma and intrasellar arachnoid invagination. AJR 122:508–510
922. Suwanwela C, Hongsaprabhas C (1966) Fronto-ethmoidal encephalomeningocele. J Neurosurg 25:172–182
923. Svab V (1937) Ostéodystrophia fibrosa cystica generalisata. ROEFO 55:450–457
924. Svien HJ (1965) Surgical experiences with craniopharyngiomas. J Neurosurg 23:148–155
925. Svien HJ, Love JG (1942) Abscess within the sella turcica simulating pituitary tumor: surgical cure. Proc Staff Meet Mayo Clin 17:497–501
926. Svolos DG (1969) Craniopharyngiomas. A study based on 107 verified cases. Acta Chir Scand [Suppl] 403:1–44
927. Swanson HA, du Boulay G (1975) Borderline variants of the normal pituitary fossa. Br J Radiol 48:366–369
928. Swischuk LE (1973) Radiology of the newborn and young infant. Williams & Wilkins, Baltimore
929. Swischuk LE (1974) The normal skull. Semin Roentgenol 9:101–113
930. Swischuk LE, Bryan RN (1974) Double midline intracranial atypical teratomas. A recognizable neuroendocrinologic syndrome. AJR 122:517–523
931. Swischuk LE, Sarwar M (1977) The sella in childhood hypothyroidism. Pediatr Radiol 6:1–3
932. Swischuk LE, Sarwar M (1978) The sella turcica (some lesser known dynamic features). CRC Crit Rev Diagn Imaging 11:37–55
933. Symon L, Ganz JC, Chir B, Burston J (1971) Granular cell myoblastoma of the neurohypophysis. J Neurosurg 35:82–89
934. Szécsi K (1978) Selle turcique vide et oedème papillaire. Klin Monatsbl Augenheilkd 172:500–505
935. Syverstsen A, Haughton VM, Williams AL, Cusick J (1979) The computed tomographic appearance of the normal pituitary gland and pituitary microadenoma. Radiology 133:385–391
936. Tänzer A (1961) Veränderung im Bereich des Processus clinoidei anteriores durch expansiv wachsende Tumoren in ihrer Umgebung. ROEFO 94:85–95
937. Tänzer A (1970) Die Röntgendiagnostik der intrasellaren und suprasellaren raumbeschränkenden Prozesse im Nativbild. Radiologe 10:440–445
938. Takeuchi J, Handa H, Nagata I (1978) Suprasellar germinoma. J Neurosurg 49:41–48
939. Takeuchi J, Handa H, Otsuka S, Takebe Y (1979) Neuroradiological aspects of suprasellar germinoma. Neuroradiology 17:153–159
940. Taptas JN (1949) La loge du sinus caverneux. Rev Otoneuroophtalmol 21:193–199
941. Taveras JM, Wood EH (1976) Diagnostic neuroradiology, 2nd edn. Williams & Wilkins, Baltimore
942. Taybi H (1962) Generalized skeletal dysplasia with multiple anomalies. A note on Pyle's disease. AJR 88:450–457
943. Taybi H, Kane PE (1971) Hypophosphatasia and hyperphosphatasia. In: Newton TH, Potts DG (eds) Radiology of the skull and brain, vol 1, book 2. 674–677 Mosby, Saint-Louis
944. Terwinghe G, Terwinghe-Nieus C, Baleriaux-Waha D, Brihaya J (1977) Chute de l'acuité visuelle comme

premier symptome d'un mucocèle du sinus sphénoïdal. Bull Soc Belge Ophtalmol 178:101–108
945. Thiebaut F, Philippides D, Buchheit F, Kurtz D (1969) Adénomes hypophysaires à expansion pharyngée. Rev Otoneuroophtalmol 41:14–18
946. Thomson JLG (1955) Enlargement of the sella turcica. A report on 27 cases. Br J Radiol 28:454–461
947. Tibbs PA, Challa V, Mortara R (1978) Isolated histiocytosis X of the hypothalamus. Case report. J Neurosurg 49:929–934
948. Tiberin P, Goldberg GM, Schwartz A (1958) Craniopharyngiomas in the aged. Neurology (Minneap) 8:51–54
949. Tillier AM (1977) La pneumatisation du massif cranio-facial chez les hommes actuels et fossiles. Bull Mem Soc Anthropol Paris 4:177–189, 287–316
950. Timme W (1921) The mongolian idiot. A preliminary note on the sella turcica finding. Arch Neurol Psychiatry 5:568–571
951. Tindall GT, Collins WF, Kirchner JA (1978) Unilateral septal technique for transsphenoidal microsurgical approach of the sella turcica. Technical note. J Neurosurg 49:138–142
952. Tindall GT, MacLanahan CS, Christy JH (1978) Transsphenoidal microsurgery for pituitary tumors associated with hyperprolactinemia. J Neurosurg 48:849–860
953. Tönnis W, Friedmann G (1964) Das Röntgenbild des Schädels bei intrakranieller Drucksteigerung im Wachstumsalter. Springer, Berlin Heidelberg New York
954. Tönnis W, Schiefer W, Rausch FJ (1954) Sellaveränderungen bei gesteigertem Schädelinnendruck. Dtsch Z Nervenheilkd 171:351–369
955. Tönnis W, Friedmann G, Albrecht H (1957) Veränderungen der Sella Turcica bei sellanahen Tumoren und Tumoren der Schädelbasis. ROEFO 87:686–692
956. Tönnis W, Friedmann G, Albrecht H (1957) Zur röntgenologischen Differentialdiagnose der Hypophysenadenome (unter besonderer Berücksichtigung der primären und sekundären Sellaveränderungen). ROEFO 87:677–686
957. Tourniaire J, Guinet P, Bachelot I, Pousset G, Pallo D, Sassolas G, Trouillas J, L'Hermite M (1974) Les adénomes à prolactine. A propos de 9 observations personnelles. Rev Otoneuroophtalmol 47:303–320
958. Traum E (1925) Beiträge zur Innervation der Dura Mater cerebri. Z Ges Anat 77:488–492
959. Trokoudes KM, Walfish PG, Holgate RC, Pritzner KPH, Schwartz ML, Kovacs K (1978) Sellar enlargement with hyperprolactinemia and a Rathke's pouch cyst. JAMA 240:471–473
960. Trouillas P, Claustrat B, Lheritier M, Goutelle A, Tourniaire J (1979) An FSH and LH – producing pituitary adenoma. 2nd European Workshop on pituitary adenomas, Paris, September 20–22, 1979
961. Tsai FY, Lisella RS, Lee KF, Roach JF (1975) Ostéosclerosis of base of skull as a manifestation of tumor invasion. AJR 124:256–264
962. Tucker RL, Holman CB, MacCarty CS, Dockerty MB (1959) The roentgenologic manifestations of meningiomas in the region of the tuberculum sellae. Radiology 72:348–355
963. Tyrrell JB, Brooks RM, Fitzegerald PA, Cofoid PB, Forsham PH, Wilson CB (1978) Cushing's disease. Selective trans-sphenoidal resection of pituitary microadenomas. N Engl J Med 298:753–758
964. Udvarhelyi GB, Khodadoust AD, Walsh FB (1966) Gliomas of the optic nerve and chiasm in children. An unusual series of cases. Clin Neurosurg 12:204–207
965. Uihlein A, Weyand RD (1953) Meningiomas of anterior clinoid process as a cause of unilateral loss of vision: surgical considerations. Arch Ophtalmol 49:261–270
966. Underwood LE, Radcliffe WB, Strickland AL, van Wyk JJ (1973) Assessment of the sella turcica volume in dwarfed children. J Clin Endocrinol Metab 36:734–741
967. Underwood LE, Radcliffe WB, Guinto FC (1976) New standards for assessment of sella turcica volume in children. Radiology 119:651–654
968. Unger SM, Roswit B (1959) Restoration of the sella turcica after treatment of pituitary adenomas. AJR 81:967–971
969. Unger JD, Unger GF, Nora D (1969) Carotid cavernous fistula presenting as a nasopharyngeal mass. AJR 106:376–380
970. Untea G, Bercu G (1969) Les modifications radiologiques de la selle turcique chez les vieillards. Rev Fr Gerontol 15:137–140
971. Urbach E, Wiethe C (1929) Lipoidosis cutis et mucosae. Virchows Arch [Pathol Anat] 273:285–319
972. Uria JC, Carrillo R, Serrano P, Bravo G (1979) Empty sella and rhinorrhea. A report of eight treated cases. J Neurosurg 50:466–471
973. Utne JR, Pugh DG (1955) The roentgenologic aspects of chordomas. AJR 74:593–608
974. Vagenakis AG, Dole K, Braverman LE (1976) Pituitary enlargement, pituitary failure and primary hypothyroidism. Ann Intern Med 85:195–198
975. Valvassori GE, Hard GE (1968) Traumatic sinus disease. Semin Roentgenol 3:160–171
976. Valvassori GE, Putterman AM (1973) Ophtalmological and radiological findings in sphenoidal mucoceles. Arch Ophthalmol 90:456–459
977. Van Alyea OE (1941) Sphenoidal sinus. Anatomic study with consideration of the clinical significance of the structural characteristics of the sphenoid sinus. Arch Otolaryngol 34:225–253
978. Van Dalen JTW, Greve EL (1977) Rapid deterioration of visual fields during bromocriptine – induced pregnancy in a patient with a pituitary adenoma. Br J Ophtalmol 61:729–733
979. Van Nouhuys JM, Bruyn GW (1964) Nasopharyngeal trans-sphenoidal encephalocele craterlike hole in the optic disc and agenesis of the corpus callosum: pneumoencephalographic visualization in a case. Psychiatr Neurol Neurochir 67:243–258
980. Van't Hoff A, Hornabrook RW, Marks V (1961) Hypopituitarism associated with intracranial aneurysm. Br Med J 4:1190–1194

981. Van Wagenen WP (1935) Chordoblastoma of the basilar plate of the skull and ecchordosis physaliphora spheno-occipitalis. Arch Neurol Psychiatry 34:548–563
982. Van Wyk JJ, Grumbach MM (1960) Syndrome of precocious menstruation and galactorrhea in juvenile hypothyroidism: an example of hormonal overlap in pituitary feedback. J Pediatr 57:416–422
983. Vesely DL, Maldonodo A, Levey GS (1977) Partial hypopituitarism and possible hypothalamic involvement in sarcoidosis – Report of a case and review of the literature. Am J Med 62:425–431
984. Vezina JL (1978) Prolacting-secreting pituitary adenomas: radiologic diagnosis. In: Robyn C, Harter M (eds) Progress in prolactin physiology and pathology. Elsevier North – Holland Biomedical Press
985. Vezina JL, Maltais R (1973) La selle turcique dans l'acromégalie. Neurochirurgie, [Suppl 2] 19:33–56
986. Vezina JL, Sutton TJ (1974) Prolacting – secreting pituitary microadenomas: Roentgenologic diagnosis. AJR 120:46–54
987. Vidic B, Radojević S (1963) Note sur le prolongement du sinus sphénoidal dans la lame quadrilatère et l'apophyse clinoide postérieure. Acta Anat 54:368–372
988. Vidic B, Stom DR (1968) The post natal development of the sphenoidal sinus and its spread into the dorsum sellae and posterior clinoid processes. AJR 104:177–183
989. Villiers HH (1956) Abscess formation in the pituitary fossa associated with a pituitary adenoma. J Neurosurg 13:208–210
990. Virchow R (1857) Untersuchungen über die Entwickelung des Schädelgrundes. Reiner, Berlin, p 51
991. Vittini F, Weber M, Montaut J, Arnould G (1974) Les tumeurs hypophysaires du sujet âgé. Rev Otoneuroophtalmol 47:277–286
992. Vogt JH (1977) Spontaneous regression of enlargement of the sella turcica and of associated panhypopituitary symptoms. Acta Endocrinol (Copenh) 85:684–691
993. Wackenheim A (1960) Neuroradiologie – Radioanatomie normale et pathologique du crâne. Doin, Paris
994. Wackenheim A, Escudero L (1969) Neuroradiologie, encéphalographie gazeuse fractionnée normale et pathologique. Doin, Paris
995. Wallace S (1973) Radiologic changes in pituitary lesions. In: Endocrine and non endocrine hormone producing tumors. Year Book Medical Publishers, Chicago, pp 63–78
996. Wallace S, Goldberg H, Leeds N, Miskin M (1967) The cavernous branches of the internal carotid artery. AJR 101:34–36
997. Walsh FB, Hoyt WF (1969) Clinical neuro-ophtalmology, 3rd ed. Williams & Wilkins, Baltimore
998. Walton JN, Warrick CK (1954) Osseous changes in myopathy. Br J Radiol 27:1–15
999. Waltz TA, Brownell B (1966) Sarcoma: a possible result of effective radiation therapy for pituitary adenoma. (report of two cases). J Neurosurg 24:901–907
1000. Warder FR, Chikes PG, Hudson WR (1975) Aspergillosis of the paranasal sinuses. Arch Otolaryngol 101:683–685
1001. Weber EL, Vogel FS, Odom GL (1970) Cysts of sella turcica. J Neurosurg 33:48–53
1002. Weed LH (1915) The formation of the cranial subarachnoid spaces. Anat Rec 10:475–482
1003. Weidner W, Rosen L, Hanafee W (1965) Neuroradiology of tumors of pituitary gland. AJR 95:884–889
1004. Weidner WA, Wenzi JE, Swischuk LE (1970) Roentgenographic findings in lipoid proteinosis. A case report. AJR 110:457–461
1005. Weinstein M, Tyrrell B, Newton TH (1976) The sella turcica in Nelson's syndrome. Radiology 118:363–365
1006. Weisberg LA (1975) Asymptomatic enlargement of the sella turcica. Arch Neurol 32:483–485
1007. Weisberg LA, Housepian EM, Saur DP (1973) Empty sella syndrom as complication of benign intracranial hypertension. J Neurosurg 43:177–180
1008. Weisberg LA, Zimmerman EA, Frantz AG (1976) Diagnosis and evaluation of patients with an enlarged sella turcica. Am J Med 61:590–596
1009. Welbourn RB, Montgomery DAD, Kennedy DL (1971) Natural history of treated Cushing's syndrome. Br J Surg 58:1–16
1010. Welch K, Stears JC (1971) Chiasmapexy for the correction of traction on the optic nerves and chiasm associated with their descent into an empty sella turcica. Case report. J Neurosurg 35:760–764
1011. Wells HG (1922) Intracranial aneurysm of the vertebral artery. Arch Neurol Psychiatry 7:311–320
1012. Wermer P (1954) Genetic aspects of adenomatosis of endocrine glands. Am J Med 16:363–371
1013. Wermer P (1963) Endocrine adenomatosis and peptic ulcer in a large kindred. Am J Med 35:205–212
1014. Werner A (1960) Two cases of intrasellar chordoma. Acta Neurochir (Wien) 8:436–443
1015. Weyand RD, Camp JD (1954) Roentgenographic examination in meningioma of the tuberculum sellae or olfactory groove. AJR 71:947–951
1016. Whalley N (1952) Abscess formation in a pituitary adenoma. J Neurol Neurosurg Psychiatry 15:66–67
1017. White J, Ballantine HT Jr (1961) Intrasellar aneurysms simulating hypophyseal tumours. J Neurosurg 18:34–50
1018. Wiederholt WC, Siekert RG (1965) Neurological manifestations of sarcoidosis. Neurology (Minneap) 15:1147–1151
1019. Wiese GM, Kempf LG, Hammon WM (1972) Transsphenoidal meningohydroencephalocele. Case report. J Neurosurg 37:475–478
1020. Wieser W (1933) Zur Entwicklung der kindlichen Sella unter normalen und pathologischen Verhältnissen. Wiener Klin Wochenschr 41:1220–1224
1021. Wiggli U, Oberson R (1975) Pneumosinus dilatans and hyperostosis: early signs of meningiomas of the anterior chiasmatic angle. Neuroradiology 8:217–221
1022. Williams JP, Slimach NM, Fowler FW (1972) Ra-

diographic hippocampal calcifications. Neuroradiology 4:159–161
1023. Williams JP, Shawker TH, Lora J (1975) Pneumosinus dilatans of the sphenoid sinus. Bull Los Angeles Neurol Soc 40:45–48
1024. Williams LW (1908) The later development of the notochord in mammals. Am J Anat 8:251–282
1025. Wilson CB, Dempsey LC (1978) Transsphenoidal microsurgical removal of 250 pituitary adenomas. J Neurosurg 48:13–22
1026. Wilson GH, Hanafee WN (1969) Angiographic findings in 16 patients with juvenile nasopharyngeal angiofibroma. Radiology 92:279–284
1027. Winchester PH, Grossman H (1971) Congenital skull dysplasias. In: Newton TH, Potts DG (eds) Radiology of the skull and brain, vol 1, book 2. 653–664, Mosby, Saint-Louis
1028. Winckelman HW (1930) Hyperostosis and tumor infiltration of base of skull associated with overlying meningeal fibroblastoma. Arch Neurol Psychiatry 23:494–501
1029. Wirt TC, Hester RW (1978) Suprasellar arachnoid cyst. Surg Neurol 9:322–323
1030. Wislocki GB (1937) The meningeal relations of the hypophysis cerebri. II. An embryological study of the meninges and blood vessels of the human hypophysis. Am J Anat 61:95–130
1031. Witcombe JB, Cremin BJ (1978) Tuberculous erosion of the sphenoid bone. Br J Radiol 51:347–350
1032. Wohlfart G (1951) Dystrophia myotonica and myotonica congenita. Histopathologic studies with special reference to changes in muscles. J Neuropathol Exp Neurol 10:109–124
1033. Wolf LM, Houdent C, Peugnet JP, Courtois H, Peillon F (1975) Adénome à prolactine révélé par une grossesse gémellaire après traitement par clomiphène. Ann Endocrinol (Paris) 36:107–108
1034. Wollschlaeger G, Wollschlaeger PB, Lopez VF, Zemel HJ (1970) A rare cause of occlusion of the internal carotid artery. Neuroradiology 1:32–38
1035. Wolman L (1956) Pituitary necrosis in raised intracranial pressure. J Pathol Bacteriol 72:575–586
1036. Wood EH, Himadi GM (1950) Chordomas: a roentgenological study of 16 cases previously unreported. Radiology 54:706–716
1037. Wren MWG (1969) Significance of the so-called j-shaped sella in the diagnosis of intracranial aneurysm. Br J Ophtalmol 53:307–309
1038. Wright D (1967) Nasopharyngeal and cervical chordoma: some aspects of their development and treatment. J Laryngol Otol 81:1355–1377
1039. Xuereb GP, Prichard MML, Daniel PM (1954) The hypophyseal portal system of vessels in man. J Exp Physiol 39:219–230
1040. Yamada T, Tsukui T, Ikejiri K, Yukimura Y, Kotani M (1976) Volume of sella turcica in normal subjects and in patients with primary hypothyroidism and hyperthyroidism. J Clin Endocrinol Metab 42:817–822
1041. Yoshida J, Kobayashi T, Kageyama N, Kanzaki M (1977) Symptomatic Rathke's cleft cyst. Morphological study with light and electron microscopy and tissue culture. J Neurosurg 47:451–458
1042. Young LW, Lim GHK, Forbes GB, Bryson MF (1976) Postadrenalectomy pituitary adenoma (Nelson's syndrome) in childhood: clinical and roentgenologic detection. AJR 126:550–559
1043. Yune HY, Holden RW, Smith JA (1975) Normal variations and lesions of the sphenoid sinus. AJR 124:129–138
1044. Zacche E, Musco G (1975) The anatomo-radiological syndrome of empty sella turcica. Minerva Med 66:4399–4402
1045. Zahra M, Sengupta RP (1977) An unusual case of chromophobe adenoma with conspicuous calcification. Acta Neurochir (Wien) 38:279–284
1046. Zander R (1896) Beiträge zur Kenntnis der mittleren Schädelgrube mit besonderer Berücksichtigung der Lage des Chiasma opticum. Anat Anz 12:457–467
1047. Zatz LM, Janon EA, Newton TH (1969) The enlarged sella and the intrasellar cistern. Radiology 93:1085–1091
1048. Zingesser LH, Schechter MM (1964) The radiology of masses lying within and adjacent to the tentorial hiatus. Br J Radiol 37:486–510
1049. Zizmor J, Noyek AM (1978) An atlas of otolaryngologic radiology. Saunders, Philadelphia
1050. Zollinger R, Cutler EC (1933) Aneurysm of the internal carotid artery. Report of a case simulating tumor of the pituitary. Arch Neurol Psychiatry 30:607–611
1051. Zülch KJ (1965) Brain tumors: their biology and pathology. Springer, Berlin Heidelberg New York

Subject Index

Italic page numbers refer to the figure captions

MIX
Papier aus verantwortungsvollen Quellen
Paper from responsible sources
FSC® C105338

If you have any concerns about our products,
you can contact us on
ProductSafety@springernature.com

In case Publisher is established outside the EU,
the EU authorized representative is:
Springer Nature Customer Service Center GmbH
Europaplatz 3, 69115 Heidelberg, Germany

Printed by Libri Plureos GmbH
in Hamburg, Germany